Interdisciplinary Applied Mathematics

Volume 50

More information about this series at http://www.springer.com/series/1390

Jeff D. Eldredge

Mathematical Modeling of Unsteady Inviscid Flows

Springer

Jeff D. Eldredge
Mechanical and Aerospace Engineering
University of California, Los Angeles
Los Angeles, CA, USA

ISSN 0939-6047 ISSN 2196-9973 (electronic)
Interdisciplinary Applied Mathematics
ISBN 978-3-030-18318-9 ISBN 978-3-030-18319-6 (eBook)
https://doi.org/10.1007/978-3-030-18319-6

Mathematics Subject Classification: 76G25, 76B99

This Springer imprint is published by the registered company Springer Nature Switzerland AG.
The registered company address is: Gewerbestrasse 11, 6330 Cham, Switzerland

For Mina, Alex, and Daphne.

Preface

In this modern era of ready access to computational resources, both for research and for instruction, potential flow theory has become somewhat of an endangered species. Though the subject persists in most undergraduate- and graduate-level aerospace and mechanical engineering curricula, a typical course covers only the most basic aspects and generally culminates in steady flows past streamlined planar bodies, setting the foundation for classical (steady) aerodynamics instruction. The associated coursework tends to consist of contrived problems adapted from magically attached flows past circular cylinders that even the instructor is at a loss to explain the relevance of. It is no surprise, then, that a student often leaves such a class wondering about the modern relevance of potential flow. Indeed, even experienced researchers tend to dismiss it as a historical legacy. However, it is precisely because of its truncated treatment that even well-respected researchers and practitioners of fluid dynamics have adopted a rather dim view of the subject of inviscid flows.

But the true utility of inviscid fluid dynamics—both to illuminate fundamental concepts and to serve as the foundation for practical modeling—only emerges just past this point of truncation, when the subject is extended to *unsteady flows*. In that context, the flow past a circular cylinder finally displays one of its virtues: outfitted with conformal transformations, we can develop this benign flow into the flow past *any* closed planar body. By adding a few simple tools in the circle plane and the transport rules for vortex elements, we can construct an endless variety of physically plausible unsteady models. Even the potential flow generated by the cylinder has an important physical role, for if the cylinder is accelerated from rest in a quiescent fluid, the flow that it develops at its first instant is entirely potential. Furthermore, in unsteady inviscid flows, we have a powerful milieu from which to understand and distinguish a variety of fundamental concepts in flow physics: flows induced by moving bodies, the influence of added mass, vortex–vortex and vortex–body dynamics, and the associated forces, moments, and energetics.

It must be emphasized that the subject of unsteady inviscid flows is as relevant as ever, providing a foundation from which to analyze the dynamics of agile aerial and aquatic vehicles, the performance of wind turbines and other devices that harvest energy from fluid flows, and the locomotion of large creatures in fluid media. Furthermore, the tools we develop in this subject enable unique insight into interacting *groups* of such systems—schools, swarms, and wind farms.

The manner in which inviscid flow models are used has evolved in recent years. In the 1970s–1990s, numerical codes based on inviscid models were in common use in the design of aircraft, based on complex panelizations of the fuselage and wing system. This is still the case, but there is now a heavier reliance on higher-fidelity analysis from computational fluid dynamics (CFD). In contrast, inviscid flow models are increasingly used in a *data-driven* capacity, wherein the flow model serves as a *dynamical template*: the model describes the underlying dynamical evolution of the flow, but unspecified parameters of the model are obtained from experiments or CFD by means of data assimilation or machine learning. The resulting reduced-order model is therefore a distillation of the real flow physics and might serve as the backbone of a flow estimation and/or control strategy, for example.

Therefore, the purpose of this book is to extend the instruction of potential flow theory to this richer modern context. My objective in writing the book has been to make this extended treatment accessible to as broad an audience as possible and to highlight the powerful set of tools that it provides for exploring fluid dynamics. The book consolidates various mathematical tools and theoretical frameworks for describing the physics of unsteady inviscid flows, particularly in the context of lifting and propulsive surfaces. Though many of these tools are classical, others are relatively recent. It is important to add that this book focuses exclusively on bodies immersed in unbounded homogeneous fluid media, generating (or subjected to) incompressible flow; it does not contain detailed treatments of internal flows, compressible flows, or flows with free surfaces or fluid-fluid interfaces, aside from their basic governing equations.

In many university courses, three-dimensional inviscid flows, even the steady ones, are treated only briefly (e.g., to establish the basis for lifting-line theory) or skipped altogether due to time constraints. However, it is illuminating to see the relationships between two-dimensional and three-dimensional concepts, so wherever possible, I simultaneously present concepts in both contexts. With a slightly liberalized notation, it is quite easy to write equations that hold in both dimensions.

For flows in the plane, I have devoted a particular attention to unifying the vector and complex approaches. Most books choose one or the other and thereby lose the opportunity to highlight the inherent connections between these approaches. Many students who feel comfortable with a vector-based presentation of potential flow tend to be a bit apprehensive of the complex analysis of the problem. Thus, I have taken special care throughout the book to demonstrate how to move fluidly from one perspective to the other. Wherever possible, formulas are presented in both forms.

Expected Level of Preparation

In a perfect world, an instructor would expand the potential flow modeling section of a class to accommodate this extended material. However, such an expansion is practically impossible, since it would require trimming some other essential subject to fit within a fixed academic term. Thus, my primary purpose has been to create a resource for self-directed study, one which provides an opportunity for students to go at their own pace beyond the typical graduate-level class (or advanced undergraduate class) on potential flow.

While writing this book, I have imagined the reader to be a mature graduate student who has already completed a graduate-level course on inviscid flows. In other words, the reader is already familiar with the foundations of potential flow, but has probably only seen it applied to simple steady (two-dimensional) flows past bodies and perhaps a few elementary unsteady motions (e.g., of point vortices). A reader with an aerodynamics background will likely be accustomed to a bit more, such as thin airfoil theory and lifting lines, but again, this is likely to be in a steady flow setting. It is possible for a reader to proceed directly to this book from an intermediate undergraduate fluid mechanics class. However, most of the foundational concepts are only summarized in Chap. 3, with the presumption that this only serves as a review.

The reader should have the mathematical preparation typical of an undergraduate engineering curriculum. Though the mathematical analysis is extensive in this book, I have tried to ensure that it is accessible to a reader who is only modestly inclined toward mathematics. I have also striven to make it as self-contained as possible, so that the reader can avoid looking up or learning about mathematical topics from other sources. Therefore, the Appendix of the book contains an extensive summary and brief discussion of all of the mathematical results essential for this book. These include basic theorems from vector calculus and their consequences, definitions, and concepts from complex analysis and some useful integrals. They also include some nonstandard fare, such as time differentiation of volume, surface, and line integrals.

Some Notes on Notation

Because of the extensive collection of symbols used in this book, I have taken particular care to organize the notation along a few rules. The variables of real and complex scalar quantities are denoted with italics, such as x and z. Vector- and tensor-valued quantities are represented with bold italics, like $\boldsymbol{x}$, $\boldsymbol{v}$, or $\boldsymbol{M}$. The arrays of their components in some coordinate system are, in contrast, denoted with bold sans serif: $\mathbf{\mathsf{x}}$ and $\mathbf{\mathsf{v}}$, for example.

Other Sources

There is a variety of references from which I have drawn the material presented in this book. For basic inviscid flow theory, I recommend the book of Karamcheti [38]. For a much deeper source of material on the subject, particularly in its use of complex analysis, I am quite fond of the books by Milne-Thomson [53, 52]. Lamb's classic book [43] is an absolute treasure of useful results, many of which are sadly neglected by modern textbooks. For a thoughtful discussion of the foundations of steady aerodynamics, particularly on the suitability of the Kutta condition (or, by his attribution, the Joukowski hypothesis), I suggest Glauert's book [26]. Anderson's text [4] has become a modern classic and provides an accessible introduction to aerodynamics. Aeroelasticity theory has been, historically, the most significant motivation for studying unsteady aerodynamics, and I recommend the classic book by Bisplinghoff, Ashley, and Halfman for this subject and its thorough treatment [8]. On the subject of vortex dynamics, both inviscid and viscous, I highly recommend the compact treatment of Saffman's book [63]. The book by Katz and Plotkin [39] provides an excellent resource on the use of numerical methods for potential flows and should be regarded as a practitioner's guide to complement the more mathematical treatment of the present book. Vortex particle methods have also matured into an attractive tool for high-fidelity numerical simulation of viscous external flows, and the book by Cottet and Koumoutsakos [14] serves as a very good reference for their implementation.

Los Angeles, CA, USA
April 2019

Jeff D. Eldredge

Acknowledgments

There are a number of people whose efforts have been essential for the writing of this book. I have been fortunate to work with several excellent students whose research has formed some of the book's material: Ratnesh Shukla, Maziar Hemati, Chengjie Wang, and Darwin Darakananda; it was during Dr. Darakananda's tenure as a student that I first started working on the book, and some of his research constitutes the more advanced chapters. I also wish to thank my collaborators, Tim Colonius and David Williams, with whom I have enjoyed several years of engaging research. It was this collaboration, and the highly appreciated support of my research by Dr. Douglas Smith via the US Air Force Office of Scientific Research, that has made this work possible.

Most importantly, I wish to thank my family for their patience, particularly for all of the times I was perched on the bed of one of my kids, working on the book, when I was meant to be playing with them. Book writing encourages bad parenting!

Contents

1 Introduction ... 1
1.1 A Motivating Example Problem ... 1
1.1.1 What Would Viscosity Do? ... 2
1.1.2 The Inviscid Model ... 3
1.2 Organization of the Book ... 5

2 Reference Frames, Body Motion and Notation ... 7
2.1 Reference Frames ... 7
2.2 Body Definitions ... 8
2.3 Vector Notation ... 10
2.4 Complex Notation ... 16
2.5 Plücker Notation ... 17
Note 2.5.1: Plücker Transform Matrices in Two Dimensions ... 20
Note 2.5.2: Setting the Pivot Axis of a Translating and Rotating Body ... 21

3 Foundational Concepts ... 25
3.1 Flow Field Definitions and General Relations ... 26
3.1.1 The Velocity Field and Flow Potentials ... 26
Note 3.1.1: Vector Potential in a Moving Rigid Body ... 27
Result 3.1: Volume Flow Rate and Vector Potential ... 27
3.1.2 The No-Penetration Condition ... 28
3.1.3 Vorticity Transport Equation ... 29
3.1.4 Kinetic Energy and Rate of Work ... 30
Result 3.2: A Form of the Kinetic Energy in Three Dimensions ... 31
Note 3.1.2: Another Form of the Rotational Part of Kinetic Energy ... 32

3.1.5 Circulation and Its Invariance 34
Result 3.3: Kelvin's Circulation Theorem 36
Note 3.1.3: Other Invariants of Unbounded Inviscid Flow 37
3.1.6 Helmholtz' Theorems 39
3.2 Elements of Potential Flow 43
3.2.1 Pressure and the Bernoulli Equation 44
Note 3.2.1: Pressure in the Inertial and Windtunnel Frames 45
3.2.2 Two-Dimensional Flows and the Complex Potential 46
Note 3.2.2: Multi-Valuedness and the Branch Cut 50
Note 3.2.3: Velocity Behavior in Corner Flows 53
3.2.3 Three-Dimensional Flows 56
3.3 Vortex Structures 58
3.3.1 Point Vortex 59
3.3.2 Vortex Filament 60
Note 3.3.1: The Scalar Potential Field of a Closed Vortex Filament 63
Result 3.4: Volume Flow Rate Induced by a Closed Filament 65
3.3.3 Vortex Sheet 66
Result 3.5: Strength of a Vortex Sheet 66
Note 3.3.2: Parameterization of a Vortex Sheet 67
3.4 Other Surface Distributions of Singularities 72
3.5 Relationships Between Singularity Distributions 74
3.5.1 Two-Dimensional Vortex Sheet as a Set of Point Vortices 74
3.5.2 Vortex Filament or Vortex Pair as a Double-Layer Potential 76
3.5.3 Vortex Sheet as a Double-Layer Potential 77
3.5.4 Three-Dimensional Vortex Sheet as a Mesh of Vortex Filaments 78
3.6 Free and Bound Vortex Sheets 80
3.6.1 Free Vortex Sheets 81
Note 3.6.1: Parameterization of a Vortex Sheet by Circulation . 82
Result 3.6: Characteristics of a Free Vortex Sheet 84
3.6.2 Bound Vortex Sheets 85
Note 3.6.2: The Surface Vortex Sheet and the No-Penetration Condition 85
Note 3.6.3: Complex Form of the No-Penetration Condition ... 86
Note 3.6.4: No-Penetration Condition on a Sheet Immersed in the Fluid 87
Note 3.6.5: A Note on Point Vortices Near Thin Plates 89

4 General Results of Incompressible Flow About a Body ... 91
4.1 The Basic Potential Flow Problem ... 92
4.2 Velocity Field Outside a Moving Body ... 94
4.2.1 Vector Form ... 94
Result 4.1: Velocity Field of an Incompressible Flow in the Presence of a Body ... 96
4.2.2 Complex Form ... 97
Result 4.2: Velocity Field of a Two-Dimensional Flow in the Presence of a Body ... 98
4.2.3 Infinitely-Thin Plate ... 98
4.2.4 Alternative Surface Formulations: Source and Dipole Distributions ... 99
4.3 The Integral Equation for an Impenetrable Body ... 101
4.3.1 Vector Form ... 102
Result 4.3: General Integral Equation for Bound Vortex Sheet ... 103
4.3.2 Complex Form ... 104
4.3.3 Infinitely-Thin Plate ... 105
4.3.4 The Double Layer Formulation ... 106
4.3.5 General Notes on These Integral Equations ... 107
4.4 Solution of Two-Dimensional Problem by Conformal Mapping ... 107
Note 4.4.1: Notational Convention for Functions and Mappings ... 109
4.4.1 Moving Bodies in Irrotational Flow ... 111
Note 4.4.2: Solution for Rigid-Body Motion by Conformal Mapping ... 111
4.4.2 Stationary Body in a Known Potential Flow: The Circle Theorem ... 113
Note 4.4.3: Point Vortex Outside a Stationary Body ... 114
Note 4.4.4: Uniform Flow Past a Stationary Body ... 115
4.4.3 Solution via the Schwarz–Christoffel Transformation ... 116
Note 4.4.5: Velocity Field Near a Corner, Revisited ... 117
Result 4.4: Signed Intensity of a Corner in Schwarz–Christoffel Mapping ... 120
Note 4.4.6: Corner Intensity and the Bound Vortex Sheet Strength ... 120
4.5 The Non-uniqueness of Two-Dimensional Potential Flow ... 121
4.6 Decomposition of the Flow into Basis Fields ... 125
4.6.1 Complex Form, via Conformal Mapping Solution ... 128
Result 4.5: Two-Dimensional Velocity Field via Conformal Mapping ... 129

4.6.2 Vector Form ... 135
Note 4.6.1: Useful Properties of the Basis Scalar Potential Fields ... 144
Note 4.6.2: Useful Property of the Basis Vector Potential Fields ... 145
4.6.3 Flow Decomposition and the Kinetic Energy: Added Mass ... 145
Result 4.6: Kinetic Energy of the Flow About a Rigid Body ... 148
Note 4.6.3: Positive Semi-definiteness of Added Mass ... 149
4.7 Multipole Expansion of the Flow Field ... 149
4.7.1 Two-Dimensional Flow ... 150
Result 4.7: Multipole Expansion of the Two-Dimensional Flow Field ... 152
4.7.2 Three-Dimensional Flow ... 153
Result 4.8: Multipole Expansion of the Three-Dimensional Flow Field ... 155
Note 4.7.1: Multipole Coefficients in Terms of Body Vorticity ... 156
Note 4.7.2: Alternate Form of Dipole Coefficients ... 157
4.7.3 Two-Dimensional Expansion in Complex Form ... 157
Note 4.7.3: Multipole Coefficients for Plates ... 159
Note 4.7.4: Multipole Coefficients Under Conformal Mapping ... 159

5 Edge Conditions ... 161
5.1 The Kutta Condition ... 162
Result 5.1: The Kutta Condition ... 162
Result 5.2: The Kutta Condition, Alternatively Stated ... 163
5.1.1 Steady Flow ... 164
5.1.2 Unsteady Flow ... 165
Result 5.3: Continuity of Vortex Sheet Strengths ... 168
Result 5.4: Giesing–Maskell Extension of the Kutta Condition ... 170
Note 5.1.1: The Kutta Condition and Kelvin's Circulation Theorem ... 172
Note 5.1.2: The Kutta Condition and the Release of Vortex Elements ... 172
5.2 Application of the Kutta Condition ... 173
Result 5.5: The Constraint Form of the Kutta Condition ... 174
5.3 Traditional Enforcement of the Kutta Condition: A Contrast ... 176
5.4 Generalized Edge Conditions ... 178
Result 5.6: Critical Edge Suction Condition ... 179

6 Force and Moment on a Body 183
6.1 Force and Moment via Surface Traction 184
Result 6.1: Surface Integral Form 1 of Force and Moment 184
Result 6.2: Surface Integral Form 2 of Force and Moment 185
Note 6.1.1: Force and Moment in the Windtunnel Frame 186
6.2 Force and Moment via Vorticity Impulse 187
6.2.1 Vectorial Forms 188
Result 6.3: Force as Rate of Change of Linear Impulse 190
Result 6.4: Moment as Rate of Change of Angular Impulse 191
Note 6.2.1: A Vorticity Form of the Impulse 192
Note 6.2.2: Another Form of the Moment's Relationship with Impulse 193
Note 6.2.3: Impulse Defined About Points Other than the Origin 194
Note 6.2.4: Impulse Formulas in the Presence of a Uniform Flow 195
6.2.2 Complex Form for Planar Applications 196
Result 6.5: Complex Form of Force as Rate of Change of Linear Impulse 196
Result 6.6: Complex Form of Moment as Rate of Change of Angular Impulse 197
Note 6.2.5: Impulses Obtained Through Conformal Mapping 198
6.3 Reconciliation of Force via Traction and via Impulse 199
6.3.1 The Force and Moment on a Region of Vorticity 200
Result 6.7: The Force and Moment on a Vortex 202
6.3.2 The Spurious Force and Moment on Vorticity 204
6.3.3 Spurious Force and Moment on a Vortex in the Presence of a Body 207
Result 6.8: Spurious Force and Moment on a Vortex Element Near a Body 210
6.3.4 Revisiting the Traction Force and Moment on the Body 211
Result 6.9: Force and Moment on a Body via Traction over Its Surface 212
6.4 Edge Suction 213
Result 6.10: Edge Suction and the Suction Coefficient 215
6.5 Decomposition of the Force into Contributors 216
6.5.1 Complex Form, via Conformal Mapping Solution 217
6.5.2 Decomposed Force and Moment, Generalized 220
Result 6.11: Decomposed Force and Moment on a Rigid Body 222
Note 6.5.1: Plücker Added Mass Matrices in Two Dimensions . 223

6.6 The Contribution of Fluid Vorticity to Force and Moment 226
6.6.1 Basic Definitions of the Vorticity-Induced Impulses 227
6.6.2 The Rates of Change of the Vorticity-Induced Impulses 228
Result 6.12: Vorticity-Induced Impulse and the Basis Vector Potentials 229
Note 6.6.1: The Elemental Contribution to Vorticity-Induced Impulses 232
Result 6.13: Time Derivative of the Vorticity-Induced Impulses 233
Result 6.14: The Force and Moment on a Rigid Body 235
6.6.3 The Rate of Change of Impulse for Singular Vortex Elements 237
6.6.4 Alternative Derivation of the Force and Moment 239
6.7 The Mechanical Energy Equation, Revisited 243

7 Transport of Vortex Elements 245
7.1 Planar Vortex Elements 246
Result 7.1: Transport of a Point Vortex in the Physical Plane 247
7.1.1 Justification for Ignoring the Self-induction of a Point Vortex 248
7.1.2 Vortex Transport in the Conformal Mapping Plane: The Routh Correction 249
Result 7.2: The Routh Correction 250
Result 7.3: Vortex Transport in the Mapped Plane 252
7.1.3 Vortex Clouds and Sheets 252
Note 7.1.1: Efficient Calculation of Vortex Cloud Transport ... 253
Result 7.4: Evolution Equation for a Free Vortex Sheet in Two Dimensions 254
Result 7.5: The Birkhoff–Rott Equation 255
7.1.4 Variable-Strength Point Vortices 255
7.1.5 Blob Regularization 260
7.2 Vortex Filaments 263
Result 7.6: Evolution Equation for a Vortex Filament 265

8 Flow About a Two-Dimensional Flat Plate 269
8.1 Basic Notation 270
8.2 Three Approaches to Solving for the Flow 273
8.2.1 Approach 1: Conformal Transformation from the Circle Plane 274
8.2.2 Approach 2: Inversion of the Cauchy Integral 280
Result 8.1: Cauchy Integral Equation for a Rigid Flat Plate 281

8.2.3 Approach 3: Fourier–Chebyshev Expansion 287
Result 8.2: Fourier–Chebyshev Form of the Bound Vortex Sheet Strength 289
Note 8.2.1: Fourier–Chebyshev for Wavy Plate or Sinusoidal Gust 291
Result 8.3: Fourier–Chebyshev Form of the Velocity and Potential Fields 293
8.3 Force and Moment 295
Result 8.4: Force and Moment on a Flat Plate 296
Result 8.5: Local Pressure Jump Across a Flat Plate 299
Note 8.3.1: Edge Suction on a Two-Dimensional Flat Plate 300
8.4 Application of the Kutta Condition 302
8.4.1 At the Trailing Edge: Thin Airfoil Theory 302
8.4.2 At Both Edges 309
8.5 Classical Results 309
8.5.1 Steady Flow at Fixed Small Angle of Attack 310
Note 8.5.1: Alternative Enforcement of the Trailing-Edge Kutta Condition 311
8.5.2 General Background on Classical Unsteady Results 313
8.5.3 Oscillatory Motion: Theodorsen 316
Result 8.6: Low-Amplitude Oscillatory Motion of a Flat Plate 320
8.5.4 Impulsive Change of Motion: Wagner 321
Result 8.7: Force and Moment due to a Sudden Change of Motion 323
Note 8.5.2: The Wagner Function 325
8.5.5 Sinusoidal Gust Response 327
8.5.6 Sharp-Edge Gust 330
8.6 Generalized Edge Conditions and Their Interpretation 331
8.7 A Deforming Plate 334

9 Examples of Two-Dimensional Flow Modeling 341
9.1 Time Marching 342
Result 9.1: State Update 343
9.2 Co-rotating Vortex Patches 344
9.3 Interaction of Vortex Patches with a Bluff Body 348
9.4 A Translating Flat Plate at Large Angle of Attack 352
9.4.1 Free Vortex Sheet Model 352
Result 9.2: The Minimum Physical Length Scale in an Inviscid Flow 359
9.4.2 Variable-Strength Vortex Model 361
9.5 Flow Past a NACA Airfoil 364

10 Rigid Motion of an Ellipsoidal Body 369
10.1 Ellipsoidal Coordinates and Harmonics 369
10.2 Translation of a General Ellipsoidal Body 372
10.3 Rotational Motion of a General Ellipsoidal Body 379
10.4 Added Mass 380
Result 10.1: The Added Mass of an Ellipsoid 382

Mathematical Tools 389
A.1 Some Vector Calculus Results 389
A.1.1 Cartesian Index Notation 389
A.1.2 Derivatives of Fundamental Solutions of Laplace's Equation 392
A.1.3 The Divergence Theorem and Some Relevant Uses 393
Result A.1: Scalar Form of Green's Theorem 400
Result A.2: Vector Form of Green's Theorem 401
Result A.3: Scalar Form of Extended Green's Theorem 402
Result A.4: Vector Form of Extended Green's Theorem 403
A.1.4 Stokes' Theorem and Some Relevant Uses 405
Result A.5: Continuity of Tangent Component of Vector Fields 408
Result A.6: Equivalence of Two Forms of Surface Singularity Distributions 409
A.1.5 Change of Variables Theorem 411
A.1.6 Field Quantities and Their Rate of Change 411
A.1.7 Time Differentiation of Spatial Integrals 412
A.2 Useful Tools from Complex Analysis 416
A.2.1 Basic Properties 417
A.2.2 The Cauchy Integral and Residue Theorem 420
A.2.3 Conformal Mapping 425
A.2.4 The Joukowski and Kármán–Trefftz Airfoils 431
A.2.5 The Schwarz–Christoffel Transformation 436
Result A.7: Power Series Representation of Schwarz–Christoffel Map 438
Note A.2.1: Schwarz–Christoffel Transformation for Rectangular Bodies 441
A.3 Mathematical Results for the Infinitely-Thin Plate 443
A.3.1 Notes on an Important Factor 443
A.3.2 Properties of Chebyshev Polynomials 444
A.3.3 Contour Integrals of Interest 452

References 455

Index 459

Chapter 1
Introduction

Contents

1.1 A Motivating Example Problem 1
  1.1.1 What Would Viscosity Do? 2
  1.1.2 The Inviscid Model 3
1.2 Organization of the Book 5

This book will cover many aspects of inviscid incompressible flows. But rather than proceed directly, let us first provide ourselves a roadmap of sorts in order to clarify the role of each ensuing chapter in the journey. The best such roadmap is an illustrative example containing nearly all of the components of an unsteady inviscid flow.

1.1 A Motivating Example Problem

Let us imagine a flat plate of length c and much longer span that sits at rest in a quiescent fluid of uniform density ρ. At some instant, the plate is suddenly accelerated from rest to constant speed U at 60° angle of attack relative to the direction of motion.

We seek to construct an inviscid model for the flow generated by this sudden motion of the plate. As is typical of studies of this simple geometry, we will suppose the plate to have infinitesimal thickness and the flow to be confined to a plane and independent of the dimension along the plate's span. The reader who has studied aerodynamics will recall that these approximations also underpin **thin airfoil theory**. But in conventional thin airfoil theory, one imagines that the plate has already been in constant motion for so long that the flow can be considered steady, and furthermore, that this motion is at such a small angle of attack that the flow remains attached along the plate. Neither of these assumptions is valid in the current scenario. Indeed, this scenario deviates substantially from the classical aerodynamics paradigm. The

J. D. Eldredge, *Mathematical Modeling of Unsteady Inviscid Flows*, Interdisciplinary Applied Mathematics 50, https://doi.org/10.1007/978-3-030-18319-6_1

essential components of an inviscid model that describes such a departure from steady attached flow constitute the major content of this book.

So what are these essential components? Because the fluid was initially at rest and was then set into motion, we very clearly must expect the ensuing flow to be **unsteady**. For ascertaining the more specific requirements of a model, let us consider the sequence of events that transpire after this plate is set in motion in a Newtonian viscous fluid.

1.1.1 What Would Viscosity Do?

Immediately after starting (at time $t = 0$, say), the plate's new velocity differs from that of the surrounding quiescent fluid. However, in order to prevent it from penetrating the plate, the adjacent fluid must be set in motion with the same normal component of velocity. This new motion imparted to one region of the fluid is communicated to the rest of it by means of pressure. In real fluids the motion would be communicated at the speed of sound. However, we may reasonably suppose, at the low Mach numbers typical of the target contexts of this book, that the fluid medium is strictly incompressible. Thus, the pressure transmits this motion's influence instantaneously, and establishes an irrotational flow throughout the fluid that vanishes at large distances from the plate.

It is at this point in the sequence, at $t = 0^+$, that the viscosity of the fluid first asserts itself. The irrotational flow established by the plate's motion inevitably leaves a discontinuity between the fluid's tangential velocity and that of the plate—a spurious slip velocity. This slip velocity, which we will describe later in the book as a **bound vortex sheet**, cannot persist in a viscous fluid, but is instead diffused into the neighboring fluid. The result is a flow that now satisfies both the no-penetration and the no-slip conditions on the plate surface; it is a flow that is no longer entirely irrotational, but is now adorned with a thin layer of vorticity adjacent to the plate.

How does the flow proceed from this point? The vorticity does not remain stagnant, but convects with the local velocity of the fluid and diffuses further by action of viscosity. It is most illuminating to imagine its subsequent evolution as a sequence of vanishingly-small time increments that follow from this first instant. The vorticity's convection and diffusion over the next small time increment modify the fluid velocity field, violating—as in the initial instant, but now less egregiously—the boundary conditions on the plate surface. The fluid responds to this violation as it did at $t = 0$: first, by developing a new irrotational flow, added to the existing rotational flow, that prevents the fluid's penetration of the plate, but which leaves a spurious slip on the surface; then, by diffusing this new bound vortex sheet into the surrounding fluid to annihilate the slip. The steps are repeated in the next small increment, and thence, in every increment that follows. This manner of viewing the flow development as a sequence of infinitesimal steps, each consisting of vorticity convection and diffusion, followed by bound vortex sheet development and diffusion, is sometimes referred to as the **Lighthill vorticity creation mechanism** [47].

The vorticity's behavior after it enters the fluid depends strongly on the circumstances of the flow. Let us first agree that ***we are devoting our attention in this book***

to modeling moderate and high Reynolds number flows, in which inertial forces in the fluid dominate viscous traction, or, stated another way, in which convective processes occur more quickly than those due to viscous diffusion. In this regime, the vorticity developed about a flat plate moving at small angle of attack remains confined to thin boundary layers on either side of the plate. These developing boundary layers meet just aft of the trailing edge, where, by virtue of the relatively stronger vorticity on the lower side, the merged boundary layers roll up into a **starting vortex** that convects away from the edge.

At the much larger angle of attack of our current scenario, this starting vortex develops with significantly more strength than at small angle. Furthermore, the flow about the leading edge of the plate, which remains attached and unremarkable at small angle of attack, now encounters a strongly adverse pressure gradient. In general, an adverse pressure gradient tends to rob the viscous boundary layer of momentum, and if it does so to such an extent that the flow can no longer continue in that direction, the flow must **separate** from the body. The pressure gradient that greets the flow as it negotiates the turn about the sharp leading edge not only induces the flow to separate, but also ensures that the point of separation remains fixed at this leading edge. This separated flow rolls up into a **leading-edge vortex, or 'LEV'**, which grows for some time and then ***sheds*** from the plate [23]. In the language of aerodynamics, the plate is experiencing the onset of stall. The LEV tends to induce the development of a new vortex at the trailing edge, and after shedding, a new LEV develops in its place. The collective shedding from both edges of the stalled plate forms a wide vortical wake behind the translating plate.

1.1.2 The Inviscid Model

Now let us return to the inviscid model and the essential components it must embody. By definition, an inviscid model lacks the mechanism for describing the diffusive components in the sequential process described above. Note that viscous diffusion enters the actual process in two roles: first, in the creation of vorticity at the plate surface, and second, in the diffusion of vorticity already in the fluid. The second of these roles, which gradually spreads the extent of existing vorticity, we can reasonably argue is nonessential, and the model's effectiveness will not suffer significantly without it.

The omission of the first role, however, seems more problematic, given that viscous diffusion is the means by which vorticity actually enters the fluid. But if we adopt a global view of the process we have just described—of the viscous boundary layers' development, their separation from the edges of the plate, and the formation of **coherent vortices** from this separated vorticity—then we need only capture the process' net effect on the flow's evolution. If we return once again to the two steps comprising the Lighthill creation mechanism, and imagine that the second of these steps—the diffusion of the bound vortex sheet—is skipped and that the slip velocity is allowed to persist, then the boundary layer has effectively been confined to zero thickness. The boundary layer's vorticity, however, still lives—approximately—in

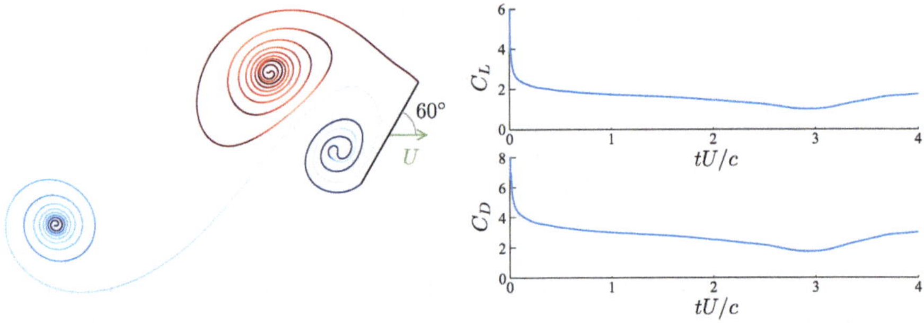

Fig. 1.1 Inviscid model of a two-dimensional flat plate of length c accelerated impulsively from rest to constant speed U to the right at 60° angle of attack in a fluid of density ρ. (Left) The vortex sheets originating from the two edges at time $tU/c = 4$ (i.e., after four chord-lengths of travel). The plate is depicted in black. The red hues in the sheet denote positive vorticity, and the blue hues represent negative vorticity; darker hues denote stronger vorticity. (Right) The coefficients of lift, $C_L := 2F_L/\rho U^2 c$, and drag, $C_D := 2F_D/\rho U^2 c$, exerted on the plate. Lift is defined as the upward component of force, perpendicular to the direction of motion; drag is the component parallel to and opposite the direction of motion

the form of the bound vortex sheet. This suggests that, in order to model the process by which this boundary layer vorticity separates from the edges of the plate, without explicit reference to the role of viscosity, we need only develop a means of extending this sheet from its edges into the fluid.

Figure 1.1 depicts the result of this inviscid approach to vorticity creation in our example of a plate at 60° angle of attack. Vorticity is shown here after the plate has traveled four chord-lengths from its initial position (i.e., at time $t = 4c/U$). This vorticity has been continuously introduced into the fluid from each edge in the form of a **free vortex sheet**, and the progressive action of the flow has caused these sheets to roll up into coherent vortex structures. What components must comprise such an inviscid model? By reflecting on the result in Fig. 1.1, it is clear that we need

- **A means of introducing vorticity into the fluid.** In the absence of viscosity, is there a criterion or constraint that we can apply at the edge of the plate (or other body) to serve as a suitable surrogate for the action of viscosity? We hinted at the idea of extending the bound vortex sheet on the plate into the fluid. How would this work?
- **A representation of vorticity in the fluid.** This representation must, of course, be compatible with the inviscid setting. We have already referred multiple times to the notion of a 'sheet' of vorticity. What does this mean? And is this the only compatible form of vorticity?
- **A transport model for this fluid vorticity.** We noted that vorticity is convected by the fluid velocity field. As we mentioned, the plate's motion induces motion in the fluid. However, vorticity is not simply a passive rider of this flow, but actively contributes to the velocity field. Thus, we need a means of constructing the overall velocity field that includes the vorticity's contribution and also satisfies the boundary conditions on the surface of the moving body.

Generally, we are not only interested in the flow's development, but also in its effect on the body. Figure 1.1 also depicts the force exerted by the fluid on the plate, in the form of the lift and drag coefficients, during the first four chord-lengths of travel. Thus, we also need

- **Mathematical tools for calculating force and moment on a body.** This force and moment are exerted directly by the action of pressure in the adjacent fluid, but could we also perhaps obtain them by observing the changes in momentum in the fluid? Can we distinguish the contributions of vorticity and body motion in this force and moment?

1.2 Organization of the Book

The rest of this book is organized into a sequence that addresses the essential model components outlined above. The next few chapters prepare for this sequence. Chapter 2 introduces the reference frames and notation to be used throughout the book. Chapter 3 reviews the foundations of inviscid flow theory, and particularly, the definitions of most of the important flow quantities and the governing equations of the flow. Then, Chaps. 4 through 7 constitute the core material of the book, and each of these focuses on a key aspect of the modeling of a general inviscid flow problem. First, Chap. 4 presents the tools for obtaining the fluid velocity field about a body in arbitrary motion. Chapter 5 describes the details of the model component for introducing vorticity into the fluid, via conditions applied at the sharp edges of a body. Then Chap. 6 develops the framework for computing force and moment on the body. And finally, Chap. 7 provides the means of evolving the flow, by describing the transport of vortex elements in the fluid.

The next chapter, Chap. 8, ties these three aspects together by demonstrating them on the solution of unsteady inviscid flow about a two-dimensional flat plate, as we have already illustrated in this chapter. This topic is so important that we devote considerable attention to its details and provide multiple perspectives on its treatment. We also demonstrate our framework in obtaining the classical results from unsteady aerodynamics. In Chap. 9 we go beyond the simple analytical results of a flat plate to illustrate the means of solving more interesting problems in two dimensions with the assistance of numerical calculation. Inviscid flows about three-dimensional bodies are discussed briefly in Chap. 10.

A pervasive concept throughout the book is the **flow contributor**, defined as an entity, such as a moving body or a vortex element, that contributes to the overall motion in the fluid. This concept is quite useful because many important measures of fluid motion can be linearly decomposed into influences from each such contributor. We will first introduce the concept in Chap. 4, along with the accompanying notion of a **basis field**, and return to them frequently throughout the book.

Chapter 2
Reference Frames, Body Motion and Notation

Contents

2.1	Reference Frames	7
2.2	Body Definitions	8
2.3	Vector Notation	10
2.4	Complex Notation	16
2.5	Plücker Notation	17

In this chapter, we establish some notation and conventions to be followed throughout the book regarding a body and its motion. We rely on three different types of notation to describe concepts and solve problems in this book: vector notation, generally useful in both two- and three-dimensional contexts; complex notation, for problems in the plane; and a less familiar notation called Plücker notation that facilitates our analyses of rigid bodies. These are described, and related to each other, in the following sections. First, we discuss the reference frames we will utilize in the book.

2.1 Reference Frames

Figure 2.1 illustrates the notation on a two-dimensional body; the conventions for a three-dimensional body follow naturally, as depicted on the right. The interior of the body is denoted by $\mathcal{V}_b$; this is enclosed by a surface, $\mathcal{S}_b$, that serves as the interface with the surrounding fluid in region $\mathcal{V}_f$. The unit normal, $\boldsymbol{n}$, on $\mathcal{S}_b$ is directed into the fluid. Note that, in this definition, as well as many others in this book, we use the notations $\mathcal{V}$ and $\mathcal{S}$ liberally, so that they also, respectively, denote the interior region and enclosing surface in a two-dimensional context. We will stretch the notation even further by using the symbol $\mathcal{V}$ to denote the volume (and in two dimensions, the area, or strictly, volume per unit depth) of the region $\mathcal{V}$, and $\mathcal{S}$ to denote the

J. D. Eldredge, *Mathematical Modeling of Unsteady Inviscid Flows*, Interdisciplinary Applied Mathematics 50, https://doi.org/10.1007/978-3-030-18319-6_2

area (and in two dimensions, the perimeter length, or surface area per unit depth) of the surface $\mathcal{S}$. The following volume and surface integrals express this liberalized notation in both dimensions:

$$\mathcal{V} = \int_{\mathcal{V}} \mathrm{d}V, \qquad \mathcal{S} = \int_{\mathcal{S}} \mathrm{d}S. \tag{2.1}$$

Figure 2.1 depicts two sets of coordinate systems that will be in common use in this book. In addition to the coordinate system associated with the inertial frame of reference, we will also rely on coordinate axes that are rigidly attached to the body, with their origin at the body's reference point. These axes are also shown in Fig. 2.1, labeled by a $\tilde{(\cdot)}$. Throughout this book, we will refer to these two reference frames as the **inertial frame** and the **body-fixed frame**, respectively, and their associated coordinate systems will be described similarly.

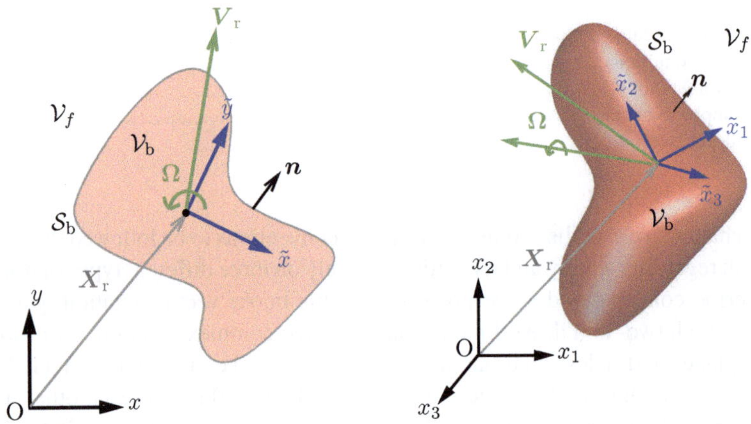

Fig. 2.1 Schematic of a moving body immersed in a fluid. (Left) A two-dimensional example. (Right) A three-dimensional example

It is important to point out that ***we will always assume that, in the inertial frame, the fluid at infinity is at rest***. This will help us avoid many complications later. However, for occasions in which we wish to pose the problem with a uniform flow at infinity, we will employ a third reference frame, which we will call the **windtunnel frame**. Coordinates and quantities in this reference frame will be denoted by $(\cdot)^{\dagger}$. For a stationary body in a uniform flow, the windtunnel frame is equivalent to the body-fixed frame. An illustration of the windtunnel frame and its relationship to the inertial frame is given in Fig. 2.2. Note that this windtunnel frame will be non-inertial whenever the uniform flow varies in time.

2.2 Body Definitions

The centroid of the body relative to the origin of the inertial coordinate system, denoted by $\boldsymbol{X}_c$, is defined as

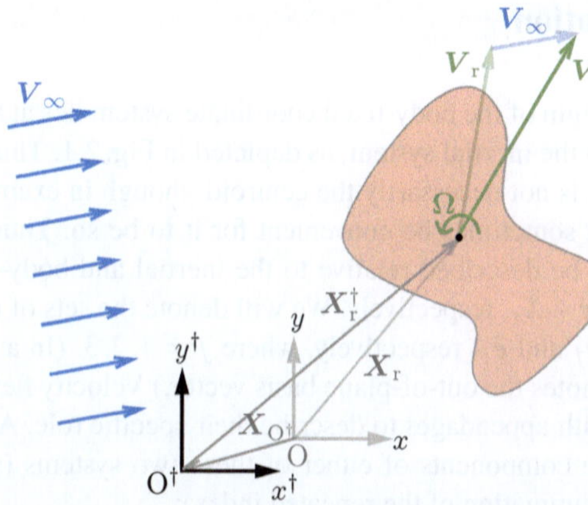

Fig. 2.2 The windtunnel reference frame. The associated coordinate system and quantities are denoted with $(\cdot)^\dagger$. The inertial system and its corresponding motion vectors are also included in lighter format

$$\boldsymbol{X}_c := \frac{1}{\mathcal{V}_b} \int_{\mathcal{V}_b} \boldsymbol{x}\, dV. \tag{2.2}$$

In this book, we will assume for simplicity that the body's mass is uniformly distributed, so that its density ρ_b is a constant, its mass is simply $M = \rho_b \mathcal{V}_b$, and its centroid is equivalent to its center of mass. The moment of inertia tensor of the body, $\boldsymbol{I}$, represents the linear operator acting on the angular velocity of the body to provide the body's angular momentum. As such, it is defined about the origin of the inertial coordinate system as

$$\boldsymbol{I}_O := \rho_b \int_{\mathcal{V}_b} \left(|\boldsymbol{x}|^2 \mathbf{1} - \boldsymbol{x}\boldsymbol{x} \right) dV, \tag{2.3}$$

where $\mathbf{1}$ is the identity tensor, and the notation $\boldsymbol{x}\boldsymbol{x}$ denotes the dyadic product of the vector $\boldsymbol{x}$ with itself. In Cartesian index notation, which we review briefly in Sect. A.1.1, the ijth component of this tensor would be

$$I_{O,ij} = \rho_b \int_{\mathcal{V}_b} \left(x_k x_k \delta_{ij} - x_i x_j \right) dV, \tag{2.4}$$

in which summation is implied for repeated (i.e., dummy) indices. For a two-dimensional problem, only the 33 component of the moment of inertia is invoked, and the definition above reverts gracefully to this special case provided that it is remembered that the indices on the components of $\boldsymbol{x}$ only take the values 1 and 2. This component, without the density factor, is referred to as the second area moment of the two-dimensional shape, denoted by $\mathcal{J}$.

2.3 Vector Notation

Suppose that the origin of the body-fixed coordinate system lies at a **body reference point** $\boldsymbol{X}_{\mathrm{r}}$ relative to the inertial system, as depicted in Fig. 2.1. This reference point, affixed to the body, is not necessarily the centroid, though in examples with simple body shapes it will sometimes be convenient for it to be so. Thus, the position of a certain point can be described relative to the inertial and body-fixed systems by vectors $\boldsymbol{x}$ and $\tilde{\boldsymbol{x}} = \boldsymbol{x} - \boldsymbol{X}_{\mathrm{r}}$, respectively. We will denote the sets of unit basis vectors of the systems as $\boldsymbol{e}_j$ and $\tilde{\boldsymbol{e}}_j$, respectively, where $j = 1, 2, 3$. (In a two-dimensional setting, $\boldsymbol{e}_3 \equiv \tilde{\boldsymbol{e}}_3$ denotes the out-of-plane basis vector.) Velocity fields will generally be denoted by $\boldsymbol{v}$, with appendages to describe their specific role. Any vector, $\boldsymbol{u}$, can be expressed in the components of either of these two systems (where, again, we stress the implied summation of the repeated index):

$$\boldsymbol{u} = u_j \boldsymbol{e}_j = \tilde{u}_j \tilde{\boldsymbol{e}}_j. \tag{2.5}$$

Following our notation for the axes and unit vectors, we will distinguish the components of any vector in these two systems by the use of $\tilde{(\cdot)}$ for body-fixed components.

Transformation Between Inertial and Body-Fixed Coordinates In order to transform between the inertial and body-fixed reference frames, we simply express the basis vectors of one system in terms of those of the other:

$$\boldsymbol{e}_j = (\boldsymbol{e}_j \cdot \tilde{\boldsymbol{e}}_k)\tilde{\boldsymbol{e}}_k = \mathsf{R}_{jk}\tilde{\boldsymbol{e}}_k, \quad j = 1, 2, 3. \tag{2.6}$$

Here, we have defined the components of the **rotation matrix, R**—sometimes called the direction cosine matrix—as $\mathsf{R}_{jk} := \boldsymbol{e}_j \cdot \tilde{\boldsymbol{e}}_k$. As exhibited by the definition, R_{jk} specifies the kth component of the jth inertial basis vector in the body-fixed basis.

The inverse of the rotation matrix transforms inertial unit vectors into body-fixed unit vectors, $\tilde{\boldsymbol{e}}_j = \mathsf{R}^{-1}_{jk}\boldsymbol{e}_k$. But we can also obtain this transformation by projecting each body-fixed vector onto the set of inertial unit vectors:

$$\tilde{\boldsymbol{e}}_j = (\tilde{\boldsymbol{e}}_j \cdot \boldsymbol{e}_k)\boldsymbol{e}_k = \mathsf{R}_{kj}\boldsymbol{e}_k. \tag{2.7}$$

By inspection, it is clear that the inverse of the rotation matrix is simply its transpose, $\mathsf{R}^{-1} = \mathsf{R}^{\mathsf{T}}$. Thus, R_{jk} also specifies the jth component of the kth body-fixed basis vector in the inertial basis. The components of any vector $\boldsymbol{u}$ can then be transformed from one system to the other through the use of this rotation matrix:

$$u_j = \mathsf{R}_{jk}\tilde{u}_k, \qquad \tilde{u}_k = \mathsf{R}_{jk}u_j. \tag{2.8}$$

Relative position vectors are described in their respective coordinate systems, e.g., x_j to describe the coordinates of $\boldsymbol{x}$ in the inertial system, and $\tilde{x}_j$ to describe $\tilde{\boldsymbol{x}} = \boldsymbol{x} - \boldsymbol{X}_{\mathrm{r}}$ in the body-fixed system. These are related by

$$x_j = X_{\mathrm{r},j} + \mathsf{R}_{jk}\tilde{x}_k. \tag{2.9}$$

Index notation will be used extensively in this book, since it allows us to compactly express derivations of vector identities. However, occasionally even this notation is too cumbersome, particularly if the indices interfere with some other notation. Therefore, we will occasionally assemble components into a matrix notation, denoted with bold sans serif. For example, the components of vector $\boldsymbol{u}$ in the inertial and body-fixed systems will each be assembled into a 3×1 column array:

$$\mathsf{u} = \begin{pmatrix} u_1 \\ u_2 \\ u_3 \end{pmatrix}, \quad \tilde{\mathsf{u}} = \begin{pmatrix} \tilde{u}_1 \\ \tilde{u}_2 \\ \tilde{u}_3 \end{pmatrix}. \tag{2.10}$$

This allows us to alternatively express (2.8) and (2.9) in the matrix forms

$$\mathsf{u} = \mathsf{R}\tilde{\mathsf{u}}, \qquad \tilde{\mathsf{u}} = \mathsf{R}^{\mathsf{T}}\mathsf{u}, \qquad \mathsf{x} = \mathsf{X}_{\mathrm{r}} + \mathsf{R}\tilde{\mathsf{x}}, \tag{2.11}$$

where X_{r} denotes the components of the body reference point in the inertial system.

Basis vectors in the body-fixed frame change with time, due to the rotation of that frame with respect to the inertial frame. The rate of change of any such basis vector is given by

$$\frac{\mathrm{d}\tilde{\boldsymbol{e}}_j}{\mathrm{d}t} = \boldsymbol{\Omega} \times \tilde{\boldsymbol{e}}_j, \tag{2.12}$$

where $\boldsymbol{\Omega}$ is the angular velocity of the body-fixed frame. We note that Eq. (2.12) shows that $\tilde{\boldsymbol{e}}_3$ remains invariant in a two-dimensional problem—as it clearly must—since the rotational axis is aligned with this out-of-plane vector. It then follows from (2.12) that the rate of change of the rotation matrix is given (in terms of its components) by

$$\frac{\mathrm{d}\mathsf{R}_{jk}}{\mathrm{d}t} = \epsilon_{jml}\Omega_m \mathsf{R}_{lk}, \tag{2.13}$$

where ϵ_{ijk} are the components of the **permutation symbol** (A.3). This enables us to compute the rate of change of a vector's components in the body-fixed system:

$$\frac{\mathrm{d}\tilde{u}_k}{\mathrm{d}t} = \mathsf{R}_{jk}\left(\frac{\mathrm{d}u_j}{\mathrm{d}t} - \epsilon_{jml}\Omega_m u_l\right). \tag{2.14}$$

Thus, the change of such a vector component arises from two sources: from time dependence in the underlying vector (the first term) and from the rate of rotation of the coordinate frame (the second term).

We can represent relations (2.13) and (2.14) in matrix notation if we define one more useful (and commonly-used) notation. Let $\mathsf{a}^{\times}$ denote the skew-symmetric 3×3 matrix that performs the equivalent of the cross product of $\boldsymbol{a}$ with another vector. Specifically, define

$$\mathsf{a}^{\times} = \begin{bmatrix} 0 & -a_3 & a_2 \\ a_3 & 0 & -a_1 \\ -a_2 & a_1 & 0 \end{bmatrix}, \tag{2.15}$$

where a_j, for $j = 1, 2, 3$, are the components of vector $\boldsymbol{a}$ in the inertial coordinate system. Then, for two vectors $\boldsymbol{a}$ and $\boldsymbol{b}$, the matrix-vector product $\mathbf{a}^\times \mathbf{b}$ computes the components of the cross product $\boldsymbol{a} \times \boldsymbol{b}$ in the inertial system. Note that $(\mathbf{a}^\times)^\mathsf{T} = -\mathbf{a}^\times$.

Now, we can express (2.13) and (2.14) in the more compact forms

$$\dot{\mathbf{R}} = \boldsymbol{\Omega}^\times \mathbf{R} \tag{2.16}$$

and

$$\dot{\tilde{\mathbf{u}}} = \mathbf{R}^\mathsf{T} \left(\dot{\mathbf{u}} - \boldsymbol{\Omega}^\times \mathbf{u} \right), \tag{2.17}$$

respectively, where $\dot{()}$ denotes the time derivative.

Motion of Rigid Bodies Suppose a point $\tilde{\boldsymbol{x}}$ is fixed relative to the body reference frame. Therefore, its position $\boldsymbol{x}$ relative to the inertial system, given by (2.9), must vary in time if the body itself is moving. Generically, we can write this functional relationship as $\boldsymbol{x} = \boldsymbol{x}(\tilde{\boldsymbol{x}}, t)$, and the velocity of this point relative to the inertial system, $\boldsymbol{V}_\mathrm{b}$, can be computed from $\boldsymbol{V}_\mathrm{b}(\tilde{\boldsymbol{x}}, t) = \partial \boldsymbol{x}(\tilde{\boldsymbol{x}}, t)/\partial t$. That is, we simply differentiate (2.9) while holding the body-fixed coordinates constant. Using our previous results, it is easy to show that, in component form, this leads to

$$V_{\mathrm{b},j} = \frac{\partial x_j}{\partial t} = \frac{\mathrm{d}X_{\mathrm{r},j}}{\mathrm{d}t} + \epsilon_{jml} \Omega_m \mathsf{R}_{lk} \tilde{x}_k, \tag{2.18}$$

or, in matrix notation,

$$\mathbf{V}_\mathrm{b} := \frac{\partial \mathbf{x}}{\partial t} = \frac{\mathrm{d}\mathbf{X}_\mathrm{r}}{\mathrm{d}t} + \boldsymbol{\Omega}^\times \mathbf{R} \tilde{\mathbf{x}}. \tag{2.19}$$

Denoting the rate of translation of the body's reference point by $\boldsymbol{V}_\mathrm{r} := \mathrm{d}\boldsymbol{X}_\mathrm{r}/\mathrm{d}t$, Eq. (2.18) can be written in vector form as

$$\boldsymbol{V}_\mathrm{b} = \boldsymbol{V}_\mathrm{r} + \boldsymbol{\Omega} \times (\boldsymbol{x} - \boldsymbol{X}_\mathrm{r}). \tag{2.20}$$

In particular, Eq. (2.20) clearly holds for any **material point** of a rigid body—i.e., a point that moves with the body. In other words, if a body is rigid, its *configuration* in the inertial reference frame is completely described by the rotation $\mathbf{R}$ and the position $\boldsymbol{X}_\mathrm{r}$, and its *motion* relative to that frame is thus described by the rates of change of these: its angular velocity, $\boldsymbol{\Omega}$, and translational velocity, $\boldsymbol{V}_\mathrm{r}$, as depicted in Fig. 2.1. We make three notes:

- For a two-dimensional problem, note that $\tilde{\boldsymbol{e}}_3 = \boldsymbol{e}_3$, that $V_{\mathrm{r},3} \equiv 0$ (and $\tilde{V}_{\mathrm{r},3} \equiv 0$), and that $\boldsymbol{\Omega} = \Omega \boldsymbol{e}_3$, where Ω is the rate of change of the body's sole orientation angle, α, defined as the angle between the $\tilde{x}_j$ axis and x_j axis (for $j = 1$ or 2). In our matrix notation, we can generally specialize to two-dimensional problems by zeroing the out-of-plane components of motion (and eventually, force) vectors.
- For a single body, it is generally possible to assume—in situations where no generality is lost—that the reference point instantaneously coincides with the origin O of the inertial system, so that $\boldsymbol{X}_\mathrm{r} = 0$. This is because only the rate of change of this point's position is relevant to the fluid dynamics.

- The pair $(\boldsymbol{\Omega}, \boldsymbol{V}_{\mathrm{r}})$ is the unique specification of a body's motion based on the reference point $\boldsymbol{X}_{\mathrm{r}}$. However, the same motion can also be described about any other point—say, $\boldsymbol{x}_P$. The angular velocity based on this new point will be the same, $\boldsymbol{\Omega}$, and the translational velocity will be

$$\boldsymbol{V}_P = \boldsymbol{V}_{\mathrm{r}} + \boldsymbol{\Omega} \times (\boldsymbol{x}_P - \boldsymbol{X}_{\mathrm{r}}). \tag{2.21}$$

Therefore, one could also equivalently specify the motion of the body by the pair $(\boldsymbol{\Omega}, \boldsymbol{V}_P)$. In general, the point for which the translational motion is defined must be explicitly described.

Field Quantities in Different Reference Frames Since we will often deal with both the inertial and the body-fixed reference frame in a given problem, it is important to understand how to reconcile spatial field quantities between the two frames. In Sect. A.1.6, we describe the basic notation for a generic field quantity f as a function of position in inertial space and of time: $f(\boldsymbol{x}, t)$. We can alternatively regard that same field as a function of position $\tilde{\boldsymbol{x}}$ in the body-fixed reference frame: $\tilde{f}(\tilde{\boldsymbol{x}}, t)$. To relate these two different views of the same field, let's assume (without loss of generality) that at $t = 0$, the coordinate systems of the two frames coincide. Again, let's consider a point whose position $\tilde{\boldsymbol{x}}$ does not change relative to the body-fixed reference frame. At $t = 0$, $\boldsymbol{x}(\tilde{\boldsymbol{x}}, 0) = \tilde{\boldsymbol{x}}$, since the two systems coincide. As time proceeds, this point necessarily moves in inertial space, at a rate described by the rigid-body velocity $\boldsymbol{V}_{\mathrm{b}}(\tilde{\boldsymbol{x}}) = \boldsymbol{V}_{\mathrm{r}} + \boldsymbol{\Omega} \times \tilde{\boldsymbol{x}}$. In fact, we can think of this point as ***mapped*** to new positions in inertial space, by a time-parameterized flow map $\boldsymbol{\phi}_t$ constructed from the integral curves of this rigid-body velocity field: $\boldsymbol{x}(\tilde{\boldsymbol{x}}, t) = \boldsymbol{\phi}_t(\tilde{\boldsymbol{x}})$. In other words, the time rate of change of this flow map is the rigid-body velocity, $\boldsymbol{V}_{\mathrm{b}}(\tilde{\boldsymbol{x}})$. Using this map formalism, we can reconcile the field f as a function of corresponding points in each reference frame:

$$f(\boldsymbol{\phi}_t(\tilde{\boldsymbol{x}}), t) \equiv \tilde{f}(\tilde{\boldsymbol{x}}, t). \tag{2.22}$$

Since this equivalence holds at all instants, it is straightforward to relate their rates of change. In particular, the rate of change at the fixed position $\tilde{\boldsymbol{x}}$ in the body-fixed reference frame is simply $\partial \tilde{f}(\tilde{\boldsymbol{x}}, t)/\partial t$. When viewed in inertial space, this rate of change is described by Eq. (A.83), which here is specialized to

$$\frac{\partial f}{\partial t}(\boldsymbol{\phi}_t(\tilde{\boldsymbol{x}}), t) + \boldsymbol{V}_{\mathrm{b}}(\tilde{\boldsymbol{x}}) \cdot \nabla f(\boldsymbol{\phi}_t(\tilde{\boldsymbol{x}}), t) \equiv \frac{\partial \tilde{f}}{\partial t}(\tilde{\boldsymbol{x}}, t). \tag{2.23}$$

This equation is particularly useful when the field is invariant in one of the two reference frames, since that invariance implies that the corresponding partial derivative with respect to time in that frame is identically zero. The partial derivative in the other reference frame can then be computed in terms of the spatial gradient.

Suppose, instead, that we are interested in the time rate of change of a vector field. As with the scalar field, the vector field can be viewed from either frame of reference, and the field is equivalent at corresponding points between the two frames. However, we now have the additional complication that the field's components can be expressed in either coordinate system, and the basis vectors in the body-fixed frame vary in time with respect to the inertial frame. Let us consider a vector $\boldsymbol{u} = u_j \boldsymbol{e}_j = \tilde{u}_j \tilde{\boldsymbol{e}}_j$ and its rate of change. The derivatives of the corresponding perspectives of $\tilde{\boldsymbol{x}}$ give us

$$\frac{\mathrm{d}\boldsymbol{u}}{\mathrm{d}t}(\boldsymbol{\phi}_t(\tilde{\boldsymbol{x}}), t) = \frac{\mathrm{d}(u_j \boldsymbol{e}_j)}{\mathrm{d}t}(\boldsymbol{\phi}_t(\tilde{\boldsymbol{x}}), t) = \frac{\partial \boldsymbol{u}}{\partial t}(\tilde{\boldsymbol{x}}, t) = \frac{\partial(\tilde{u}_j \tilde{\boldsymbol{e}}_j)}{\partial t}(\tilde{\boldsymbol{x}}, t) \tag{2.24}$$

For the time rate of change of components in the inertial system, we can apply (2.23), and the time derivative of the unit vectors of the body-fixed system follow from Eq. (2.12):

$$\frac{\partial u_j}{\partial t}(\boldsymbol{\phi}_t(\tilde{\boldsymbol{x}}), t)\boldsymbol{e}_j + \boldsymbol{V}_\mathrm{b}(\tilde{\boldsymbol{x}}) \cdot \nabla u_j(\boldsymbol{\phi}_t(\tilde{\boldsymbol{x}}), t)\boldsymbol{e}_j = \frac{\partial \tilde{u}_j}{\partial t}(\tilde{\boldsymbol{x}}, t)\tilde{\boldsymbol{e}}_j + \tilde{u}_j(\tilde{\boldsymbol{x}}, t)\boldsymbol{\Omega} \times \tilde{\boldsymbol{e}}_j. \tag{2.25}$$

The unit vectors are spatially uniform and can therefore be pulled inside the gradient without consequence. Furthermore, it is easy to show that

$$\boldsymbol{\Omega} \times \boldsymbol{u} \equiv \boldsymbol{u} \cdot \nabla \boldsymbol{V}_\mathrm{b}. \tag{2.26}$$

Thus, after applying a vector identity on the collected terms, we arrive at the following:

$$\frac{\partial \boldsymbol{u}}{\partial t}(\boldsymbol{\phi}_t(\tilde{\boldsymbol{x}}), t) - \nabla \times (\boldsymbol{V}_\mathrm{b}(\tilde{\boldsymbol{x}}) \times \boldsymbol{u}(\boldsymbol{\phi}_t(\tilde{\boldsymbol{x}}), t)) \equiv \frac{\partial \tilde{u}_j}{\partial t}(\tilde{\boldsymbol{x}}, t)\tilde{\boldsymbol{e}}_j. \tag{2.27}$$

When we state that such a vector field is invariant in the body-fixed reference frame, this necessarily means that its components $\tilde{u}_j$ in that coordinate system are independent of time when observed at a point $\tilde{\boldsymbol{x}}$ fixed in that frame. In other words, the right-hand side of (2.27) vanishes.

Deformable Bodies If the body is deformable or has a flux through its surface, then we add to (2.20) a deformation velocity field, $\boldsymbol{u}_\mathrm{b}$, which accounts for deviations of the body's shape from rigid (or, alternatively, flux through the surface of a rigid body):

$$\boldsymbol{V}_\mathrm{b} = \boldsymbol{V}_\mathrm{r} + \boldsymbol{\Omega} \times (\boldsymbol{x}_\mathrm{b} - \boldsymbol{X}_\mathrm{r}) + \boldsymbol{u}_\mathrm{b}. \tag{2.28}$$

Throughout this book, we will assume that the body's motion preserves its volume. This is expressed mathematically as a constraint on the surface integral of the body's normal component of velocity,

$$\int_{S_\mathrm{b}} \boldsymbol{V}_\mathrm{b} \cdot \boldsymbol{n} \, \mathrm{d}S = 0. \tag{2.29}$$

This, of course, is trivially satisfied for a rigid body. For a deforming body (or a body with surface flux), it requires us to place the same constraint on the deformation velocity, $\boldsymbol{u}_\mathrm{b}$.

The Windtunnel Reference Frame An example of a windtunnel reference frame is depicted in Fig. 2.2. In this reference frame, the fluid is in motion at infinity, at uniform (but possibly unsteady) velocity $\boldsymbol{V}_\infty$. Remember that, in this book, we regard this reference frame as non-inertial in general. The origin of its coordinate system, initially coincident with that of the inertial system, is at position $\boldsymbol{X}_{O^\dagger}(t) := -\int_0^t \boldsymbol{V}_\infty(\tau)\,\mathrm{d}\tau$ at time t. Thus, any position $\boldsymbol{x}^\dagger$ in the windtunnel coordinate system is related to position $\boldsymbol{x}$ in the inertial coordinate system by

$$\boldsymbol{x} = \boldsymbol{x}^\dagger + \boldsymbol{X}_{O^\dagger}(t). \tag{2.30}$$

Note that, at all times, the basis vectors of the inertial and windtunnel coordinate systems are identical and time invariant.

A velocity $\boldsymbol{v}$ in the inertial reference frame is related to the corresponding velocity, $\boldsymbol{v}^\dagger$, measured in the windtunnel frame by $\boldsymbol{v}^\dagger = \boldsymbol{v} + \boldsymbol{V}_\infty$. For example, consider the simple (and common) case of a body at rest in the windtunnel frame. A stationary observer in the inertial frame moves with the bulk fluid, so, to such an observer, this body appears to move at velocity $-\boldsymbol{V}_\infty$. Since we will refer most analyses to the inertial frame of reference, any problem initially posed in a windtunnel reference frame of uniform flow $\boldsymbol{V}_\infty$, consisting of a body in rigid motion with translational velocity $\boldsymbol{V}_\mathrm{r}^\dagger$ and angular velocity $\boldsymbol{\Omega}^\dagger$, can be equivalently expressed in the inertial reference frame with translational velocity $\boldsymbol{V}_\mathrm{r} = \boldsymbol{V}_\mathrm{r}^\dagger - \boldsymbol{V}_\infty$ and angular velocity $\boldsymbol{\Omega} = \boldsymbol{\Omega}^\dagger$. If the uniform flow is time varying, there will generally be a difference in the pressures and forces observed in the two frames. This will be addressed later in the book.

Tensors We will also have occasion to use (rank-2) tensors in this book. Like vectors, these can be expressed in either of the two coordinate systems. For tensor $\boldsymbol{M}$,

$$\boldsymbol{M} = M_{ij}\boldsymbol{e}_i\boldsymbol{e}_j = \tilde{M}_{ij}\tilde{\boldsymbol{e}}_i\tilde{\boldsymbol{e}}_j. \tag{2.31}$$

Clearly, we can relate the components in these two systems by the same means as for vectors, via the rotation matrix:

$$\tilde{M}_{kl} = \mathsf{R}_{ik}\mathsf{R}_{jl}M_{ij}, \tag{2.32}$$

or compactly, in matrix notation, by

$$\tilde{\mathbf{M}} = \mathbf{R}^\mathsf{T}\mathbf{M}\mathbf{R}. \tag{2.33}$$

The rates of change of the tensorial components are easily related by differentiating this transformation, with the help of (2.13):

$$\frac{\mathrm{d}\tilde{M}_{kl}}{\mathrm{d}t} = \mathsf{R}_{ik}\mathsf{R}_{jl}\left(\frac{\mathrm{d}M_{ij}}{\mathrm{d}t} - \epsilon_{imn}\Omega_m M_{nj} - \epsilon_{jmn}\Omega_m M_{in}\right), \tag{2.34}$$

which in matrix notation is

$$\dot{\tilde{\mathbf{M}}} = \mathbf{R}^\mathsf{T}\left(\dot{\mathbf{M}} - \mathbf{\Omega}^\times\mathbf{M} + \mathbf{M}\mathbf{\Omega}^\times\right)\mathbf{R}, \tag{2.35}$$

where again we have used $\dot{()}$ to denote time derivative.

2.4 Complex Notation

In two-dimensional problems, we will make extensive use of complex analysis. In this context, we will denote the coordinate directions as x and y rather than x_1 and x_2, and the Cartesian components of the velocity as u and v rather than v_1 and v_2. The complex position in the inertial system is denoted by $z = x + \mathrm{i}y$, and the **complex conjugate velocity** (often referred to as simply the **complex velocity**) is described by $w = u - \mathrm{i}v$. The reason for the minus sign will be clarified when we introduce the complex potential in Chap. 3. We denote the equivalence between w and the vector form $\boldsymbol{v}$ using notation introduced in Sect. A.2:

$$\boldsymbol{v} \Longleftrightarrow w^*, \tag{2.36}$$

where $(\cdot)^*$ denotes complex conjugate. This simply formalizes the fact that the real and imaginary parts of w^* are equal to the components of $\boldsymbol{v}$ in the inertial coordinate system.

In complex form, the inertial (z) and body-fixed ($\tilde{z}$) coordinate systems are related to each other through the position of the reference point, Z_r, and the rotation operator, $\mathrm{e}^{\mathrm{i}\alpha}$, based on the current angle, α, of the $\tilde{x}$ axis with respect to the x axis:

$$z = Z_\mathrm{r} + \tilde{z}\mathrm{e}^{\mathrm{i}\alpha}. \tag{2.37}$$

This is the complex equivalent of (2.9). Note that any other complex vector can be transformed from body-fixed to inertial components by multiplying by $\mathrm{e}^{\mathrm{i}\alpha}$. For the complex conjugate velocity, however, we need the conjugate of this rotation, so $\mathrm{e}^{\mathrm{i}\alpha}$ is used to transform from the inertial to the body-fixed system:

$$\tilde{w} = w\mathrm{e}^{\mathrm{i}\alpha}. \tag{2.38}$$

If a point $\tilde{z}$ is fixed in the body reference frame, then its coordinates z in the inertial system must change in time as the body moves. Its velocity relative to the inertial frame is found by differentiating (2.37). Denoting this by the complex velocity $W_\mathrm{b} := U_\mathrm{b} - \mathrm{i}V_\mathrm{b}$, the result is

$$W_\mathrm{b} = W_\mathrm{r} - \mathrm{i}\Omega(z^* - Z_\mathrm{r}^*), \tag{2.39}$$

where $W_\mathrm{r} := \dot{Z}_\mathrm{r}^*$ denotes the complex velocity of the body's reference point, Z_r. The notation is easily reconciled with the vector components: for example, $\mathrm{Re}(W_\mathrm{r}) \equiv U_\mathrm{r} \equiv V_{\mathrm{r},1}$ and $\mathrm{Im}(W_\mathrm{r}) \equiv -V_\mathrm{r} \equiv -V_{\mathrm{r},2}$. Thus, (2.39) describes the velocity of any material point on a rigid body, and, by comparing with (2.20), the following vector/complex equivalence is noted:

$$\boldsymbol{V}_\mathrm{b} \Longleftrightarrow W_\mathrm{b}^*. \tag{2.40}$$

As mentioned in the vectorial context, we will often assume $Z_\mathrm{r} = 0$ wherever generality is not sacrificed.

In body-fixed coordinates, we can write the velocity of a point $\tilde{z}_\mathrm{b}$ on a rigid body as

$$\tilde{W}_\mathrm{b} := \tilde{U}_\mathrm{b} - \mathrm{i}\tilde{V}_\mathrm{b} = W_\mathrm{b}\mathrm{e}^{\mathrm{i}\alpha} = \tilde{W}_\mathrm{r} - \mathrm{i}\Omega\tilde{z}_\mathrm{b}^*, \tag{2.41}$$

where $\tilde{W}_r := \tilde{U}_r - i\tilde{V}_r \equiv W_r e^{i\alpha}$ describes the components of the reference point's velocity in body-fixed coordinates. The rate of change of this reference point's velocity is easily determined by differentiating its relationship with its inertial form,

$$\frac{d\tilde{W}_r}{dt} = \left(\dot{W}_r + i\Omega W_r\right) e^{i\alpha}. \tag{2.42}$$

This, of course, can be generalized to any complex velocity, and is clearly the complex version of the equation for the vector components (2.14), applied to the reference point velocity W_r.

We insist that the body's enclosed area is preserved by constraining its motion so that

$$\mathrm{Im} \int_{C_b} W_b(z)\, dz = 0, \tag{2.43}$$

where C_b is the contour that tightly encloses the body. This contour and the body surface S_b are effectively equivalent in two-dimensional problems (with the latter interpreted per unit depth). In fact, (2.43) is simply the complex form of (2.29), as can be verified from the notes on complex analysis in Sect. A.2 of the Appendix. The left-hand side of Eq. (2.43) vanishes identically for rigid-body motion. As in the vector case, we can account for body deformation with an additional deformation velocity field, denoted here by its conjugate form, $w_b(z)$, which is constrained by (2.43).

2.5 Plücker Notation

Analysis of rigid-body dynamics, which forms an inseparable part of the unsteady flows pursued in this book, is greatly assisted by a notation in which the components of angular and linear motion and of moment and force are represented by six-dimensional column 'vectors', sometimes called **Plücker coordinates**. Motion vectors, such as velocity and acceleration, are distinguished from force vectors, such as moment/force, momentum and impulse.

For example, the velocity of the body (in inertial coordinates), defined about the body's reference point, is written (and subsequently shortened to two 3×1 arrays) as

$$\begin{pmatrix} \Omega_1 \\ \Omega_2 \\ \Omega_3 \\ V_{r,1} \\ V_{r,2} \\ V_{r,3} \end{pmatrix} := \begin{pmatrix} \boldsymbol{\Omega} \\ \mathbf{V}_r \end{pmatrix}. \tag{2.44}$$

Similarly, the Plücker vector for the moment and force on the body, written in the components of the body-fixed coordinate system, is

$$\begin{pmatrix} \tilde{m}_{r,1} \\ \tilde{m}_{r,2} \\ \tilde{m}_{r,3} \\ \tilde{f}_1 \\ \tilde{f}_2 \\ \tilde{f}_3 \end{pmatrix} := \begin{pmatrix} \tilde{\mathbf{m}}_r \\ \tilde{\mathbf{f}} \end{pmatrix}. \tag{2.45}$$

Other Plücker vectors, in any coordinate system, follow this same template. Note that, in all cases, the angular components are listed above the linear components. In two-dimensional problems, only the third, fourth and fifth entry in a vector are non-zero.

Operations on these Plücker vectors are performed with 6×6 matrices: 2×2 matrices of blocks, each of size 3×3. For motion or force vectors, the rotation from the body-fixed coordinate system to inertial coordinates is performed with the rotation matrix (2.11), now assembled into a 6×6 matrix to transform both the angular and linear parts of the Plücker array, e.g.,

$$\begin{pmatrix} \mathbf{m}_r \\ \mathbf{f} \end{pmatrix} = \begin{bmatrix} \mathbf{R} & 0 \\ 0 & \mathbf{R} \end{bmatrix} \begin{pmatrix} \tilde{\mathbf{m}}_r \\ \tilde{\mathbf{f}} \end{pmatrix}. \tag{2.46}$$

Recall that the inverse of rotation $\mathbf{R}$—i.e., to rotate from inertial to body-fixed components—is given by its transpose $\mathbf{R}^T$. For example,

$$\begin{pmatrix} \tilde{\boldsymbol{\Omega}} \\ \tilde{\mathbf{V}}_r \end{pmatrix} = \begin{bmatrix} \mathbf{R}^T & 0 \\ 0 & \mathbf{R}^T \end{bmatrix} \begin{pmatrix} \boldsymbol{\Omega} \\ \mathbf{V}_r \end{pmatrix}. \tag{2.47}$$

Remember that the moment vector can be evaluated about another point P with the help of the translation formula

$$\boldsymbol{m}_P = \boldsymbol{m}_O - \boldsymbol{x}_P \times \boldsymbol{f}, \tag{2.48}$$

where $\boldsymbol{m}_O$ is the moment about the origin of the inertial coordinate system, and $\boldsymbol{x}_P$ is the vector position of the point P relative to this origin. Thus, to translate a Plücker force vector (or momentum or impulse) between, say, the body's reference point and the origin, we use the transformation given by

$$\begin{pmatrix} \mathbf{m}_O \\ \mathbf{f} \end{pmatrix} = \begin{bmatrix} \mathbf{1} & \mathbf{X}_r^\times \\ 0 & \mathbf{1} \end{bmatrix} \begin{pmatrix} \mathbf{m}_r \\ \mathbf{f} \end{pmatrix}, \tag{2.49}$$

where $\mathbf{1}$ is the 3×3 identity matrix and $\mathbf{X}_r^\times$ is the cross-product matrix (2.15) containing the inertial coordinates of the body's reference point, $\boldsymbol{X}_r$. In order to invert the translation, we simply switch the sign of the $\mathbf{X}_r^\times$ entry: e.g., to obtain the moment about the reference point from the moment about the origin,

$$\begin{pmatrix} \mathbf{m}_r \\ \mathbf{f} \end{pmatrix} = \begin{bmatrix} \mathbf{1} & -\mathbf{X}_r^\times \\ 0 & \mathbf{1} \end{bmatrix} \begin{pmatrix} \mathbf{m}_O \\ \mathbf{f} \end{pmatrix}. \tag{2.50}$$

Clearly, the transformations (2.46) and (2.49) can be composed into a single transformation, ${}^{\mathrm{i}}\mathcal{T}_{\mathrm{b}}^{(\mathrm{f})}$, from the body-fixed components of a force vector about the reference point of the body, to inertial components about the origin. This transformation, called the **body-to-inertial force transformation operator**, is

$$ {}^{\mathrm{i}}\mathcal{T}_{\mathrm{b}}^{(\mathrm{f})} := \begin{bmatrix} \mathbf{1} & \mathbf{X}_{\mathrm{r}}^{\times} \\ 0 & \mathbf{1} \end{bmatrix} \begin{bmatrix} \mathbf{R} & 0 \\ 0 & \mathbf{R} \end{bmatrix}. \tag{2.51} $$

With this operator, we can now transform Plücker coordinates in the body frame to coordinates in the inertial frame, e.g.,

$$ \begin{pmatrix} \mathbf{m}_{\mathrm{O}} \\ \mathbf{f} \end{pmatrix} = {}^{\mathrm{i}}\mathcal{T}_{\mathrm{b}}^{(\mathrm{f})} \begin{pmatrix} \tilde{\mathbf{m}}_{\mathrm{r}} \\ \tilde{\mathbf{f}} \end{pmatrix}. \tag{2.52} $$

The inverse, ${}^{\mathrm{b}}\mathcal{T}_{\mathrm{i}}^{(\mathrm{f})}$, is simply composed of the individual inverses, applied in reverse order:

$$ {}^{\mathrm{b}}\mathcal{T}_{\mathrm{i}}^{(\mathrm{f})} := \begin{bmatrix} \mathbf{R}^{\mathrm{T}} & 0 \\ 0 & \mathbf{R}^{\mathrm{T}} \end{bmatrix} \begin{bmatrix} \mathbf{1} & -\mathbf{X}_{\mathrm{r}}^{\times} \\ 0 & \mathbf{1} \end{bmatrix}. \tag{2.53} $$

The operator ${}^{\mathrm{i}}\mathcal{T}_{\mathrm{b}}^{(\mathrm{f})}$ can be generalized to allow transformations from any coordinate system to any another. To construct such an operator, from system A to system B, one writes

$$ {}^{B}\mathcal{T}_{A}^{(\mathrm{f})} := \begin{bmatrix} \mathbf{1} & -{}^{B}\mathbf{X}_{A}^{\times} \\ 0 & \mathbf{1} \end{bmatrix} \begin{bmatrix} {}^{B}\mathbf{R}_{A} & 0 \\ 0 & {}^{B}\mathbf{R}_{A} \end{bmatrix}. \tag{2.54} $$

where ${}^{B}\mathbf{X}_{A}^{\times}$ contains the position of the origin of B relative to the origin of A, expressed in system B's coordinates. The entries in the rotation matrix ${}^{B}\mathbf{R}_{A}$ are given by dot products of the unit vectors of each system, as follows:

$$ {}^{B}\mathsf{R}_{A,jk} := \boldsymbol{e}_{j}^{B} \cdot \boldsymbol{e}_{k}^{A}. \tag{2.55} $$

Motion vectors can be translated from one reference point to another in similar fashion, based on Eq. (2.21). For example, to translate the inertial components of velocity from the body reference point to the inertial system's origin, we write

$$ \begin{pmatrix} \mathbf{\Omega} \\ \mathbf{V}_{\mathrm{O}} \end{pmatrix} = \begin{bmatrix} \mathbf{1} & 0 \\ \mathbf{X}_{\mathrm{r}}^{\times} & \mathbf{1} \end{bmatrix} \begin{pmatrix} \mathbf{\Omega} \\ \mathbf{V}_{\mathrm{r}} \end{pmatrix}. \tag{2.56} $$

The composition of rotation and translation leads to the **body-to-inertial motion transformation operator**,

$$ {}^{\mathrm{i}}\mathcal{T}_{\mathrm{b}}^{(\mathrm{m})} := \begin{bmatrix} \mathbf{1} & 0 \\ \mathbf{X}_{\mathrm{r}}^{\times} & \mathbf{1} \end{bmatrix} \begin{bmatrix} \mathbf{R} & 0 \\ 0 & \mathbf{R} \end{bmatrix}, \tag{2.57} $$

which allows us to write

$$ \begin{pmatrix} \mathbf{\Omega} \\ \mathbf{V}_{\mathrm{O}} \end{pmatrix} = {}^{\mathrm{i}}\mathcal{T}_{\mathrm{b}}^{(\mathrm{m})} \begin{pmatrix} \tilde{\mathbf{\Omega}} \\ \tilde{\mathbf{V}}_{\mathrm{r}} \end{pmatrix}. \tag{2.58} $$

The inverse is,

$$ {}^{\mathrm{b}}\mathcal{T}_{\mathrm{i}}^{(\mathrm{m})} := \begin{bmatrix} \mathbf{R}^{\mathsf{T}} & 0 \\ 0 & \mathbf{R}^{\mathsf{T}} \end{bmatrix} \begin{bmatrix} \mathbf{1} & 0 \\ -\mathbf{X}_{\mathrm{r}}^{\times} & \mathbf{1} \end{bmatrix}. \tag{2.59} $$

The general motion transformation operator from system A to system B is given by

$$ {}^{B}\mathcal{T}_{A}^{(\mathrm{m})} := \begin{bmatrix} \mathbf{1} & 0 \\ -{}^{B}\mathbf{X}_{A}^{\times} & \mathbf{1} \end{bmatrix} \begin{bmatrix} {}^{B}\mathbf{R}_{A} & 0 \\ 0 & {}^{B}\mathbf{R}_{A} \end{bmatrix}, \tag{2.60} $$

where ${}^{B}\mathbf{X}_{A}^{\times}$ has the same meaning as in (2.54). In general, the force and motion transformation operators are related by ${}^{B}\mathcal{T}_{A}^{(\mathrm{f})} = ({}^{A}\mathcal{T}_{B}^{(\mathrm{m})})^{\mathsf{T}}$ and ${}^{B}\mathcal{T}_{A}^{(\mathrm{m})} = ({}^{A}\mathcal{T}_{B}^{(\mathrm{f})})^{\mathsf{T}}$.

Note 2.5.1: Plücker Transform Matrices in Two Dimensions

As we noted earlier, only the third, fourth and fifth entry in a Plücker vector—the sole angular component and the two planar linear components—are non-zero in two-dimensional applications. Thus, the transform matrices in such contexts consist only of the 3×3 square sub-matrix consisting of those rows and columns. For example, the rotation transform consists of the following boxed portion

$$ \begin{bmatrix} \mathbf{R} & 0 \\ 0 & \mathbf{R} \end{bmatrix} = \left[\begin{array}{cc|ccc|c} \mathsf{R}_{11} & \mathsf{R}_{12} & \mathsf{R}_{13} & 0 & 0 & 0 \\ \mathsf{R}_{21} & \mathsf{R}_{22} & \mathsf{R}_{23} & 0 & 0 & 0 \\ \hline \mathsf{R}_{31} & \mathsf{R}_{32} & \mathsf{R}_{33} & 0 & 0 & 0 \\ 0 & 0 & 0 & \mathsf{R}_{11} & \mathsf{R}_{12} & \mathsf{R}_{13} \\ 0 & 0 & 0 & \mathsf{R}_{21} & \mathsf{R}_{22} & \mathsf{R}_{23} \\ \hline 0 & 0 & 0 & \mathsf{R}_{31} & \mathsf{R}_{32} & \mathsf{R}_{33} \end{array}\right], \tag{2.61} $$

where $\mathsf{R}_{33} = 1$. Analogously, the translation transform reduces to the corresponding boxed sub-matrix, shown here for a translation of a force-type vector:

$$ \begin{bmatrix} \mathbf{1} & -\mathbf{X}^{\times} \\ 0 & \mathbf{1} \end{bmatrix} = \left[\begin{array}{cc|ccc|c} 1 & 0 & 0 & 0 & X_3 & -X_2 \\ 0 & 1 & 0 & -X_3 & 0 & X_1 \\ \hline 0 & 0 & 1 & X_2 & -X_1 & 0 \\ 0 & 0 & 0 & 1 & 0 & 0 \\ 0 & 0 & 0 & 0 & 1 & 0 \\ \hline 0 & 0 & 0 & 0 & 0 & 1 \end{array}\right]. \tag{2.62} $$

The overall transformation operator is simply the composition of these boxed sub-matrices.

Later in this book, we will need the rate of change of the body-to-inertial transformation operator (2.51), with the result inverted back to body-fixed components about the reference point. Using the product rule and the rate of change of the rotation matrix (2.16), one can verify that, for the force transformation, this leads to

$$ {}^{\mathrm{b}}\mathcal{T}_{\mathrm{i}}^{(\mathrm{f})} \frac{\mathrm{d}\,{}^{\mathrm{i}}\mathcal{T}_{\mathrm{b}}^{(\mathrm{f})}}{\mathrm{d}t} = \begin{bmatrix} \mathbf{R}^{\mathrm{T}} & 0 \\ 0 & \mathbf{R}^{\mathrm{T}} \end{bmatrix} \begin{bmatrix} \boldsymbol{\Omega}^{\times} & \mathbf{V}_{\mathrm{r}}^{\times} \\ 0 & \boldsymbol{\Omega}^{\times} \end{bmatrix} \begin{bmatrix} \mathbf{R} & 0 \\ 0 & \mathbf{R} \end{bmatrix} = \begin{bmatrix} \tilde{\boldsymbol{\Omega}}^{\times} & \tilde{\mathbf{V}}_{\mathrm{r}}^{\times} \\ 0 & \tilde{\boldsymbol{\Omega}}^{\times} \end{bmatrix}. \tag{2.63} $$

In the last result, the vectors in the cross-product operators have been expressed in body-fixed components. Similarly, for the motion transformation operator,

$$ {}^{\mathrm{b}}\mathcal{T}_{\mathrm{i}}^{(\mathrm{m})} \frac{\mathrm{d}\,{}^{\mathrm{i}}\mathcal{T}_{\mathrm{b}}^{(\mathrm{m})}}{\mathrm{d}t} = \begin{bmatrix} \tilde{\boldsymbol{\Omega}}^{\times} & 0 \\ \tilde{\mathbf{V}}_{\mathrm{r}}^{\times} & \tilde{\boldsymbol{\Omega}}^{\times} \end{bmatrix}. \tag{2.64} $$

We can use this last result to compute the body's acceleration (relative to the inertial reference frame), which is defined as the rate of change of the body's velocity when viewed from the inertial reference frame. This is embodied by the time derivative of (2.58), which, when combined with (2.64), leads to

$$ \begin{pmatrix} \dot{\boldsymbol{\Omega}} \\ \dot{\mathbf{V}}_{\mathrm{O}} \end{pmatrix} = \frac{\mathrm{d}}{\mathrm{d}t}\left[{}^{\mathrm{i}}\mathcal{T}_{\mathrm{b}}^{(\mathrm{m})} \begin{pmatrix} \tilde{\boldsymbol{\Omega}} \\ \tilde{\mathbf{V}}_{\mathrm{r}} \end{pmatrix} \right] = {}^{\mathrm{i}}\mathcal{T}_{\mathrm{b}}^{(\mathrm{m})} \begin{pmatrix} \dot{\tilde{\boldsymbol{\Omega}}} \\ \dot{\tilde{\mathbf{V}}}_{\mathrm{r}} \end{pmatrix} + {}^{\mathrm{i}}\mathcal{T}_{\mathrm{b}}^{(\mathrm{m})} \begin{bmatrix} \tilde{\boldsymbol{\Omega}}^{\times} & 0 \\ \tilde{\mathbf{V}}_{\mathrm{r}}^{\times} & \tilde{\boldsymbol{\Omega}}^{\times} \end{bmatrix} \begin{pmatrix} \tilde{\boldsymbol{\Omega}} \\ \tilde{\mathbf{V}}_{\mathrm{r}} \end{pmatrix}. \tag{2.65} $$

But the last term on the right-hand side is identically zero. Thus, the acceleration of the body is transformed in the same manner as velocity:

$$ \begin{pmatrix} \dot{\boldsymbol{\Omega}} \\ \dot{\mathbf{V}}_{\mathrm{O}} \end{pmatrix} = {}^{\mathrm{i}}\mathcal{T}_{\mathrm{b}}^{(\mathrm{m})} \begin{pmatrix} \dot{\tilde{\boldsymbol{\Omega}}} \\ \dot{\tilde{\mathbf{V}}}_{\mathrm{r}} \end{pmatrix}, \quad \begin{pmatrix} \dot{\tilde{\boldsymbol{\Omega}}} \\ \dot{\tilde{\mathbf{V}}}_{\mathrm{r}} \end{pmatrix} = {}^{\mathrm{b}}\mathcal{T}_{\mathrm{i}}^{(\mathrm{m})} \begin{pmatrix} \dot{\boldsymbol{\Omega}} \\ \dot{\mathbf{V}}_{\mathrm{O}} \end{pmatrix}. \tag{2.66} $$

Note that this does *not* represent the acceleration of any material point affixed to the body, as would be true of the classical acceleration. Rather, it is a measure of the change in the flow of material points through a fixed observation location (i.e., using fluid dynamics terminology, it is the Eulerian derivative of velocity, not the Lagrangian). The acceleration as defined here is the more relevant measure in rigid-body dynamics, for it describes the overall acceleration of the body; the definition ensures that it is a true vector, with components that transform accordingly.

Note 2.5.2: Setting the Pivot Axis of a Translating and Rotating Body

It is often the case in unsteady aerodynamics that a body rotates about a certain pivot axis, whose own motion is relatively simple. However, much of the analysis that we perform in this book is predicated on the body's reference point, $\boldsymbol{X}_{\mathrm{r}}$, which, in general, does not coincide with the pivot. The motion transformation described in (2.60) gives us an easy way to write the velocity of this pivoting in the form we need it.

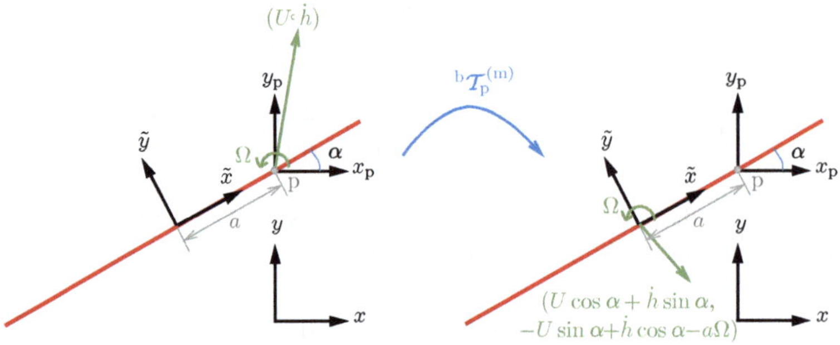

Fig. 2.3 Transform of the motion of a pivoting plate from the pivot coordinate system (left) to the body-fixed system (right)

Let's consider a two-dimensional example of this, illustrated in Fig. 2.3, in which the pivot axis is in the out-of-plane direction and located at an invariant point $(\tilde{x}_p, \tilde{y}_p) = (a, 0)$ in the body-fixed coordinate system. Let the angle of the body relative to the inertial x axis be denoted by α, and its rate of change by Ω. We will suppose that the pivot point's motion is described in inertial coordinates by velocity $\boldsymbol{V}_p = U\boldsymbol{e}_1 + \dot{h}\boldsymbol{e}_2$.

There are three different coordinate systems depicted in the diagram in Fig. 2.3: the usual two—the inertial system and the body-fixed system—and an additional one, whose origin is attached to the pivot point p, and whose axes are always aligned with those of the inertial system. The pivot coordinate system has been constructed to make it easy to describe the body's velocity in those coordinates. Remember, the Plücker vectors in a two-dimensional problem only have three elements: the angular velocity and the two linear velocity components:

$$\begin{pmatrix} \boldsymbol{\Omega} \\ \mathbf{V}_p \end{pmatrix} = \begin{pmatrix} \Omega \\ U \\ \dot{h} \end{pmatrix}. \tag{2.67}$$

Now, to transform the body's velocity into its body-fixed form, we need to construct the pivot-to-body transformation operator, ${}^{b}\mathcal{T}_{p}^{(m)}$. It is easy to construct the following:

$${}^{b}\mathbf{X}_{p}^{\times} = \begin{bmatrix} 0 & 0 & 0 \\ 0 & 0 & -a \\ 0 & a & 0 \end{bmatrix}, \quad {}^{b}\mathbf{R}_{p} = \begin{bmatrix} \cos\alpha & \sin\alpha & 0 \\ -\sin\alpha & \cos\alpha & 0 \\ 0 & 0 & 1 \end{bmatrix}. \tag{2.68}$$

Note that the components of the first are expressed in the body-fixed coordinate system: The rotation operator acts first, rotating a vector from the pivot axes to the body-fixed axes, and then the translation operator shifts it to the reference point. Then, following the general template given by (2.60), and restricting them to the sub-matrices of a two-dimensional transform, as exhibited in Note 2.5.1, the motion transform is

$$ {}^{\mathrm{b}}\mathcal{T}_{\mathrm{p}}^{(\mathrm{m})} = \begin{bmatrix} 1 & 0 & 0 \\ 0 & \cos\alpha & \sin\alpha \\ -a & -\sin\alpha & \cos\alpha \end{bmatrix}, \tag{2.69} $$

and the body's motion in its own coordinate system follows easily:

$$ \begin{pmatrix} \tilde{\boldsymbol{\Omega}} \\ \tilde{\mathbf{V}}_{\mathrm{r}} \end{pmatrix} = \begin{bmatrix} 1 & 0 & 0 \\ 0 & \cos\alpha & \sin\alpha \\ -a & -\sin\alpha & \cos\alpha \end{bmatrix} \begin{pmatrix} \Omega \\ U \\ \dot{h} \end{pmatrix} = \begin{pmatrix} \Omega \\ U\cos\alpha + \dot{h}\sin\alpha \\ -U\sin\alpha + \dot{h}\cos\alpha - a\Omega \end{pmatrix}. \tag{2.70} $$

The body's acceleration, in body-fixed coordinates, then follows by simply differentiating this expression:

$$ \begin{pmatrix} \dot{\tilde{\boldsymbol{\Omega}}} \\ \dot{\tilde{\mathbf{V}}}_{\mathrm{r}} \end{pmatrix} = \begin{pmatrix} \dot{\Omega} \\ \dot{U}\cos\alpha + \ddot{h}\sin\alpha - U\Omega\sin\alpha + \dot{h}\Omega\cos\alpha \\ -\dot{U}\sin\alpha + \ddot{h}\cos\alpha - a\dot{\Omega} - U\Omega\cos\alpha - \dot{h}\Omega\sin\alpha \end{pmatrix}. \tag{2.71} $$

Finally, we note that, if we desire to express the force and moment on the body in the pivot axes (e.g., to obtain the moment about the pivot point), then we would utilize the operator ${}^{\mathrm{p}}\mathcal{T}_{\mathrm{b}}^{(\mathrm{f})}$, which is simply the transpose of ${}^{\mathrm{b}}\mathcal{T}_{\mathrm{p}}^{(\mathrm{m})}$ in (2.69).

Chapter 3
Foundational Concepts

Contents

3.1 Flow Field Definitions and General Relations 26
  3.1.1 The Velocity Field and Flow Potentials 26
  3.1.2 The No-Penetration Condition 28
  3.1.3 Vorticity Transport Equation 29
  3.1.4 Kinetic Energy and Rate of Work 30
  3.1.5 Circulation and Its Invariance 34
  3.1.6 Helmholtz' Theorems 39
3.2 Elements of Potential Flow 43
  3.2.1 Pressure and the Bernoulli Equation 44
  3.2.2 Two-Dimensional Flows and the Complex Potential 46
  3.2.3 Three-Dimensional Flows 56
3.3 Vortex Structures 58
  3.3.1 Point Vortex 59
  3.3.2 Vortex Filament 60
  3.3.3 Vortex Sheet 66
3.4 Other Surface Distributions of Singularities 72
3.5 Relationships Between Singularity Distributions 74
  3.5.1 Two-Dimensional Vortex Sheet as a Set of Point Vortices 74
  3.5.2 Vortex Filament or Vortex Pair as a Double-Layer Potential 76
  3.5.3 Vortex Sheet as a Double-Layer Potential 77
  3.5.4 Three-Dimensional Vortex Sheet as a Mesh of Vortex Filaments 78
3.6 Free and Bound Vortex Sheets 80
  3.6.1 Free Vortex Sheets 81
  3.6.2 Bound Vortex Sheets 85

This book is focused on inviscid, incompressible flows generated by or interacting with impenetrable bodies in motion. This chapter is focused on presenting many of the basic concepts and physical laws on which the rest of this book depends. Some of these topics are standard fare in a graduate-level text on fluid dynamics. However, there are also some topics that are less commonly treated, so these are presented in somewhat more detail.

J. D. Eldredge, *Mathematical Modeling of Unsteady Inviscid Flows*, Interdisciplinary Applied Mathematics 50, https://doi.org/10.1007/978-3-030-18319-6_3

3.1 Flow Field Definitions and General Relations

3.1.1 The Velocity Field and Flow Potentials

A general differentiable velocity field, $\boldsymbol{v}(\boldsymbol{x}, t)$ (or, for that matter, any vector field), can be derived from two *potentials fields*,

$$\boldsymbol{v} = \nabla\varphi + \nabla \times \boldsymbol{\Psi}. \tag{3.1}$$

This is usually known as the Helmholtz decomposition (or, in geometric mechanics, the Hodge decomposition), and is proved to exist in Sect. A.1.3 in the Appendix. The first potential, φ, is called the **scalar potential**, and the second, $\boldsymbol{\Psi}$, is called the **vector potential**. It should be noted that we could add the gradient of any scalar field to $\boldsymbol{\Psi}$ without affecting the velocity field, since the curl of the gradient is always zero. To avoid this non-uniqueness, we generally place the additional constraint on $\boldsymbol{\Psi}$ that $\nabla \cdot \boldsymbol{\Psi} = 0$. This constraint, which simplifies the relationships below, is more important in the field of electrodynamics, where it is called **setting the gauge**.

Each of these potentials satisfies a Poisson equation. To show this, let us first take the divergence of (3.1). The divergence of the curl of a vector is always identically zero, so the vector potential term vanishes; however, the divergence of the gradient is the Laplacian operator, so we end up with

$$\nabla^2\varphi = \Theta := \nabla \cdot \boldsymbol{v}, \tag{3.2}$$

where we have defined the **rate of dilatation**, Θ, as the divergence of the velocity field. In contrast, when we apply the curl to (3.1), then the scalar potential term vanishes because the curl of the gradient is also identically zero; the curl of the curl of the vector potential, however, satisfies the following vector identity, easily verified using index notation:

$$\nabla \times (\nabla \times \boldsymbol{\Psi}) = -\nabla^2\boldsymbol{\Psi} + \nabla(\nabla \cdot \boldsymbol{\Psi}). \tag{3.3}$$

But since we have constrained the vector potential to be divergence free, we are left with

$$\nabla^2\boldsymbol{\Psi} = -\boldsymbol{\omega} := -\nabla \times \boldsymbol{v}, \tag{3.4}$$

where the **vorticity**, $\boldsymbol{\omega}$, is defined as the curl of the velocity field. Note that ***vorticity, defined in this manner, is always divergence free***. It is also important to note that these governing equations for φ and $\boldsymbol{\Psi}$ are entirely ***kinematic***: they express relationships between the various fields describing fluid motion, and hold regardless of the types of forces acting on the fluid.

A flow is called **incompressible** (or volume-preserving) if $\Theta = 0$ everywhere and **irrotational** if $\boldsymbol{\omega} = 0$ everywhere. In two-dimensional flows, the vorticity and the vector potential each have only a single component in the out-of-plane direction, $\boldsymbol{\omega} = \omega\boldsymbol{e}_3$ and $\boldsymbol{\Psi} = \psi\boldsymbol{e}_3$, where ψ is called the **streamfunction**. The two-dimensional form of Eq. (3.4) is simply

$$\nabla^2\psi = -\omega. \tag{3.5}$$

Note 3.1.1: Vector Potential in a Moving Rigid Body

For a rigid body in motion, it is easy to verify that the vector potential inside the body is given by

$$\boldsymbol{\Psi}_{\mathrm{b}}(\boldsymbol{x}) = \boldsymbol{\Psi}_0 + \frac{1}{n_{\mathrm{d}}-1}\boldsymbol{V}_{\mathrm{r}} \times \boldsymbol{x} + \frac{1}{n_{\mathrm{d}}}(\boldsymbol{\Omega} \times \boldsymbol{x}) \times \boldsymbol{x}, \tag{3.6}$$

where n_{d} is the number of spatial dimensions and $\boldsymbol{\Psi}_0$ is an arbitrary constant vector with no effect on the velocity field. When the Laplacian of this vector potential is evaluated, the vorticity is seen to be $\boldsymbol{\omega}_{\mathrm{b}} = 2\boldsymbol{\Omega}$. In the planar case, the vector potential is simply the streamfunction,

$$\psi_{\mathrm{b}}(x, y) = \psi_0 + U_{\mathrm{r}}y - V_{\mathrm{r}}x - \frac{1}{2}\Omega(x^2 + y^2). \tag{3.7}$$

We also can write (3.7) in a complex form,

$$\psi_{\mathrm{b}}(z, z^*) = \psi_0 + \frac{\mathrm{i}}{2}(W_{\mathrm{r}}^* z^* - W_{\mathrm{r}} z) - \frac{1}{2}\Omega z z^*. \tag{3.8}$$

The reader might recall some insight regarding the streamfunction: The streamfunction is constant along any streamline, and the difference between the values of streamfunction on two streamlines is equal to the volume flow rate per unit depth between these streamlines. Here, we extend this insight to the vector potential field. First, it should be noted that, in the plane, we can choose any two points in the plane and, provided that neither is a stagnation point, obtain an instantaneous conduit between the unique pair of streamlines that pass through these points. The volume flow rate per unit depth is uniform along this conduit. In a full three-dimensional flow, we can analogously choose any closed loop, and the bundle of streamlines that pass through this loop also form a conduit. The volume flow rate through this conduit is given by the following:

Result 3.1: Volume Flow Rate and Vector Potential

For a velocity field $\boldsymbol{v}$ derived entirely from a vector potential field, $\boldsymbol{\Psi}$, the volume flow rate Q through a surface S enclosed by contour C is given by

$$Q = \int_C \boldsymbol{\Psi} \cdot \mathrm{d}\boldsymbol{l}. \tag{3.9}$$

Proof The volume flow rate is defined as

$$Q = \int_S \boldsymbol{n} \cdot \boldsymbol{v} \,\mathrm{d}S. \tag{3.10}$$

The desired result follows easily by applying Stokes' theorem (A.60) with $\boldsymbol{v} = \nabla \times \boldsymbol{\Psi}$. □

3.1.2 The No-Penetration Condition

In inviscid flows, the basic kinematic constraint at the interface, $\mathcal{S}_\mathrm{b}$, between an impenetrable body and the surrounding fluid is the **no-penetration condition**, which requires that the normal components of their velocities be equal on the interface,

$$\boldsymbol{n} \cdot \boldsymbol{v} = \boldsymbol{n} \cdot \boldsymbol{V}_\mathrm{b}, \tag{3.11}$$

at all points $\boldsymbol{x} \in \mathcal{S}_\mathrm{b}$. As we discussed in the previous chapter, we constrain the body motion in this book to be volume-preserving, so that $\boldsymbol{V}_\mathrm{b}$ satisfies the constraint (2.29).

Let us also write this condition in complex form, for later reference. To do so, we will use the vector/complex equivalences written in Sect. A.2.1 for the dot product (A.104) and the normal vector (A.107). On the bounding contour of the body, $\mathcal{C}_\mathrm{b}$, the complex conjugate velocity w in the fluid must satisfy

$$\mathrm{Im}\left(w \frac{\mathrm{d}z}{\mathrm{d}s}\right) = \mathrm{Im}\left(W_\mathrm{b} \frac{\mathrm{d}z}{\mathrm{d}s}\right), \tag{3.12}$$

where $\mathrm{d}z/\mathrm{d}s$ denotes the rate of change of position along $\mathcal{C}_\mathrm{b}$ with increasing arc length, s.

When the fluid velocity can be derived from a scalar potential field, condition (3.11) becomes a Neumann-type condition on this potential on surface $\mathcal{S}_\mathrm{b}$:

$$\boldsymbol{n} \cdot \nabla\varphi \equiv \frac{\partial \varphi}{\partial n} = \boldsymbol{n} \cdot \boldsymbol{V}_\mathrm{b}. \tag{3.13}$$

If, instead, the velocity can be obtained from a vector potential field, then we would place a condition on this field on $\mathcal{S}_\mathrm{b}$:

$$\boldsymbol{n} \cdot (\nabla \times \boldsymbol{\Psi}) = \boldsymbol{n} \cdot \boldsymbol{V}_\mathrm{b}. \tag{3.14}$$

As discussed in Note 3.1.1, rigid-body motion can be described by its own vector potential, $\boldsymbol{\Psi}_b$. Thus, with the help of Stokes' theorem (A.60) applied on arbitrary patches on the surface, we show in Result A.5 that (3.14) is automatically satisfied if ***the corresponding tangential components of the vector potential in the fluid and the body match at the interface,*** $\mathcal{S}_b$,

$$\boldsymbol{n} \times \boldsymbol{\Psi} = \boldsymbol{n} \times \boldsymbol{\Psi}_b. \tag{3.15}$$

This means that the tangential components of the potential given by (3.6) represent the expected form of the corresponding components of the fluid's vector potential when the surface of the body is approached. In two dimensions, this simply means that the streamfunction in the fluid is equal to the streamfunction given by Eq. (3.7), evaluated on the body surface. As we note below Result A.5, these tangential components of the vector potential need only match to within an arbitrary divergence-free vector field in $\mathcal{S}_b$. For example, a constant difference between the fluid and body vector potentials would make zero contribution to the velocity.

3.1.3 Vorticity Transport Equation

Inviscid flows are governed by the **Euler equations**, which describe the local conservation of momentum in a fluid medium that can only exert normal forces; tangential forces are impossible without viscosity. Ignoring gravity, these equations can be written as

$$\rho\left(\frac{\partial \boldsymbol{v}}{\partial t} + \boldsymbol{v} \cdot \nabla \boldsymbol{v}\right) = -\nabla p, \tag{3.16}$$

where ρ is the fluid density and p is the pressure. The operator in parentheses acting on velocity is the **material derivative**, generally denoted by D/Dt, and defined in Eq. (A.84). The derivation of these equations can be found in most standard texts on fluid dynamics.

These equations can be written in a form more conducive to our needs in this book by rewriting the convective term with the vector identity $\boldsymbol{v}\cdot\nabla\boldsymbol{v} = \nabla|\boldsymbol{v}|^2/2 - \boldsymbol{v}\times(\nabla\times\boldsymbol{v})$. If we assume the flow to be **barotropic**, in which density is only a function of pressure, then we can bring the density inside the gradient:

$$\frac{\partial \boldsymbol{v}}{\partial t} + \nabla\left(\int \frac{dp}{\rho(p)} + \frac{1}{2}|\boldsymbol{v}|^2\right) = \boldsymbol{v} \times \boldsymbol{\omega}. \tag{3.17}$$

We will be more restrictive in this book and assume throughout that ***the flow is strictly incompressible and of uniform density***; this is a subset of the barotropic class of flows. In incompressible flows, the equation of conservation of mass is expressed by the kinematic constraint that the velocity remain divergence free,

$$\nabla \cdot \boldsymbol{v} = 0. \tag{3.18}$$

Indeed, the role of pressure in such an incompressible medium is that of a constraint force, or **Lagrange multiplier**; attempts to alter the volume of fluid particles are prevented by reactions from the pressure field. Lanczos [44] provides a nice treatment of the Euler equations that clarifies this role.

Let us also replace the velocity in the time derivative term of (3.17) by its Helmholtz decomposition (3.1). We thus have

$$\frac{\partial}{\partial t}(\nabla \times \boldsymbol{\Psi}) - \boldsymbol{v} \times \omega + \nabla\left(\frac{p}{\rho} + \frac{1}{2}|\boldsymbol{v}|^2 + \frac{\partial \varphi}{\partial t}\right) = 0. \tag{3.19}$$

This form of the Euler equations serves two purposes. Later, in Sect. 3.2.1, we will specify its use for potential flow in order to obtain the Bernoulli equation. However, more generally, by taking the curl we eliminate all of the terms under the gradient operator. This gives rise to the so-called **vorticity transport equation**, which is

$$\frac{\partial \omega}{\partial t} = \nabla \times (\boldsymbol{v} \times \omega). \tag{3.20}$$

Applying the standard vector identity (A.7) and using the fact that both velocity and the vorticity are divergence free, we arrive at another form of these equations,

$$\frac{D\omega}{Dt} = \omega \cdot \nabla \boldsymbol{v}. \tag{3.21}$$

The left-hand side of this equation is the material derivative of the vorticity. The right-hand side brings about stretching and tilting of vortex lines; it is identically zero in two-dimensional flows, where vorticity is orthogonal to velocity gradients.

3.1.4 Kinetic Energy and Rate of Work

The kinetic energy in the fluid is defined as

$$T_f := \frac{1}{2}\int_{\mathcal{V}_f} \rho|\boldsymbol{v}|^2 \,\mathrm{d}V, \tag{3.22}$$

where $\boldsymbol{v}$ is the usual velocity field in the inertial reference frame, so that, by definition, it goes to zero at infinite distance. This is necessary to ensure that the integral is convergent.

From this basic definition, we can obtain some interesting alternative forms. Let us write the velocity field as $\boldsymbol{v} = \boldsymbol{v}_{\mathrm{v}} + \nabla\varphi$, where we will soon identify $\boldsymbol{v}_{\mathrm{v}}$ as the curl of a vector potential as in (3.1); clearly, this part of the velocity describes the fluid's rotational motion, $\nabla \times \boldsymbol{v}_{\mathrm{v}} = \omega$, if any. Then, when we substitute this decomposed velocity into the kinetic energy, we get

$$T_f = \frac{1}{2}\rho \int\limits_{\mathcal{V}_f} \nabla\varphi \cdot \nabla\varphi \, \mathrm{d}V + \frac{1}{2}\rho \int\limits_{\mathcal{V}_f} |\boldsymbol{v}_\mathrm{v}|^2 \, \mathrm{d}V + \rho \int\limits_{\mathcal{V}_f} \boldsymbol{v}_\mathrm{v} \cdot \nabla\varphi \, \mathrm{d}V. \tag{3.23}$$

Let us suppose, without loss of generality, that any surface motion of a body is to be matched by the part of $\boldsymbol{v}$ derived from the scalar potential field, φ, so that (3.13) holds, and furthermore, $\boldsymbol{n} \cdot \boldsymbol{v}_\mathrm{v} = 0$ on $\mathcal{S}_\mathrm{b}$. In other words, $\boldsymbol{v}_\mathrm{v}$ not only contains the velocity induced directly by the vorticity, but also its modification due to the presence of the impenetrable body. We will have much more to say about this modification in future chapters.

Since $\boldsymbol{v}_\mathrm{v}$ is divergence free by construction, we must have that $\nabla \cdot (\boldsymbol{v}_\mathrm{v}\varphi) = \boldsymbol{v}_\mathrm{v} \cdot \nabla\varphi$. Furthermore, let us assume that this rotational part of the velocity field decays at least as fast as $1/r^3$ (or $1/r^2$ in two dimensions) and the scalar potential field like $1/r^2$ (or $1/r$ in two dimensions) with increasing distance r from the origin. As we will discuss at length in the following chapters, particularly Sect. 4.7, every flow field of interest to us in this work will obey this assumption. Then, by applying the divergence theorem (A.19) to the fluid region bounded internally by $\mathcal{S}_\mathrm{b}$ and externally by a stationary surface $\mathcal{S}_R$ that encloses all fluid vorticity, we can show that the last integral in (3.23) must vanish identically as the surface $\mathcal{S}_R$ is pushed out to infinity. A similar set of steps can be applied to the first integral, with the assumption that the flow is incompressible, and we obtain the enlightening result,

$$T_f = -\frac{1}{2}\rho \int\limits_{\mathcal{S}_\mathrm{b}} \varphi \frac{\partial\varphi}{\partial n} \, \mathrm{d}S + \frac{1}{2}\rho \int\limits_{\mathcal{V}_f} |\boldsymbol{v}_\mathrm{v}|^2 \, \mathrm{d}V. \tag{3.24}$$

The first term in (3.24), which we will denote by $T_{f,\mathrm{ir}}$, represents the kinetic energy of the irrotational motion—verifiably positive, in spite of the negative sign—and the second term, denoted by $T_{f,\mathrm{v}}$, contains any additional (rotational) components of the fluid motion. The most obvious conclusion we can draw from (3.24) is that the irrotational and rotational contributions to the kinetic energy are decoupled. But more importantly, the rotational contribution clearly increases the overall kinetic energy. Thus, we have just proved **Kelvin's minimum energy theorem**: The kinetic energy of an entirely irrotational motion of a fluid has less kinetic energy than any other fluid motion satisfying the same boundary conditions.

Let us go a step further with our manipulation of the rotational term in the kinetic energy. The following form will be useful in some of our subsequent analysis:

Result 3.2: A Form of the Kinetic Energy in Three Dimensions

In three-dimensional flow, with velocity field $\boldsymbol{v} = \nabla\varphi + \boldsymbol{v}_\mathrm{v}$, the kinetic energy in the fluid can be written as

$$T_f = -\frac{1}{2}\rho \int\limits_{\mathcal{S}_b} \varphi \frac{\partial \varphi}{\partial n}\, dS + \rho \int\limits_{\mathcal{V}_f} \boldsymbol{v}_v \cdot (\boldsymbol{x} \times \boldsymbol{\omega})\, dV + \frac{1}{2}\rho \int\limits_{\mathcal{S}_b} \boldsymbol{v}_v \cdot [\boldsymbol{x} \times (\boldsymbol{n} \times \boldsymbol{v}_v)]\, dS, \tag{3.25}$$

where φ is a scalar potential field generated by motion of the surface $\mathcal{S}_b$, and $\boldsymbol{v}_v$ is a velocity field induced by fluid vorticity and obeying the condition $\boldsymbol{v}_v \cdot \boldsymbol{n} = 0$ on $\mathcal{S}_b$.

Proof The starting point for the derivation of Result 3.2 is the vector identity

$$\boldsymbol{u} = \frac{1}{n_d - 1} [\boldsymbol{x} \times (\nabla \times \boldsymbol{u}) - \nabla(\boldsymbol{x} \cdot \boldsymbol{u}) + \nabla \cdot (\boldsymbol{x}\boldsymbol{u})], \tag{3.26}$$

where $\boldsymbol{u}$ is any continuously differentiable vector field, and again, n_d is the number of spatial dimensions. This identity, which the reader is invited to prove, holds for the full fluid velocity, $\boldsymbol{v}$, but also holds for just $\boldsymbol{v}_v$—in both cases, the curl of the vector field is the vorticity, $\boldsymbol{\omega}$. Now, suppose that we take the dot product of this latter version of the identity with $\boldsymbol{v}_v$. Then, we have

$$\boldsymbol{v}_v \cdot \boldsymbol{v}_v = \frac{1}{n_d - 1} [\boldsymbol{v}_v \cdot (\boldsymbol{x} \times \boldsymbol{\omega}) - \boldsymbol{v}_v \cdot \nabla(\boldsymbol{x} \cdot \boldsymbol{v}_v) + \nabla \cdot (\boldsymbol{x}\boldsymbol{v}_v) \cdot \boldsymbol{v}_v]. \tag{3.27}$$

We can work $\boldsymbol{v}_v$ inside the spatial derivatives in the last two terms: Since $\boldsymbol{v}_v$ is divergence free, then $\boldsymbol{v}_v \cdot \nabla(\boldsymbol{x} \cdot \boldsymbol{v}_v) = \nabla \cdot (\boldsymbol{v}_v\boldsymbol{x} \cdot \boldsymbol{v}_v)$. Also, it can easily verified that

$$\nabla \cdot (\boldsymbol{x}\boldsymbol{v}_v) \cdot \boldsymbol{v}_v = \frac{1}{2}\nabla \cdot (\boldsymbol{x}\boldsymbol{v}_v \cdot \boldsymbol{v}_v) + \frac{n_d}{2}\boldsymbol{v}_v \cdot \boldsymbol{v}_v. \tag{3.28}$$

Thus, in three dimensions,

$$\frac{1}{2}|\boldsymbol{v}_v|^2 = \boldsymbol{v}_v \cdot (\boldsymbol{x} \times \boldsymbol{\omega}) - \nabla \cdot (\boldsymbol{v}_v\boldsymbol{x} \cdot \boldsymbol{v}_v) + \frac{1}{2}\nabla \cdot (\boldsymbol{x}\boldsymbol{v}_v \cdot \boldsymbol{v}_v). \tag{3.29}$$

If we integrate this over the fluid region, $\mathcal{V}_f$, apply the divergence theorem to the last two terms, and use once again the fact that $\boldsymbol{n} \cdot \boldsymbol{v}_v = 0$, then we obtain our desired result. □

Note 3.1.2: Another Form of the Rotational Part of Kinetic Energy

As we mentioned earlier, the component $\boldsymbol{v}_v$ can be derived from a vector potential, $\boldsymbol{v}_v = \nabla \times \boldsymbol{\Psi}_v$. Then, by applying a rotational version of the divergence theorem (A.20), we can manipulate the final term in (3.24) to show that

$$T_f = -\frac{1}{2}\rho \int_{S_b} \varphi \frac{\partial \varphi}{\partial n}\, dS + \frac{1}{2}\rho \int_{\mathcal{V}_f} \boldsymbol{\Psi}_v \cdot \boldsymbol{\omega}\, dV - \frac{1}{2}\rho \int_{S_b} \boldsymbol{\Psi}_v \cdot (\boldsymbol{n} \times \boldsymbol{v}_v)\, dS. \quad (3.30)$$

We note that, in the second surface integral of (3.30), only the components of $\boldsymbol{\Psi}_v$ tangent to the surface are invoked, and, from our discussion in Sect. 3.1.2, these must match the corresponding tangential components inside the body. But since we have assumed that any surface motion is described by the scalar potential field, this vector potential necessarily has a uniform value throughout the body's interior; thus, the vector potential can be taken outside of this integral. A thorough interpretation of the two final terms of (3.30) is postponed to a later section.

What is the governing equation for the kinetic energy? In incompressible flow, mechanical and thermodynamic energies are decoupled from each other, and the equation governing the mechanical energy is really just a restatement of the momentum equation. Let us develop that equation here. We start by taking the time derivative of the kinetic energy, defined in its original form (3.22), and then use our mathematical tools for differentiating volume integrals in Sect. A.1.7, specifically Eq. (A.88) to interchange the time derivative with the integral over $\mathcal{V}_f$. Since this integration region is a material volume, deformed by the fluid velocity itself, and the velocity field is divergence free,

$$\frac{dT_f}{dt} = \int_{\mathcal{V}_f} \frac{D}{Dt}\left(\frac{1}{2}\rho|\boldsymbol{v}|^2\right) dV. \quad (3.31)$$

The density is assumed to be uniform. The material derivative of $|\boldsymbol{v}|^2/2$ is, by the product rule, $\boldsymbol{v} \cdot D\boldsymbol{v}/Dt$. From the Euler equations (3.16), the material derivative of velocity can be replaced by the pressure gradient, so we get

$$\frac{dT_f}{dt} = -\int_{\mathcal{V}_f} \boldsymbol{v} \cdot \nabla p\, dV. \quad (3.32)$$

We can write the integrand as $\nabla \cdot (\boldsymbol{v}p)$, since the velocity is divergence free. This form enables us to use the divergence theorem (A.18) in order to transform the volume integral over $\mathcal{V}_f$ into an integral over the body surface. To be careful, we might apply the theorem to a region bounded externally by S_R, as we did before. It is easy to show that the contribution from that surface vanishes as it is moved outward to infinity. We also apply the no-penetration condition (3.11) on the body surface, and end up with the following integral energy equation,

$$\frac{dT_f}{dt} = \int_{S_b} p\boldsymbol{V}_b \cdot \boldsymbol{n}\, dS. \quad (3.33)$$

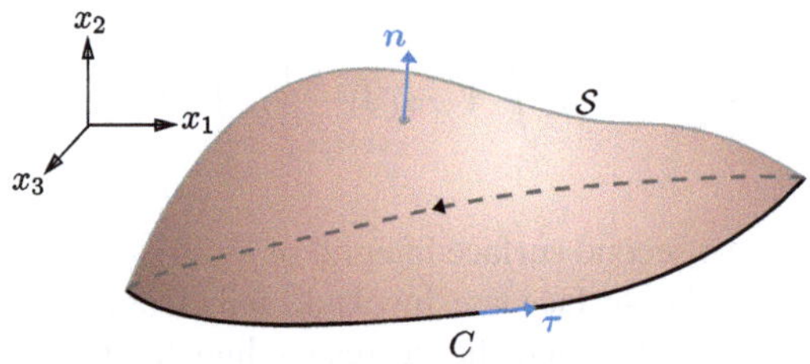

Fig. 3.1 Closed contour C and an enclosed surface S

The expression on the right-hand side is the rate of work done on the fluid along the body's surface. It is important to note that, in the absence of viscosity and gravity, the only means of affecting the kinetic energy is through the rate of work done by the body on the fluid. This rate of work might be negative, of course: the interchange of energy between the fluid and body goes both ways. However, once the motion of the body stops, the kinetic energy in the fluid remains constant. As we will see later in Sect. 4.6.3, any energy that persists in the fluid while a body remains stationary must be borne by the part of (3.24) associated with vorticity.

3.1.5 Circulation and Its Invariance

In flows with vorticity, a quantity of substantial importance is the **circulation**. Around a closed curve, C, as depicted in Fig. 3.1, the circulation is defined as

$$\Gamma := \int_{C} \boldsymbol{v} \cdot \mathrm{d}\boldsymbol{l}, \tag{3.34}$$

where $\mathrm{d}\boldsymbol{l} := \boldsymbol{\tau}\mathrm{d}s$ is the vector differential element along the curve, of length $\mathrm{d}s$ in the direction of the unit tangent $\boldsymbol{\tau}$. By Stokes' theorem (A.62), the circulation can also be written in terms of the total vorticity piercing any surface S enclosed by C:

$$\Gamma = \int_{S} \boldsymbol{\omega} \cdot \boldsymbol{n}\, \mathrm{d}S. \tag{3.35}$$

In two-dimensional problems, the surface in (3.35) is in the plane of the flow and its normal is $\boldsymbol{e}_3$, in the same direction as the single component of vorticity. Thus, the circulation can be written as

$$\Gamma = \int_{\mathcal{V}} \omega\, \mathrm{d}V. \tag{3.36}$$

Here, we have used our liberalized notation, in which a planar region is denoted by a volume $\mathcal{V}$ (of which the region is a cross-sectional slice), and the volume integral should be interpreted per unit depth. Alternatively, the circulation in two-

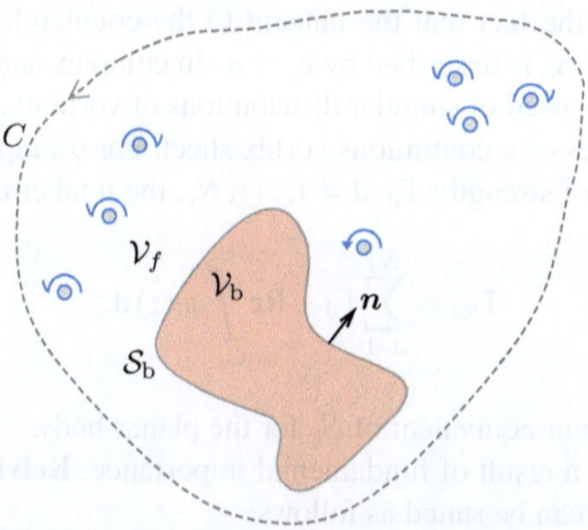

Fig. 3.2 Contour enclosing a body and fluid vorticity, for the definition of Γ_{tot}

dimensional problems can be written in a complex version of (3.34) from a contour integral of the complex velocity, w:

$$\Gamma = \mathrm{Re}\left(\int_C w(z)\,\mathrm{d}z \right). \tag{3.37}$$

(Note that the imaginary part of the integral is zero, provided there is no net source or sink of mass contained within C.)

The circulation in problems with bodies require some additional comments. In three-dimensional problems involving bodies without holes, any closed contour in the fluid is **reducible**, regardless of whether there is a body or not. In other words, the contour can be topologically shrunk to a point while remaining entirely in the fluid. Furthermore, for a given contour, we can always find an enclosed surface that does not intersect a body. Thus, the circulation in such problems can always be expressed entirely in terms of vorticity in the fluid, through (3.35). This also holds for contours in two-dimensional problems that do not encircle the body. However, contours that do surround the body are irreducible, and thus, the circulation about these contours necessarily includes the body's contribution. In particular, let us consider a contour C that encloses the body and all of the vorticity in the fluid, as shown in Fig. 3.2. Again, using our notation in which the planar fluid region is denoted by $\mathcal{V}_f$ and the body perimeter by S_b—both interpreted per unit depth—then, by applying the divergence theorem (A.20) to $\nabla \times \boldsymbol{v}$ in $\mathcal{V}_f$ (or, equivalently, by applying Stokes' theorem (A.62) to a reducible contour formed by connecting C with another that wraps tightly around S_b via a 'barrier' with canceling segments), the circulation about C—which we will henceforth call the **total circulation** of the flow—can be written as

$$\Gamma_{tot} = \int_{\mathcal{V}_f} \omega\,\mathrm{d}V + \int_{S_b} (\boldsymbol{n} \times \boldsymbol{v}) \cdot \boldsymbol{e}_3\,\mathrm{d}S. \tag{3.38}$$

Here, we have used the fact that the tangent to the counterclockwise contour that wraps tightly around $\mathcal{S}_b$ is described by $\boldsymbol{e}_3 \times \boldsymbol{n}$. In complex analysis of the flow, the fluid vorticity is composed of singular distributions of vorticity, in the form of either discrete point vortices or a continuous vortex sheet. For example, considering a set of N_v point vortices of strengths Γ_J, $J = 1, \ldots, N_v$, the total circulation is given by

$$\Gamma_{tot} = \sum_{J=1}^{N_v} \Gamma_J + \text{Re} \int_{C_b} w(z)\, dz, \tag{3.39}$$

where C_b is the contour equivalent of $\mathcal{S}_b$ for the planar body.

Let us now prove a result of fundamental importance: **Kelvin's circulation theorem**. This theorem can be stated as follows:

Result 3.3: Kelvin's Circulation Theorem

If C is a **material loop**—that is, a closed curve composed of points that move with the local fluid velocity—then the rate of change of circulation, Γ, about C is zero in inviscid barotropic flow with conservative body forces:

$$\frac{d\Gamma}{dt} = 0. \tag{3.40}$$

We can prove this result by taking the time derivative of either the contour or the surface expression for Γ. Starting with the surface form (3.35), let us apply identity (A.90) for the time derivative of the flux of a vector field through a material surface. Note that, by requiring that S be a material surface, we are making only one particular choice out of many for how to evolve a surface enclosed by the material loop. But, by Stokes' theorem, the result for $d\Gamma/dt$ for this surface must hold for the loop itself, and thus, for any other surface enclosed by the loop. Since vorticity is identically divergence free, this results in

$$\frac{d\Gamma}{dt} = \int_S \left(\frac{\partial \boldsymbol{\omega}}{\partial t} - \nabla \times (\boldsymbol{v} \times \boldsymbol{\omega}) \right) dS. \tag{3.41}$$

However, by the vorticity transport equation (3.20), the integrand is identically zero, which completes the proof.

We can apply Kelvin's circulation theorem immediately to show that, in inviscid two-dimensional flow, Γ_{tot} is constant. This is obvious, because we can always interpret the contour we considered earlier in defining Γ_{tot} as a material loop, and thus, this contour will always enclose the body and all fluid vorticity, even if we include a mechanism for generating new vorticity at the body.

Note 3.1.3: Other Invariants of Unbounded Inviscid Flow

Inviscid flows that are **unbounded**—that is, for which there is no internal or external rigid boundary—possess other invariants, and we will discuss a few here. In three-dimensional flows, the **total vorticity**—the vorticity integrated over all of $\mathbb{R}^3$ is invariant, and in fact, is identically zero,

$$\int_{\mathbb{R}^3} \boldsymbol{\omega}\, \mathrm{d}V = 0. \tag{3.42}$$

Note that the two-dimensional version of total vorticity is equivalent to the **total circulation** in the fluid (times the unit vector $\boldsymbol{e}_3$), which need *not* be zero. However, by Kelvin's circulation theorem (3.40), this total circulation must remain invariant. Later in this book we will be more particular about the total circulation in flows with a rigid body; this will be discussed in Sect. 4.5.

The first integral moment of vorticity is also invariant, in both two and three dimensions. We will consider a form of this moment that will be very important later in the book:

$$\boldsymbol{P} = \frac{1}{n_\mathrm{d} - 1} \int_{\mathbb{R}^{n_\mathrm{d}}} \boldsymbol{x} \times \boldsymbol{\omega}\, \mathrm{d}V, \tag{3.43}$$

where n_d denotes, as usual, the number of spatial dimensions. Finally, the second integral moment of vorticity is also time invariant. As with the first moment, we will consider this moment in a particular form in anticipation of its later importance:

$$\boldsymbol{\Pi}_\mathrm{O} = \frac{1}{n_\mathrm{d}} \int_{\mathbb{R}^{n_\mathrm{d}}} \boldsymbol{x} \times (\boldsymbol{x} \times \boldsymbol{\omega})\, \mathrm{d}V, \tag{3.44}$$

where the subscript $\boldsymbol{\Pi}_\mathrm{O}$ denotes that the moment is taken about the origin.

Proof The vanishing total vorticity can be proved by applying the divergence theorem in the form (A.20) to the volume integral, with $\boldsymbol{\omega}$ replaced by its definition, $\nabla \times \boldsymbol{v}$. To be cautious, we apply this theorem to a region $\mathcal{V}_R$ that contains all of the vorticity in the fluid and is bounded externally by a closed surface $\mathcal{S}_R$ that we will allow to approach infinite distance. Since the velocity is entirely derivable from the gradient of a scalar potential on this surface, then we can apply Stokes theorem on this surface in the form given by (A.63). But the surface $\mathcal{S}_R$ is closed and has no bounding curve, so the resulting integral is zero and the identity is proved. The total vorticity, identically zero, is clearly invariant with time.

Now, to prove the invariance of $\boldsymbol{P}$, we will use the same artifice as we did for total vorticity, by considering the integration over $\mathcal{V}_R$ bounded externally by $\mathcal{S}_R$. Clearly, $\boldsymbol{P}$ itself is unaffected by this change of integration region. The region is constant in time, so the time derivative and the integral can be interchanged directly:

$$\frac{\mathrm{d}\boldsymbol{P}}{\mathrm{d}t} = \frac{1}{n_{\mathrm{d}} - 1} \int_{\mathcal{V}_R} \boldsymbol{x} \times \frac{\partial \boldsymbol{\omega}}{\partial t} \,\mathrm{d}V. \tag{3.45}$$

Now we can introduce the vorticity transport equation (3.20), and note that we can apply identity (3.26) with $\boldsymbol{u}$ replaced by $\boldsymbol{v} \times \boldsymbol{\omega}$ to obtain

$$\frac{\mathrm{d}\boldsymbol{P}}{\mathrm{d}t} = \int_{\mathcal{V}_R} \boldsymbol{v}\times\boldsymbol{\omega} \,\mathrm{d}V + \frac{1}{n_{\mathrm{d}} - 1} \left[\int_{\mathcal{V}_R} \nabla(\boldsymbol{x} \cdot (\boldsymbol{v} \times \boldsymbol{\omega}))\,\mathrm{d}V - \int_{\mathcal{V}_R} \nabla \cdot (\boldsymbol{x}(\boldsymbol{v} \times \boldsymbol{\omega}))\,\mathrm{d}V \right]. \tag{3.46}$$

The last two integrals can be transformed to surface integrals over $\mathcal{S}_R$ with the use of the divergence theorem (A.18). However, these surface integrals are clearly both zero, since the $\boldsymbol{\omega}$ field is completely contained within $\mathcal{V}_R$. Thus, we are left to contend with the integral of $\boldsymbol{v} \times \boldsymbol{\omega}$. This, too, can be shown to vanish; we will do so in Lemma 6.3 later in the book.

The steps to prove that the second moment is time invariant also start with identity (3.26); a closely related identity is

$$\boldsymbol{x} \times \boldsymbol{u} = \frac{1}{n_{\mathrm{d}}} \left[\boldsymbol{x} \times (\boldsymbol{x} \times (\nabla \times \boldsymbol{u})) + \nabla \times (\boldsymbol{x}\boldsymbol{x} \cdot \boldsymbol{u}) + \nabla \cdot (\boldsymbol{x}\boldsymbol{x} \times \boldsymbol{u}) \right]. \tag{3.47}$$

Thus, with the use of the divergence theorem on $\mathcal{V}_R$, it is easy to show that

$$\frac{\mathrm{d}\boldsymbol{\Pi}_{\mathrm{O}}}{\mathrm{d}t} = \int_{\mathcal{V}_R} \boldsymbol{x} \times (\boldsymbol{v} \times \boldsymbol{\omega})\,\mathrm{d}V. \tag{3.48}$$

This integral will also be shown to vanish in Lemma 6.3. Thus, the second moment is also invariant. □

In Chap. 6, in which we address the force and moment on a body, we will return to $\boldsymbol{P}$ and $\boldsymbol{\Pi}_{\mathrm{O}}$ and rechristen them the **linear** and **angular impulse**, respectively, and modify them to account for the body's presence. In that context, they will generally not remain invariant.

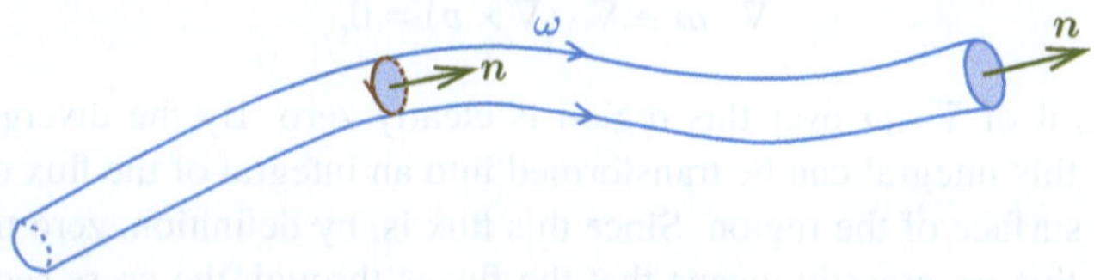

Fig. 3.3 A vortex tube. The wall of a vortex tube is composed from all of the vortex lines that pass through a closed loop

3.1.6 Helmholtz' Theorems

The assumptions used to prove Kelvin's circulation theorem—inviscid barotropic flow under the influence of conservative body forces, or, what we will refer to for brevity as **circulation-preserving flow**—can be used to obtain other, farther reaching conclusions about such a flow. These are collectively known as Helmholtz' theorems [30]. Helmholtz published these theorems around a decade before Lord Kelvin made his own remarkable contributions [69]. The publication of Helmholtz' work led to a vibrant correspondence between the two men, and much of our fundamental understanding of vortex dynamics today stems from their contributions. Before we proceed to describe these theorems, it is important to first define a few concepts. A **vortex line** is a space curve that is everywhere tangent to the local vorticity vector field. It is analogous to a streamline for the velocity field. From this, we can define a **vortex tube**, whose wall is composed from the collection of all vortex lines that pass through a closed loop. An example of such a tube is depicted in Fig. 3.3.

It should be noted that the numbering of the theorems below is somewhat at odds with Helmholtz' own numbering, and, in some cases, the theorems have been expanded in breadth since their original presentation. Let us consider the flux of vorticity through the cross-section of a vortex tube, equivalent to the circulation about the loop forming the perimeter of this cross-section. The circulation can, of course, be computed about any loop enclosing the vortex tube, and we will take this circulation to be a local strength of the tube. In other words, ***the local strength of a vortex tube is defined as the flux of vorticity through its cross section***. The first Helmholtz theorem then states that this strength is the same for any such cross-section of a given tube:

Helmholtz' Theorem 1 The strength of a vortex tube is uniform along its length.

This can be proved by first noting that, for any two closed loops that envelop the tube and lie in the tube wall, one can devise a volumetric region bounded by the wall of the tube between these loops and, at either end, by a cross-sectional surface of the tube enclosed by one of the loops. The vorticity is always divergence free,

$$\nabla \cdot \boldsymbol{\omega} \equiv \nabla \cdot (\nabla \times \boldsymbol{v}) = 0, \tag{3.49}$$

so the integral of $\nabla \cdot \boldsymbol{\omega}$ over this region is clearly zero. By the divergence theorem (A.19), this integral can be transformed into an integral of the flux of vorticity through the surface of the region. Since this flux is, by definition, zero through the tube's wall, this necessarily means that the fluxes through the cross-sectional surfaces must be equal, and the theorem is proved. As a consequence of this theorem, since the strength of a tube cannot vary along its length, then it necessarily follows that ***a vortex tube cannot end within the fluid. It must either be closed or extend to infinity.*** This is sometimes given as a separate theorem of Helmholtz, although it originally appeared together with the first.

Thus, in this first theorem, we have established a single scalar-valued property—the strength—of a vortex tube, and we will have more to say about this later. But first, we must prove the next theorem, which provides an essential rule for the behavior of vorticity in circulation-preserving flows; its importance cannot be overstated.

Helmholtz' Theorem 2 In a circulation-preserving flow, vortex lines are material lines.

Stated in other words, this theorem tells us that ***vorticity is transported by the flow***. To prove the theorem, it is sufficient to consider the subsequent behavior of a vortex line and a material line that are initially coincident. If the material line is described parametrically by the space curve $\boldsymbol{X}(\xi, t)$, where ξ is a scalar parameter that uniquely identifies each material point on the curve and increases monotonically along it, then the initial coincidence of these two lines is described mathematically by the requirement that their tangents—$\partial \boldsymbol{X}/\partial \xi$ and $\boldsymbol{\omega}$, respectively—are everywhere parallel to one another, $\partial \boldsymbol{X}/\partial \xi \times \boldsymbol{\omega} = 0$.[1] Another way to write this is that the two vectors are initially identical to within a scalar factor: $\partial \boldsymbol{X}/\partial \xi = c\boldsymbol{\omega}$. To prove that the vortex line is a material line, we must therefore show that this parallel condition is preserved as time proceeds, viz.

$$\frac{D}{Dt}\left(\frac{\partial \boldsymbol{X}}{\partial \xi} \times \boldsymbol{\omega}\right) = 0. \tag{3.50}$$

By the product rule, this time derivative can be expanded:

$$\frac{D}{Dt}\left(\frac{\partial \boldsymbol{X}}{\partial \xi} \times \boldsymbol{\omega}\right) = \frac{D}{Dt}\left(\frac{\partial \boldsymbol{X}}{\partial \xi}\right) \times \boldsymbol{\omega} + \frac{\partial \boldsymbol{X}}{\partial \xi} \times \frac{D\boldsymbol{\omega}}{Dt}. \tag{3.51}$$

The material derivative of the tangent vector is simply its partial derivative with respect to time with the material coordinate, ξ, held fixed, and this derivative can be interchanged with the one with respect to ξ:

$$\frac{D}{Dt}\left(\frac{\partial \boldsymbol{X}}{\partial \xi}\right) \equiv \frac{\partial}{\partial \xi}\left(\frac{\partial \boldsymbol{X}}{\partial t}\right) = \frac{\partial \boldsymbol{v}(\xi, t)}{\partial \xi}. \tag{3.52}$$

[1] Note that ξ is not necessarily arc length, so $\partial \boldsymbol{X}/\partial \xi$ does not necessarily have unit length.

The right-hand side of this last equality expresses the rate of stretching along the material line. Writing velocity in its usual Eulerian form, $\boldsymbol{v}(\boldsymbol{x}, t)$, we can use the chain rule to express this same quantity as

$$\frac{D}{Dt}\left(\frac{\partial \boldsymbol{X}}{\partial \xi}\right) = \frac{\partial \boldsymbol{X}}{\partial \xi} \cdot \nabla \boldsymbol{v}. \tag{3.53}$$

Now, let us use this last result in our expansion of the time derivative above, and furthermore, replace all instances of the material line's tangent vector with $c\boldsymbol{\omega}$, since these are equivalent at the moment in question. When we do so, and collect common factors, we get

$$\frac{D}{Dt}\left(\frac{\partial \boldsymbol{X}}{\partial \xi} \times \boldsymbol{\omega}\right) = c\boldsymbol{\omega} \times \left(\frac{D\boldsymbol{\omega}}{Dt} - \boldsymbol{\omega} \cdot \nabla \boldsymbol{v}\right). \tag{3.54}$$

However, we already know that, for a circulation-preserving flow, the expression in parentheses vanishes because of the vorticity transport equation (3.21). Thus, the theorem is proved.

Now, with the third theorem, we return to the strength of a vortex tube. Not only does the strength remain uniform along the entire tube, but it also remains constant in time:

Helmholtz' Theorem 3 The strength of a vortex tube in a circulation-preserving flow is time invariant.

This is easy to prove: By the second theorem, the vortex lines that constitute the wall of the vortex tube are all material lines. Thus, by assumption, the circulation about a material loop that surrounds any such vortex line will remain constant in time, and this must clearly hold for any material loop that encloses the entire tube.

In particular, this third theorem must hold for an infinitely slender vortex tube, whose axis must be a material line, which we will describe again by the curve $\boldsymbol{X}(\xi, t)$. Then any cross section of this tube is an infinitesimal material surface, with area δS, whose normal is simply the unit tangent of the axis curve, $\partial \boldsymbol{X}/\partial s = (\partial \boldsymbol{X}/\partial \xi)/|\partial \boldsymbol{X}/\partial \xi|$, where s denotes the arc length along the curve. This unit tangent is aligned with the local vorticity vector: that is, $\partial \boldsymbol{X}/\partial s = c(s)\boldsymbol{\omega}$, where $c(s)$ is a scalar function. What is the value of this function, c? Provided the curve is differentiable, then the tangent exists and has unit length, so for such a case, $c = \pm|\boldsymbol{\omega}|^{-1}$. We must also consider the possibility that $\boldsymbol{\omega} = 0$. However, in such a case—a vorticity 'stagnation point'—the axis of the vortex tube is not uniquely defined, and, as we will see in the next passage, the original assumption of infinitesimal cross section is contradicted.

The strength of the narrow tube is given by the flux through this infinitesimal cross section, $\boldsymbol{\omega} \cdot (\partial \boldsymbol{X}/\partial s)\delta S = \pm|\boldsymbol{\omega}|\delta S$. Remember that, by the first Helmholtz theorem, this strength is uniform along the tube, so only one of the two signs is possible; the positive sign can be chosen without loss of generality. The magnitude of the vorticity may vary spatially along the tube's axis, since the cross-sectional surface's area may also vary along the tube. By the third theorem, the tube's strength is also invariant in time, and the cross-sectional area and vorticity magnitude at a certain material point along the tube axis must vary in inverse proportion to ensure this invariance. In particular, consider a short cylindrical material element of length δs lying along the axis of the tube, whose ends constitute the cross-sections of the tube. Then, a velocity gradient along the tube will tend to stretch or compress the length of the element, and the cross-sectional area must vary inversely in order to preserve the element's volume, which in turn proportionally changes the vorticity's magnitude. Mathematically, this element's volume is described by $\delta S \delta s$. Since both the element's volume and the strength $|\omega|\delta S$ must remain constant, then $|\omega|$ changes inversely with δS and proportionally with δs.

The proportionality between vorticity magnitude and incremental arc length is clearly constant in space and time, since it is composed from the element volume and tube strength. Indeed, since $\omega = |\omega|\partial \boldsymbol{X}/\partial s$, then the vorticity ω is also invariantly proportional to $\delta \boldsymbol{X} := (\partial \boldsymbol{X}/\partial s)\delta s$, an incremental length vector along the tube axis. In other words, ***if we track the stretching or compression of the curve that constitutes the axis of a narrow vortex tube, then we automatically track the changes to the vorticity along the tube***. This is an important result that has tremendous value in computational modeling of vortex dynamics.

There are other versions of Helmholtz theorems that are consequences of results previously stated. For example, the next follows from the vorticity transport equation (3.21):

Helmholtz' Theorem 4 In circulating-preserving flow, fluid particles that are initially free of vorticity remain free of vorticity at all times.

This must be true, for, if vorticity is zero at the location of a particle at any instant, then the right-hand side of Eq. (3.21) is zero, and thus, the material derivative of vorticity—i.e., the rate of change of the vorticity associated with a material element—must also be zero.

The collection of Helmholtz theorems—along with Kelvin's circulation theorem—provide the foundations for vorticity dynamics. It should be noted that, in two-dimensional flows, all vortex lines and tubes extend infinitely and uniformly in the out of plane direction. Thus, only the second and third Helmholtz theorems have meaningful consequences in this context. (And the third is simply a restatement of Kelvin's theorem.) The term 'dynamics' is somewhat of a misnomer, because the theorems reveal that ***vorticity transport is entirely kinematic***, computable without regard for the forces in the fluid. Later, when we have developed the tools to determine the velocity field from a given vorticity field in Chap. 4, we will have everything we need to compute these kinematics. This will be the subject of Chap. 7. But first, we must discuss the basic elements of an inviscid incompressible flow.

3.2 Elements of Potential Flow

A flow field is defined as a **potential flow** if it is both incompressible and irrotational.[2] Actually, we qualify this definition with the mathematician's phrase 'almost everywhere', for we allow it to also describe flows in which vorticity is concentrated on a set of points, lines, and surfaces—that is, concentrated on a set of measure zero. We can also allow for distributed rate of dilatation, Θ, on such a zero-volume set. To ensure that the resulting flow field is incompressible, Eq. (3.2) shows that the scalar potential must be the solution of Laplace's equation almost everywhere,

$$\nabla^2 \varphi = 0. \tag{3.55}$$

And, since we require here that any vorticity in this flow is concentrated in a set of zero volume, Eq. (3.4) shows that the vector potential satisfies Laplace's equation almost everywhere. But lest one imagine that only trivial solutions exist for φ and $\boldsymbol{\Psi}$, rest assured: Laplace's equation has a very rich set of homogeneous solutions, as we will discuss in this section.

It is important to stress that, in regions in which the flow is truly potential, the velocity field can be derived entirely from a scalar potential

$$\boldsymbol{v} = \nabla \varphi, \tag{3.56}$$

or entirely from a vector potential field,

$$\boldsymbol{v} = \nabla \times \boldsymbol{\Psi}, \tag{3.57}$$

or from some mix of the two. This means that, in such regions, for every scalar potential field there is a corresponding vector potential field that produces the same flow, and their equivalence can be established up to arbitrary constants. Which potential we use to describe a flow is a matter of choice, dictated by convenience and practical considerations.

In this book, we are primarily concerned with **external flows**—that is, flows that are formed in regions that may be bounded internally by the surface of one or more bodies, but which extend infinitely far in every direction. We have already discussed the boundary condition of inviscid flow on an impenetrable surface (3.11). In an external flow, we must also insist that the flow remain bounded at infinitely large distances,

$$|\boldsymbol{v}| < \infty, \quad \text{as } |\boldsymbol{x}| \longrightarrow \infty. \tag{3.58}$$

The most basic potential flow is a **uniform flow**, whose uniform velocity we denote vectorially by $\boldsymbol{V}_\infty$, and whose jth velocity component in the inertial basis we denote by $V_{\infty,j}$. This flow clearly satisfies the boundedness condition (3.58). The

---

[2] This definition is more restrictive than other sources, such as [46], which define potential flow simply as irrotational flow, regardless of whether it is incompressible or not. We will consistently use the more restrictive definition in this book.

scalar potential field and equivalent vector potential field associated with a uniform flow are, respectively,

$$\varphi = \boldsymbol{V}_\infty \cdot \boldsymbol{x}, \qquad \boldsymbol{\Psi} = \frac{1}{n_\mathrm{d} - 1} \boldsymbol{V}_\infty \times \boldsymbol{x}. \tag{3.59}$$

As with any vector, the components of $\boldsymbol{V}_\infty$ can also be expressed in the body-fixed coordinate system, with the jth such component denoted by $\tilde{V}_{\infty,j}$; these components are related to those in the inertial system by the transformation (2.8).

3.2.1 Pressure and the Bernoulli Equation

In regions of potential flow, the pressure, p, at any point can be computed with the unsteady Bernoulli equation, which can be obtained from the Euler equations, written in the form (3.19) but now under the more restrictive condition that the flow is incompressible and the density uniform. The velocity field in potential flow, as we now know, can be expressed entirely as the gradient of the scalar potential at all points except singularities. Thus, the Euler equations are equivalent to the scalar-valued equation

$$p + \frac{1}{2}\rho|\boldsymbol{v}|^2 + \rho\frac{\partial\varphi}{\partial t} = B(t), \tag{3.60}$$

where the Bernoulli 'constant', B, is a spatially-uniform value. Note that we have ignored gravity here, though it is easy to account for it.

When evaluating the pressure at a moving point $\boldsymbol{x}_e(t)$, such as on the surface of a moving body, this form is somewhat awkward since the time derivative of the scalar potential corresponds to a fixed evaluation point. To derive a more appropriate form, let us take the time derivative of the potential as we follow the evaluation point, $\varphi(\boldsymbol{x}_e(t), t)$. By the chain rule, this is

$$\frac{\mathrm{d}\varphi}{\mathrm{d}t}(\boldsymbol{x}_e(t), t) = \frac{\partial\varphi}{\partial t}(\boldsymbol{x}_e(t), t) + \dot{\boldsymbol{x}}_e \cdot \nabla\varphi(\boldsymbol{x}_e(t), t), \tag{3.61}$$

where $\dot{\boldsymbol{x}}_e$ is the velocity of the evaluation point. The first term on the right-hand side comprises the rate of change of the potential at the fixed location currently occupied by the evaluation point; this is the same term that appears in (3.60) when that equation is evaluated at $\boldsymbol{x}_e(t)$. Furthermore, we can replace the gradient of the potential in (3.61) with the fluid velocity. Substituting this into (3.60) and manipulating further, we arrive at

$$p - \frac{1}{2}\rho|\dot{\boldsymbol{x}}_e|^2 + \frac{1}{2}\rho|\boldsymbol{v} - \dot{\boldsymbol{x}}_e|^2 + \rho\frac{\mathrm{d}\varphi}{\mathrm{d}t} = B(t), \tag{3.62}$$

where each spatially-dependent term is evaluated at $\boldsymbol{x}_e(t)$. Thus, this form of the Bernoulli equation accounts for the moving evaluation point, since the fluid velocity is replaced by the velocity relative to this point, and the time derivative is taken with respect to the moving point.

Note 3.2.1: Pressure in the Inertial and Windtunnel Frames

Form (3.62) of the Bernoulli equation allows us to reconcile pressure in two reference frames translating relative to one another, for example, between the inertial reference frame and the windtunnel reference frame. Consider the first reference frame as our usual inertial reference frame, which is always constructed in this book so that the fluid at infinity is at rest. As we defined in Sect. 2.3, $\boldsymbol{v}$ denotes the fluid velocity field measured with respect to this reference frame. The associated scalar potential field, φ, thus tends to a constant value at infinity, which we can take simply as zero for simplicity.

Now, consider the second reference frame as our windtunnel reference frame, as illustrated in Fig. 2.2. In this frame, the fluid at infinity moves as a (possibly time-varying) uniform flow $\boldsymbol{V}_\infty$. The fluid velocity field measured relative to this reference frame, which we denote by $\boldsymbol{v}^\dagger$, is related to the other simply by $\boldsymbol{v}^\dagger = \boldsymbol{v} + \boldsymbol{V}_\infty$. Positions $\boldsymbol{x}$ and $\boldsymbol{x}^\dagger$ in the respective reference frames were related in (2.30); time proceeds the same in both reference frames, so $t^\dagger = t$. For example, this relationship between positions can be interpreted as the time-varying position of an evaluation point, $\boldsymbol{x}_e(\boldsymbol{x}^\dagger, t^\dagger)$, in the inertial frame whose position is fixed at $\boldsymbol{x}^\dagger$ in the windtunnel frame. Clearly, this evaluation point moves at velocity $\dot{\boldsymbol{x}}_e = -\boldsymbol{V}_\infty$ relative to the inertial frame. Thus, Eq. (3.62) for such a point becomes

$$p - \frac{1}{2}\rho|\boldsymbol{V}_\infty|^2 + \frac{1}{2}\rho|\boldsymbol{v}^\dagger|^2 + \rho\frac{\mathrm{d}\varphi}{\mathrm{d}t} = B(t). \tag{3.63}$$

Remember that the time derivative in this equation is taken with respect to an observer moving with velocity $\dot{\boldsymbol{x}}_e$—that is, at a fixed point $\boldsymbol{x}^\dagger$ in the windtunnel frame. Thus, if one interprets any field quantity as a function of the independent variables of the windtunnel frame, $\boldsymbol{x}^\dagger$ and $t^\dagger$, then this time derivative is equivalent to $\partial/\partial t^\dagger$ of the field.

The scalar potential field in the windtunnel frame, $\varphi^\dagger$, defined so that $\boldsymbol{v}^\dagger = \nabla\varphi^\dagger$, tends to $\boldsymbol{V}_\infty \cdot \boldsymbol{x}^\dagger + C(t)$ at infinite distances, where $C(t)$ is a uniform value. Thus, it is easy to verify that this potential field is related to the one in the inertial frame by $\varphi^\dagger = \varphi + \boldsymbol{V}_\infty \cdot \boldsymbol{x}^\dagger + C(t)$. When φ in our Bernoulli equation is substituted with this relationship, we can write the equation in a form appropriate for the windtunnel frame,

$$p - \dot{\boldsymbol{V}}_\infty \cdot \boldsymbol{x}^\dagger + \frac{1}{2}\rho|\boldsymbol{v}^\dagger|^2 + \rho\frac{\partial\varphi^\dagger}{\partial t^\dagger} = B^\dagger(t), \tag{3.64}$$

in which all spatially uniform terms have been collected on the right-hand side in $B^\dagger(t)$. It is easy to see that, to within a uniform constant, the first two terms comprise the pressure, $p^{\dagger}$, measured in the windtunnel frame; that is,

$$p^{\dagger} = p - \rho \dot{V}_{\infty} \cdot \boldsymbol{x}^{\dagger}. \tag{3.65}$$

This additional term, which only arises when the uniform flow is accelerating, is identical in form to the hydrostatic pressure encountered in the presence of a gravitational field. As we will see later in this book when we explore the force on bodies, this term gives rise to a type of 'buoyant' force, analogous to the classical form of buoyancy attributed to hydrostatic pressure.

3.2.2 Two-Dimensional Flows and the Complex Potential

The Basic Singularities and Their Associated Velocity Many potential flows of interest in fluid dynamics are composed from singular solutions of Laplace's equation—that is, solutions that satisfy the equation everywhere except at isolated points—for the streamfunction or the scalar potential. Among these singular solutions, there is one for each equation that is referred to as the **fundamental solution**. Each of these is based on the Green's function of the (negative) Laplace operator, i.e., the solution of the equation

$$-\nabla^2 G = \delta(\boldsymbol{x}) \tag{3.66}$$

in an unbounded region, where δ is the Dirac delta function. In two dimensions, this Green's function is

$$G(\boldsymbol{x}) = -\frac{1}{2\pi} \log |\boldsymbol{x}|. \tag{3.67}$$

Thus, a fundamental solution of the streamfunction, the solution of $\nabla^2 \psi = -\Gamma \delta(\boldsymbol{x})$, is

$$\psi(\boldsymbol{x}) = \Gamma G(\boldsymbol{x}). \tag{3.68}$$

This constitutes a planar **point vortex** of strength Γ centered at the origin (or, when viewed in a three-dimensional context, a line vortex whose axis passes through the origin and is perpendicular to the plane of the flow). The strength Γ has the same units as the streamfunction—velocity times length—and represents the circulation about a contour that encloses the vortex. Recall that the general relationship between the streamfunction and the vorticity was given in Eq. (3.5). By simple comparison of the right-hand side with our particular case of the point vortex, it is clear that its vorticity field is $\omega(\boldsymbol{x}) = \Gamma \delta(\boldsymbol{x})$—infinitely large at the center and zero everywhere else. Remember from Eq. (3.36) that the circulation about a closed loop in the plane is equivalent to the vorticity enclosed by the loop. With vorticity of the point vortex concentrated at its center, we are thus assured that the circulation is the same for any contour that encircles this center.

The velocity associated with this vortex is easily found from Eq. (3.57), where $\boldsymbol{\Psi} = \psi \boldsymbol{e}_3$ as usual. Its components in the two Cartesian coordinate directions are

$$v_1(\boldsymbol{x}) = -\frac{\Gamma}{2\pi} \frac{x_2}{|\boldsymbol{x}|^2}, \quad v_2(\boldsymbol{x}) = \frac{\Gamma}{2\pi} \frac{x_1}{|\boldsymbol{x}|^2}, \tag{3.69}$$

producing a counterclockwise motion around the origin when Γ is positive. We will often find it preferable to write the velocity in vector form as

$$\boldsymbol{v}(\boldsymbol{x}) = \boldsymbol{K}(\boldsymbol{x}) \times \Gamma \boldsymbol{e}_3, \tag{3.70}$$

where $\boldsymbol{K}$ is called the **velocity kernel**, defined as the gradient of the Green's function,

$$\boldsymbol{K}(\boldsymbol{x}) := \nabla G = -\frac{\boldsymbol{x}}{2\pi |\boldsymbol{x}|^2}. \tag{3.71}$$

Analogous to the streamfunction, the solution of $\nabla^2 \varphi = Q\delta(\boldsymbol{x})$ gives rise to a fundamental solution for the scalar potential field,

$$\varphi(\boldsymbol{x}) = -QG(\boldsymbol{x}). \tag{3.72}$$

This constitutes a **monopole source** centered at the origin (or a line source when viewed in three dimensions), with a strength Q—again, with units of velocity times length—that represents the volume flow rate per unit depth in the out-of-plane direction through a closed contour enclosing the source. In fact, the reader can verify that, as the point vortex is associated with singular vorticity, the monopole source corresponds to a rate of dilatation concentrated at its center, $\Theta(\boldsymbol{x}) = Q\delta(\boldsymbol{x})$. The velocity components of the source can be determined from Eq. (3.56),

$$v_1(\boldsymbol{x}) = \frac{Q}{2\pi} \frac{x_1}{|\boldsymbol{x}|^2}, \quad v_2(\boldsymbol{x}) = \frac{Q}{2\pi} \frac{x_2}{|\boldsymbol{x}|^2}. \tag{3.73}$$

These indicate a radial outflow from the origin when Q is positive. The velocity can be written in vector form as

$$\boldsymbol{v}(\boldsymbol{x}) = -Q\boldsymbol{K}(\boldsymbol{x}), \tag{3.74}$$

where $\boldsymbol{K}$ is still defined by (3.71).

As we discussed earlier, any potential flow can be equivalently derived from either a scalar or a vector potential field (a streamfunction in two dimensions). These fundamental solutions are no different. The point vortex, for example, has an associated scalar potential field given by

$$\varphi(\boldsymbol{x}) = \frac{\Gamma}{2\pi} \arctan(x_2, x_1), \tag{3.75}$$

where we have used the notation $\arctan(x_2, x_1)$—sometimes called 'atan2' in mathematical software—to convey that this arc tangent respects the quadrant in which the point (x_1, x_2) lies. That is, it returns the angle between the positive x_1 axis and the point (x_1, x_2), in a range of angles of length 2π. This function and its ramifications will be discussed in more detail in Note 3.2.2. For now, let us note that the gradient of the function is

$$\frac{\partial}{\partial x_1}\arctan(x_2, x_1) = -\frac{x_2}{|\boldsymbol{x}|^2}, \quad \frac{\partial}{\partial x_2}\arctan(x_2, x_1) = \frac{x_1}{|\boldsymbol{x}|^2}. \tag{3.76}$$

The reader can thus easily verify that the gradient of the scalar potential field (3.75) is the velocity field (3.69). Similarly, the monopole source's velocity field (3.73) can be equivalently derived from the streamfunction field

$$\psi(\boldsymbol{x}) = \frac{Q}{2\pi}\arctan(x_2, x_1). \tag{3.77}$$

These companion potential fields will be important below when we discuss the description of these flows in complex variables.

To shift the center of the singularity from the origin to some other point, $\boldsymbol{y}$, the Green's function and velocity kernel are simply written as $G(\boldsymbol{x} - \boldsymbol{y})$ and $\boldsymbol{K}(\boldsymbol{x} - \boldsymbol{y})$, respectively. Throughout this work, we will refer to $\boldsymbol{y}$ and $\boldsymbol{x}$ as the source and evaluation points, respectively. It is clear from the definitions that the Green's function G is symmetric with respect to an interchange of these points, whereas the velocity kernel $\boldsymbol{K}$ is anti-symmetric:

$$G(\boldsymbol{y} - \boldsymbol{x}) = G(\boldsymbol{x} - \boldsymbol{y}), \quad \boldsymbol{K}(\boldsymbol{y} - \boldsymbol{x}) = -\boldsymbol{K}(\boldsymbol{x} - \boldsymbol{y}). \tag{3.78}$$

Derivatives of the fundamental solution are also singular solutions of Laplace's equation; we call these **higher-order singularities**. For example, the scalar potential field of the form

$$\varphi(\boldsymbol{x}) = \boldsymbol{Q} \cdot \nabla G(\boldsymbol{x}) \equiv -\frac{\boldsymbol{Q} \cdot \boldsymbol{x}}{2\pi|\boldsymbol{x}|^2} \tag{3.79}$$

constitutes a **dipole**, or doublet; its strength, $\boldsymbol{Q}$, is vector-valued, describing both the dipole's magnitude and orientation. Similarly, one can define a dipole of the streamfunction field,

$$\psi(\boldsymbol{x}) = -\boldsymbol{S} \cdot \nabla G(\boldsymbol{x}) \equiv \frac{\boldsymbol{S} \cdot \boldsymbol{x}}{2\pi|\boldsymbol{x}|^2}, \tag{3.80}$$

with strength $\boldsymbol{S}$. But, since potential flow fields in the plane are equivalently describable by either a scalar potential or a streamfunction, their dipole fields can be shown to be equivalent. Indeed, one can confirm that they produce identical flow fields when their strengths are related by

$$\boldsymbol{S} = \boldsymbol{e}_3 \times \boldsymbol{Q}. \tag{3.81}$$

We will revisit higher singularities in our discussion of multipole expansions in Chap. 4.

Thus far in this discussion of potential flows, we have relied on the vectorial notation to introduce the basic flows, and this facilitates direct comparison with the three-dimensional flows in the next section. However, it is foolish to venture further into a discussion of potential flows in the plane without introducing the very powerful tools of complex analysis.

Equivalence of Vectorial and Complex Variable Forms of Equations It is well known that a complex-valued function of z whose real and imaginary parts each satisfy Laplace's equation in the plane is **holomorphic** (sometimes called **analytic**). This property brings with it several useful mathematical tools for further analysis, which are discussed in Sect. A.2. Thus, since both φ and ψ individually satisfy Laplace's equation in potential flow, we are assured that a **complex potential**, F, defined as

$$F := \varphi + \mathrm{i}\psi, \tag{3.82}$$

is a holomorphic function (except at singular points). This means, in particular, that the complex potential is differentiable with respect to z, and that its derivative with respect to z^* is exactly zero.

The derivative of the complex potential is the complex velocity, denoted by w:

$$w := u - \mathrm{i}v := \frac{\mathrm{d}F}{\mathrm{d}z}. \tag{3.83}$$

The minus sign is necessary to ensure that each velocity component is correctly related to the corresponding partial derivative of the scalar potential and streamfunction in the Cauchy–Riemann equations (A.102). Sometimes it is useful to write this velocity in polar components, v_r and v_θ, in correspondence to the polar form of z: $z = r\mathrm{e}^{\mathrm{i}\theta}$. These can be found by rotating w into the local r–θ system:

$$v_r - \mathrm{i}v_\theta = w\mathrm{e}^{\mathrm{i}\theta}. \tag{3.84}$$

The complex form of the uniform flow's velocity field is described by

$$W_\infty = U_\infty - \mathrm{i}V_\infty. \tag{3.85}$$

Note that the vectorial and complex notations are related by $\mathrm{Re}(W_\infty) \equiv U_\infty \equiv V_{\infty,1}$ and $\mathrm{Im}(W_\infty) \equiv -V_\infty \equiv -V_{\infty,2}$. The associated complex potential is easily seen to be

$$F(z) = W_\infty z. \tag{3.86}$$

The components of the uniform flow in the body-fixed coordinate system are related to the inertial components through the transformation exhibited by (2.38):

$$\tilde{W}_\infty := \tilde{U}_\infty - \mathrm{i}\tilde{V}_\infty = W_\infty \mathrm{e}^{\mathrm{i}\alpha}. \tag{3.87}$$

The vortex and monopole source discussed above can be represented collectively as a single singularity with complex strength,

$$F(z) = \frac{Q - \mathrm{i}\Gamma}{2\pi} \log z, \tag{3.88}$$

with an associated complex velocity,

$$w(z) = \frac{Q - \mathrm{i}\Gamma}{2\pi z}. \tag{3.89}$$

It is easy to verify that this is equivalent to the sum of a point vortex of strength Γ and a monopole source of strength Q, each centered at the origin. Either singularity on its own is easily recovered by simply setting the other strength to zero. The center of the singularity can be easily shifted to a point z_0 by replacing z with $z - z_0$. Note that the functions $-1/(2\pi)\log z$ and $-1/(2\pi z)$ are the complex equivalents of the Green's function (3.67) and velocity kernel (3.71), respectively. This complex Green's function, however, includes somewhat more information, so that F in (3.88) not only contains the streamfunction of the point vortex (3.68), but also its associated scalar potential (3.75), and similarly, describes both the scalar potential (3.72) and the streamfunction (3.77) of the monopole source.

Higher-order singularities, starting with the dipole (3.79), have the form

$$F(z) = \frac{a_n}{2\pi z^n}, \tag{3.90}$$

where a_n is a complex strength coefficient and n is any integer greater than or equal to 1. This form is holomorphic throughout the complex plane, except at the origin. If, for $n = 1$, we compare the real part of this complex potential with the expression (3.79) for the scalar potential of the dipole, we find that the real and imaginary parts of a_1 correspond, respectively, to the negative of the x and y components of the vector strength $\boldsymbol{Q}$. We represent this correspondence here, as in Sect. A.2, by the notation

$$\boldsymbol{Q} \Longleftrightarrow -a_1. \tag{3.91}$$

It is useful to note that, even for singularities of order higher than 1, the strength is completely described by only two numbers: the real and imaginary parts of a_n.

Note 3.2.2: Multi-Valuedness and the Branch Cut

The complex form of the vortex and source potential described by (3.88) reveals an important aspect of this flow. We will discuss this in the context of an isolated vortex (i.e., with $Q = 0$), the case most relevant to our interests in this book. The logarithmic dependence on z carries with it an associated multi-valuedness, borne by the scalar potential in the case of a vortex. This can be seen by writing z in its polar form, $z = |z|\mathrm{e}^{\mathrm{i}\arg z}$, so that $\log z = \log|z| + \mathrm{i}\arg z$. The imaginary part of the log is equal to the argument of z, which increases by 2π each time the singular point (the origin in this case) is encircled. For a vortex, it is easy to see that

$$\varphi = \frac{\Gamma}{2\pi} \arg z, \qquad \psi = -\frac{\Gamma}{2\pi} \log |z|, \tag{3.92}$$

so the scalar potential's value must be established by constructing a **branch cut**—that is, a cut that renders the function single-valued—and deciding the **branch** of the log to which we are restricting ourselves. For example, we could place the branch cut on the ray extending along the negative real axis, and decide that the function is restricted to the branch $-\pi \leq \arg z < \pi$. Some notes are in order here:

- In this note, we have presented the vortex and the associated multi-valuedness of the scalar potential, primarily because it is most relevant to our work in this book. However, we could also have illustrated the concept on the monopole source, whose streamfunction is multi-valued.
- The velocity field itself is always single valued and continuous, even if the potential is not.
- It is important to stress that we have complete authority over how the branch cut is constructed, provided that it originates at the center of the singularity. For example, it need not be straight.
- A vortex singularity is always paired with another of equal but opposite strength, and the branch cut can always be constructed so that it joins these two singularities in some arbitrary manner. In some cases, one of the members in this pair is at infinity, so the branch cut extends accordingly. However, in this work, we will restrict ourselves to cases in which both members of the pair are a finite distance from the origin, so that one can construct a closed contour that encloses all of the branch cuts in the flow.
- Branch cuts arise only in two-dimensional flows. However, in Sect. 3.3.2, we will see a three-dimensional singularity (a closed vortex filament) that presents a three-dimensional manifestation of a branch cut.

Corner and Wedge Flows In the previous discussion we found that higher-order singularities have the form z^{-n}, where n is a positive integer. But we do not need to restrict ourselves to integer powers to obtain useful flows. And, if we temporarily ignore the fact that it does not satisfy the boundary condition (3.58) at infinity, many of these useful flows involve a positive power of z. Thus, let us consider complex potentials whose z dependence is $F(z) \sim z^{\delta}$, where δ is a real number, but not necessarily an integer. Here, we will discuss these in the general form

$$F(z) = \nu S L^{1-1/\nu} z^{1/\nu}, \tag{3.93}$$

where S is a complex constant, ν a real-valued constant, and L is a characteristic length scale (and thus, also a real-valued constant). The associated complex velocity is

$$w(z) = S\left(\frac{L}{z}\right)^{1-1/\nu}. \tag{3.94}$$

This shows clearly that S has units of velocity.

To see what kind of flow this represents, we write z and S in their polar forms, $z = re^{i\theta}$ and $S = |S|e^{i\phi}$, and substitute into (3.93). Then, the polar form of the complex potential is $F(r, \theta) = \nu|S|r^{1/\nu}e^{i(\theta/\nu+\phi)}$, and the streamfunction is just the imaginary part of this:

$$\psi(r, \theta) = \nu|S|L^{1-1/\nu}r^{1/\nu}\sin(\theta/\nu + \phi). \tag{3.95}$$

The flow is most easily revealed by noting the streamlines corresponding to $\psi = 0$, which are generated by setting $\theta/\nu + \phi = m\pi$, for m an arbitrary integer. These are rays extending from the origin at angles $\theta_m = \nu(m\pi - \phi)$; the rays are regularly spaced by angle $\pi\nu$. Since we are free to interpret any streamline as an impenetrable wall, the most obvious interpretation of (3.93) is that of a **corner flow**: a flow confined to the sector of angle

$$\theta_{m+1} - \theta_m = \nu\pi \tag{3.96}$$

between two adjacent streamline rays, θ_m and θ_{m+1}. In this context, we only permit values of ν that ensure that the **exterior angle** of the corner—the angle of the flow sector—is between 0 and 2π. That is, we require

$$0 < \nu \leq 2. \tag{3.97}$$

At the upper end of this range, $\nu = 2$, the entire plane constitutes the flow region, and the body is limited to a single ray with **interior angle**—the angle of the body sector—equal to zero. In contrast, as ν approaches 0, the flow is confined to a vanishingly small sector of the plane, and the interior angle is 2π.

It should be noted that the phase ϕ of coefficient S sets the orientation of the corner. The corner has an associated normal vector, n_0, that bisects the flow sector; its angle is simply the average of the two bounding rays, $(\theta_m + \theta_{m+1})/2 = \nu(m\pi + \pi/2 - \phi)$. We can choose ϕ to set this angle to a desired value. For all of the results shown in Fig. 3.4, the normal is at angle $\pi/2$.

We have to be a little bit careful about the multi-valuedness of this flow. If we encircle the origin once (i.e., add 2π to θ), then the streamfunction (3.95) will generally have a different value than it started with. In fact, except for integer values of $1/\nu$, the streamfunction, the scalar potential and the velocity are all multi-valued for this flow. Thus, we must restrict the flow to a branch, with a corresponding branch cut that does not interfere with the flow. A natural choice is to restrict the solution to the branch $\beta \leq \theta < \beta + 2\pi$, where β is an angle that bisects the interior angle of the corner (i.e., inside the body sector):

$$\beta := \nu(m\pi + \pi/2 - \phi) - \pi. \tag{3.98}$$

The directionality of the flow can be established by inspecting the velocity (3.94) along the confining streamlines. For this, it is more useful to write the velocity in its polar form (3.84), since v_r represents the tangent velocity along these streamlines. Throughout the flow, these polar components are

$$v_r = |S| \left(\frac{L}{r}\right)^{1-1/\nu} \cos(\theta/\nu + \phi), \quad v_\theta = -|S| \left(\frac{L}{r}\right)^{1-1/\nu} \sin(\theta/\nu + \phi), \tag{3.99}$$

and in particular, on the streamline along ray θ_m,

$$v_r = (-1)^m |S| \left(\frac{L}{r}\right)^{1-1/\nu}, \quad v_\theta = 0. \tag{3.100}$$

Thus, we see that the direction of the flow is opposite along the two bounding rays of the flow sector, at angles θ_m and θ_{m+1}. This is not surprising, since the flow must satisfy conservation of mass across any surface that extends between the two rays. Examples of the streamlines generated for an even value of m and various choices of ν are depicted in Fig. 3.4; we will refer to each of these as a positive corner flow. If we choose instead an odd value for m, the sense of the streamlines is reversed, and we will refer to this as a negative corner flow.

From the examples depicted in Fig. 3.4, we can see that $\nu = 1$ is the threshold between two different types of corner flow. For $1 < \nu \leq 2$, the interior angle of the body is less than π, so the body's corner is convex and the flow comprises more than half of the plane; for $0 \leq \nu < 1$, the interior angle exceeds π, so the body's corner is concave and less than half of the plane is filled with fluid. These two types of corner are different in another important respect, easily observed in (3.99):

Note 3.2.3: Velocity Behavior in Corner Flows

Consider a corner whose exterior angle—the angle swept by the flow region—is $\pi\nu$. In a concave corner, for which $0 \leq \nu < 1$, the corner itself is a stagnation point, $\boldsymbol{v} = 0$, while at a convex corner, $1 < \nu \leq 2$, the flow is singular, and its velocity goes like $|\boldsymbol{v}| \sim (L/r)^{1-1/\nu}$ as the corner is approached. The value $\nu = 1$ corresponds to a flat wall with non-zero but finite velocity.

In particular, for the case $\nu = 2$—a thin plate—the velocity goes like the inverse square root of distance. This fact will be crucially important in many parts of this book.

The singular behavior of the flow past a convex corner makes it difficult to physically interpret the strength of such a flow. We want to ask, "How fast is the flow around a singular corner?" It seems reasonable that the answer is dependent upon $|S|$, but what does $|S|$ physically represent? At this point, all we know is that $|S|$ has units of velocity. Here, we will provide one answer to this that will serve us later. The definition of circulation (3.34) strictly requires a contour C that is completely closed. However, suppose we extend this definition to a non-closed contour such

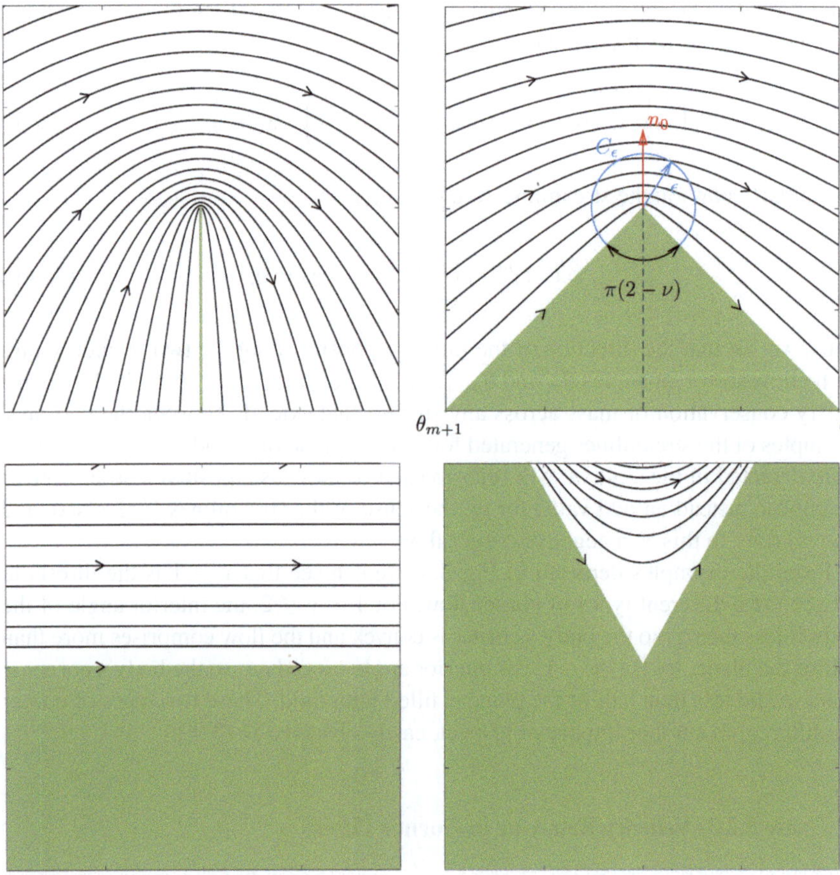

Fig. 3.4 Examples of positive corner flows with $\nu = 2$ (upper left), $\nu = 3/2$ (upper right), $\nu = 1$ (lower left) and $\nu = 1/3$ (lower right) and an even value of m for the flow sector. The interior angle is denoted in the upper right panel, as is the contour C_ϵ used to define the partial circulation and the corner flow intensity (3.103). The branch cut for the flow depicted in this panel is shown as a dashed line. The corner normal, n_0, is also illustrated

as C_ϵ, depicted in the upper right panel of Fig. 3.4, which extends only from one bounding streamline at θ_m to the other at angle θ_{m+1}:

$$\int_{C_\epsilon} \boldsymbol{v} \cdot \mathrm{d}\boldsymbol{l}. \tag{3.101}$$

We will call this the **partial circulation** on this contour. It is easy to show that any two contours between the same two endpoints will lead to the same value of partial circulation, as long as no vorticity is enclosed by the closed circuit formed by joining the contours. Intuitively, we expect this to provide some measure of the intensity of the flow about the corner point. The complex form of this partial circulation follows from (3.37). Let us note that

$$\int_{C_\epsilon} w(z)\,\mathrm{d}z = \int_{C_\epsilon} \mathrm{d}F = F(\epsilon, \theta_{m+1}) - F(\epsilon, \theta_m). \tag{3.102}$$

The imaginary part, associated with the difference in streamfunction on the two rays, is obviously zero, since these streamlines were each identified by setting $\psi = 0$. (Note that this is also a consequence of the conservation of mass, since this imaginary part also represents the net volume flow rate per unit depth across C_ϵ.) For the remaining real part, it is easy to verify that the partial circulation on C_ϵ is $\mp 2\nu|S|L^{1-1/\nu}\epsilon^{1/\nu}$, where the negative sign corresponds to a positive corner flow and the opposite sign to a negative corner flow. This partial circulation clearly vanishes—for any permissible value of ν—as the radius of the contour is shrunk to zero. So even in the case of the convex corner, though the velocity becomes infinite, its integrated value along the vanishingly small arc approaches zero. But the form of this partial circulation suggests we can define a quantity that remains finite and non-zero in this limit:

$$\sigma := -\lim_{\epsilon \to 0} \frac{1}{2\nu L^{1-1/\nu}\epsilon^{1/\nu}} \int_{C_\epsilon} \boldsymbol{v} \cdot \mathrm{d}\boldsymbol{l}. \tag{3.103}$$

We have given this definition here in a vectorial form that can be applied in more general settings, such as along sharp edges of a three-dimensional body. The negative sign ensures that a positive corner flow returns a positive value of σ, the factor $1/(2\nu)$ is just provided for convenience, and L is again the characteristic length scale of the geometry to which the corner belongs. Since $\boldsymbol{v} = \nabla\varphi$ in the vicinity of the corner, then this quantity can also be calculated from the difference in scalar potential at either end of the contour:

$$\sigma = -\lim_{\epsilon \to 0} \frac{\varphi(\epsilon, \theta_{m+1}) - \varphi(\epsilon, \theta_m)}{2\nu L^{1-1/\nu}\epsilon^{1/\nu}}. \tag{3.104}$$

If we apply L'Hôpital's rule to the limit, then we can also write σ as a limit on the difference in velocity tangent to the bounding streamlines:

$$\sigma = -\lim_{\epsilon \to 0} \frac{1}{2} \left(\frac{\epsilon}{L}\right)^{1-1/\nu} \left(v_r(\epsilon, \theta_{m+1}) - v_r(\epsilon, \theta_m)\right). \tag{3.105}$$

These formulas for σ are provided for later use, when we will ascribe another role for σ as a measure of edge suction. But based on its adaptation from the partial circulation, we need no calculation to see that $\sigma = \pm|S|$. In other words, at any corner, this finite non-zero quantity σ incorporates both the strength of the flow, as measured by $|S|$, as well as its directionality, determined by the choice of sign—positive for a positive corner flow, or negative for a negative corner flow. For this reason, we will call σ the corner flow's **signed intensity**. This quantity, which, like S, has units of velocity, will arise several times in this book. In fact, as the reader can verify, we can write our coefficient S in a manner that easily exhibits its role in setting the strength and orientation of the corner flow:

$$S = \mathrm{i}\sigma n_0^{-1/\nu}, \tag{3.106}$$

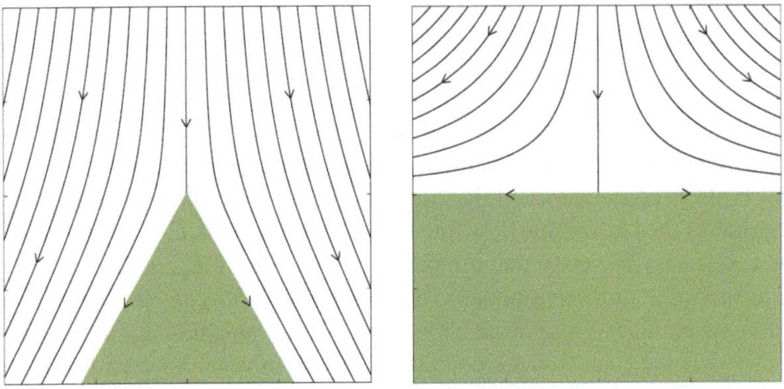

Fig. 3.5 Examples of positive wedge flows with $\nu = 5/6$ (left) and $\nu = 1/2$ (right), and an even value of m for the flow sector. The angle of the wedge in the left panel is $\pi/3$

where n_0 is the corner's normal, as illustrated in Fig. 3.4.

In the corner flow, we have restricted the flow domain to the region between two adjacent streamline rays. However, another reasonable interpretation of (3.93) arises from combining two neighboring sectors, so that the flow region extends between the rays θ_m and θ_{m+2}:

$$\theta_{m+2} - \theta_m = 2\nu\pi. \tag{3.107}$$

This is called a **wedge flow**, since the flow is in the same direction on the two bounding streamlines. A few examples are depicted in Fig. 3.5. The angle of the 'wedge'—the body sector—is $2\pi(1 - \nu)$, so clearly the range of permissible values of ν is slightly more restricted than for the corner flow,

$$0 \le \nu \le 1. \tag{3.108}$$

The upper bound corresponds to uniform flow, as in the corner flow context, but now with the entire plane composed of fluid. The case $\nu = 1/2$, shown in the right panel in Fig. 3.5, is the well-known **stagnation flow**.

Let us conclude with one final note. It is natural to wonder about the utility of the corner or wedge flow in the larger context of external inviscid flows. In particular, all but the convex corners produce an unbounded velocity field at large distances from the origin, which fails to satisfy the boundary condition (3.58). In contrast to the basic singularities we discussed earlier, corner and wedge flows are not building blocks that we can use to systematically develop more sophisticated flows fields. Rather, they describe the dominant flow behavior in the immediate vicinity of these geometric features on an impenetrable boundary. Thus, even in the most complicated flows, they allow us to easily abstract and develop insight into the local behavior.

3.2.3 Three-Dimensional Flows

As in planar flows, the point vortex is a fundamental singularity in three-dimensional flows. In this context, we not only must specify the strength of the vortex, but also its direction. Thus, we provide it with a vectorial strength $\boldsymbol{\alpha}$, whose units are vorticity times volume. The vector potential associated with a vortex at the origin is

$$\boldsymbol{\Psi}(\boldsymbol{x}) = \boldsymbol{\alpha} G(\boldsymbol{x}), \tag{3.109}$$

where, as in the two-dimensional case, $G(\boldsymbol{x})$ is the Green's function of the negative Laplace operator, i.e., the solution of $-\nabla^2 G = -\delta(\boldsymbol{x})$. In this three dimensional context, this is

$$G(\boldsymbol{x}) = \frac{1}{4\pi|\boldsymbol{x}|}. \tag{3.110}$$

The velocity field of this vector potential is calculated by taking the curl, as usual. This produces

$$\boldsymbol{v}(\boldsymbol{x}) = \boldsymbol{K}(\boldsymbol{x}) \times \boldsymbol{\alpha}, \tag{3.111}$$

where we have defined the three-dimensional velocity kernel, $\boldsymbol{K}$, analogously to the two-dimensional kernel[3]: namely, as the gradient of the fundamental solution of the negative Laplacian,

$$\boldsymbol{K}(\boldsymbol{x}) := \nabla G = -\frac{\boldsymbol{x}}{4\pi|\boldsymbol{x}|^3}. \tag{3.112}$$

Equation (3.111) indicates that the velocity field is perpendicular to the axis of the vortex and to the line that joins the vortex with the point of evaluation, $\boldsymbol{x}$. The three-dimensional Green's function and velocity kernel, like their two-dimensional counterparts, have the same symmetric and anti-symmetric properties (3.78).

The monopole source (or sink) is also a fundamental singularity: in this case, of Laplace's equation for the scalar potential. Its form is

$$\varphi(\boldsymbol{x}) = -QG(\boldsymbol{x}). \tag{3.113}$$

It is easy to see that its associated velocity, obtained by taking the gradient of the scalar potential, is

$$\boldsymbol{v}(\boldsymbol{x}) = -Q\boldsymbol{K}(\boldsymbol{x}). \tag{3.114}$$

This velocity field is directed radially outward from (or, if $Q < 0$, inward toward) the singular point.

As in two dimensions, higher-order singularities are formed from derivatives of the fundamental solution. A dipole, for example, has scalar potential field

$$\varphi(\boldsymbol{x}) = \boldsymbol{Q} \cdot \nabla G(\boldsymbol{x}) \equiv -\frac{\boldsymbol{Q} \cdot \boldsymbol{x}}{4\pi|\boldsymbol{x}|^3}, \tag{3.115}$$

where $\boldsymbol{Q}$ is, as in the planar case, the vector-valued strength of the dipole. There must be an equivalent description of this flow in terms of the vector potential; one can verify that the following produces the same velocity field as the scalar potential in (3.115):

[3] We have used the same symbol to denote the Green's function and velocity kernel in two-dimensional and three-dimensional flows in order to unify some of the discussion that follows; whether its specific forms are (3.67) and (3.71) or (3.110) and (3.112) will be clear from the context in which it is used.

$$\Psi(x) = -Q \times \nabla G(x) \equiv \frac{Q \times x}{4\pi |x|^3}. \tag{3.116}$$

Singularities of higher order will arise in our discussion of multipole expansion in Sect. 4.7.

3.3 Vortex Structures

In a general viscous flow, vorticity is often distributed throughout an extensive region of space. For the purposes of determining its associated velocity field, we can regard the vorticity field as an infinitely dense collection of point vortices spread over the vortical region, each with strength $\omega\,\delta V$ given by the vorticity's local value and the infinitesimal volume δV it occupies. Each point vortex's contribution to velocity is given by (3.70) or (3.111), with its center shifted to the vortex's location. The overall velocity field—in two or three dimensions—is then given by the collective contribution from all of these point vortices,

$$v(x) = \int_{\mathcal{V}} K(x - y) \times \omega(y)\, \mathrm{d}V(y), \tag{3.117}$$

where, as usual, we interpret this as an area integral in two-dimensional calculations; the velocity kernel K is given by either (3.71) or (3.112). This equation is generally known as the **Biot–Savart integral**, and formally, it represents the particular solution of the identity $-\nabla^2 v = \nabla \times \omega$ in unbounded incompressible flow. As we will discuss more extensively in Chap. 4 (and as we show in detail in Sect. A.1.3 in the Appendix), the velocity field of (3.117) is one part of the overall velocity field in a general flow.[4]

Vorticity is also spatially distributed in inviscid flows, and (3.117) holds regardless of the presence of viscosity, of course. But the absence of diffusion limits the manner in which the vorticity can be spread, as embodied in Helmholtz' theorems discussed earlier. In fact, under the assumptions that underlie the analysis in this book—inviscid incompressible flow with conservative body forces—there is strictly no mechanism by which vorticity can be generated in the first place. It is only through our relaxation of these assumptions at singular points, generally on a body–fluid interface, that we can provide a means for vorticity to enter the fluid, as will be illustrated in Sect. 5.1. At these singular transfer points, fluid vorticity is generated in an inherently concentrated form.

Thus, when we mathematically model an inviscid flow, it is natural—and often advantageous—for vorticity to be represented by lower-dimensional subsets of the fluid region. For example, in a two-dimensional flow, vorticity might be non-zero only

[4] It is common to describe the vorticity as 'inducing' the velocity given by the Biot–Savart integral. This terminology is convenient, but misleading because it implies a causal relationship between the vorticity and velocity. The vorticity cannot 'cause' the velocity field, of course. Rather, these fields only hold a consistent relationship with each other, expressed either by $\omega = \nabla \times v$ or its corollary (3.117).

at discrete points or along curves; in three-dimensional flow, such lower-dimensional representations might consist of points, curves or surfaces. The total volume (or area in two dimensions) occupied by fluid vorticity will remain zero for all of these lower-dimensional structures. For each of these lower-dimensional vortex structures, the equation for its associated velocity field can be obtained from a specialization of the more general result, the Biot–Savart integral (3.117).

3.3.1 Point Vortex

In Sects. 3.2.2 and 3.2.3 we presented the point vortex, the lowest-dimensional form of a vortex structure, characterized only by its position and strength. Suppose we have a collection of N_v such point vortices, where the position of the Jth vortex is denoted by $\boldsymbol{x}_J$ and its strength is given by Γ_J in two-dimensions or $\boldsymbol{\alpha}_J$ in three dimensions. We will often refer to such a set as a **vortex cloud**. Then the form of the velocity field (3.117) induced by this collection can be immediately obtained by superposition of the velocity fields of each point vortex, (3.70) or (3.111) in two or three dimensions, which we represent in unified form as

$$\boldsymbol{v}(\boldsymbol{x}) = \sum_{J=1}^{N_v} \boldsymbol{K}(\boldsymbol{x} - \boldsymbol{x}_J) \times \boldsymbol{\alpha}_J, \tag{3.118}$$

with the understanding that $\boldsymbol{\alpha}_J = \Gamma_J \boldsymbol{e}_3$ in two-dimensional flow. Note that in the two-dimensional context, under the conditions we have assumed for these flows (inviscid, incompressible), the strength of each point vortex must be time invariant by Kelvin's circulation theorem (3.40). In three dimensions, this strength is modified by vortex stretching and tilting.

In complex analysis of a two-dimensional flow, the position of the Jth point vortex is denoted by z_J, and the velocity field of the collection is obtained from superposition of (3.89) (with the source strength set to zero):

$$w(z) = \frac{1}{2\pi i} \sum_{J=1}^{N_v} \frac{\Gamma_J}{z - z_J}. \tag{3.119}$$

Earlier in this section, we developed the Biot–Savart integral (3.117) in an intuitive fashion, through a limit obtained from a collection of point vortices whose strengths were allowed to shrink to zero. However, we could have followed this path in the opposite direction: by starting with (3.117) for a general vorticity distribution, and then specializing this and affiliated results to a collection of singular point vortices. This process is facilitated by noting that a point vortex can be represented as a vorticity field in the form of a **Dirac delta function**:

$$\omega(\boldsymbol{x}) = \sum_{J=1}^{N_v} \boldsymbol{\alpha}_J \delta(\boldsymbol{x} - \boldsymbol{x}_J). \tag{3.120}$$

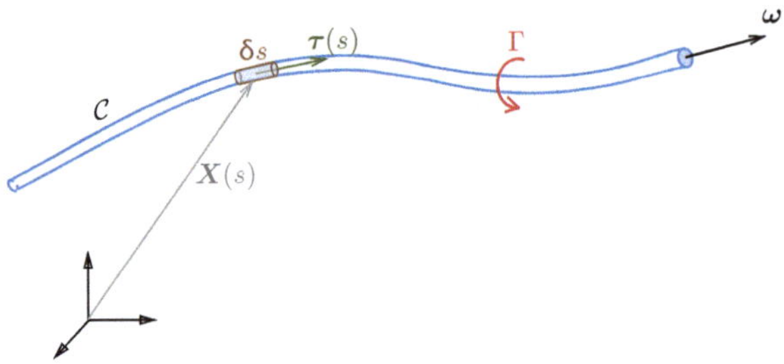

Fig. 3.6 Schematic of a portion of a vortex filament C of strength Γ

We need only recall the basic property (A.32) of this delta function. Using this property, then, for example, Eq. (3.118) follows immediately from (3.117) when the distribution (3.120) is substituted for the vorticity field. We will use this same process in several other places in this book in order to specialize general results on vorticity fields to collections of point vortices.

The velocity field associated with a point vortex is obviously unbounded as the evaluation point approaches the position of the vortex, and this leads to important questions about the vortex' transport. As we will discuss in detail in Chap. 7, we can rigorously argue that a point vortex does not induce velocity on itself. Furthermore, we will also show in that chapter that we can smooth the flow field in the immediate vicinity of a vortex by using **regularization** techniques, essential for practical computations. These regularization techniques can be thought of as a replacement of the singular representation (3.120) by a collection of smooth **blobs** of vorticity.

In the following sections, we define other vortex structures, in the forms of curves (vortex filaments) or surfaces (vortex sheets), and provide some mathematical background on their geometry.

3.3.2 Vortex Filament

A **vortex filament** is an entity in which vorticity is concentrated in a vortex tube of very small cross-sectional area. This property allows us to (usually) think of a vortex filament as a space curve that is everywhere tangent to the vorticity in the filament, as depicted in Fig. 3.6. However, it is distinguished from a vortex line by the fact that the vorticity is zero immediately outside of the narrow region surrounding the curve. Indeed, in order for the strength of this filament to be finite, the vorticity must be of very large magnitude. Since a vortex filament is a special example of a vortex tube, when it is immersed in a fluid it must obey Helmholtz' theorems: it is transported by the flow; it either forms a closed loop or ends at infinity; and its strength, Γ, is uniform along its length and constant in time. As usual for a tube, this strength represents the circulation about any contour enclosing the filament.

Let us describe a filament's curve C in space by the parameterization $\boldsymbol{X}(s)$, where s is the arc length measured from some arbitrary position on the curve. The curve's local unit tangent, $\boldsymbol{\tau}(s)$, parallel to the vorticity, is given by the derivative $\boldsymbol{\tau}(s) := \partial \boldsymbol{X}/\partial s$. Because a filament's configuration will evolve, $\boldsymbol{X}$ and its derivatives

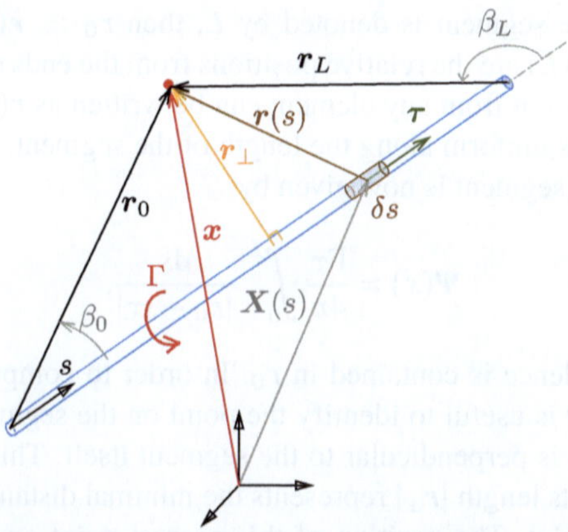

Fig. 3.7 Straight segment of a vortex filament. The evaluation point, x, is depicted in red

are actually functions of both s and time, t, but we will generally omit the time dependence for brevity. We will denote the local velocity of the filament by $\dot{X}(s) := \partial X(s,t)/\partial t$; this velocity is itself time dependent in general circumstances.

As with any singular distribution of vorticity, a vortex filament induces a potential flow around it. This flow can be derived by regarding the continuous filament as composed of infinitesimal segments oriented along the curve, $\delta l = (\partial X/\partial s)\delta s = \tau\delta s$, each of which is associated with a three-dimensional point vortex of vectorial strength $\delta\alpha = \Gamma\delta l$. The flow field is then given by the collective contribution from all of these point vortices. Based on the results for a single vortex, (3.109) and (3.111), the vector potential field induced by the filament at an evaluation point $x \notin C$ is given by

$$\Psi(x) = \Gamma \int_C G(x - X(s))\,\mathrm{d}l(s), \qquad (3.121)$$

and the velocity field at x by

$$v(x) = \Gamma \int_C K(x - X(s)) \times \mathrm{d}l(s), \qquad (3.122)$$

where the three-dimensional Green's function (3.110) and velocity kernel (3.112) are used. It is important to note that the integrals are determined only by the geometry of the filament and the relative position of the evaluation point.

For a filament of general shape, these integrals must be evaluated numerically. In later chapters, particularly in our discussions of numerical modeling, it will be useful to discretize a filament into a connected set of straight segments of finite length. For any such segment, the integrals in (3.121) and (3.122) can be evaluated analytically. To do so, let us denote the position of the evaluation point relative to an infinitesimal element on the filament by $r(s) := x - X(s)$, as depicted in Fig. 3.7, where s remains an arc length parameter, defined here from the starting point of the segment. If the overall length of the segment is denoted by L, then $\boldsymbol{r}_0 := \boldsymbol{r}(0) = \boldsymbol{x} - \boldsymbol{X}(0)$ and $\boldsymbol{r}_L := \boldsymbol{r}(L) = \boldsymbol{x} - \boldsymbol{X}(L)$ are the relative positions from the ends of the segment. Note that the relative position from any element can be written as $\boldsymbol{r}(s) = \boldsymbol{r}_0 - s\boldsymbol{\tau}$, where here, the tangent $\boldsymbol{\tau}$ is uniform along the length of the segment. The vector potential field induced by the segment is now given by

$$\boldsymbol{\Psi}(\boldsymbol{x}) = \frac{\Gamma\boldsymbol{\tau}}{4\pi}\int_0^L \frac{\mathrm{d}s}{|\boldsymbol{r}_0 - s\boldsymbol{\tau}|}, \tag{3.123}$$

where the $\boldsymbol{x}$ dependence is contained in $\boldsymbol{r}_0$. In order to compute this integral and others that follow, it is useful to identify the point on the segment axis from which the relative position is perpendicular to the segment itself. This relative position is denoted by $\boldsymbol{r}_\perp$ and its length $|\boldsymbol{r}_\perp|$ represents the minimal distance from the filament to the evaluation point. The position of this nearest point on the segment axis is given by the arc length parameter $s_\perp = \boldsymbol{r}_0 \cdot \boldsymbol{\tau}$; it need not lie on the segment itself. If we shift the origin of the segment parameterization to $s_\perp$, then the integral is easily evaluated:

$$\boldsymbol{\Psi}(\boldsymbol{x}) = \frac{\Gamma\boldsymbol{\tau}}{4\pi}\left[\log\left(|\boldsymbol{r}_L| - \boldsymbol{r}_L \cdot \boldsymbol{\tau}\right) - \log\left(|\boldsymbol{r}_0| - \boldsymbol{r}_0 \cdot \boldsymbol{\tau}\right)\right] \tag{3.124}$$

The velocity field can be derived by similar steps,

$$\boldsymbol{v}(\boldsymbol{x}) = \frac{\Gamma}{4\pi|\boldsymbol{r}_\perp|}\boldsymbol{\tau} \cdot \left(\frac{\boldsymbol{r}_0}{|\boldsymbol{r}_0|} - \frac{\boldsymbol{r}_L}{|\boldsymbol{r}_L|}\right)\boldsymbol{\tau} \times \frac{\boldsymbol{r}_\perp}{|\boldsymbol{r}_\perp|}. \tag{3.125}$$

The first part of this expression provides the magnitude of the velocity, while the trailing factor provides the directionality: perpendicular to the filament segment's axis and to the relative position of the evaluation point. It is useful to note that expressions for the vector potential and velocity fields given by (3.124) and (3.125) show that these fields are determined entirely by the relative positions from the endpoints, $\boldsymbol{r}_0$ and $\boldsymbol{r}_L$, since both $\boldsymbol{\tau}$ and $\boldsymbol{r}_\perp$ can be written in terms of these relative positions. However, to orient these fields in inertial space, we also need the position of one of these endpoints or of the evaluation point itself in the inertial coordinate system. Thus, we can denote the entire dependency of the vector potential and velocity as $\boldsymbol{\Psi}(\boldsymbol{x}, \boldsymbol{r}_0, \boldsymbol{r}_L)$ and $\boldsymbol{v}(\boldsymbol{x}, \boldsymbol{r}_0, \boldsymbol{r}_L)$, respectively.

The formulas derived here for the straight filament segment are often written in terms of the angles β_0 and β_L formed between the relative positions from the ends and the axis of the segment, as defined in Fig. 3.7. In terms of these angles, these formulas are

$$\boldsymbol{\Psi}(\boldsymbol{x}) = \frac{\Gamma\boldsymbol{\tau}}{8\pi}\left[\log\left(\frac{1-\cos\beta_L}{1+\cos\beta_L}\right) - \log\left(\frac{1-\cos\beta_0}{1+\cos\beta_0}\right)\right]. \tag{3.126}$$

and

$$\boldsymbol{v}(\boldsymbol{x}) = \frac{\Gamma}{4\pi|\boldsymbol{r}_\perp|}\left(\cos\beta_0 - \cos\beta_L\right)\boldsymbol{\tau} \times \frac{\boldsymbol{r}_\perp}{|\boldsymbol{r}_\perp|}. \tag{3.127}$$

When the segment becomes infinitely long, these reduce, respectively, to the two-dimensional results (3.68) and (3.70) (in a plane oriented perpendicularly to the filament axis).[5]

In our derivation of these preceding equations, we had no reason to think of the filament as anything more than vorticity concentrated into an infinitesimal cross-section along a space curve—in other words, a **line vortex** (not to be confused with a vortex line). In fact, this interpretation is reasonable when we wish to evaluate the filament's induced flow field at points off of this curve. To be sure, it is obvious that the velocity field is singular as one approaches the axis of the filament. After all, the point vortex also has this singular behavior, and a planar point vortex is simply the cross-section of a rectilinear line vortex. However, on a *curved* vortex filament, the constituent elements are able to induce velocity on one another, a mechanism that is absent in a perfectly straight filament. If we regard the filament as simply a line vortex, then this self-induced velocity is logarithmically divergent, as we will discuss in Chap. 7. In other words, there is more subtlety to the singular behavior of a vortex filament than there is for a point vortex. ***The filament's transport therefore requires that we pay some consideration to its finite cross-sectional structure.*** This will be elaborated in Chap. 7.

Note 3.3.1: The Scalar Potential Field of a Closed Vortex Filament

The flow field associated with a vortex filament is irrotational apart from on the filament itself. Thus, the velocity field (3.122) should also be derivable from the gradient of a scalar potential field. Let us derive that potential field here when the filament C forms a closed loop. First, we use the generalized Stokes' theorem (A.62) to transform the integral over C to another over a surface S enclosed by this filament loop, as illustrated by Fig. A.3. In index notation, this integral over S can be written as

$$v_i(\boldsymbol{x}) = \Gamma \int_S \epsilon_{lmk} \frac{\partial}{\partial y_m} \left(\epsilon_{ijk} K_j(\boldsymbol{x} - \boldsymbol{y})\right) n_{y,l} \, \mathrm{d}S(\boldsymbol{y}), \tag{3.128}$$

where $\boldsymbol{n}_y$ is the unit normal of S at the source point $\boldsymbol{y} \in S$, directed in the right-handed sense of Stokes' theorem. The velocity kernel is anti-symmetric with respect to the source and evaluation points (3.78).

[5] This limit only is sensible if the filament is closed. For example, it can be taken on a filament with a straight section that extends from $-R$ to R, closed by a semi-circular portion of radius R, with R allowed to become infinitely large. Note that, in this limit, the vector potential field retains uniform terms, one of them proportional to $\log R$. These terms are simply ignored, since they are of no consequence on the velocity field.

Furthermore, recall that the velocity kernel is the gradient of the Green's function for the negative Laplacian, so $\partial K_j(\boldsymbol{x}-\boldsymbol{y})/\partial x_j = \nabla^2 G(\boldsymbol{x}-\boldsymbol{y}) = -\delta(\boldsymbol{x}-\boldsymbol{y})$, which vanishes for any evaluation point $\boldsymbol{x}$ not on the surface $\mathcal{S}$. Thus, the velocity field due to the filament can be written as

$$v_i(\boldsymbol{x}) = \Gamma \frac{\partial}{\partial x_i} \int_S n_{y,j} K_j(\boldsymbol{x}-\boldsymbol{y})\,\mathrm{d}S(\boldsymbol{y}), \tag{3.129}$$

and the associated scalar potential field is easily identified:

$$\varphi(\boldsymbol{x}) = \Gamma \int_S \boldsymbol{n}_y \cdot \boldsymbol{K}(\boldsymbol{x}-\boldsymbol{y})\,\mathrm{d}S(\boldsymbol{y}). \tag{3.130}$$

The velocity field induced by the filament is spatially continuous, aside from on the singular filament itself. The scalar potential field of (3.130), however, is discontinuous across the surface $\mathcal{S}$, as we will show here. Suppose we evaluate this potential field at points that approach the surface from either side, the positive side taken as the side into which the normal vector $\boldsymbol{n}$ points. Equation (A.46) provides the general behavior of an integral of the velocity kernel in such a limit. Applying this result to (3.130), and denoting the respective limiting values of the scalar potential by $\varphi^{\pm}$, we obtain

$$\varphi^{\pm}(\boldsymbol{x}) = \mp\frac{1}{2}\Gamma + \Gamma \fint_S \boldsymbol{n}_y \cdot \boldsymbol{K}(\boldsymbol{x}-\boldsymbol{y})\,\mathrm{d}S(\boldsymbol{y}). \tag{3.131}$$

The first term in this expression indicates the difference in limiting values in φ; the second term is the principal value of the integral, the average of the limiting values. Thus, at any points $\boldsymbol{x}$ on the surface $\mathcal{S}$, the jump in the scalar potential across the surface is (uniformly) equal to the negative of the filament strength:

$$\varphi^{+}(\boldsymbol{x}) - \varphi^{-}(\boldsymbol{x}) = -\Gamma. \tag{3.132}$$

This could have been obtained by simpler means by evaluating the circulation (3.34) about any closed loop that encircles the filament once, replacing the velocity in the contour integral by $\nabla\varphi$. It is important to stress that, in either means of obtaining (3.132) for a closed filament, the location of the jump—embodied by the surface $\mathcal{S}$ enclosed by C or by the beginning and end of the closed circulation contour—is arbitrary.

Furthermore, by inspection of the general expression (A.36) for a scalar field in terms of surface and volume integrals, any closed vortex filament of strength Γ can be equivalently regarded as a discontinuity of φ, with the spatially uniform value $-\Gamma$, distributed over an enclosed surface. We will discuss this further in Sect. 3.5.2.

If the filament obeys Helmholtz' third theorem, its strength Γ must be time invariant, and so must be the rate of change of the jump in scalar potential across the enclosed surface. Thus, by the unsteady Bernoulli equation (3.60), the pressure must be continuous across any surface enclosed by a closed filament of time-invariant strength. However, if Helmholtz' theorem is relaxed, then there must be a corresponding jump in pressure across the enclosed surface equal to $[p]_-^+ = \rho\dot{\Gamma}$.

It is important to note that this surface of discontinuity in the scalar potential field is the three-dimensional manifestation of the branch cut between a pair of point vortices of equal but opposite strength, as discussed in Sect. 3.2.2. Indeed, this pair of vortices can be regarded as the planar cross-section of a closed filament with two anti-parallel segments of infinite extent in the out-of-plane direction, closed at $\pm\infty$. If this pair of vortices have time-varying strength, then there must be a discontinuity in pressure, $\rho\dot{\Gamma}$, just as there is for a variable-strength filament.

Result 3.4: Volume Flow Rate Induced by a Closed Filament

The volume flow rate, Q, through a surface $\tilde{S}$ enclosed by curve $\tilde{C}$ due to a vortex filament C with strength Γ is given by

$$Q = \Gamma \int_{\tilde{C}} \int_{C} G(\tilde{\boldsymbol{X}}(\tilde{s}) - \boldsymbol{X}(s))\, \mathrm{d}\boldsymbol{l}(s) \cdot \mathrm{d}\tilde{\boldsymbol{l}}(\tilde{s}). \qquad (3.133)$$

Proof In Result 3.1 we derived the expression for the volume flow rate through a surface in terms of vector potential, when the velocity field can be entirely derived from this potential. The desired result follows directly by applying expression (3.121) for the vector potential of a filament. □

The symmetry of the expression for volume flow rate given by Result 3.4 is pleasing, particularly because the Green's function G is symmetric with respect to the interchange of source and target. In fact, if the strength of the filament is set to unity, then the roles of the source filament and the target curve can be interchanged between C and $\tilde{C}$ without effect: curve $\tilde{C}$ induces the same volume flow rate through the surface enclosed by C. This symmetric form can be useful for numerical methods.

3.3.3 Vortex Sheet

A **vortex sheet** is a surface in the fluid across which there exists a discontinuity in tangential velocity. This surface can equivalently be regarded as a singular distribution of vorticity, in which that vorticity is everywhere tangent to the surface. A vortex sheet is characterized by its strength distribution, or strength density, $\boldsymbol{\gamma}(\boldsymbol{x})$, which can be defined from the local jump in tangential velocity, as we will show below. We will often refer to this quantity as the **strength** of the sheet. Equivalently, this strength can be interpreted as the limit of the product of the vanishing thickness of a volumetric layer and the vorticity confined to that layer. Since the strength is finite, the vorticity in the sheet formed from this limit has infinite magnitude. Let us discuss the definition of this vortex sheet strength in more detail first:

Result 3.5: Strength of a Vortex Sheet

In a vortex sheet, the local strength, $\boldsymbol{\gamma}$, is tangent to the sheet itself; in fact, this strength, along with the sheet's unit normal vector, $\boldsymbol{n}$—directed toward the + side of the sheet—and the jump in velocities $\boldsymbol{v}^{+} - \boldsymbol{v}^{-}$, form a locally orthogonal right-handed coordinate system,

$$\boldsymbol{\gamma} \times \boldsymbol{n} = \boldsymbol{v}^{+} - \boldsymbol{v}^{-}, \qquad \boldsymbol{\gamma} = \boldsymbol{n} \times (\boldsymbol{v}^{+} - \boldsymbol{v}^{-}), \tag{3.134}$$

where $\boldsymbol{v}^{\pm}$ denote the fluid velocities as the vortex sheet is approached from the respective sides. In two dimensions, $\boldsymbol{\gamma}$ has only a single component in the out-of-plane direction, $\boldsymbol{\gamma} = \gamma \boldsymbol{e}_3$.

In a two-dimensional vortex sheet, we can also define a complex form of the strength, g, at any point in the sheet, equal to the difference in complex conjugate velocities on either side of the sheet:

$$g := w^{+} - w^{-}. \tag{3.135}$$

The definition of this complex strength can be easily connected to the real-valued strength γ, using the tools in Sect. A.2 and noting that the complex forms of the unit tangent τ and unit normal n—defined as in Fig. 3.8—are related by $n = \mathrm{i}\tau$:

$$\gamma = -\mathrm{Re}(g\tau), \qquad \mathrm{Im}(g\tau) = 0. \tag{3.136}$$

The second of these expresses the fact that the normal velocity across the sheet is continuous. Note that these are equivalent to writing

$$g = -\gamma\tau^{*}, \tag{3.137}$$

which implies that (the conjugate of) g is locally parallel to (and in the plane of) the vortex sheet.

The edges of a vortex sheet may be at finite distances or may lie infinitely far away. In order to be consistent with Helmholtz' theorems, any such edge must necessarily coincide with a vortex line. In fact, as we will discuss later, vortex sheets can be interpreted as a dense and continuous network of vortex filaments, in which some filaments comprise the edges of the sheet. The filaments either form closed loops of vanishingly small size or extend to infinity; in the latter case, the sheet itself must extend to infinity in that direction. Thus, just as vortex filaments are regarded as a collection of point vortices distributed along a curve, vortex sheets are interpreted similarly as an accumulation of lower-dimensional vortex elements. We will discuss this further in Sect. 3.5.

Here, we discuss the structure of a vortex sheet and some basic relationships. First, let us clarify our notation:

Note 3.3.2: Parameterization of a Vortex Sheet

Since the sheet is topologically a surface entity, its configuration in Euclidean space can be described parametrically by $\boldsymbol{X}(\boldsymbol{\xi})$, where $\boldsymbol{\xi}$ is a unique set of surface coordinates—a pair of coordinates (ξ, η) in three dimensions; or, in two dimensions, in which the sheet is a curve, a single coordinate ξ—for all points in the sheet. Since a vortex sheet is generally in motion, the relationship $\boldsymbol{X}(\boldsymbol{\xi})$ is time dependent; we omit the time argument for brevity wherever possible. In complex notation, the sheet is described parametrically by $Z(\xi)$.

In two dimensions, the arc length, s, defined as

$$\mathrm{d}s := |\partial \boldsymbol{X}/\partial \xi|\, \mathrm{d}\xi \Longrightarrow s(\xi) = \int_0^{\xi} |\partial \boldsymbol{X}/\partial \hat{\xi}|\, \mathrm{d}\hat{\xi}, \tag{3.138}$$

will be another useful parameterization of the sheet. In complex form,

$$\mathrm{d}s := |\partial Z/\partial \xi|\, \mathrm{d}\xi. \tag{3.139}$$

Based on these relationships, we will interchangeably refer to the spatial dependence of a sheet's properties by its position in Euclidean space, $\boldsymbol{x} = \boldsymbol{X}(\boldsymbol{\xi})$ (or $z = Z(\xi)$ in complex form), by its position in the sheet, $\boldsymbol{\xi}$, and, in two dimensions, by its arc length position, s. For example, the strength $\boldsymbol{\gamma}$ might be written as $\boldsymbol{\gamma}(\boldsymbol{x})$ or $\boldsymbol{\gamma}(\boldsymbol{\xi})$, or by $\gamma(s)$ in a planar sheet. All such notations refer to the same property and are easily reconciled with each other through the relations above. It is to be remembered, of course, that the sheet's definition is limited to $\boldsymbol{x} \in \mathcal{S}$.

We will discuss the two-dimensional sheet first, and then generalize this to three dimensions:

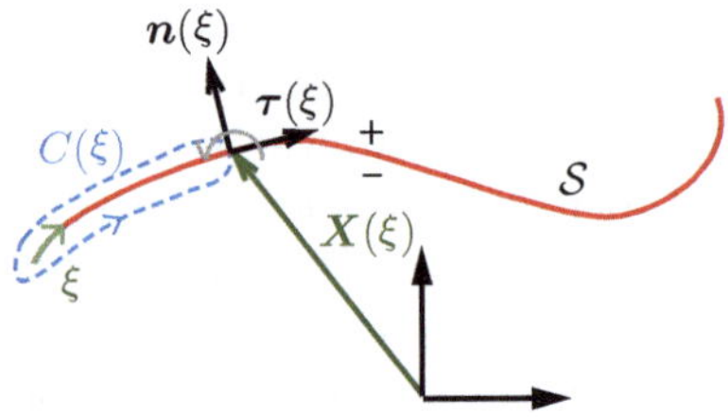

Fig. 3.8 Schematic of a vortex sheet $\mathcal{S}$ (in red) in a two-dimensional flow. Contour $C(\xi)$ is used to define the local circulation of the sheet. The parameter ξ provides a unique label (not necessarily arc length) of each point on the sheet

Two-Dimensional Sheet In the sheet $\mathcal{S}$ depicted in Fig. 3.8 in a two-dimensional flow, all of the vortex lines are perpendicular to the plane and extend to infinity in the out-of-plane direction. The sheet's strength has only a single component in the out-of-plane direction, $\boldsymbol{\gamma} = \gamma \boldsymbol{e}_3$, so that (3.134) simplify to

$$\gamma(\xi) = \boldsymbol{\tau}(\xi) \cdot \left[\boldsymbol{v}^-(\xi) - \boldsymbol{v}^+(\xi)\right]; \tag{3.140}$$

$\xi \in [0, \xi_L]$ is the monotonically-increasing parameter that uniquely labels each point on the sheet and $\boldsymbol{\tau}(\xi) \equiv (\partial \boldsymbol{X}/\partial \xi)/|\partial \boldsymbol{X}/\partial \xi|$ is the local unit tangent vector on the sheet, and $\boldsymbol{v}^{\pm}(\xi)$ are the fluid velocities evaluated as the point $\boldsymbol{X}(\xi)$ on the sheet is approached from the + or − side, as defined in Fig. 3.8:

$$\boldsymbol{v}^{\pm}(\xi) := \lim_{\boldsymbol{x} \to \boldsymbol{X}(\xi)^{\pm}} \boldsymbol{v}(\boldsymbol{x}). \tag{3.141}$$

For this sheet of finite length in a two-dimensional context, we can define a local circulation of the sheet, $\Gamma(\xi)$, from the contour $C(\xi)$ depicted in Fig. 3.8. This is based on the conventional definition when following the contour in counterclockwise fashion:

$$\Gamma(\xi) = \oint_{C(\xi)} \boldsymbol{v} \cdot \mathrm{d}\boldsymbol{l}. \tag{3.142}$$

If ξ_L denotes the largest value of ξ on the sheet, then $\Gamma(\xi_L) = \Gamma_{\mathcal{S}}$, the total circulation of the sheet. Since the vorticity is confined to the surface of the sheet, we can let this contour wrap tightly around the portion of the sheet that it encloses without any change to the circulation. In doing so, this contour integral can be written as

$$\Gamma(\xi) = \int_0^{\xi} \left(\boldsymbol{v}^-(\hat{\xi}) - \boldsymbol{v}^+(\hat{\xi})\right) \cdot \boldsymbol{\tau}(\hat{\xi}) |\partial \boldsymbol{X}/\partial \hat{\xi}| \, \mathrm{d}\hat{\xi} = \int_0^{\xi} \gamma(\hat{\xi}) |\partial \boldsymbol{X}/\partial \hat{\xi}| \, \mathrm{d}\hat{\xi}. \tag{3.143}$$

Differentiating both sides with respect to ξ, we see that the vortex sheet strength is simply the local density of circulation along the sheet,

$$\gamma(\xi) = \frac{1}{|\partial \boldsymbol{X}/\partial \xi|} \frac{\partial \Gamma}{\partial \xi}, \tag{3.144}$$

or simply, in terms of arc length (3.138),

$$\gamma(s) = \frac{\partial \Gamma}{\partial s}. \tag{3.145}$$

It is important to note, for later applications, that we could have reasonably defined the contour in Fig. 3.8 so that it instead encloses the opposite end of the vortex sheet (but retains its counterclockwise sense). The local circulation defined with this contour is equal to $\Gamma_S - \Gamma(\xi)$, and its derivative with respect to s is $-\gamma$.

Since the vortex sheet is immersed in an otherwise potential flow, the velocity field on either side of the sheet is described by the gradient of a scalar potential field, and the limiting values $\boldsymbol{v}^{\pm}$ can be obtained by the respective limits:

$$\boldsymbol{v}^{\pm}(\xi) = \lim_{\boldsymbol{x} \to \boldsymbol{X}(\xi)^{\pm}} \nabla \varphi(\boldsymbol{x}). \tag{3.146}$$

Thus, by Eq. (3.140), we can relate the vortex sheet strength to the jump in scalar potential:

$$\gamma(\xi) = -\boldsymbol{\tau}(\xi) \cdot \lim_{\boldsymbol{x} \to \boldsymbol{X}(\xi)^{\pm}} \nabla \varphi(\boldsymbol{x}), \tag{3.147}$$

or simply

$$\gamma(s) = -\frac{\partial}{\partial s} [\varphi]_{-}^{+}, \tag{3.148}$$

where we have used a standard notation for differences, $[f]_{-}^{+} = f^{+} - f^{-}$, to denote the local jump in scalar potential across the sheet. It follows, then, that the local circulation is equal to the negative of this jump in potential,

$$\Gamma(\xi) = -[\varphi]_{-}^{+}(\xi). \tag{3.149}$$

Note that, in deriving this last expression, we have relied on the fact that the contour $C(\xi)$ encloses the end of the sheet, where the scalar potential is continuous.

Thus far, we have obtained many of our results by depending on the fact that the sheet has an edge that can be enclosed by the contour. However, we will often encounter sheets that are closed, so our contour cannot possibly pierce the sheet at only a single point. To deal with such situations, we extend our approach by allowing the contour to pierce the sheet at two locations, ξ_1 and ξ_2, and denote this contour by $C(\xi_1, \xi_2)$. The circulation about this contour, $\Gamma(\xi_1, \xi_2)$, is defined in the natural way, and carrying through with our previous analysis, it is straightforward to show that

$$\Gamma(\xi_1, \xi_2) = -[\varphi]_{-}^{+}(\xi_2) + [\varphi]_{-}^{+}(\xi_1). \tag{3.150}$$

Furthermore, relation (3.145) between this circulation and the local sheet strength still holds when we take the first parameter as a fixed reference point and the gradient with respect to the second parameter.

In Eq. (3.135), we defined the complex strength, g, of a two-dimensional vortex sheet, and related it to the real-valued strength in (3.136). Let us relate the local circulation of the sheet to this complex strength. At any point $z \in \mathcal{S}$, Eq. (3.136) implies that

$$g(z) = -\frac{\partial \Gamma}{\partial z}, \tag{3.151}$$

as can be verified by the chain rule, or, denoting the contour that pierces the sheet at z by $C(z)$,

$$\Gamma(z) = -\int_{C(z)} g(\lambda)\,\mathrm{d}\lambda. \tag{3.152}$$

The flow field associated with a two-dimensional vortex sheet is easily obtained. First, we note that the sheet can be regarded as a collection of infinitesimal segments of arc length δs. Each of these segments constitutes a point vortex; the point vortex associated with the Jth segment, located at s_J, has infinitesimal strength $\delta\Gamma_\mathrm{J} = \gamma(s_\mathrm{J})\delta s$, based on Eq. (3.145). Thus, we can compose the overall sheet's induced flow field from the collective contributions (3.118) of the point vortices associated with these segments. In the limit, the velocity is given at any point $\boldsymbol{x} \notin S$ by

$$\boldsymbol{v}(\boldsymbol{x}) = \sum_\mathrm{J} \boldsymbol{K}(\boldsymbol{x} - \boldsymbol{X}(s_\mathrm{J})) \times \boldsymbol{\gamma}(s_\mathrm{J})\delta s \longrightarrow \int_S \boldsymbol{K}(\boldsymbol{x} - \boldsymbol{X}(s)) \times \boldsymbol{\gamma}(s)\,\mathrm{d}s, \tag{3.153}$$

where the velocity kernel $\boldsymbol{K}$ has the form (3.71), and $\boldsymbol{\gamma}(s) = \gamma(s)\boldsymbol{e}_3$. The associated streamfunction field follows from a similar process, starting from (3.68) for a single point vortex:

$$\psi(\boldsymbol{x}) = \int_S G(\boldsymbol{x} - \boldsymbol{X}(s))\gamma(s)\,\mathrm{d}s. \tag{3.154}$$

In complex form, starting from (3.119), and using the relationship (3.151), the complex velocity field of the sheet is

$$w(z) = \frac{1}{2\pi \mathrm{i}} \int_S \frac{g(\lambda)}{\lambda - z}\,\mathrm{d}\lambda; \tag{3.155}$$

The complex potential field is clearly

$$F(z) = -\frac{1}{2\pi \mathrm{i}} \int_S g(\lambda) \log(z - \lambda)\,\mathrm{d}\lambda. \tag{3.156}$$

Both forms of the velocity (3.153) and (3.155) can be applied to points on the sheet itself, but the evaluation process requires care, since the integrand is singular. This will be addressed in Chap. 4.

Three-Dimensional Sheet The results developed for the two-dimensional vortex sheet can be naturally extended to a three-dimensional vortex sheet, depicted schematically in Fig. 3.9. The relationship between the strength of a vortex sheet and the local jump in velocity was given by Eq. (3.134). As we described in Note 3.3.2, the sheet is parameterized by the pair of coordinates, $\boldsymbol{\xi} = (\xi, \eta)$.

Analogous to the two-dimensional example, a contour $C(\xi, \eta)$ that pierces the sheet at a single point can be used to define the local circulation,

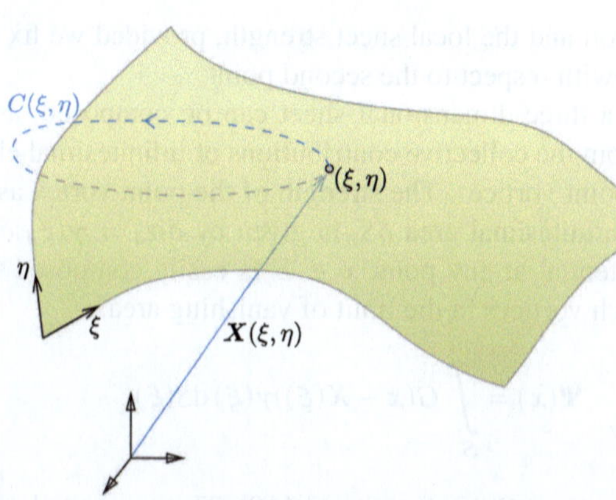

Fig. 3.9 Schematic of a three-dimensional vortex sheet S. The surface is parameterized by a coordinate system (ξ, η). Contour $C(\xi, \eta)$ is used to define the local circulation of the sheet

$$\Gamma(\xi,\eta) = \oint_{C(\xi,\eta)} \boldsymbol{v} \cdot \mathrm{d}\boldsymbol{l}. \tag{3.157}$$

It is important to stress that the circulation is insensitive to how the contour is otherwise configured; it depends only on where it pierces the sheet. A special consequence of this is that, ***if the evaluation point itself lies on any edge of the sheet, then the local circulation must vanish***, so long as the contour used to calculate this circulation can be deformed into one that encloses no part of the sheet. This is *not* true for a two-dimensional vortex sheet, since it runs to negative and positive infinity in the out-of-plane direction, so a contour that envelops the entire sheet cannot be deformed into one that encloses no vorticity. This means that it can have non-zero overall circulation.

By using standard results from the differential geometry of surfaces, we can develop a local relationship between the circulation and the vortex sheet strength, analogous to (3.145). These relations can be expressed succinctly in vector form:

$$\nabla\Gamma \times \boldsymbol{n} = \boldsymbol{\gamma}, \qquad \nabla\Gamma = \boldsymbol{n} \times \boldsymbol{\gamma}, \tag{3.158}$$

where it is to be understood that the gradient is restricted to the sheet. As in the two-dimensional case, we can use this to relate Γ to the jump in scalar potential across the sheet: we combine (3.158) with (3.134) and with the facts that $\boldsymbol{v}^{\pm} = \nabla\varphi^{\pm}$ and that the local circulation vanishes at the edges of the sheet to arrive at the relation

$$\Gamma(\boldsymbol{\xi}) = -[\varphi]_{-}^{+}(\boldsymbol{\xi}). \tag{3.159}$$

As in the two-dimensional case, we can extend the definition of circulation to contours that pierce the sheet in two locations, making it possible to describe the circulation in closed vortex sheets. Let us denote such a contour by $C(\boldsymbol{\xi}_1, \boldsymbol{\xi}_2)$, and the circulation about this contour by $\Gamma(\boldsymbol{\xi}_1, \boldsymbol{\xi}_2)$. The relationship (3.158) still holds between this circulation and the local sheet strength, provided we fix the first point and take the gradient with respect to the second point.

The flow field of a three-dimensional sheet can be composed, as for the two-dimensional sheet, from the collective contributions of infinitesimal elements of the sheet interpreted as point vortices. The strength of the point vortex associated with the Jth element, of infinitesimal area δS, is given by $\delta\alpha_{\mathrm{J}} = \gamma(\xi_{\mathrm{J}})\delta S$. From this, the overall vector potential at any point $x \notin S$ is easily composed from the sum of (3.109) over all such vortices in the limit of vanishing area:

$$\boldsymbol{\Psi}(\boldsymbol{x}) = \int_S G(\boldsymbol{x} - \boldsymbol{X}(\boldsymbol{\xi}))\boldsymbol{\gamma}(\boldsymbol{\xi})\,\mathrm{d}S(\boldsymbol{\xi}), \tag{3.160}$$

where the Green's function, G, is given by (3.110). The associated velocity field is obtained from the curl of this potential:

$$\boldsymbol{v}(\boldsymbol{x}) = \nabla \times \int_S G(\boldsymbol{x} - \boldsymbol{X}(\boldsymbol{\xi}))\boldsymbol{\gamma}(\boldsymbol{\xi})\,\mathrm{d}S(\boldsymbol{\xi}) \equiv \int_S \boldsymbol{K}(\boldsymbol{x} - \boldsymbol{X}(\boldsymbol{\xi})) \times \boldsymbol{\gamma}(\boldsymbol{\xi})\,\mathrm{d}S(\boldsymbol{\xi}), \tag{3.161}$$

where the velocity kernel $\boldsymbol{K}$ takes its three-dimensional form (3.112). Note that (3.161) can also be evaluated at a point on the sheet itself, but we must take care with the singular integrand. This will be discussed in Chap. 4.

3.4 Other Surface Distributions of Singularities

The vortex sheet defined in the previous section is not the only entity characterized by a set of singularities distributed over a surface. Indeed, one can also conceive surface distributions of monopole sources and of dipoles (or 'doublets'). The equations that describe their associated flow fields are easily constructed by the same process we followed for a vortex sheet: we imagine a discrete set of singularities on the surface, of increasing number and vanishingly small strength. Our notation for the spatial dependence of the properties of these singularity distributions follows the same pattern discussed in Note 3.3.2 for the vortex sheet.

Single-Layer Potential Let us consider a surface $\mathcal{S}$, and construct expressions for distributions on that surface. The scalar potential at any point $x \notin \mathcal{S}$ due to a monopole source distribution on $\mathcal{S}$, also known as a **single-**, or **simple-layer potential**, is obtained from the superposition of contributions (3.113) from sources at locations y and infinitesimal strengths $\delta Q = q(y)\delta S$, where q is the local monopole source density:

$$\varphi(\boldsymbol{x}) = -\int_S q(\boldsymbol{y})G(\boldsymbol{x} - \boldsymbol{y})\,\mathrm{d}S(\boldsymbol{y}). \tag{3.162}$$

The associated velocity field at $\boldsymbol{x} \notin S$ is simply the gradient of this:

$$\boldsymbol{v}(\boldsymbol{x}) = -\int_S q(\boldsymbol{y}) \nabla G(\boldsymbol{x}-\boldsymbol{y})\,\mathrm{d}S(\boldsymbol{y}) \equiv -\int_S q(\boldsymbol{y}) \boldsymbol{K}(\boldsymbol{x}-\boldsymbol{y})\,\mathrm{d}S(\boldsymbol{y}). \tag{3.163}$$

These expressions hold for either two- or three-dimensional contexts, provided the appropriate Green's function—(3.67) or (3.110)—and velocity kernel—(3.71) or (3.112)—are used. In two dimensions, we can also describe such a field in complex form, starting from (3.88) for a source of strength $\delta Q = q(s)\,\mathrm{d}s$ at position $\lambda(s)$ on the surface (or rather, curve) S, where s denotes the arc length along the curve. The complex potential of this field at $z \notin S$ is

$$F(z) = \frac{1}{2\pi}\int_S q(s)\log(z-\lambda(s))\,\mathrm{d}s. \tag{3.164}$$

Remember that only the real part of this describes the scalar potential field; the imaginary part is the associated streamfunction field. The complex conjugate velocity field is then simply

$$w(z) = \frac{1}{2\pi}\int_S \frac{q(s)}{z-\lambda(s)}\,\mathrm{d}s. \tag{3.165}$$

Double-Layer Potential A surface distribution of dipoles, each oriented normal to the surface, represents a **double-layer potential**; its scalar potential field at $\boldsymbol{x} \notin S$ is formed from a collection of dipoles of infinitesimal strength $\delta \boldsymbol{Q} = \mu(\boldsymbol{y})\boldsymbol{n}_y\,\delta S$, where μ is the dipole strength density and $\boldsymbol{n}_y$ is the local outward normal at position $\boldsymbol{y} \in S$. Starting from the potential of each such dipole (3.115), the overall collection's potential field must be

$$\varphi(\boldsymbol{x}) = \int_S \mu(\boldsymbol{y})\boldsymbol{n}_y \cdot \nabla G(\boldsymbol{x}-\boldsymbol{y})\,\mathrm{d}S(\boldsymbol{y}). \tag{3.166}$$

Again, the velocity field is just the gradient of this potential field,

$$\boldsymbol{v}(\boldsymbol{x}) = \nabla \int_S \mu(\boldsymbol{y})\boldsymbol{n}_y \cdot \nabla G(\boldsymbol{x}-\boldsymbol{y})\,\mathrm{d}S(\boldsymbol{y}). \tag{3.167}$$

As for the single-layer potential, these formulas hold in both two- and three-dimensional contexts. We can also express the two-dimensional flow of the double layer in complex form, of course; its complex potential follows directly from (3.90) for $n = 1$, with strength $-a_1 = \mathrm{i}\mu(\lambda)\,\mathrm{d}\lambda$ at all positions $\lambda \in S$:

$$F(z) = \frac{1}{2\pi\mathrm{i}}\int_S \frac{\mu(\lambda)}{z-\lambda}\,\mathrm{d}\lambda; \tag{3.168}$$

its corresponding conjugate velocity field is

$$w(z) = \frac{1}{2\pi i}\frac{d}{dz}\int_S \frac{\mu(\lambda)}{z-\lambda}\,d\lambda. \tag{3.169}$$

3.5 Relationships Between Singularity Distributions

In the chapters ahead, we will present many results in which the influence of fluid vorticity is expressed as a sum over vortex filaments (in three dimensions) or point vortices (in the plane). Though it might appear at first glance that we have neglected other types of elements (principally, vortex sheets) in these results, rest assured that we have not. We need only remind ourselves of the manner in which we have derived the flow field induced by vortex sheets in Sect. 3.3.3: as a limiting process of a set of point vortices of infinitesimal strength. By this same process, we can adapt any result that we obtain in the future for such discrete elements to an alternative form appropriate for a vortex sheet. We can also obtain representations in terms of other singularities, and these may be useful for alternative perspectives. Let us start with a simple observation:

3.5.1 Two-Dimensional Vortex Sheet as a Set of Point Vortices

Following the approach we used to develop the velocity field induced by a vortex sheet, we can adapt any result obtained for N_v point vortices in the plane, with strengths Γ_J and positions $\boldsymbol{x}_J$, into a form corresponding to a free vortex sheet, $\mathcal{S}$, described by position $\boldsymbol{X}(s)$ and scalar-valued strength $\gamma_{\mathcal{S}}(s)$, where s is the arc length along the sheet:

$$\sum_{J=1}^{N_v}\Gamma_J f(\boldsymbol{x}_J) \longrightarrow \int_{\mathcal{S}} f(\boldsymbol{X}(s))\,\gamma_{\mathcal{S}}(s)\,ds, \tag{3.170}$$

where f is any continuous (scalar- or vector-valued) function of position of the vortex elements. In complex form, an equivalent expression of this adaptation is

$$\sum_{J=1}^{N_v}\Gamma_J f(z_J) \longrightarrow -\int_{\mathcal{S}} f(z)g_{\mathcal{S}}(z)\,dz. \tag{3.171}$$

Both of these forms are obtained by identifying the Jth segment of the sheet with a point vortex of infinitesimal strength: $\delta\Gamma_J = \gamma_{\mathcal{S}}(s_J)\delta s = -g_{\mathcal{S}}(z_J)\delta z$.

As we have observed before, a vortex filament is the natural three-dimensional extension of a point vortex in the plane. Might there be a generalization of (3.170) that accounts for three-dimensional sheets, as well? In Sect. 3.3.2, we discussed the fact that a closed vortex filament of strength Γ consists of an enclosed surface across which the scalar potential jumps uniformly by $-\Gamma$. We have also seen that both two- and three-dimensional vortex sheets possess the same jump locally. So it seems that the key to generalizing (3.170) lies in exploring this connection further. Let us first take note of identity (A.76),

$$-\nabla \times \int_S G(\boldsymbol{x}-\boldsymbol{y})\boldsymbol{n}_y \times \nabla_y \mu \, \mathrm{d}S(\boldsymbol{y}) =$$
$$-\int_C \mu(\boldsymbol{y}) \nabla G(\boldsymbol{x}-\boldsymbol{y}) \times \mathrm{d}\boldsymbol{l}(\boldsymbol{y}) + \nabla \int_S \mu(\boldsymbol{y}) \boldsymbol{n}_y \cdot \nabla G(\boldsymbol{x}-\boldsymbol{y}) \, \mathrm{d}S(\boldsymbol{y}), \quad \text{(A.76)}$$

which relates two forms of surface distributions. The complex form of Eq. (A.76), applied to S in the form of a planar curve as shown in the right panel in Fig. A.3, reduces simply to an integration by parts:

$$\frac{1}{2\pi \mathrm{i}} \frac{\mathrm{d}}{\mathrm{d}z} \int_S \log(z-\lambda) \, \mathrm{d}\mu(\lambda) = \frac{1}{2\pi \mathrm{i}} \left[\frac{\mu(\lambda)}{z-\lambda} \right]_{z_f}^{z_i} + \frac{1}{2\pi \mathrm{i}} \frac{\mathrm{d}}{\mathrm{d}z} \int_S \frac{\mu(\lambda)}{\lambda - z} \, \mathrm{d}\lambda. \quad (3.172)$$

Note that z_i and z_f are labeled so that they are consistent with the definition of these points in Fig. A.3, but to ensure consistency with the predominant convention, the curve is actually integrated from z_f to z_i, with the normal directed to the left.

What is the significance of Eq. (A.76), or its complex equivalent (3.172)? Let us inspect them term by term. In the complex version, the left-hand side takes the form of the velocity (that is, the derivative of a complex potential) induced by a distribution of point vortices along S, each of infinitesimal strength $\mathrm{d}\mu$. In other words, the left-hand side is the velocity field due to a vortex sheet. The vector form (A.76) simply generalizes this to all dimensions, with the left-hand side taking the form of the curl of the vector potential from a vortex sheet of strength $\nabla\mu \times \boldsymbol{n}$, similar to the form (3.158) we discussed earlier. Thus, we can think of μ as the local circulation, Γ.

What about the right-hand side of the equations? Remember from our discussion of vortex sheets that the local circulation is equal to the negative of the jump in scalar potential, $-[\varphi]_-^+$, where, as usual, the positive side of S is the one to which $\boldsymbol{n}$ is directed. Thus, μ can equivalently be thought of as the negative of this jump in scalar potential. The second term on the right-hand side of either Eq. (A.76) or (3.172) is in the form of the gradient of a scalar potential field. In particular, if we compare these, respectively, with (3.167) or (3.169), they each represent the velocity induced by a distribution of dipoles in S—that is, a double-layer potential—of strength density $\mu = -[\varphi]_-^+$.

The difference between the velocities induced by these two singularity distributions—vortices and dipoles—lies in the first term on the right-hand side. In the complex version (3.172), this term takes the form of the velocity induced by two point vortices—one at either end of S, with strength equal to the local value of μ (at z_i) or $-\mu$ (at z_f). In the vector form (A.76), this term is a line integral about the enclosing contour C, closely resembling the velocity field induced by a vortex filament (3.122), except for the fact that $[\varphi]_-^+$ generally varies along the contour in three dimensions.

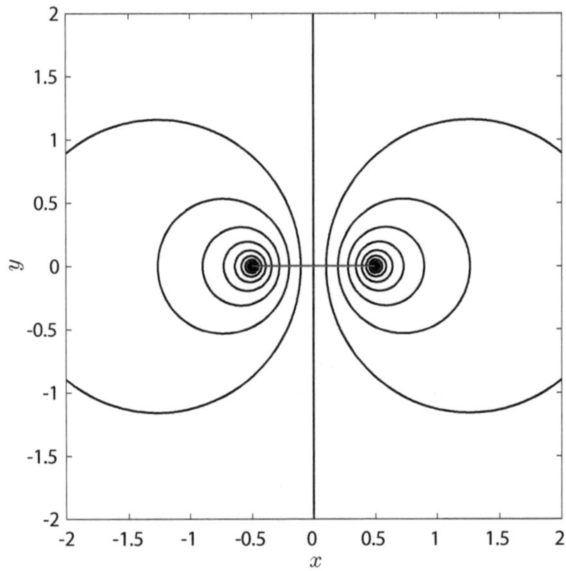

Fig. 3.10 A pair of point vortices of strengths Γ and $-\Gamma$ at $(0, -1/2)$ and $(0, 1/2)$, respectively, and the equivalent dipole distribution of uniform strength Γ along the branch cut (in red) between them. The streamlines of the induced flow field are depicted as black contours

Thus, Eqs. (A.76) and (3.172) provide us with a tool for reinterpreting the distribution of surface singularities that gives rise to a certain velocity field in the fluid. There are a few ways that we can use this tool. The first is perhaps the most obvious.

3.5.2 Vortex Filament or Vortex Pair as a Double-Layer Potential

If μ (or rather, Γ) is uniform, then there is no vortex sheet, and the identities reduce to

$$\Gamma \int_C \boldsymbol{K}(\boldsymbol{x}-\boldsymbol{y}) \times \mathrm{d}\boldsymbol{l}(\boldsymbol{y}) = \Gamma \nabla \int_S \boldsymbol{n}_y \cdot \nabla G(\boldsymbol{x}-\boldsymbol{y})\, \mathrm{d}S(\boldsymbol{y}), \tag{3.173}$$

or, in complex form in two dimensions,

$$\frac{\Gamma}{2\pi \mathrm{i}} \left(\frac{1}{z-z_f} - \frac{1}{z-z_i} \right) = \frac{\Gamma}{2\pi \mathrm{i}} \frac{\partial}{\partial z} \int_S \frac{1}{z-\lambda}\, \mathrm{d}\lambda. \tag{3.174}$$

These confirm that the velocity induced by a vortex filament (or, in two dimensions, by a pair of point vortices of equal and opposite strength) is equivalent to that from a double-layer potential on an enclosed surface of the filament (resp., a branch cut

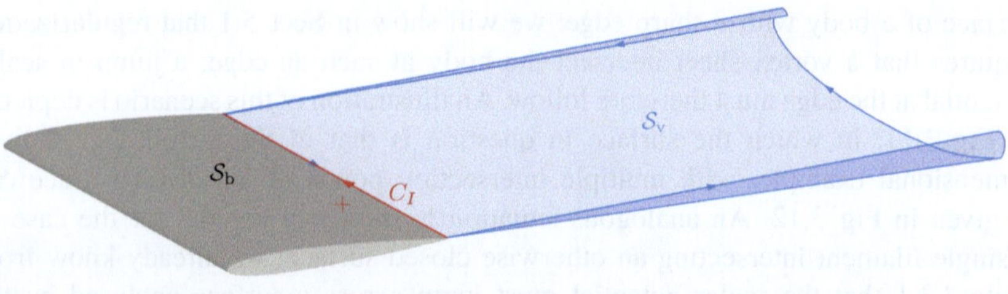

Fig. 3.11 An example of a common intersection of two entities: the surface of an airfoil, S_b, intersected by a wake vortex sheet S_v at the airfoil's trailing edge. The contour of intersection, C_I, in S_b is oriented as shown, and the positive side of the contour in S_b is labeled +

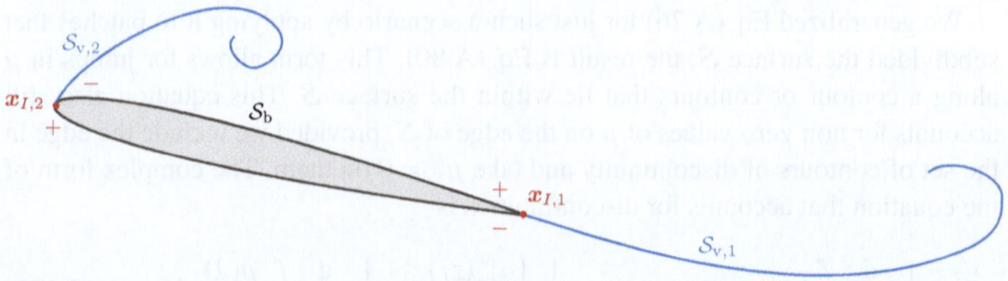

Fig. 3.12 A two-dimensional example of an intersection. In this example, there are two intersections: between the surface of an airfoil, S_b, and vortex sheets $S_{v,1}$ and $S_{v,2}$ at the airfoil's trailing and leading edges, respectively. The points of intersection, $\boldsymbol{x}_{I,1}$ and $\boldsymbol{x}_{I,2}$, are labeled, as are their respective positive and negative sides

between the point vortices). As an illustration, consider the planar configuration depicted in Fig. 3.10, consisting of two point vortices of strength Γ and $-\Gamma$. The velocity field induced by this pair is equivalently effected by a distribution of dipoles between them, with uniform strength density Γ. Note that we could have set this dipole distribution on *any* contour that extends between the two point vortices and still achieved the same flow.

3.5.3 Vortex Sheet as a Double-Layer Potential

Let us return to the identities (A.76) and (3.172). If the surface S is closed, so that the contour C—or the analogous endpoints in two dimensions—does not exist, and if μ is continuous throughout the surface, then the vortex sheet in a surface can be equivalently regarded as a distribution of dipoles in that same surface. Even if the surface is not closed, but still represents an entire vortex sheet, then we know from our previous discussion that $[\varphi]_-^+$ vanishes along the sheet's edge C. Thus, this dual interpretation as vortex sheet and dipole distribution is still available for a sheet with edges.

Let us make this interpretation a bit more general by supposing that μ is *not* continuous in S. This will be important, for example, when we consider S to be the surface of a body with a sharp edge: we will show in Sect. 5.1 that regularization requires that a vortex sheet intersect the body at such an edge; a jump in scalar potential at the edge must therefore follow. An illustration of this scenario is depicted in Fig. 3.11, in which the surface in question is that of the airfoil, $\mathcal{S}_b$. A two-dimensional example, with multiple intersection points of an airfoil surface $\mathcal{S}_b$, is given in Fig. 3.12. An analogous situation is shown in Fig. 6.2 for the case of a single filament intersecting an otherwise closed surface; we already know from Note 3.3.1 that the scalar potential must jump across a surface enclosed by the filament. Regardless of whether the surface $\mathcal{S}$ is intersected by a vortex sheet or a filament, or by the branch cut of a point vortex in two dimensions, the intersection gives rise to a jump in μ in $\mathcal{S}$.

We generalized Eq. (A.76) for just such a scenario by applying it to patches that subdivided the surface $\mathcal{S}$; the result is Eq. (A.80). This form allows for jumps in μ along a contour or contours that lie within the surface $\mathcal{S}$. This equation also still accounts for non-zero values of μ on the edge of $\mathcal{S}$, provided we include the edge in the set of contours of discontinuity and take $\mu^- = 0$ on them. The complex form of the equation that accounts for discontinuities is

$$\frac{1}{2\pi \mathrm{i}} \frac{\mathrm{d}}{\mathrm{d}z} \int_{\mathcal{S}} \log(z - \lambda)\, \mathrm{d}\mu(\lambda) = \frac{1}{2\pi \mathrm{i}} \frac{[\mu]_-^+(z_I)}{z - z_I} + \frac{1}{2\pi \mathrm{i}} \frac{\mathrm{d}}{\mathrm{d}z} \int_{\mathcal{S}} \frac{\mu(\lambda)}{\lambda - z}\, \mathrm{d}\lambda, \tag{3.175}$$

where z_I is a point of discontinuity along $\mathcal{S}$, and its $-$ and $+$ sides are just before and just after this point is encountered as the contour $\mathcal{S}$ is traversed in its positive direction; see, for example, Fig. 3.12. If the surface contour $\mathcal{S}$ is finite in extent, its end points can also be included in the list of discontinuities; we would take $\mu^- = 0$ at the initial point of the contour and $\mu^+ = 0$ at the final point.

This leads us to another reinterpretation: ***The vortex sheet in a surface can be equivalently regarded as a distribution of dipoles in that same surface, plus an account for discontinuities in the dipole distribution due to intersections with other vortex sheets or filaments.*** It should be added that the contribution from an intersection contour vanishes if both of the intersecting entities are represented in the same form: i.e., both in vortex form or both in dipole form. This is because both of these entities will have the same intersection contour in their dipole formulation, but these contours will run in opposite directions and therefore cancel each other out, as illustrated in Fig. 3.11, for example. This cancellation clearly requires that the entities' strengths match along their intersection; this is assured by continuity of the scalar potential field.

3.5.4 Three-Dimensional Vortex Sheet as a Mesh of Vortex Filaments

There is also a fourth, less obvious, reinterpretation of the surface distribution, which extends our first interpretation a bit farther. We have already observed that a vortex sheet can be reformulated as a double-layer potential, replacing the strength $\nabla \mu \times \boldsymbol{n}$ in the surface representation with μ itself. Suppose that we divide $\mathcal{S}$ into patches, as we have done before, but now with the goal of ensuring that μ (that is, the local circulation, Γ) is approximately uniform in each patch. Its value can still vary from patch to patch, but the main consequence is that $\nabla\Gamma$ is zero within any given patch, so that the left-hand side of the equations—the contribution from the vortex sheet within the patch—vanishes. The remaining terms are the dipole distribution, which now has a uniform strength in each patch; and the contour integral, which now, with uniform Γ, is truly in the form of the velocity induced by a vortex filament of strength Γ aligned with the patch's contour, C. That is, ***when a surface patch's dipole strength density*** Γ ***is uniform, its associated flow field is equivalently effected by a vortex filament of strength*** Γ ***along its bounding contour***.

Let us denote the division of surface S into a non-intersecting set of patches by

$$S = \bigcup_{\mathsf{P}=1}^{N_P} S_{\mathsf{P}}. \tag{3.176}$$

Then, using notation for the vortex sheet, Eq. (A.76) can be written exactly as

$$\int_S \boldsymbol{K}(\boldsymbol{x}-\boldsymbol{y}) \times \boldsymbol{\gamma}(\boldsymbol{y})\, \mathrm{d}S(\boldsymbol{y}) =$$

$$-\int_C \Gamma(\boldsymbol{y}) \boldsymbol{K}(\boldsymbol{x}-\boldsymbol{y}) \times \mathrm{d}\boldsymbol{l}(\boldsymbol{y}) + \sum_{\mathsf{P}=1}^{N_P} \nabla \int_{S_{\mathsf{P}}} \Gamma(\boldsymbol{y}) \boldsymbol{n}_y \cdot \boldsymbol{K}(\boldsymbol{x}-\boldsymbol{y})\, \mathrm{d}S(\boldsymbol{y}). \tag{3.177}$$

Now, we will approximate Γ in patch P by a uniform value Γ_{P}. When we apply Eq. (A.76) to the surface integral over S_{P} inside the summation, with its strength taken as uniform, it is clear that this integral is equivalent to an integral over C_{P}, the closed contour enclosing the patch:

$$\int_S \boldsymbol{K}(\boldsymbol{x}-\boldsymbol{y}) \times \boldsymbol{\gamma}(\boldsymbol{y})\, \mathrm{d}S(\boldsymbol{y}) \approx$$

$$-\int_C \Gamma(\boldsymbol{y}) \boldsymbol{K}(\boldsymbol{x}-\boldsymbol{y}) \times \mathrm{d}\boldsymbol{l}(\boldsymbol{y}) + \sum_{\mathsf{P}=1}^{N_P} \Gamma_{\mathsf{P}} \int_{C_{\mathsf{P}}} \boldsymbol{K}(\boldsymbol{x}-\boldsymbol{y}) \times \mathrm{d}\boldsymbol{l}(\boldsymbol{y}). \tag{3.178}$$

Note that we have not eliminated the vortex sheet by assuming uniform Γ in each patch. Rather, we have simply replaced its effect with an equivalent effect from each patch's contour. In other words, at any evaluation point, ***a vortex sheet of strength*** $\boldsymbol{\gamma} = \nabla\Gamma \times \boldsymbol{n}$ ***can be regarded as a mesh of closed vortex filaments, adjacent to one another, and each enclosing a patch of vanishingly small area***. The strength of each filament is given by the local circulation, Γ. An illustration of this interpretation of a sheet is provided in Fig. 3.13. Adjacent closed filaments share a common segment, though with opposite orientation. In the illustration, in which adjacent filaments of respective strengths $\Gamma(\xi, \eta)$ and $\Gamma(\xi + \mathrm{d}\xi, \eta)$ are labeled, the net strength of the common filament segment is $\Gamma(\xi, \eta) - \Gamma(\xi + \mathrm{d}\xi, \eta)$. Similar differences characterize all such filament segments in the interior of the mesh. As we discussed earlier, the

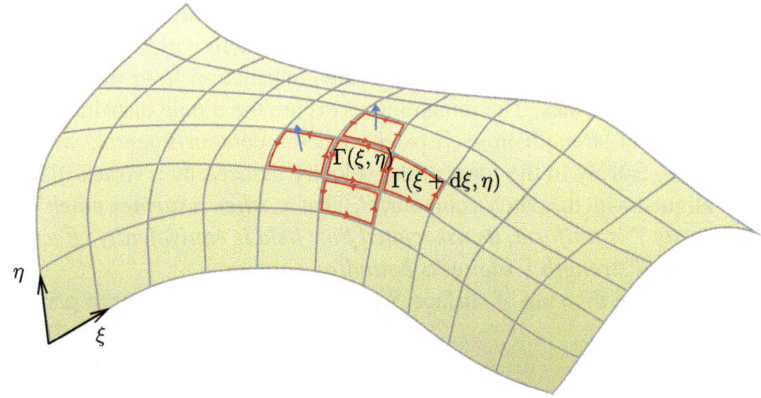

Fig. 3.13 A vortex sheet as a mesh of filaments, with a sample of the filaments depicted in red. Unit normals are depicted in blue

strength Γ on the curve C enclosing $\mathcal{S}$ is zero wherever the contour does not comprise an intersection with another surface. However, in those sections where Γ is not zero, the contour is naturally subdivided into segments formed by the adjacent patches; the negative sign on the integral over C in Eq. (3.178) indicates that its local value of Γ subtracts from the value of Γ_P on the patch. This interpretation is very useful when developing numerical approximations of the influence of a vortex sheet.

This is precisely the generalization of the two-dimensional scenario we were seeking: in a planar case, the adjacent filaments along a contour reduce to a single point vortex whose strength is equal to the difference in the value of Γ in neighboring patches. Indeed, the complex form (3.175) explicitly exhibits these point vortices in the first term on the right-hand side. In Note 3.6.1 below, we will take this interpretation a bit farther by showing that Γ can often be used to parameterize the sheet in a two-dimensional flow.

Of course, this interpretation of a vortex sheet as a network of filaments or point vortices is only approximate if the patches have finite area. The error of the approximation depends on the degree to which the local circulation in the patch departs from uniform. But this error is easily controlled by using smaller patches or by only relying on the filament form when evaluating at distant points. However, our discussion has at least provided convincing evidence that ***vortex filaments, like point vortices in two dimensions, are quite fundamental structures, and we lose no generality in our analysis by assuming that the fluid vorticity takes such form***.

3.6 Free and Bound Vortex Sheets

In Sect. 3.3, we defined a vortex sheet as one of three fundamental vortex structures in inviscid flow, and discussed some of its characteristics. In this section we provide further background on vortex sheets and derive some basic results to prepare for their important roles in forthcoming chapters.

Throughout this book, we will distinguish between two types of vortex sheets: a **free vortex sheet**, which is a subset of the fluid vorticity and is a material surface by Helmholtz' second theorem (i.e., is transported with the fluid); and a **bound vortex sheet**, which lies in and is intrinsically part of a fluid–body interface, and is therefore kinematically constrained to follow this interface's motion. Now, let us discuss the distinct properties of these two types of sheets, and, for the first time in this book, explicitly acknowledge their time dependence.

3.6.1 Free Vortex Sheets

As a material surface, a free vortex sheet is composed of a collection of infinitesimal elements, each with invariant material coordinates provided by the parameterizations discussed above (ξ in two dimensions or (ξ, η) in three dimensions, and now denoted in all dimensions by vector $\boldsymbol{\xi}$), so that the instantaneous configuration of the sheet is described by $\boldsymbol{X}(\boldsymbol{\xi}, t)$. By Helmholtz' second theorem, these elements move with the local fluid velocity. But what is this local velocity? This question, a general one pertaining to all vortex elements, constitutes much of our discussion in Chap. 7. In the case of a vortex sheet, ambiguity arises because there is a relative slip between the fluid elements on either side of the sheet. We can unambiguously require that the ***normal component*** of the sheet's velocity match those of the fluid on either side; the fluid components are necessarily equal to each other by conservation of mass. However, what should the sheet's tangential velocity be? Let us propose that it take the mean value of the fluid's corresponding components, so that the overall transport of the sheet by

$$\frac{\partial \boldsymbol{X}(\boldsymbol{\xi}, t)}{\partial t} = \frac{1}{2}\left(\boldsymbol{v}^{+}(\boldsymbol{\xi}, t) + \boldsymbol{v}^{-}(\boldsymbol{\xi}, t)\right), \tag{3.179}$$

where $\boldsymbol{v}^{\pm}$ have been defined in (3.141). This choice of transport does not uniquely satisfy the constraint on the normal velocity. We could also have chosen, for example, that the sheet move with velocity $\boldsymbol{v}^{+}$. We will provide a firmer justification for the choice (3.179) below.

Furthermore, in a free vortex sheet, the contour $C(\boldsymbol{\xi}, t)$ is readily interpreted as a material loop, so by Kelvin's circulation theorem (3.40), $\Gamma(\boldsymbol{\xi}, t)$ is constant for fixed value of the material coordinates $\boldsymbol{\xi}$:

$$\frac{\partial \Gamma(\boldsymbol{\xi}, t)}{\partial t} = 0. \tag{3.180}$$

In other words, the local circulation is, like the material coordinates, invariant for each element of a free vortex sheet. In fact, in two-dimensional vortex sheets, this aspect of the local circulation makes it a natural parameter for the sheet under some conditions, as the following note describes.

Note 3.6.1: Parameterization of a Vortex Sheet by Circulation

In some two-dimensional problems, the strength γ maintains the same sign throughout the length of the sheet. Thus, by Eq. (3.143), the local circulation varies monotonically along the sheet and can be used to uniquely identify points along it, i.e., it serves as a parameterization of the sheet, $\boldsymbol{X}(\Gamma)$. Furthermore, Γ is also a ***material coordinate*** of the sheet, since its value at a material point on the sheet does not change in time, as Eq. (3.180) indicates.

It is useful to note that, under such a parameterization, the vortex sheet strength, γ, is the Jacobian of the mapping between the alternative coordinate systems—from arc length to local circulation:

$$\mathrm{d}\Gamma = \gamma(s)\,\mathrm{d}s. \tag{3.181}$$

Alternatively, we can view the sheet strength as a function of local circulation, $\gamma(\Gamma)$, in which case

$$\gamma(\Gamma) = \frac{1}{|\partial \boldsymbol{X}/\partial\Gamma|}, \qquad \frac{\mathrm{d}s}{\mathrm{d}\Gamma} = \frac{1}{\gamma(\Gamma)}, \qquad s(\Gamma) = \int_0^{\Gamma} \frac{\mathrm{d}\hat{\Gamma}}{\gamma(\hat{\Gamma})}. \tag{3.182}$$

In complex form, where the complex sheet strength is related to the infinitesimal change in circulation by $\mathrm{d}\Gamma = -g\,\mathrm{d}z$, we can parameterize the sheet's position Z and strength g by Γ, and write

$$\frac{\mathrm{d}Z}{\mathrm{d}\Gamma} = -\frac{1}{g(\Gamma)}, \qquad Z(\Gamma) = -\int_0^{\Gamma} \frac{\mathrm{d}\hat{\Gamma}}{g(\hat{\Gamma})}. \tag{3.183}$$

In such circumstances, the velocity field (3.153) associated with the vortex sheet can be written as

$$\boldsymbol{v}(\boldsymbol{x}) = \int_0^{\Gamma_S} \boldsymbol{K}(\boldsymbol{x} - \boldsymbol{X}(\hat{\Gamma}))\,\mathrm{d}\hat{\Gamma} \times \boldsymbol{e}_3, \tag{3.184}$$

or, in complex form, the velocity (3.155) as

$$w(z) = \frac{1}{2\pi\mathrm{i}} \int_0^{\Gamma_S} \frac{\mathrm{d}\hat{\Gamma}}{z - Z(\hat{\Gamma})}. \tag{3.185}$$

It might appear quite limiting to only allow this parameterization for vortex sheets in which Γ varies monotonically. However, to circumvent this restriction in sheets along which Γ is not monotonic, we can simply interpret the sheet as multiple **subsheets**, formed by splitting the original sheet at points where $\gamma = 0$. Then, Γ varies monotonically along each of these subsheets and serves as an adequate local parameterization.

It is useful to have a measure of **local strain** on a vortex sheet—that is, the change of length of an infinitesimal portion of the sheet, relative to length of the sheet in some reference state (e.g., at an earlier time). We can write this as $\mathrm{d}s/\mathrm{d}s_0$, where s_0 is the arc length coordinate of the sheet in its reference state. In a two-dimensional sheet for which Γ is a suitable parameterization—that is, for which Γ varies monotonically along the sheet—this strain can be related to the local sheet strength, γ, by equations (3.182):

$$\frac{\mathrm{d}s}{\mathrm{d}s_0} = \frac{\gamma_0(\Gamma)}{\gamma(\Gamma)}, \tag{3.186}$$

where we have denoted the local strength of the sheet at its reference state by $\gamma_0(\Gamma)$.

The two characteristics of a free vortex sheet discussed thus far—that the elements of the sheet are transported by (3.179) and that they retain the same value of circulation—imply a third: ***the pressure is continuous across a free vortex sheet.*** This is easily proved by using the Bernoulli equation in its convected form (3.62), at an evaluation point moving with a given element of the sheet, $\boldsymbol{X}(\boldsymbol{\xi}, t)$, with fixed material coordinates $\boldsymbol{\xi}$. Relating the terms in this equation on either side of the sheet through their common value $B(t)$,

$$p^+ - \frac{1}{2}\rho|\dot{\boldsymbol{X}}(\boldsymbol{\xi},t)|^2 + \frac{1}{2}\rho|\boldsymbol{v}^+ - \dot{\boldsymbol{X}}(\boldsymbol{\xi},t)|^2 + \rho\frac{\mathrm{d}\varphi^+}{\mathrm{d}t}$$
$$= p^- - \frac{1}{2}\rho|\dot{\boldsymbol{X}}(\boldsymbol{\xi},t)|^2 + \frac{1}{2}\rho|\boldsymbol{v}^- - \dot{\boldsymbol{X}}(\boldsymbol{\xi},t)|^2 + \rho\frac{\mathrm{d}\varphi^-}{\mathrm{d}t}, \tag{3.187}$$

where we have used $\dot{\boldsymbol{X}}(\boldsymbol{\xi},t) := \partial \boldsymbol{X}(\boldsymbol{\xi},t)/\partial t$ to denote the velocity of the sheet element, described by (3.179). By that equation, the velocity of the fluid relative to the sheet itself has the same magnitude on both sides of the sheet,

$$|\boldsymbol{v}^\pm - \dot{\boldsymbol{X}}(\boldsymbol{\xi},t)| = \frac{1}{2}|\boldsymbol{v}^+ - \boldsymbol{v}^-|. \tag{3.188}$$

Using this fact, canceling common terms, and relating the local circulation to the jump in scalar potential by (3.159), we end up with

$$[p]_-^+(\boldsymbol{\xi},t) = \rho\frac{\partial\Gamma(\boldsymbol{\xi},t)}{\partial t} = 0, \tag{3.189}$$

where the final result follows from Eq. (3.180). Let us emphasize these important facts by summarizing them in a slightly different manner:

Result 3.6: Characteristics of a Free Vortex Sheet

In an inviscid barotropic (e.g., incompressible) flow, consider a free vortex sheet that satisfies Helmholtz' second and Kelvin's circulation theorems, namely, that a point labeled by $\boldsymbol{\xi}$ on the sheet moves with the material (fluid), and that the local circulation at any material point $\boldsymbol{\xi}$ does not vary in time, $\partial\Gamma(\boldsymbol{\xi}, t)/\partial t = 0$. The pressure is continuous across the sheet, $[p]_-^+ = 0$, if and only if the transport of the elements of the sheet is described by

$$\frac{\partial \boldsymbol{X}(\boldsymbol{\xi}, t)}{\partial t} = \frac{1}{2}\left(\boldsymbol{v}^+(\boldsymbol{\xi}, t) + \boldsymbol{v}^-(\boldsymbol{\xi}, t)\right), \tag{3.190}$$

where $\boldsymbol{v}^\pm$ denote the velocity in the fluid immediately adjacent to the sheet on either side. We will discuss the detailed forms of these limiting values of velocity in Chap. 4, and return to this transport equation in Chap. 7.

In proving (3.189), we specifically chose to follow a material element of the sheet in order to cancel each of the terms that potentially contributes to pressure difference. Now that we have proved that pressure is continuous at all points along a free vortex sheet, we can use it to obtain another useful result. Here, we will again use the convected form of the Bernoulli equation, but rather than allowing the evaluation point on the sheet—denoted by $\boldsymbol{x}_e(t)$—to move with a given material element, we will keep its motion arbitrary (but still confined to the sheet). Note that the pressure is still continuous at $\boldsymbol{x}_e$, regardless of its motion. Thus, the Bernoulli equation provides a relationship between the rate of change of local circulation at $\boldsymbol{x}_e$ and the jump in velocity across the sheet:

$$\frac{\mathrm{d}\Gamma_e}{\mathrm{d}t} = \frac{1}{2}\left(|\boldsymbol{v}^+ - \dot{\boldsymbol{x}}_e|^2 - |\boldsymbol{v}^- - \dot{\boldsymbol{x}}_e|^2\right). \tag{3.191}$$

We have denoted the local circulation at $\boldsymbol{x}_e$ by Γ_e and used an ordinary derivative to denote its time derivative while moving with the evaluation point. This local circulation represents, as before, the circulation about a closed contour that pierces the sheet at $\boldsymbol{x}_e(t)$ (and along which one passes from the – to the + side of the sheet as it is traversed). Thus, its rate of change must be due to a net flux of circulation into this non-material contour. The right-hand side indicates that this flux arises due to differences in the relative velocity between the fluid and the contour. A slightly more intuitive form of this difference can be obtained by expanding the squared magnitudes with the identity $|\boldsymbol{a}|^2 - |\boldsymbol{b}|^2 = (\boldsymbol{a} - \boldsymbol{b}) \cdot (\boldsymbol{a} + \boldsymbol{b})$ and using the definition of the vector strength of the vortex sheet. This results in

$$\frac{\mathrm{d}\Gamma_e}{\mathrm{d}t} = \left[\frac{1}{2}(\boldsymbol{v}^+ + \boldsymbol{v}^-) - \dot{\boldsymbol{x}}_e\right] \cdot (\boldsymbol{\gamma} \times \boldsymbol{n}). \tag{3.192}$$

The expression in square brackets measures the mean difference between the motions of the fluid and the evaluation point; this represents the net convection across the contour at $\boldsymbol{x}_e$. The expression $\boldsymbol{\gamma} \times \boldsymbol{n}$ is the negative of the local gradient of the circulation, by (3.158). Thus, the right-hand side is the convective flux of circulation into the region enclosed by the contour. It is important to note that both $\boldsymbol{\gamma}$ and the fluid velocity relative to the evaluation point are parallel to the sheet; the latter must be so because of conservation of mass. We will have use for (3.192) later in this book.

3.6.2 Bound Vortex Sheets

Bound vortex sheets have quite different properties from free vortex sheets due to their constrained kinematics. For example, let us consider a bound vortex sheet of finite extent. (Such a sheet will be identified with infinitely-thin bodies later in this book.) An element of this sheet described by parametric coordinates $\boldsymbol{\xi}$ is constrained to move with the local velocity of the fluid–body interface, $\boldsymbol{V}_{\mathrm{b}}$. The local circulation of this element is computed about a contour $C(\boldsymbol{\xi}, t)$ that encloses the edge of the sheet, as before. However, the contour is no longer a material loop because of the constrained kinematics of $\boldsymbol{X}(\boldsymbol{\xi}, t)$, so Kelvin's circulation theorem does not hold. Thus, a bound vortex sheet admits a pressure jump across it. It is straightforward to show, using a similar set of steps that led to (3.192), that this pressure jump is given by

$$[p]_{-}^{+} = -\rho \left[\frac{1}{2}(\boldsymbol{v}^{+} + \boldsymbol{v}^{-}) - \boldsymbol{V}_{\mathrm{b}} \right] \cdot (\boldsymbol{\gamma} \times \boldsymbol{n}) + \rho \frac{\mathrm{d}\Gamma}{\mathrm{d}t}, \tag{3.193}$$

where we have denoted the time derivative of the local vortex sheet strength Γ as an ordinary derivative, understood to be taken at fixed $\boldsymbol{\xi}$. We will go into more detail on the circulation in this expression when we use it later in this book. In complex form, this can be expressed as

$$[p]_{-}^{+} = \rho\gamma \operatorname{Re}\left[\left(\frac{1}{2}(w^{+} + w^{-}) - W_{\mathrm{b}} \right) \tau \right] + \rho \frac{\mathrm{d}\Gamma}{\mathrm{d}t}, \tag{3.194}$$

where τ is the unit tangent on the bound vortex sheet, defined such that the + side is to the *left* (in contrast to how it is conventionally defined on a closed counterclockwise contour). As will be shown in Note 3.6.4 below, the imaginary part of the term in square brackets is identically zero due to the no-penetration condition.

Note 3.6.2: The Surface Vortex Sheet and the No-Penetration Condition

The enforcement of the no-penetration condition, discussed in Sect. 3.1.2, leads to a mismatch only in the tangential component(s) of velocity between the body and the surrounding fluid. Therefore, this condition is equivalent to the presence of a bound vortex sheet on the body surface, whose strength is equal to this mismatch:

$$\boldsymbol{\gamma} = \boldsymbol{n} \times (\boldsymbol{v} - \boldsymbol{V}_\mathrm{b}). \tag{3.195}$$

Thus, one can view the no-penetration condition as consistent with a concentration of singular vorticity distributed on the body–fluid interface. We will call this the **tangency form** of the no-penetration condition, and (3.11) the **direct form**. This perspective is a useful one, and we will make extensive use of it in later sections and chapters. It is useful to note that, if we take the cross-product of (3.195) with the unit normal, then, with the help of a standard vector identity, the right-hand side becomes $(\boldsymbol{v} - \boldsymbol{V}_\mathrm{b}) - \boldsymbol{n}[\boldsymbol{n} \cdot (\boldsymbol{v} - \boldsymbol{V}_\mathrm{b})]$. But the last term vanishes due to the direct form of the no-penetration condition. Thus, another version of the tangency form of the condition is

$$\boldsymbol{v} - \boldsymbol{V}_\mathrm{b} = \boldsymbol{\gamma} \times \boldsymbol{n}. \tag{3.196}$$

This shows that the jump in velocity between the body and fluid is not only tangent to the surface, but also orthogonal to the local strength of the vortex sheet.

Note 3.6.3: Complex Form of the No-Penetration Condition

In Eq. (3.12) we wrote the complex version of the direct form of the no-penetration condition for a two-dimensional flow. Let us write the associated vortex sheet relationship. As before, denote by $W_\mathrm{b}(z)$ the complex velocity of the body at a point z on the body surface. Then, we can use tools from Sect. A.2.1 to transform the vector expression (3.195) into complex form. The body surface is identified with a counterclockwise contour, C_b, on which the normal vector is given by expression (A.107). Collecting the direct and tangency forms, we obtain, for $z \in C_\mathrm{b}$,

$$\mathrm{Im}\left(w(z)\frac{\mathrm{d}z}{\mathrm{d}s}\right) = \mathrm{Im}\left(W_\mathrm{b}(z)\frac{\mathrm{d}z}{\mathrm{d}s}\right), \quad \gamma(z) = \mathrm{Re}\left(w(z)\frac{\mathrm{d}z}{\mathrm{d}s}\right) - \mathrm{Re}\left(W_\mathrm{b}(z)\frac{\mathrm{d}z}{\mathrm{d}s}\right), \tag{3.197}$$

where s is an arc length parameter on the surface and $\mathrm{d}z/\mathrm{d}s$ is the unit tangent, τ, on C_b. Note that this definition of the tangent is opposite to that defined for a vortex sheet in Fig. 3.8: here the tangent is directed along the counterclockwise contour C_b with the normal directed to the right (rather than to the left).

These conditions can be written in a different form if we define, as we did in (3.135), the complex sheet strength, g. Here, this is defined as the difference between the fluid and body velocity on the surface:

$$g(z) := w(z) - W_b(z), \qquad z \in C_b. \tag{3.198}$$

Then, the no-penetration condition (3.197) can be written in terms of g:

$$\operatorname{Im}(g\tau) = 0, \qquad \gamma = \operatorname{Re}(g\tau). \tag{3.199}$$

Since $|dz/ds| = 1$, conditions (3.199) indicate that g and γ are compactly related by

$$g = \gamma\tau^*, \tag{3.200}$$

which implies that (the conjugate of) g is locally parallel to the surface. In other words, equations (3.199) are a **flow tangency** condition: they specify that the difference in velocity between the fluid and the body is tangent to the surface, consistent with a vortex sheet. The function g can be interpreted as the **complex strength** of the vortex sheet, since it embodies the same information as γ, but also contains the surface's local directionality. The sign difference between (3.200) and (3.137) is due to the difference in orientation of the tangent relative to the normal.

Another complex form of the no-penetration condition arises from matching the streamfunction at the interface between body and fluid, as described in the discussion following Eq. (3.14). This is equivalent to the first of equations (3.197), integrated along the enclosing contour. The resulting form is expressed in terms of complex potential in the fluid:

$$\operatorname{Im}(F(z)) = \psi_b(z, z^*), \tag{3.201}$$

where $\psi_b(z, z^*)$ (for a rigid body) is the real-valued function given by (3.8). Note that the boundary condition is still satisfied if the left- and right-hand sides of (3.201) are different by a constant, since a constant streamfunction makes no contribution to the velocity field.

Note 3.6.4: No-Penetration Condition on a Sheet Immersed in the Fluid

On a free or bound vortex sheet immersed entirely in the fluid—denoted by $\mathcal{S}$ in either dimension—the configuration contrasts slightly with those considered in Notes 3.6.2 and 3.6.3: here, both sides of the sheet are adjacent to the fluid. But the no-penetration condition still holds: the normal component of the fluid velocity must match that of the sheet in the absence of singular sources or sinks of mass, whether or not the sheet is bound to a plate or free to move in the fluid. We already stated this point for the free vortex sheet in our discussion of its transport, and here, we generalize the point.

Let us define the positive side of the sheet as the direction into which the normal vector points, as in Figs. 3.8 and 3.9, and the negative side as its opposite counterpart. As a result, the sheet has vector strength given by the second of equations (3.134), namely

$$\boldsymbol{\gamma} := \boldsymbol{n} \times (\boldsymbol{v}^+ - \boldsymbol{v}^-), \tag{3.202}$$

or, in complex form in two dimensions, by Eqs. (3.135) and (3.137),

$$g := w^+ - w^- \equiv -\gamma\tau^*. \tag{3.203}$$

Equations (3.202) and (3.136) are clearly analogous to Eqs. (3.195) and (3.199). However, it is important to note that this vortex sheet is not quite the same as those in Notes 3.6.2 and 3.6.3. First of all, the sign of the second of equations (3.136) is different from (3.199). This is because it is more conventional to define the tangent of such a sheet so that the positive side of the sheet is to the left, as in Fig. 3.8, in contrast to the counterclockwise orientation of a closed contour. Furthermore, in those previous equations, the sheet constituted the jump in tangential velocity between the fluid and the body. Here, in (3.202) and (3.203), the sheet represents instead a surface of discontinuity of tangential velocity immersed entirely in the fluid. Since the sheet itself has its own (local) intrinsic velocity, $\boldsymbol{V}_\mathrm{b}$ (or W_b in complex form), which might be the velocity of the plate it is bound to or the convection velocity of a free vortex sheet, then it is helpful to think of this sheet as actually composed of two sheets: one associated with either side. If we denote the strengths of the individual sheets on either side as $\boldsymbol{\gamma}^+$ and $\boldsymbol{\gamma}^-$, respectively defined as

$$\boldsymbol{\gamma}^\pm := \pm\boldsymbol{n} \times (\boldsymbol{v}^\pm - \boldsymbol{V}_\mathrm{b}), \tag{3.204}$$

then it is easy to see that the overall vortex sheet is simply the sum, $\boldsymbol{\gamma} = \boldsymbol{\gamma}^+ + \boldsymbol{\gamma}^-$. Correspondingly in the complex form, if we define

$$g^\pm := w^\pm - W_\mathrm{b}, \tag{3.205}$$

then $g = g^+ - g^-$, where subtraction replaces addition because the complex sheet strength also embodies the local direction of the contour, which is opposite on either side.

In two dimensions, equations (3.136) hold for each of these 'side' sheets. A useful form of the no-penetration condition can thus be formed by taking the average of the two:

$$\frac{1}{2}\text{Im}\left[\left(g^{+}+g^{-}\right)\tau\right] \equiv \text{Im}\left[\frac{1}{2}\left(w^{+}+w^{-}\right)\tau - W_{b}\tau\right] = 0. \tag{3.206}$$

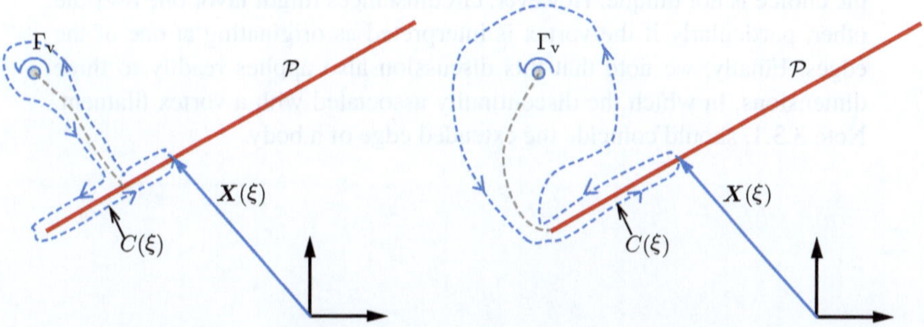

Fig. 3.14 Point vortex near an infinitely thin plate $\mathcal{P}$ with two possible choices of branch cut. The contour $C(\xi)$ used to compute the local circulation at ξ is also shown

Note 3.6.5: A Note on Point Vortices Near Thin Plates

Later in this book, we will encounter scenarios in which an impenetrable plate is in the midst of one or more point vortices. As we have just discussed in Note 3.6.4, the enforcement of the no-penetration condition on the plate gives rise to a bound vortex sheet along the plate. Furthermore, any point vortex is necessarily accompanied by a branch cut in the scalar potential field, as we described in Note 3.2.2. By Kelvin's circulation theorem, this point vortex's strength must be matched by equal and opposite strength in the bound vortex sheet, so the branch cut—which, remember, is arbitrarily configurable—can be designed to intersect and end somewhere in the plate. Is there a preferable choice for this branch cut's configuration? In Fig. 3.14, we depict two possible choices of this branch cut for a single point vortex of strength Γ_v near a plate $\mathcal{P}$. In both cases, the contour $C(\xi)$ used to define the local circulation of the bound vortex sheet, $\Gamma(\xi)$, is deformed around the point vortex in order to account for that vortex's circulation in the total. When the branch cut intersects the plate somewhere along the plate's length, the local circulation starts at zero and jumps discontinuously by Γ_v as the intersection point is passed. If Γ_v is time-varying, then the pressure is awkwardly discontinuous at this point, too. On the other hand, when the branch cut intersects the plate at its endpoint, as in the right panel, the local circulation starts at Γ_v and varies continuously along the plate's length, as does the pressure. This configuration of the branch cut—intersecting the plate at one of its edges—is clearly preferable, and we will utilize it in discussions to follow. Of course, since there are two endpoints, the choice is not unique. However, circumstances might favor one over the other, particularly if the vortex is interpreted as originating at one of the edges. Finally, we note that this discussion also applies readily to three dimensions, in which the discontinuity associated with a vortex filament, Note 3.3.1, should coincide the extended edge of a body.

Chapter 4
General Results of Incompressible Flow About a Body

Contents

4.1 The Basic Potential Flow Problem 92
4.2 Velocity Field Outside a Moving Body 94
4.2.1 Vector Form 94
4.2.2 Complex Form 97
4.2.3 Infinitely-Thin Plate 98
4.2.4 Alternative Surface Formulations: Source and Dipole Distributions 99
4.3 The Integral Equation for an Impenetrable Body 101
4.3.1 Vector Form 102
4.3.2 Complex Form 104
4.3.3 Infinitely-Thin Plate 105
4.3.4 The Double Layer Formulation 106
4.3.5 General Notes on These Integral Equations 107
4.4 Solution of Two-Dimensional Problem by Conformal Mapping 107
4.4.1 Moving Bodies in Irrotational Flow 111
4.4.2 Stationary Body in a Known Potential Flow: The Circle Theorem 113
4.4.3 Solution via the Schwarz–Christoffel Transformation 116
4.5 The Non-uniqueness of Two-Dimensional Potential Flow 121
4.6 Decomposition of the Flow into Basis Fields 125
4.6.1 Complex Form, via Conformal Mapping Solution 128
4.6.2 Vector Form 135
4.6.3 Flow Decomposition and the Kinetic Energy: Added Mass 145
4.7 Multipole Expansion of the Flow Field 149
4.7.1 Two-Dimensional Flow 150
4.7.2 Three-Dimensional Flow 153
4.7.3 Two-Dimensional Expansion in Complex Form 157

The fundamental solutions described in Sects. 3.2.2 and 3.2.3 serve as the foundation for calculating more sophisticated incompressible flows. This is because the governing equations for the potentials—Laplace's equation, or strictly speaking, Poisson's equation—are linear, and therefore admit superposition of the elemental solutions. We have already made use of this in the previous chapter in our construction of flows induced by distributions of singularities in a surface (the vortex sheet, and single- and double-layer potentials).

J. D. Eldredge, *Mathematical Modeling of Unsteady Inviscid Flows*, Interdisciplinary Applied Mathematics 50, https://doi.org/10.1007/978-3-030-18319-6_4

In this chapter, we will make more definite use of this superposition principle in order to obtain a general form of the incompressible flow field about a body. To focus our presentation, let us return to the geometry of a moving body surrounded by an infinite region of fluid, depicted in Fig. 2.1. To keep the results reasonably general, we will suppose for now that the surface $\mathcal{S}_b$ is impenetrable, but not necessarily rigid. Both the two-dimensional and three-dimensional contexts will be discussed, in as unified a manner as possible.

For the most part, we will present results that correspond to the inertial frame of reference, in which the flow is at rest at infinity. However, these results can easily be adapted from the inertial to the windtunnel reference frame by simply substituting their relationship: $\boldsymbol{v} = \boldsymbol{v}^{\dagger} - \boldsymbol{V}_{\infty}$ to relate their fluid velocities, or analogously to relate their body velocities. We will provide details of this adaptation throughout.

4.1 The Basic Potential Flow Problem

The formulations in the next few sections will provide us with a toolbox for constructing velocity fields from the basic contributors of the flow: fluid vorticity, body motion, and uniform flow. Before we discuss this procedure in the following sections, it is important to take note of the underlying mathematical problem we seek to solve. Throughout this book, we are interested in the velocity field associated with a singular vorticity distribution outside one or more impenetrable bodies. Let us write this velocity field in the Helmholtz decomposition (3.1). Remember that the Helmholtz decomposition for a potential flow affords us considerable freedom in describing the various contributors to a flow field either via the scalar potential or by an equivalent vector potential (or streamfunction in two dimensions). Here, we can use this freedom to group contributions from vorticity in the leading term, implicitly derived from the curl of a vector potential, so that the velocity at any point in the fluid is

$$\boldsymbol{v} = \overline{\boldsymbol{v}}_{\omega} + \nabla\varphi, \tag{4.1}$$

where we obtain the vorticity's contribution from the Biot–Savart integral (3.117):

$$\overline{\boldsymbol{v}}_{\omega}(\boldsymbol{x}) := \int_{\mathcal{V}_f \cup \mathcal{V}_b} \boldsymbol{K}(\boldsymbol{x} - \boldsymbol{y}) \times \boldsymbol{\omega}(\boldsymbol{y})\,\mathrm{d}V(\boldsymbol{y}). \tag{4.2}$$

It is important to note that $\mathcal{V}_f \cup \mathcal{V}_b$ signifies that the volume integral extends into a fictitious flow region occupied by the body. The vorticity in this region does not correspond to the body's own motion. So what is the significance of the vorticity in this region? The reader is probably familiar with this in two-dimensional flows from the idea of the ***image vortex***. This image vortex, which has strength equal and opposite to the fluid vortex with which it is paired, is often strategically placed in order to enforce the no-penetration condition on $\mathcal{S}_b$. However, we will be more lenient here and simply insist that it has equal and opposite strength and lies somewhere inside the body. Generically, we refer to this as **bound vorticity**, since it is intrinsically tied to the body. The bound vortex sheet that we have already discussed at length in

Sect. 3.6.2 is also in this class. However, in the present discussion, we are focused on bound vorticity that is specifically inside the body. In three dimensions, this corresponds to closing a vortex filament in some fashion inside the body. By grouping the contribution from this bound interior vorticity with that of the real vorticity, we avoid any awkwardness associated with discontinuous scalar potential fields that intersect the body. Note that if there are multiple bodies, we only need to extend a given vortex element into one of them to avoid such a discontinuity.

The velocity field $\overline{\boldsymbol{v}}_\omega$ does not, in general, satisfy the no-penetration condition on the surface of the body. Thus, the scalar potential field φ is included in order to adjust the flow field in the fluid for the presence of the body. Its governing equation and boundary conditions are

$$\nabla^2\varphi = 0 \text{ in } \mathcal{V}_f, \quad \boldsymbol{n}\cdot\nabla\varphi = \boldsymbol{n}\cdot(\boldsymbol{V}_\mathrm{b} - \overline{\boldsymbol{v}}_\omega) \text{ on } \mathcal{S}_\mathrm{b}, \quad \varphi \to 0 \text{ as } |\boldsymbol{x}| \to \infty. \tag{4.3}$$

Note that the no-penetration condition on the surface $\mathcal{S}_\mathrm{b}$ has been adjusted to account for the effect of the vorticity. By solving (4.3), we seek to cancel the penetrating influence of the fluid vorticity on the surface and replace it with the body's own intrinsic motion. The resulting velocity field in (4.1) is thus assured of satisfying all of the boundary conditions of the problem. It should be noted that we have posed this problem in the inertial reference frame, in which there is no uniform flow at infinity; this problem can be easily transformed to the windtunnel reference frame by including the scalar potential field of the uniform flow with φ and adjusting the condition on this potential field at infinity in (4.3) so that it approaches the form in Eq. (3.59).

In Eq. (4.1) we chose to represent the non-vortical part of the velocity field as the gradient of a scalar potential. We could also have chosen to describe this part instead by the curl of a vector potential,

$$\boldsymbol{v} = \overline{\boldsymbol{v}}_\omega + \nabla\times\boldsymbol{\Psi}. \tag{4.4}$$

This vector potential field, $\boldsymbol{\Psi}$, is not associated with any fluid vorticity, but performs the same role as the scalar potential above: correct the flow field of the vorticity, $\overline{\boldsymbol{v}}_\omega$, so that the boundary conditions are satisfied. As such, it satisfies the governing equation

$$\nabla^2\boldsymbol{\Psi} = 0. \tag{4.5}$$

The no-penetration condition on $\mathcal{S}_\mathrm{b}$ can be equivalently placed on the tangential components of $\boldsymbol{\Psi}$, as we discussed in (3.15), adjusted so that it cancels the influence of vorticity:

$$\boldsymbol{n}\times\boldsymbol{\Psi} = \boldsymbol{n}\times(\boldsymbol{\Psi}_\mathrm{b} - \boldsymbol{\Psi}_\omega), \tag{4.6}$$

where $\boldsymbol{\Psi}_\mathrm{b}$ is the vector potential field inside the body, given, e.g., for a rigid body in Note 3.1.1; and $\boldsymbol{\Psi}_\omega$ is the vector potential field associated with the vorticity, $\overline{\boldsymbol{v}}_\omega = \nabla\times\boldsymbol{\Psi}_\omega$:

$$\boldsymbol{\Psi}_\omega(\boldsymbol{x}) := \int_{\mathcal{V}_f\cup\mathcal{V}_\mathrm{b}} G(\boldsymbol{x}-\boldsymbol{y})\boldsymbol{\omega}(\boldsymbol{y})\,\mathrm{d}V(\boldsymbol{y}). \tag{4.7}$$

The vector potential field $\boldsymbol{\Psi}$ should approach zero—or, at worst, a constant vector—at infinity. If the problem is posed in the windtunnel reference frame, then $\boldsymbol{\Psi}$ includes a uniform flow and should approach the form given by Eq. (3.59) at infinity.

The approaches embodied in Eqs. (4.1) and (4.4) constitute two possible ways to complete the velocity field from a given distribution of vorticity so that the boundary conditions are met on the surface of a body (or bodies). But we should note that there is no strict need to rely exclusively on a scalar or a vector potential field to complete the velocity field, and there might be some advantage to using some mix of the two. In fact, such a mix will be implicit in the formulation we describe at length in the next section.

4.2 Velocity Field Outside a Moving Body

The equations described in the previous section constitute the basic mathematical problem for flow past a moving impenetrable body. For example, we could solve Eq. (4.3), by a variety of analytical or computational means, to determine a scalar potential field that corrects for the body's presence amidst vorticity in the surrounding fluid. In the sections that follow, we will discuss another approach to this problem—an approach that we will frequently follow in this book—in which we construct a velocity field that does not explicitly acknowledge this scalar potential field. This approach proceeds from expressions for the velocity field, derived in this section, that directly account for the body via integrals over its interface with the fluid. We will find in this approach that we have shifted the burden of solving Eq. (4.3) for the full scalar potential field in the fluid to another 'corollary' problem in which we must determine the unknown strength of a singularity distribution on the surface of the body.

4.2.1 Vector Form

The velocity field of a general flow in a region can be obtained from the contributions of vorticity and rate of dilatation distributed throughout the region and the components of the velocity on the enclosing surface. This is proved mathematically in the steps leading to Eq. (A.43) in the Appendix. Consider a region $\mathcal{V}_f$, bounded internally by surface $\mathcal{S}_b$, on which the unit normal $\boldsymbol{n}$ points into $\mathcal{V}_f$ (which, we note, is opposite the direction assumed in Eq. (A.43)). At any point $\boldsymbol{x} \in \mathcal{V}_f$, the fluid velocity $\boldsymbol{v}$ of an incompressible flow, measured with respect to the inertial reference frame, can be written as

$$\boldsymbol{v}(\boldsymbol{x}) = \int_{\mathcal{V}_f} \boldsymbol{K}(\boldsymbol{x}-\boldsymbol{y}) \times \boldsymbol{\omega}(\boldsymbol{y})\, \mathrm{d}V(\boldsymbol{y}) + \int_{\mathcal{S}_b} \left[\boldsymbol{K}(\boldsymbol{x}-\boldsymbol{y}) \times (\boldsymbol{n}_y \times \boldsymbol{v}(\boldsymbol{y})) - \boldsymbol{K}(\boldsymbol{x}-\boldsymbol{y})(\boldsymbol{n}_y \cdot \boldsymbol{v}(\boldsymbol{y})) \right] \mathrm{d}S(\boldsymbol{y}), \quad (4.8)$$

where $\boldsymbol{n}_y$ denotes the normal at the point $\boldsymbol{y} \in \mathcal{S}_b$, and the velocity $\boldsymbol{v}$ on the surface $\mathcal{S}_b$ corresponds to the limit as the surface is approached from within $\mathcal{V}_f$. Equation (4.8) focuses much of the discussion in this chapter. Therefore, it is useful to make a few notes first:

- The volume integral here is only over the fluid vorticity, and does not include the extension of vorticity into the fictitious region occupied by the body in the previous section.
- It is derived without any specification of boundary conditions on the velocity at $\mathcal{S}_b$. As such, it holds even if the flow is viscous or the surface is permeable or deforming, possibly with non-zero net flow rate. (For example, $\mathcal{S}_b$ might simply be a fictitious surface used for limiting the size of the region $\mathcal{V}_f$ to a domain of interest.) We have only assumed that the velocity field is divergence free throughout the fluid.
- If there is more than one body, the surface integral over $\mathcal{S}_b$ can easily be interpreted as the sum over all bodies' enclosing surfaces.
- By comparison with the forms of singularity distributions (3.117), (3.161) and (3.163) developed in the preceding chapter, it is clear that the velocity consists of the collective contributions from vorticity elements of strength $\boldsymbol{\omega}\, \mathrm{d}V$ distributed in the fluid, and from singular distributions of vortex and source elements in the surface, of differential strengths $(\boldsymbol{n} \times \boldsymbol{v})\, \mathrm{d}S$ and $(\boldsymbol{n} \cdot \boldsymbol{v})\, \mathrm{d}S$, respectively. The velocity kernel, $\boldsymbol{K}(\boldsymbol{x}-\boldsymbol{y})$, embodies the action at the evaluation point $\boldsymbol{x}$ of each 'source' (i.e., the quantities in the integrands: fluid vorticity, surface distributions), located at point $\boldsymbol{y}$ in the fluid or on the surface.
- From Eq. (A.43), it is apparent that, when the evaluation point is inside $\mathcal{V}_b$—the region enclosed by $\mathcal{S}_b$—and thus not in $\mathcal{V}_f$ or on $\mathcal{S}_b$, the left-hand side of (4.8) is identically zero. Thus, the equation reduces to a statement about the equivalence in this region of the volume integral over the fluid vorticity and the (negative of the) surface integral over the fluid velocity.
- From our discussion in Sect. 3.5, the surface distribution of vortex elements can be reinterpreted, if desired, as a distribution of dipoles. This will be discussed further in Sect. 4.2.4.
- The equation, as it appears, is applicable in the inertial reference frame. However, it can easily be adapted to obtain the velocity $\boldsymbol{v}^{\dagger}$ in the windtunnel reference frame by making the substitution $\boldsymbol{v} = \boldsymbol{v}^{\dagger} - \boldsymbol{V}_{\infty}$ wherever $\boldsymbol{v}$ appears.
- One can always replace the surface integral of $\boldsymbol{v}$ over $\mathcal{S}_b$ by an equivalent influence from the vorticity contained within $\mathcal{V}_b$ (the actual vorticity, and not the fictitious sort described in the previous section). This can sometimes be advantageous

when that vorticity inside $\mathcal{V}_b$ is known: for example, when the body is rigid and its vorticity is twice its angular velocity, as discussed in Note 3.1.1.

- As is apparent from the integral over $\mathcal{V}_f$, the equation is valid for flows with generally distributed vorticity fields. However, it can be easily specialized to flows in which the fluid vorticity is concentrated in singular elements, such as points, filaments, or sheets, as we discussed in Sect. 3.3. In such cases, we simply replace the vorticity integral by its specialized form. The specialized form for a vortex sheet immersed in the fluid will be discussed further in Sect. 4.2.3.

In order to use Eq. (4.8), one generally must determine the distributions on the surface by solving a boundary integral problem, as we will discuss in Sect. 4.3. This process is guided by the fact that we must enforce the no-penetration condition at all points on the body surface, which requires that $\boldsymbol{n} \cdot \boldsymbol{v} = \boldsymbol{n} \cdot \boldsymbol{V}_b$ and (by Note 3.6.2) $\boldsymbol{n} \times \boldsymbol{v} = \boldsymbol{\gamma} + \boldsymbol{n} \times \boldsymbol{V}_b$, where $\boldsymbol{V}_b$ is the local velocity of the surface and $\boldsymbol{\gamma}$ is the strength of a vortex sheet bound to the surface (a **bound vortex sheet**, as defined in Sect. 3.6). Consequently, an alternative form of (4.8), more useful for our purposes, is given by the following:

Result 4.1: Velocity Field of an Incompressible Flow in the Presence of a Body

At any $\boldsymbol{x}$ in a fluid region $\mathcal{V}_f$, in the presence of fluid vorticity $\boldsymbol{\omega}$ and an impenetrable moving surface $\mathcal{S}_b$ with local velocity $\boldsymbol{V}_b$, the velocity in the inertial reference frame (i.e., at rest at infinity) is given by

$$\begin{aligned}\boldsymbol{v}(\boldsymbol{x}) = &\int_{\mathcal{V}_f} \boldsymbol{K}(\boldsymbol{x}-\boldsymbol{y}) \times \boldsymbol{\omega}(\boldsymbol{y})\,\mathrm{d}V(\boldsymbol{y}) + \int_{\mathcal{S}_b} \boldsymbol{K}(\boldsymbol{x}-\boldsymbol{y}) \times \boldsymbol{\gamma}(\boldsymbol{y})\,\mathrm{d}S(\boldsymbol{y}) \\ &+ \int_{\mathcal{S}_b} \left[\boldsymbol{K}(\boldsymbol{x}-\boldsymbol{y}) \times (\boldsymbol{n}_y \times \boldsymbol{V}_b(\boldsymbol{y})) - \boldsymbol{K}(\boldsymbol{x}-\boldsymbol{y})(\boldsymbol{n}_y \cdot \boldsymbol{V}_b(\boldsymbol{y}))\right]\,\mathrm{d}S(\boldsymbol{y}),\end{aligned} \tag{4.9}$$

where $\boldsymbol{\gamma} = \boldsymbol{n} \times (\boldsymbol{v} - \boldsymbol{V}_b)$ is the strength of a vortex sheet on $\mathcal{S}_b$, and $\boldsymbol{K}$ is the velocity kernel, given in two dimensions by

$$\boldsymbol{K}(\boldsymbol{x}) = -\frac{\boldsymbol{x}}{2\pi|\boldsymbol{x}|^2}, \tag{4.10}$$

and in three dimensions by

$$\boldsymbol{K}(\boldsymbol{x}) = -\frac{\boldsymbol{x}}{4\pi|\boldsymbol{x}|^3}. \tag{4.11}$$

Furthermore, when $\boldsymbol{x}$ is inside the region $\mathcal{V}_b$ enclosed by $\mathcal{S}_b$, the left-hand side of (4.9) is 0.

In the windtunnel reference frame, in which there is a uniform flow $\boldsymbol{V}_\infty$ at infinity, the expression for the velocity field $\boldsymbol{v}^\dagger$ in that frame can be obtained by replacing $\boldsymbol{v}$ and $\boldsymbol{V}_\mathrm{b}$ with $\boldsymbol{v}^\dagger - \boldsymbol{V}_\infty$ and $\boldsymbol{V}_\mathrm{b}^\dagger - \boldsymbol{V}_\infty$, respectively, wherever they appear in (4.9).

Equation (4.9) clearly exhibits the effect of the various **flow contributors** to the velocity field: fluid vorticity, body surface motion, and, in the windtunnel reference frame, uniform external flow. In most cases, we know the surface motion and the state of the uniform flow and fluid vorticity, so we need only to determine the bound vortex sheet's strength distribution, $\boldsymbol{\gamma}$, in order to use (4.9) to compute the velocity. This will be discussed in Sect. 4.3, where we will find that the bound vortex sheet is itself dependent on the flow contributors. This, in turn, will lead to a natural decomposition of the flow field, discussed in Sect. 4.6.

4.2.2 Complex Form

For two-dimensional flows, Eq. (4.8) has a complex equivalent that will be of value to us in some of the work in this book, including Chap. 8. Since complex analysis is of greatest utility when the velocity field is holomorphic almost everywhere, in this complex form we require that the flow is irrotational except at isolated singularities, so that the integral over $\mathcal{V}_f$ is replaced by a sum over a set of point vortices in the fluid, whose form was given in Eq. (3.119). As noted in the discussion of the volume-preserving constraint (2.43), the body surface $\mathcal{S}_\mathrm{b}$ is identified with a closed counterclockwise contour $\mathcal{C}_\mathrm{b}$. The complex version of (4.8) is actually Cauchy's integral formula (A.134), evaluated in the region bounded *internally* by $\mathcal{C}_\mathrm{b}$. A positively-oriented contour of that region would actually be clockwise. However, the sign of the integral is switched from (A.134) in order to ensure the conventional counterclockwise orientation. Then, the complex velocity at $z \in \mathcal{V}_f$ is given by

$$w(z) = \frac{1}{2\pi \mathrm{i}} \sum_{\mathrm{J}=1}^{N_\mathrm{v}} \frac{\Gamma_\mathrm{J}}{z - z_\mathrm{J}} - \frac{1}{2\pi \mathrm{i}} \int_{\mathcal{C}_\mathrm{b}} \frac{w(\lambda)}{\lambda - z}\, \mathrm{d}\lambda, \tag{4.12}$$

where $w(\lambda)$ in the integral should be interpreted as the limit of the complex velocity as $\lambda \in \mathcal{C}_\mathrm{b}$ is approached from the fluid side. As in the vector case, w is the velocity in the inertial reference frame; to obtain the velocity $w^\dagger$ in the windtunnel reference frame, simply substitute $w = w^\dagger - W_\infty$.

Thus, each element of the surface source/vortex distribution has infinitesimal strength $\mathrm{d}Q - \mathrm{i}\,\mathrm{d}\Gamma = -\mathrm{i}w(\lambda)\,\mathrm{d}\lambda$, for $w(\lambda)$ evaluated on the surface. In other words, the real part of $w\,\mathrm{d}\lambda$ represents the strength of the local vortex (equivalent to $\boldsymbol{n}\times\boldsymbol{v}\,\mathrm{d}S$ in the vector expression) and the imaginary part serves as the strength of the local source (equivalent to $\boldsymbol{n}\cdot\boldsymbol{v}\,\mathrm{d}S$). We can distinguish the roles of the body motion and the surface vortex sheet by substituting for w in the contour integral with the no-penetration condition (3.198). This leads to the following result, analogous to Result 4.1.

Result 4.2: Velocity Field of a Two-Dimensional Flow in the Presence of a Body

The complex velocity field of a two-dimensional incompressible flow, in the presence of a set of N_v point vortices with positions z_J and strengths Γ_J and an impenetrable surface $\mathcal{S}_b$ in motion with local velocity W_b, is given at any $z \in \mathcal{V}_f$ by

$$w(z) = \frac{1}{2\pi i} \sum_{J=1}^{N_v} \frac{\Gamma_J}{z - z_J} - \frac{1}{2\pi i} \int_{C_b} \frac{g(\lambda)}{\lambda - z} \, d\lambda - \frac{1}{2\pi i} \int_{C_b} \frac{W_b(\lambda)}{\lambda - z} \, d\lambda, \qquad (4.13)$$

where g is the complex strength of a vortex sheet on $\mathcal{S}_b$, and the imaginary part of $g \, d\lambda$ is identically zero. Furthermore, for $z \in \mathcal{V}_b$, the left-hand side of (4.13) vanishes identically.

In the windtunnel reference frame, in which there is a uniform flow W_∞ at infinity, the expression for the velocity field $w^\dagger$ in that frame can be obtained by replacing w and W_b with $w^\dagger - W_\infty$ and $W_b^\dagger - W_\infty$, respectively.

4.2.3 Infinitely-Thin Plate

We will also have a need in this book to compute the velocity due to a vortex sheet immersed entirely in the fluid. We already considered this notion in our discussion of free vortex sheets in Sect. 3.6. The situation also arises when we consider a flat or deformed body of infinitesimal thickness (which we will refer to as a **plate**), in which case the body's influence on the fluid motion is representable as a bound vortex sheet with the same kinematics as the body. In this case we generally are faced with the same problem of determining the unknown strength of the sheet as for a body of finite thickness. Here, we present the velocity field when the body takes the form of a plate, $\mathcal{P}$ (or multiple plates). Note that, as usual, the fluid vorticity is left general, but may take the form of points, filaments or free sheets.

Opposite sides of the bound sheet on $\mathcal{P}$ are paired in the surface integrals of (4.8) and (4.12), and the normal vectors at such paired points are in opposite directions (or, in the complex form, the contours are oppositely directed). The surface integrals can then be reformulated into new ones evaluated over the positive side of the finite-length curve (in two dimensions) or finite-area patch (in three dimensions) constituting the sheet, with surface velocity replaced by the jump in fluid velocity as the sheet is approached from the opposite sides. In Note 3.6.4, we defined this jump to be the local strength of the vortex sheet, $\boldsymbol{\gamma} := \boldsymbol{n} \times (\boldsymbol{v}^+ - \boldsymbol{v}^-)$ or $g := w^+ - w^-$, in the vector and complex forms, respectively. The vector form of the velocity (4.8) is therefore replaced by

$$\boldsymbol{v}(\boldsymbol{x}) = \int_{\mathcal{V}_f} \boldsymbol{K}(\boldsymbol{x} - \boldsymbol{y}) \times \boldsymbol{\omega}(\boldsymbol{y})\, \mathrm{d}V(\boldsymbol{y}) + \int_{\mathcal{P}} \boldsymbol{K}(\boldsymbol{x} - \boldsymbol{y}) \times \boldsymbol{\gamma}(\boldsymbol{y})\, \mathrm{d}S(\boldsymbol{y}), \tag{4.14}$$

and the complex form (4.12) by

$$w(z) = \frac{1}{2\pi \mathrm{i}} \sum_{\mathrm{J}=1}^{N_{\mathrm{v}}} \frac{\Gamma_{\mathrm{J}}}{z - z_{\mathrm{J}}} + \frac{1}{2\pi \mathrm{i}} \int_{\mathcal{P}} \frac{g(\lambda)}{\lambda - z}\, \mathrm{d}\lambda. \tag{4.15}$$

As always, these expressions can be adapted to the windtunnel reference frame by making the substitutions $\boldsymbol{v} = \boldsymbol{v}^{\dagger} - \boldsymbol{V}_{\infty}$ and $w = w^{\dagger} - W_{\infty}$, respectively.

As discussed in Note 3.6.4, the vortex sheet appearing in these integrals is not the same as the ones in (4.9) and (4.13), since it now represents the overall jump in tangential velocity on either side of the sheet, rather than the jump in velocity between the fluid and the body. The two definitions are reconciled by thinking of this overall jump as composed of two sheets—one on each side—as discussed in that Note 3.6.4. Also, the sign on the surface integral in (4.15) is switched due to the convention of defining the tangent of $\mathcal{P}$ (and hence, the integration direction) with the positive side to the left.

4.2.4 Alternative Surface Formulations: Source and Dipole Distributions

The vortex sheet is our preferred representation of the surface of the body in this book, primarily because it provides us with a formulation of the flow field that avoids any dealings with the awkward branch cuts or other discontinuities that accompany the scalar potential field. In the next section, we will develop the equation for the unknown strength of the bound vortex sheet. However, before we leave this discussion of formulations of the velocity field, let us observe that we need not rely on the vortex sheet as the exclusive entity for representing the body's presence. In this section, we discuss an alternative set of common formulations, based on distributions of sources and/or dipoles on the body surface. To construct the most generally useful form, we will make use of two flow fields: the 'real' one in the fluid region $\mathcal{V}_f$, which we will label $\boldsymbol{v}^+$; and another 'fictitious' flow in the region $\mathcal{V}_{\mathrm{b}}$ occupied by the body, denoted by $\boldsymbol{v}^-$. It is important to stress that both flow fields are meant to serve us in constructing the real flow field in $\mathcal{V}_f$; we are not seeking to describe the body's own motion with $\boldsymbol{v}^-$. Both flow fields will be written in the Helmholtz decomposition form (4.1), making use once again of the extension of the fluid vorticity into the body and the associated velocity $\overline{\boldsymbol{v}}_{\omega}$ in Eq. (4.2):

$$\boldsymbol{v}^{\pm} = \overline{\boldsymbol{v}}_{\omega} + \nabla\varphi^{\pm}. \tag{4.16}$$

The scalar potential field φ^+ is the same field φ discussed in Sect. 4.1, and is governed by (4.3). In contrast, we have some freedom to choose the role of the auxiliary potential inside the body, φ^-; generally, it is chosen to make the overall problem easier to solve. It must satisfy Laplace's equation, of course, but we can set its boundary condition on $\mathcal{S}_b$, if any, as we wish. For now, let us leave it unspecified. In fact, we can immediately write down the scalar potential field in either region, in terms of the jumps in potential and their normal derivative on $\mathcal{S}_b$, using Result A.3. The resulting velocity field in $\mathcal{V}_f$, after enforcing incompressibility, is

$$\begin{aligned}\boldsymbol{v}(\boldsymbol{x}) \equiv \boldsymbol{v}^+(\boldsymbol{x}) = &\int\limits_{\mathcal{V}_f \cup \mathcal{V}_b} \boldsymbol{K}(\boldsymbol{x}-\boldsymbol{y}) \times \boldsymbol{\omega}(\boldsymbol{y})\,\mathrm{d}V(\boldsymbol{y}) \\ &+\nabla \int\limits_{\mathcal{S}_b} \left[[\varphi]_-^+(\boldsymbol{y})\boldsymbol{n}_y \cdot \nabla_y G(\boldsymbol{x}-\boldsymbol{y}) - G(\boldsymbol{x}-\boldsymbol{y})\boldsymbol{n}_y \cdot \nabla_y [\varphi]_-^+(\boldsymbol{y}) \right] \mathrm{d}S(\boldsymbol{y}).\end{aligned} \tag{4.17}$$

Let us consider some choices for the condition satisfied by φ^- on the body surface, and the resulting formulation of the velocity field.

- First, let us set $\varphi^- = \varphi^+$ on $\mathcal{S}_b$. Note that we're only setting these potential fields to match at their interface; their normal derivatives will generally not match, since we've imposed the no-penetration condition on φ^+. In fact, let us denote the difference in their normal derivatives by $q := \boldsymbol{n} \cdot \nabla[\varphi]_-^+$. As a result, the velocity field is given by

$$\boldsymbol{v}(\boldsymbol{x}) = \int\limits_{\mathcal{V}_f \cup \mathcal{V}_b} \boldsymbol{K}(\boldsymbol{x}-\boldsymbol{y}) \times \boldsymbol{\omega}(\boldsymbol{y})\,\mathrm{d}V(\boldsymbol{y}) - \nabla \int\limits_{\mathcal{S}_b} q(\boldsymbol{y}) G(\boldsymbol{x}-\boldsymbol{y})\,\mathrm{d}S(\boldsymbol{y}). \tag{4.18}$$

  The surface integral has taken the form of a single-layer potential (3.162), with distributed sources on the surface whose strength is equal to the jump in normal derivative between the inner and outer potentials. This strength is unknown, and is to be solved for by enforcing the no-penetration condition.
  It's important to note that a single-layer potential, on its own, cannot possibly represent the flow about an infinitely thin plate. In such a case, there is no body interior, and thus, no auxiliary potential. Furthermore, there is no difference in the normal derivatives of the potential on either side of the plate, since both must satisfy the same no-penetration condition, and thus, the surface integral vanishes as body thickness goes to zero. For this reason, we will not consider this formulation any further in the book.
- Now, let us consider what emerges when we enforce the normal derivative of φ^- to match that of φ^+ on $\mathcal{S}_b$; they both then satisfy the no-penetration condition in Eq. (4.3). A difference in the values of these potentials will persist on the boundary, and we will denote this difference by $\mu := -[\varphi]_-^+$. This leads to a

velocity field obtained from a double-layer potential (3.166):

$$v(x) = \int_{\mathcal{V}_f \cup \mathcal{V}_b} K(x-y)\times\omega(y)\,\mathrm{d}V(y) + \nabla \int_{\mathcal{S}_b} \mu(y) n_y \cdot \nabla G(x-y)\,\mathrm{d}S(y). \quad (4.19)$$

As with the single-layer formulation, or with the bound vortex sheet formulation (4.9), we do not yet know the strength of the dipole distribution μ. Its solution will be discussed in the following section.
In this case, the representation remains well behaved even as the body's thickness goes to zero. For an infinitely thin plate, the dipole strength simply represents the difference in φ^+ between the lower and upper side of the plate.

Other formulations can be constructed, as well. For example, we could simply set $\varphi^- = 0$, which would lead to a combination of single- and double-layer potentials on $\mathcal{S}_b$. However, the strength of the single-layer potential is $n \cdot \nabla\varphi^+$ on $\mathcal{S}_b$, which is known from the no-penetration condition, and the problem requires the solution for the dipole strength, $\mu := -\varphi^+$ on $\mathcal{S}_b$.
These unknown strengths—q, μ, and γ—are necessarily related to each other, since they induce closely related potential flow fields for the same body in the same motion and subject to the same fluid vorticity. In particular, the vortex sheet and double-layer representations are reconciled with each other by the equivalencies of these entities we discussed in Sect. 3.5.3; the reader is encouraged to verify that, indeed, (4.8) can be obtained from (4.19).[1] In summary, we simply note that $\gamma = n \times (\overline{v}_\omega + \nabla\varphi^+ - V_b)$, which is the tangential slip velocity between the fluid flow and the body's surface motion.

4.3 The Integral Equation for an Impenetrable Body

The various formulations in Sect. 4.2 for the velocity field—Eq. (4.9), or its complex version (4.13), or the double-layer potential (4.19)—are only of value for calculation if the quantities in the integrands are known. Assuming that we know the vorticity distribution in the fluid region and the motion of the body, and that the body surface is impenetrable, then the only unknown is the strength distribution on the body surface $\mathcal{S}_b$—of the vortex sheet, $\gamma(x)$ or $g(z)$, or, alternatively, of the double layer μ. Our goal in this section is to formulate integral equations for these strengths. These equations are obtained by evaluating the velocity at a point x that approaches $\mathcal{S}_b$ and enforcing the no-penetration condition on the result.

[1] Equation (A.53) applied on v^- is useful in this procedure, since it allows us to transform the fictitious vorticity's influence in $\mathcal{V}_f$ into a surface integral.

4.3.1 Vector Form

Before we present the vector form of this integral equation, we recall the behavior of a surface integral of $\boldsymbol{K}$ as the evaluation point approaches the surface, given in the Appendix by Eq. (A.46). Using the identity (A.51), we can obtain the following integral equation for the velocity field, $\boldsymbol{v}$, at points $\boldsymbol{x} \in \mathcal{S}_b$:

$$\frac{1}{2}\boldsymbol{v}(\boldsymbol{x}) = \int_{\mathcal{V}_f} \boldsymbol{K}(\boldsymbol{x}-\boldsymbol{y}) \times \boldsymbol{\omega}(\boldsymbol{y})\,\mathrm{d}V(\boldsymbol{y}) + \fint_{\mathcal{S}_b} \left[\boldsymbol{K}(\boldsymbol{x}-\boldsymbol{y}) \times (\boldsymbol{n}_y \times \boldsymbol{v}(\boldsymbol{y})) - \boldsymbol{K}(\boldsymbol{x}-\boldsymbol{y})(\boldsymbol{n}_y \cdot \boldsymbol{v}(\boldsymbol{y}))\right] \mathrm{d}S(\boldsymbol{y}), \tag{4.20}$$

where the surface integral is evaluated at its principal value (i.e., omitting the effect from a vanishingly small circular patch surrounding $\boldsymbol{x}$). Note that, in developing this expression, we have accounted for the fact that $\boldsymbol{v}$ is divergence free, that its curl is the vorticity, and that the unit normal on $\mathcal{S}_b$ points into $\mathcal{V}_f$ rather than away from it. Now, we apply the no-penetration condition (3.11) on $\mathcal{S}_b$ and the associated bound vortex sheet identities in Note 3.6.2 to substitute each appearance of $\boldsymbol{v}$ in this equation for terms involving the vortex strength $\boldsymbol{\gamma}$ and the surface velocity $\boldsymbol{V}_b$. It can be easily verified that this leads to

$$-\frac{1}{2}\boldsymbol{n}(\boldsymbol{x}) \times \boldsymbol{\gamma}(\boldsymbol{x}) - \fint_{\mathcal{S}_b} \boldsymbol{K}(\boldsymbol{x}-\boldsymbol{y}) \times \boldsymbol{\gamma}(\boldsymbol{y})\,\mathrm{d}S(\boldsymbol{y}) = \int_{\mathcal{V}_f} \boldsymbol{K}(\boldsymbol{x}-\boldsymbol{y}) \times \boldsymbol{\omega}(\boldsymbol{y})\,\mathrm{d}V(\boldsymbol{y}) - \frac{1}{2}\boldsymbol{V}_b(\boldsymbol{x}) + \fint_{\mathcal{S}_b} \left[\boldsymbol{K}(\boldsymbol{x}-\boldsymbol{y}) \times (\boldsymbol{n}_y \times \boldsymbol{V}_b(\boldsymbol{y})) - \boldsymbol{K}(\boldsymbol{x}-\boldsymbol{y})(\boldsymbol{n}_y \cdot \boldsymbol{V}_b(\boldsymbol{y}))\right] \mathrm{d}S(\boldsymbol{y}). \tag{4.21}$$

There are a few important things to note about this equation:

- The no-penetration condition has already been applied (to obtain (4.9)), so Eq. (4.21) constitutes a vector-valued integral equation for the strength, $\boldsymbol{\gamma}$, of the vortex sheet.
- Equation (4.21), like (4.9), is applicable to general (viscous or inviscid) incompressible flows, in which vorticity may be generally distributed in the fluid. For a viscous flow, the no-slip condition is also enforced, so that $\boldsymbol{v} = \boldsymbol{V}_b$ on $\mathcal{S}_b$ and $\boldsymbol{\gamma}$ is identically zero; the remaining terms simply recover the original identity (4.20). However, since we devote our attention in this book to inviscid flows with singular distributions of vorticity, we will focus on the application of (4.21) in that more limited context.

- To adapt this equation to the windtunnel reference frame, we make the usual substitution of velocities, which in this case is simply to transform the body's surface velocity: $\boldsymbol{V}_{\mathrm{b}} = \boldsymbol{V}_{\mathrm{b}}^{\dagger} - \boldsymbol{V}_{\infty}$. The resulting contributions from $\boldsymbol{V}_{\infty}$ can be collected into a single term, equal to $\boldsymbol{V}_{\infty}$ on the right-hand side.

We will discuss methods to solve the equation in later sections and chapters. First, let us highlight the equation, and define some useful notation:

Result 4.3: General Integral Equation for Bound Vortex Sheet

Consider a fluid region $\mathcal{V}_f$ and an impenetrable surface $\mathcal{S}_{\mathrm{b}}$ moving with local velocity $\boldsymbol{V}_{\mathrm{b}}$. Define the vector-valued **surface operator**, $\mathcal{F}[\cdot](\cdot)$, as

$$\mathcal{F}[\boldsymbol{\gamma}](\boldsymbol{x}) := \frac{1}{2}\boldsymbol{n}(\boldsymbol{x}) \times \boldsymbol{\gamma}(\boldsymbol{x}) + \fint_{\mathcal{S}_{\mathrm{b}}} \boldsymbol{K}(\boldsymbol{x} - \boldsymbol{y}) \times \boldsymbol{\gamma}(\boldsymbol{y})\,\mathrm{d}S(\boldsymbol{y}). \tag{4.22}$$

The first argument of this operator is the vector-valued strength of a vortex sheet in the body surface, and the second argument is a point $\boldsymbol{x} \in \mathcal{S}_{\mathrm{b}}$. The operator returns the velocity induced by this sheet at a point on the surface when the point is approached from the negative side, inside the body. Then, the general integral equation for the bound vortex sheet strength that enforces the no-penetration condition can be written as

$$\mathcal{F}[\boldsymbol{\gamma}](\boldsymbol{x}) = \Delta\boldsymbol{V}(\boldsymbol{x}), \tag{4.23}$$

where, in the inertial reference frame, the right-hand side is defined at all points $\boldsymbol{x} \in \mathcal{S}_{\mathrm{b}}$ as

$$\begin{aligned}\Delta\boldsymbol{V}(\boldsymbol{x}) := &- \int_{\mathcal{V}_f} \boldsymbol{K}(\boldsymbol{x} - \boldsymbol{y}) \times \boldsymbol{\omega}(\boldsymbol{y})\,\mathrm{d}V(\boldsymbol{y}) \\ &+ \frac{1}{2}\boldsymbol{V}_{\mathrm{b}}(\boldsymbol{x}) - \fint_{\mathcal{S}_{\mathrm{b}}} \left[\boldsymbol{K}(\boldsymbol{x} - \boldsymbol{y}) \times (\boldsymbol{n}_{\boldsymbol{y}} \times \boldsymbol{V}_{\mathrm{b}}(\boldsymbol{y})) - \boldsymbol{K}(\boldsymbol{x} - \boldsymbol{y})(\boldsymbol{n}_{\boldsymbol{y}} \cdot \boldsymbol{V}_{\mathrm{b}}(\boldsymbol{y}))\right]\,\mathrm{d}S(\boldsymbol{y}).\end{aligned} \tag{4.24}$$

In the windtunnel reference frame, from which the body's surface velocity is observed to be $\boldsymbol{V}_{\mathrm{b}}^{\dagger} = \boldsymbol{V}_{\mathrm{b}} + \boldsymbol{V}_{\infty}$, the same right-hand side (4.24) can be expressed as

$$\Delta \boldsymbol{V}(\boldsymbol{x}) := -\boldsymbol{V}_{\infty} - \int_{\mathcal{V}_f} \boldsymbol{K}(\boldsymbol{x}-\boldsymbol{y}) \times \boldsymbol{\omega}(\boldsymbol{y})\,\mathrm{d}V(\boldsymbol{y})$$
$$+ \frac{1}{2}\boldsymbol{V}_{\mathrm{b}}^{\dagger}(\boldsymbol{x}) - \fint_{\mathcal{S}_{\mathrm{b}}} \left[\boldsymbol{K}(\boldsymbol{x}-\boldsymbol{y}) \times (\boldsymbol{n}_y \times \boldsymbol{V}_{\mathrm{b}}^{\dagger}(\boldsymbol{y})) - \boldsymbol{K}(\boldsymbol{x}-\boldsymbol{y})(\boldsymbol{n}_y \cdot \boldsymbol{V}_{\mathrm{b}}^{\dagger}(\boldsymbol{y}))\right] \mathrm{d}S(\boldsymbol{y}). \tag{4.25}$$

Both right-hand side expressions represent the discrepancy in velocity, at a point $\boldsymbol{x} \in \mathcal{S}_{\mathrm{b}}$, between the body's motion and that of the external contributors to fluid motion (uniform flow and fluid vorticity). Equation (4.23) expresses our search for a vortex sheet, with strength $\boldsymbol{\gamma}$, that induces a velocity field within the region $\mathcal{V}_{\mathrm{b}}$ which cancels the effects of any uniform flow or fluid vorticity and replaces it with the body's own velocity. We stress that, as long as $\boldsymbol{V}_{\mathrm{b}}$ and $\boldsymbol{V}_{\mathrm{b}}^{\dagger}$ are related to each other by $\boldsymbol{V}_{\infty}$, the solution $\boldsymbol{\gamma}$ to (4.23) is the same for either form of right-hand side.

Let us assume for now that Eq. (4.23) has a solution and that it is unique. Later, we will discuss the solvability of the problem, and find that the solution's uniqueness will require some additional constraint in two-dimensional problems; we postpone that discussion to Sect. 4.5. For the benefit of compact notation, define the **inverse surface operator**, $\mathcal{F}^{-1}[\cdot](\cdot)$, whose first argument is the velocity discrepancy on the surface and which returns the local value (at a point on $\mathcal{S}_{\mathrm{b}}$ given by the second argument) of the bound vortex sheet strength that annihilates this velocity discrepancy. We can then denote the formal solution of (4.23) by

$$\boldsymbol{\gamma} = \mathcal{F}^{-1}[\Delta \boldsymbol{V}](\cdot). \tag{4.26}$$

It is important to remember that $\boldsymbol{\gamma}$ only has components that are locally tangent to the body surface. Thus, though it might appear that, by solving (4.23), we are canceling all components of the velocity discrepancy on the surface, only the normal velocity difference is actually eliminated. We have, after all, only added a potential flow induced by the bound vortex sheet, and the mere presence of this sheet indicates that a tangential velocity difference (i.e., a slip motion) persists.

4.3.2 Complex Form

In two-dimensional problems, we can use (4.13) to develop a complex version of this integral equation. The analogous formulas to (A.46) are the Plemelj formulas (A.125). When z is allowed to approach $z \in C_{\mathrm{b}}$ from inside $\mathcal{V}_f$ (the $-$ side, according to the definition in (A.125)), Eq. (4.12) becomes

$$\frac{1}{2}g(z) + \frac{1}{2\pi \mathrm{i}}\fint_{C_{\mathrm{b}}} \frac{g(\lambda)}{\lambda - z}\,\mathrm{d}\lambda = \frac{1}{2\pi \mathrm{i}}\sum_{\mathrm{J}=1}^{N_{\mathrm{v}}} \frac{\Gamma_{\mathrm{J}}}{z - z_{\mathrm{J}}} - \frac{1}{2}W_{\mathrm{b}}(z) - \frac{1}{2\pi \mathrm{i}}\fint_{C_{\mathrm{b}}} \frac{W_{\mathrm{b}}(\lambda)}{\lambda - z}\,\mathrm{d}\lambda, \quad z \in C_{\mathrm{b}}. \tag{4.27}$$

(Note that we get the same result if z approaches C_b from within $\mathcal{V}_b$, since the left-hand side of (4.13) is zero when $z \in \mathcal{V}_b$.) As usual, we can adapt this to the windtunnel reference frame by replacing W_b with $W_b^\dagger - W_\infty$. Analogous to the vectorial case, we can define a complex surface integral operator,

$$\mathcal{F}[g](z) := -\frac{1}{2}g(z) - \frac{1}{2\pi i}\fint_{C_b} \frac{g(\lambda)}{\lambda - z}\,d\lambda, \tag{4.28}$$

that acts upon the complex strength, $g = \gamma\tau^*$, of a vortex sheet on the surface of the body and evaluates the velocity at a surface point, z, approached from within the body. If we denote the velocity discrepancy—the negative of the right-hand side of (4.27)—by

$$\Delta W(z) := -\frac{1}{2\pi i}\sum_{J=1}^{N_v} \frac{\Gamma_J}{z - z_J} + \frac{1}{2}W_b(z) + \frac{1}{2\pi i}\fint_{C_b} \frac{W_b(\lambda)}{\lambda - z}\,d\lambda, \tag{4.29}$$

then the integral equation for g is compactly represented by

$$\mathcal{F}[g](z) = \Delta W(z), \tag{4.30}$$

and its solution at all points $z \in C_b$ (presuming, for now, that such a solution exists) is denoted by

$$g(z) = \mathcal{F}^{-1}[\Delta W](z). \tag{4.31}$$

Note that the right-hand side can also be written in terms of the surface velocity in the windtunnel reference frame, $W_b^\dagger$, as

$$\Delta W(z) := -W_\infty - \frac{1}{2\pi i}\sum_{J=1}^{N_v} \frac{\Gamma_J}{z - z_J} + \frac{1}{2}W_b^\dagger(z) + \frac{1}{2\pi i}\fint_{C_b} \frac{W_b^\dagger(\lambda)}{\lambda - z}\,d\lambda. \tag{4.32}$$

4.3.3 Infinitely-Thin Plate

The forms of the integral equations for the vortex sheet strengths are somewhat different when the body is infinitely thin. Starting from the vector form of Eq. (4.14), we can apply the same limits (A.46) to approach the plate at $\boldsymbol{x} \in \mathcal{P}$ from either side:

$$\boldsymbol{v}^{\pm}(\boldsymbol{x}) = \int_{\mathcal{V}_f} \boldsymbol{K}(\boldsymbol{x} - \boldsymbol{y}) \times \boldsymbol{\omega}(\boldsymbol{y})\,dV(\boldsymbol{y}) \mp \frac{1}{2}\boldsymbol{n}(\boldsymbol{x}) \times \boldsymbol{\gamma}(\boldsymbol{x}) + \fint_{\mathcal{P}} \boldsymbol{K}(\boldsymbol{x} - \boldsymbol{y}) \times \boldsymbol{\gamma}(\boldsymbol{y})\,dS(\boldsymbol{y}). \tag{4.33}$$

Incidentally, this is the limiting form of the velocity at any vortex sheet $\mathcal{S}$ of strength γ; for example, it provides the relevant form for obtaining the transport of a free vortex sheet, as discussed in Result 3.6. To obtain the limiting forms of the velocity observed in the windtunnel reference frame, we simply add $\boldsymbol{V}_\infty$ to the right-hand side.

Recall from Note 3.6.4 that the fluid velocities on both sides of a vortex sheet or plate satisfy the no-penetration condition (3.11); we apply that condition here in the direct form to the average of the velocities $\boldsymbol{v}^+$ and $\boldsymbol{v}^-$ on either side of the plate. The result is the following integral equation for $\boldsymbol{\gamma}$ at all $\boldsymbol{x} \in \mathcal{P}$:

$$\boldsymbol{n}(\boldsymbol{x}) \cdot \fint_{\mathcal{P}} \boldsymbol{K}(\boldsymbol{x}-\boldsymbol{y}) \times \boldsymbol{\gamma}(\boldsymbol{y})\,\mathrm{d}S(\boldsymbol{y}) = \boldsymbol{n}(\boldsymbol{x}) \cdot \left(\boldsymbol{V}_\mathrm{b}(\boldsymbol{x}) - \int_{\mathcal{V}_f} \boldsymbol{K}(\boldsymbol{x}-\boldsymbol{y}) \times \boldsymbol{\omega}(\boldsymbol{y})\,\mathrm{d}V(\boldsymbol{y}) \right). \tag{4.34}$$

If desired, $\boldsymbol{V}_\mathrm{b}$ can be replaced with $\boldsymbol{V}_\mathrm{b}^\dagger - \boldsymbol{V}_\infty$ to adapt this equation to the windtunnel reference frame. The solution for $\boldsymbol{\gamma}$ is the same in either frame, however.

In the complex form of the two-dimensional problem, the Plemelj formulas (A.125) still hold when the finite-length contour of the plate is approached from either side. These limits, when applied to (4.15), lead to

$$w^\pm(z) = \frac{1}{2\pi\mathrm{i}} \sum_{\mathrm{J}=1}^{N_\mathrm{v}} \frac{\Gamma_\mathrm{J}}{z - z_\mathrm{J}} \pm \frac{1}{2} g(z) + \frac{1}{2\pi\mathrm{i}} \fint_{\mathcal{P}} \frac{g(\lambda)}{\lambda - z}\,\mathrm{d}\lambda; \tag{4.35}$$

the right-hand side would have an additional term W_∞ in the windtunnel reference frame. Then, since the direct form of the no-penetration condition (3.197) holds on either side of the plate, the integral equation for g can be formed by taking the sum of these two conditions (3.206), which results in

$$\mathrm{Im}\left[\frac{\tau}{2\pi\mathrm{i}} \fint_{\mathcal{P}} \frac{g(\lambda)}{\lambda - z}\,\mathrm{d}\lambda \right] = \mathrm{Im}\left[\tau \left(W_\mathrm{b}(z) - \frac{1}{2\pi\mathrm{i}} \sum_{\mathrm{J}=1}^{N_\mathrm{v}} \frac{\Gamma_\mathrm{J}}{z - z_\mathrm{J}} \right) \right], \tag{4.36}$$

where $\tau := \mathrm{d}z/\mathrm{d}s$ corresponds to the unit tangent along the plate contour (which, the reader will recall, defines the direction along the plate with respect to which the positive side is to the left). The plate velocity can be expressed in the windtunnel reference frame, if desired, by making the substitution $W_\mathrm{b} = W_\mathrm{b}^\dagger - W_\infty$. Again, it is to be remembered that the complex sheet strength g in this equation represents the jump in fluid velocity across the plate, rather than the jump between the body velocity and the fluid velocity as in (4.27). Hence, there is a factor-of-two difference between some terms in (4.36) and (4.27).

4.3.4 The Double Layer Formulation

In Sect. 4.2.4, we developed an alternative formulation for the velocity field in terms of a double-layer potential—a dipole distribution—on the body surface. Let us discuss the governing equation for this dipole distribution's strength, μ. As for the vortex sheet formulation, this governing equation is obtained by enforcing the no-penetration condition on the surface. In this case, it comes from evaluating the normal component of the velocity field in (4.19) at points $\boldsymbol{x}^+$ on the surface, approached from the positive side (the side adjacent to $\mathcal{V}_f$):

$$\lim_{\boldsymbol{x}\to\boldsymbol{x}^+} \boldsymbol{n}(\boldsymbol{x}^+)\cdot\nabla\int_{\mathcal{S}_b} \mu(\boldsymbol{y})\boldsymbol{n}_y\cdot\nabla G(\boldsymbol{x}-\boldsymbol{y})\,\mathrm{d}S(\boldsymbol{y}) =$$

$$\boldsymbol{n}(\boldsymbol{x}^+)\cdot\left(\boldsymbol{V}_b(\boldsymbol{x}^+) - \int_{\mathcal{V}_f\cup\mathcal{V}_b} \boldsymbol{K}(\boldsymbol{x}^+-\boldsymbol{y})\times\boldsymbol{\omega}(\boldsymbol{y})\,\mathrm{d}V(\boldsymbol{y})\right). \quad (4.37)$$

The right-hand side of the equation expresses the discrepancy between the normal components of the body's local motion and the velocity induced by the extended vorticity field. On the left-hand side, we have left the limit of the integral unevaluated. The derivative of a double-layer potential requires particular care when evaluated at a point on the surface. It remains convergent, in spite of the singularity, but does not have a simple result analogous to the limits (A.46).

4.3.5 General Notes on These Integral Equations

There are several general observations we can make regarding the various integral equations derived here for the strength distribution of a bound vortex sheet or double-layer potential.

- In two-dimensional problems, it is often possible to solve the problem analytically, and we will discuss a powerful set of tools for this purpose in the next section. When an analytical solution is not possible, such as in three-dimensional problems or in some two-dimensional problems (e.g., with multiple bodies), we will need to rely on polynomial expansions or a discrete approximation of the integral equation, such as with a **panel method**.
- For a two-dimensional body, the solution of the integral equation is not unique. The consequences of this issue, as well as its resolution, will be discussed in Sect. 4.5.

4.4 Solution of Two-Dimensional Problem by Conformal Mapping

Though Eq. (4.27) is, strictly speaking, the equation that governs the complex vortex sheet strength for a moving two-dimensional body in the presence of fluid vorticity, it is advantageous to take a step back to remind ourselves of the underlying problem. We seek a complex-valued function $w(z)$ that is holomorphic everywhere in the region

$\mathcal{V}_f$—except at isolated singular points—such that the no-penetration condition, in its first form in (3.199), is satisfied on the contour C_b. We can also express this in terms of complex potential: we seek a complex-valued function $F(z)$, holomorphic almost everywhere in $\mathcal{V}_f$ except at singularities and along branch cuts, such that condition (3.201) is enforced on C_b. It turns out that, in most cases, we can find a solution to this problem without direct reference to the integral equation (4.27). This is because of the special properties of holomorphic functions and, in particular, the rules they must follow when they are extended from a contour into a region. These properties, along with the Riemann mapping theorem—that any closed contour can be conformally mapped to a circle of unit radius—provide all of the tools we need to solve the problem. Conformal mapping is discussed in more detail in Sect. A.2.3 in the Appendix. Armed with a few general classes of mappings, some of which are described in Sects. A.2.4 and A.2.5, we can essentially obtain any closed shape we desire.

Let us denote the conformal mapping from the unit circle C_c (in the ζ, or **circle plane**) to the contour C_b (in the z, or **physical plane**) by $z(\zeta)$, where both contours are counterclockwise. This transformation is defined in the region $\mathcal{V}^+$ external to the circle, mapping points in this region to the region $\mathcal{V}_f$ exterior to C_b. Note that such a conformal transformation must be holomorphic, so that its derivative exists everywhere in $\mathcal{V}^+$ and on C_c, and is invertible in $\mathcal{V}_f$ at all points, with an inverse denoted by $\zeta(z)$. As a result, a given holomorphic function in one plane can be readily regarded as a holomorphic function in the other by the composition with the mapping, e.g., $\hat{F}(\zeta) = F(z(\zeta))$ or $F(z) = \hat{F}(\zeta(z))$. This ensures that, in particular, the value of the streamfunction on the circle is identical to its value at corresponding points on C_b. So if we can enforce the no-penetration condition (3.201) in the circle plane, we can be assured that it is also satisfied in the physical plane. We also note that the complex velocity in the physical plane can be related to the gradient of $\hat{F}$ in the circle plane:

$$w(z) := \frac{\mathrm{d}F}{\mathrm{d}z} = \frac{\mathrm{d}\hat{F}}{\mathrm{d}\zeta}\frac{\mathrm{d}\zeta}{\mathrm{d}z} = \frac{\hat{w}(\zeta)}{z'(\zeta)}. \tag{4.38}$$

In the last form, we have used $z'(\zeta)$ to denote the **Jacobian** of the mapping, $\mathrm{d}z/\mathrm{d}\zeta$; the complex velocity in the circle plane, $\hat{w} := \mathrm{d}\hat{F}/\mathrm{d}\zeta$, has been defined for convenience. Note that the equality $\mathrm{d}\zeta/\mathrm{d}z = 1/z'(\zeta)$ is ensured by any conformal mapping. As we note in (A.148) in the discussion on conformal mapping, the unit tangent on the body (in the physical plane) can be written as $\tau = \mathrm{i}\zeta z'(\zeta)/|z'(\zeta)|$. Thus, it is straightforward to show that the direct form of the no-penetration condition (3.12) can also be written as

$$\mathrm{Re}\,(\zeta\hat{w}(\zeta)) = \mathrm{Re}\,[\zeta z'(\zeta)W_b(z(\zeta))], \quad \zeta \in C_c. \tag{4.39}$$

Furthermore, the complex strength of the bound vortex sheet on the body surface has the form that we discussed in Note 3.6.3. Noting that the complex conjugate of the unit tangent can be written as the inverse of the tangent itself, this sheet strength can be written in terms of the mapping as

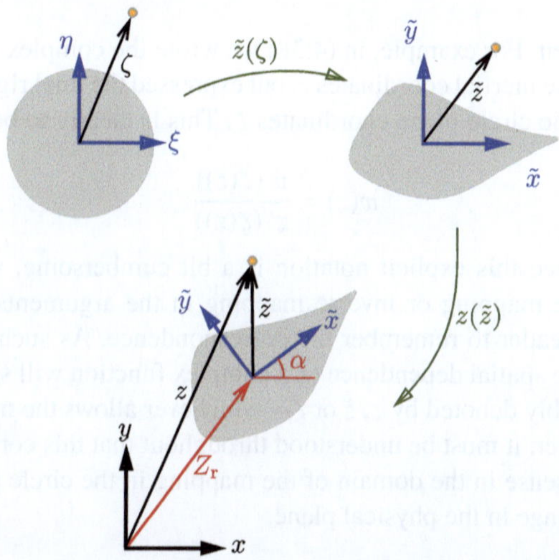

Fig. 4.1 Conformal transformation from the circle plane by a composition of two mappings: $\tilde{z}(\zeta)$ from the circle to the physical shape in the body-fixed frame of reference, and $z(\tilde{z})$ to rigidly map the shape into the inertial frame of reference. The correspondence of a given point between the three planes is indicated by the small dot

$$g(z) = \frac{1}{\mathrm{i}\zeta z'(\zeta)} |z'(\zeta)| \gamma(z), \tag{4.40}$$

where $\gamma(z)$ is the real-valued vortex sheet strength. Our notation deserves a note of clarification:

Note 4.4.1: Notational Convention for Functions and Mappings

Generally, the conformal mappings that we will consider in this book will actually be a composition of two mappings, illustrated in Fig. 4.1. The first mapping, $\tilde{z}(\zeta)$, transforms the unit circle C_c into the physical shape of the body C_b in the body-fixed frame of reference. The second mapping, $z(\tilde{z})$, then rigidly translates and rotates the body shape in the inertial reference frame. The overall mapping is simply $z(\zeta) = z(\tilde{z}(\zeta))$. The rigid transformation is defined generically by Eq. (A.152). The shape transformation, on the other hand, can be obtained from one of several families of conformal mappings; in this book, we will frequently rely on the power series mapping (A.153) or the Schwarz–Christoffel mapping (A.188).

Through the series of mappings, a point ζ in the circle plane corresponds uniquely to a point $\tilde{z}$ in the body frame and z in the physical plane. We will frequently use notation in which this correspondence is not explicitly written. For example, in (4.38) we wrote the complex velocity as a function of the inertial coordinates z, but expressed the final right-hand side in terms of the circle plane coordinates ζ. This is clearly to be understood as

$$w(z) = \frac{\hat{w}\left(\zeta(z)\right)}{z'\left(\zeta(z)\right)}. \tag{4.41}$$

However, since this explicit notation is a bit cumbersome, we generally will omit the mapping or inverse mapping in the arguments and simply rely on the reader to remember the correspondence. As such, throughout this book, the spatial dependence of a complex function will sometimes be interchangeably denoted by z, $\tilde{z}$ or ζ—whichever allows the most compact form. However, it must be understood throughout that this correspondence only makes sense in the domain of the mapping in the circle plane and its associated range in the physical plane.

From hereon in this section we will focus on the problem in the circle plane. We can rewrite (3.201) in this plane as

$$\hat{F}(\zeta) - \hat{F}(\zeta)^* = 2\mathrm{i}\psi_\mathrm{b}(z(\zeta), z(\zeta)^*), \quad \zeta \in C_c, \tag{4.42}$$

where the notation $z(\zeta)^*$ implies that the conjugate of the expression is obtained *after* it is evaluated at ζ, whereas the notation $z^*(\zeta)$, which we will encounter below, indicates that the function is conjugated *before* evaluating at the argument ζ. These distinctions are discussed in detail in Sect. A.2.1. In particular, $z(\zeta)^* \equiv z^*(\zeta^*)$.

Now, on the unit circle itself, $\zeta^* = 1/\zeta$. Thus, we can write (4.42) as

$$\hat{F}(\zeta) - \hat{F}^*(1/\zeta) = 2\mathrm{i}\psi_\mathrm{b}(z(\zeta), z^*(1/\zeta)), \quad \zeta \in C_c. \tag{4.43}$$

We can extend the definition of $\hat{F}(\zeta)$ into the region $\mathcal{V}^-$ interior to C_c by defining a function

$$\Phi(\zeta) = \begin{cases} \hat{F}(\zeta), & \zeta \in \mathcal{V}^+, \\ \hat{F}^*(1/\zeta), & \zeta \in \mathcal{V}^-. \end{cases} \tag{4.44}$$

It can be verified that, if $\hat{F}(\zeta)$ is holomorphic in $\mathcal{V}^+$, then by this definition, the function $\Phi(\zeta)$ is holomorphic in each respective region. Equation (4.44) is called the **Schwarz reflection principle**. Condition (4.43) is now equivalent to a jump condition on Φ,

$$\Phi^+(\zeta) - \Phi^-(\zeta) = B(\zeta), \quad \zeta \in C_c, \tag{4.45}$$

where

$$B(\zeta) := 2\mathrm{i}\psi_\mathrm{b}(z(\zeta), z^*(1/\zeta)) \tag{4.46}$$

is called the **surface function** and $\Phi^\pm(\zeta)$ represent the limits of $\Phi(\zeta)$ as the circle is approached from the respective sides. Thus, we seek a function $\Phi(\zeta)$ that is holomorphic in both the interior and exterior of the circle and satisfies the jump relation (4.45). This problem is a simple example of a more general class called Riemann–Hilbert problems [56]. We will consider two cases.

4.4.1 Moving Bodies in Irrotational Flow

Let us first consider the case in which the body is in rigid motion in an otherwise stagnant fluid. In this case, the streamfunction is non-zero and variable along the circle. This is an inhomogeneous Riemann–Hilbert problem, and the solution of the problem, following Eq. (A.131), is simply the Cauchy integral

$$\Phi(\zeta) = \frac{1}{2\pi\mathrm{i}} \int_{C_c} \frac{B(\lambda)}{\lambda - \zeta}\,\mathrm{d}\lambda. \tag{4.47}$$

Following Milne-Thomson [53], let us decompose the surface function $B(\zeta)$ into two parts, based on a power series expansion in ζ,

$$B(\zeta) = B_1(\zeta) + B_2(\zeta), \tag{4.48}$$

where B_1 contains all negative powers of ζ and B_2 all non-negative powers of ζ. Then, $B_1(\zeta)$ is holomorphic in all of $\mathcal{V}^+$ and $B_2(\zeta)$ is holomorphic in $\mathcal{V}^-$. By Cauchy's integral theorem, the solution in $\mathcal{V}^+$ can now be written down immediately:

$$\hat{F}(\zeta) = B_1(\zeta). \tag{4.49}$$

Note 4.4.2: Solution for Rigid-Body Motion by Conformal Mapping

A quite general class of conformal transformations is described by Eqs. (A.152) and (A.153) in Sect. A.2.3 of the Appendix, in which the power series coefficients c_{-k}, for $k \geq -1$, are chosen to generate the body shape. When we introduce this mapping into the surface function $B(\zeta)$ in (4.46), where the body streamfunction is given by (3.8), the result is

$$B(\zeta) = 2\mathrm{i}\psi_0 - W_\mathrm{r}^* \mathrm{e}^{-\mathrm{i}\alpha} \sum_{k=-1}^{\infty} c_{-k}^* \zeta^k + W_\mathrm{r} \mathrm{e}^{\mathrm{i}\alpha} \sum_{k=-1}^{\infty} \frac{c_{-k}}{\zeta^k} - \mathrm{i}\Omega \sum_{k=-1}^{\infty} \sum_{l=-1}^{\infty} \frac{c_{-k}^* c_{-l}}{\zeta^{l-k}}, \tag{4.50}$$

where we have included all terms in the mapping (A.153) inside the summation for the sake of compactness. Recall that $W_\mathrm{r}\mathrm{e}^{\mathrm{i}\alpha}$ is $\tilde{W}_\mathrm{r}$, the expression of the body's translational motion in the body-fixed coordinate system. Let us assume that $\psi_0 = 0$, as we are free to do. The portion of this surface function with negative powers of ζ can now be identified. As a result, the complex potential in $\mathcal{V}^+$ is

$$\hat{F}(\zeta) = -\tilde{W}_\mathrm{r}^* \frac{c_1^*}{\zeta} + \tilde{W}_\mathrm{r} \sum_{k=1}^{\infty} \frac{c_{-k}}{\zeta^k} - \mathrm{i}\Omega \sum_{k=1}^{\infty} \frac{d_{-k}}{\zeta^k}, \tag{4.51}$$

where the coefficients d_{-k} are defined as

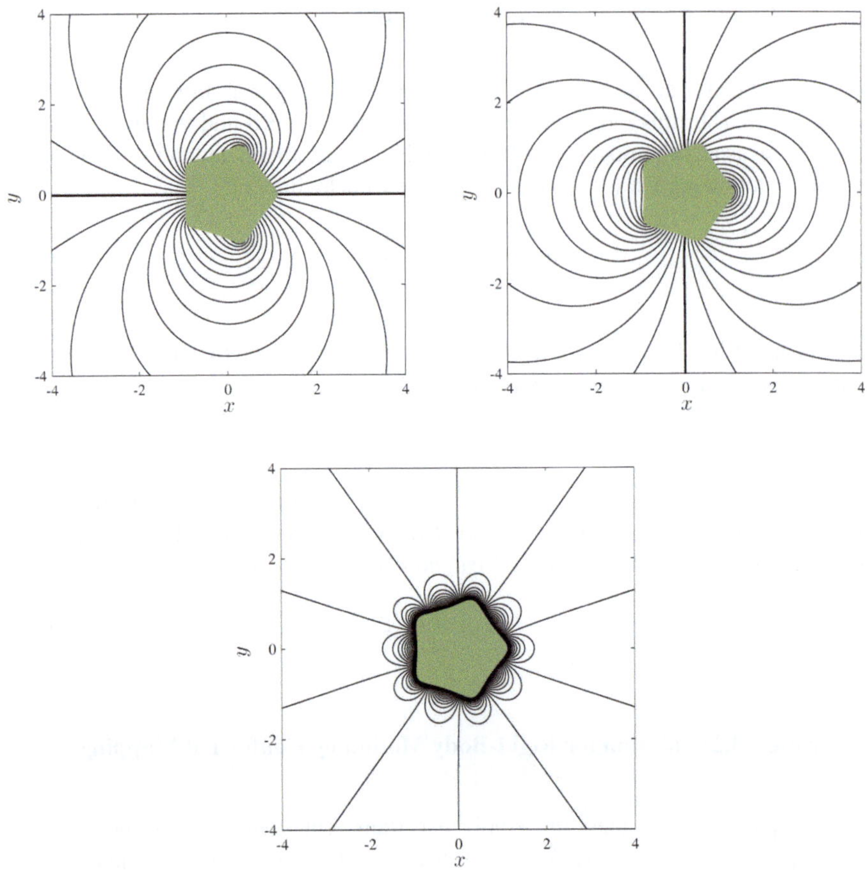

Fig. 4.2 Streamlines of the flow field generated by translational (top row) and rotational (bottom) motion of a smoothed pentagon

$$d_{-k} = \begin{cases} \sum_{l=-1}^{\infty} c_{-l}^{*} c_{-l-k}, & k \geq 0, \\ \sum_{l=-1}^{\infty} c_{-l} c_{-l+k}^{*} = d_{k}^{*}, & k \leq -1. \end{cases} \tag{A.157}$$

From this, we can evaluate the complex potential at corresponding points in the physical plane.

As an example, Fig. 4.2 depicts the streamlines (obtained from a contour plot of the imaginary part of F) of the flow fields generated by a smoothed pentagon shape, respectively, in unit translational velocity in the x and y directions and unit angular velocity about its reference point. The shape in this case was formed by setting $c_1 = 1$, $c_{-4} = 0.1$, and all other coefficients of the mapping, as well as the angle α, to zero.

4.4.2 Stationary Body in a Known Potential Flow: The Circle Theorem

Now, let us consider a flow field whose complex potential is known in the absence of the body. In this book, such a flow might be due to a point vortex or to a uniform flow at infinity. We therefore seek the potential flow to add to this so that the streamfunction on the body vanishes. In other words, we seek the holomorphic function such that the sum of this with the original potential has zero imaginary part on the circle. This is a homogeneous Riemann–Hilbert problem, and the solution is often called the **circle theorem** of Milne-Thomson [53]. Let $\hat{F}_u(\zeta)$ be the known complex potential in the absence of the body, whose only requirement is that it has no singularities within the contour of the body. Then, the function

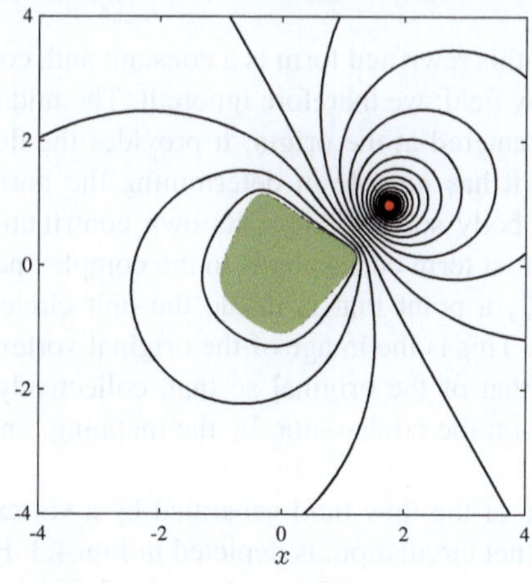

Fig. 4.3 Streamlines of the flow field generated by a vortex near a stationary planar body

$$\hat{F}(\zeta) = \hat{F}_u(\zeta) + \hat{F}_u^*(1/\zeta) \tag{4.52}$$

must be the solution to our problem. Why? First of all, it clearly satisfies condition (4.43) by symmetry. Furthermore, any singularities of $\hat{F}_u(\zeta)$—which are, by supposition, outside the circle—are singularities inside the circle for $\hat{F}_u^*(1/\zeta)$. Thus, this latter function is necessarily holomorphic outside the unit circle.

The circle theorem is a very useful tool for solving problems in two-dimensional potential flow. It represents an extension of the method of images to circular bodies, or, through conformal mapping, to any closed planar bodies. It is important to stress that this procedure is extendable by superposition to an arbitrary combination of potential flows, and particularly, to any number of point vortices.

Note 4.4.3: Point Vortex Outside a Stationary Body

Consider a point vortex of strength Γ_v at position z_v external to a stationary rigid body. Through a conformal mapping, the vortex and the body are identified with point ζ_v and the unit circle, respectively, in the circle plane. Using the circle theorem (4.52), the complex potential can be immediately constructed:

$$\hat{F}(\zeta) = \frac{\Gamma_v}{2\pi i} \log(\zeta - \zeta_v) - \frac{\Gamma_v}{2\pi i} \log(1/\zeta - \zeta_v^*). \qquad (4.53)$$

The second term, which is holomorphic in the region $\mathcal{V}^+$ external to the circle, can be rewritten slightly by noting that

$$\log(1/\zeta - \zeta_v^*) = \log(\zeta - 1/\zeta_v^*) - \log(\zeta) + \log(-\zeta_v^*).$$

The last term in this rewritten form is a constant and, consequently, has no effect on the flow field; we therefore ignore it. The middle term represents a point vortex centered at the origin. It provides the flow with an overall circulation, but it has no role in determining the normal component of velocity on the body surface, since its own contribution there is purely tangential. The first term corresponds to the complex potential of a vortex centered at $1/\zeta_v^*$, a point that is inside the unit circle and on the same radial line as ζ_v. This is the image of the original vortex, and its flow field combines with that of the original so that, collectively, they produce no normal velocity on the circle—nor, by the mapping, on the surface of the body itself.

An example of the flow field generated by a vortex near a stationary body, with zero net circulation, is depicted in Fig. 4.3. Here, we have used the same series mapping (A.153) as in Note 4.4.2. The body in this example is generated with coefficients $c_1 = 1$, $c_{-2} = 0.1$, and $c_{-3} = 0.1 \exp(i\pi/4)$. The vortex is located at $\zeta_v = 2\exp(i\pi/6)$ in the circle plane, or $z_v = 1.753 + i0.9695$ in the physical plane, and has unit strength (though the strength has no effect on the streamline pattern).

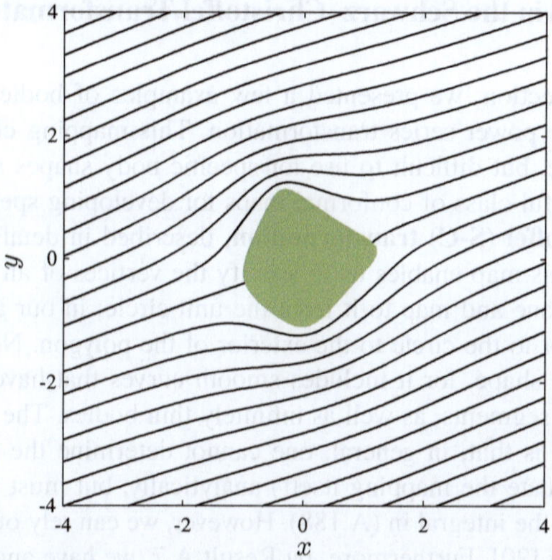

Fig. 4.4 Streamlines of the flow field generated by a uniform flow at 20° angle of incidence past a stationary planar body

Note 4.4.4: Uniform Flow Past a Stationary Body

Now consider a uniform flow with complex velocity W_∞. Once again, this flow can be mapped from the circle plane; we will again use the rigid-body/power-series mapping (A.152)–(A.153). At large distances from the unit circle, the mapping simplifies to $z \sim e^{i\alpha} c_1 \zeta$, so the complex potential must be $W_\infty e^{i\alpha} c_1 \zeta$ in order to produce the correct uniform flow in the physical plane. Recall from (3.87) that $W_\infty e^{i\alpha}$ is simply $\tilde{W}_\infty$, the uniform flow expressed in the body-fixed coordinate system. The overall complex potential in the circle plane follows from the circle theorem:

$$\hat{F}(\zeta) = \tilde{W}_\infty c_1 \zeta + \frac{\tilde{W}_\infty^* c_1^*}{\zeta}. \tag{4.54}$$

Figure 4.4 depicts an example of the streamlines in the physical plane resulting from a uniform flow of unit speed at 20° angle of incidence past a body generated with the same mapping coefficients as in Note 4.4.3.

4.4.3 Solution via the Schwarz–Christoffel Transformation

In the previous section, we presented a few examples of bodies mapped from a unit circle via the power series transformation. This mapping class is completely general, of course, but difficult to use for specific body shapes aside from simple ones. A more useful class of conformal maps for developing specific shapes is the **Schwarz–Christoffel (S-C) transformation**, described in detail in Sect. A.2.5 in the Appendix. This map enables us to specify the vertices of an arbitrary polygon in the physical plane and map to it from the unit circle; in our approach, we map the region exterior to the circle to the exterior of the polygon. Note that a polygon is a quite general shape, for it includes smooth curves that have been discretized with straight line segments, as well as infinitely thin bodies. The only drawback of the S-C mapping is that, in general, one cannot determine the parameters of the mapping (or evaluate the mapping itself) analytically, but must rely on numerical approximation of the integral in (A.188). However, we can rely on existing libraries for this, including [20]. Furthermore, by Result A.7, we have analytical expression for the coefficients of the power-series representation of the mapping. As we discuss in the Appendix, there is an infinite number of non-zero coefficients in this series for most bodies, but the higher coefficients are progressively negligible.

Armed with the properties of the S-C map in Sect. A.2.5 and the results in Notes 4.4.2–4.4.4 for transformations from the unit circle, we can construct any flow about a polygonal body. Examples of such flows are depicted in Figs. 4.5 and 4.6.

The use of the S-C transformation leads to an important, and quite general, question in inviscid fluid dynamics: ***How does the flow behave near a corner?***. The answer to this question, presented below, will figure prominently in our approach to solving flow problems, for as we will discuss in Sect. 5.2, it is crucially important for devising a means of removing or modifying this behavior.

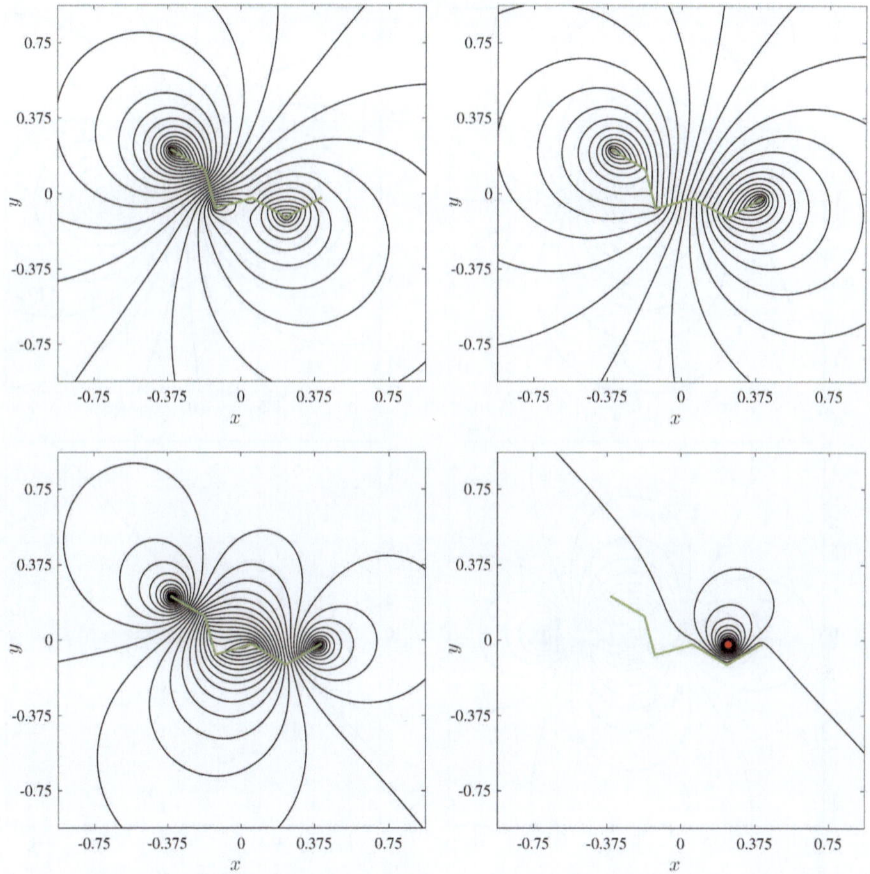

Fig. 4.5 Basis flow fields for a multi-segmented infinitely thin body, obtained by Schwarz–Christoffel transformation. Upper left: unit translational velocity in the x direction. Upper right: unit translational velocity in the y direction. Lower left: unit angular velocity about the origin. Lower right: a point vortex near the body

Note 4.4.5: Velocity Field Near a Corner, Revisited

To ensure a smooth transformation, the Jacobian $z'(\zeta)$ must not vanish anywhere in $\mathcal{V}^+$. However, as discussed in Sect. A.2.3, the Jacobian *can* vanish at discrete points on the unit circle, forming convex corners on the mapped body geometry. Indeed, this aspect is the foundation of the Schwarz–Christoffel transformation (A.188). In the vicinity of a corner of internal angle $(2-\nu)\pi$ at z_0, where $1 < \nu \leq 2$, we found in Eq. (3.94) in our discussion of basic potential flows that the velocity in this corner's vicinity

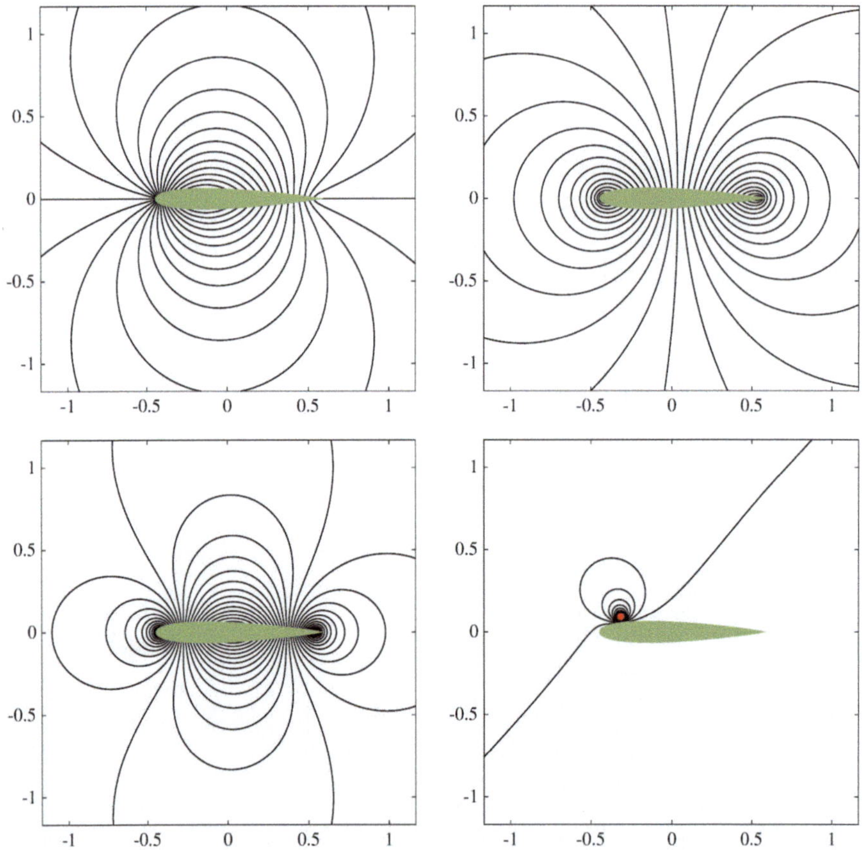

Fig. 4.6 Basis flow fields for a NACA 0012 airfoil, approximately described by a polygon with 38 straight segments. Upper left: unit translational velocity in the x direction. Upper right: unit translational velocity in the y direction. Lower left: unit angular velocity about the origin. Lower right: a point vortex near the body

has the form

$$w(z) \approx \frac{SL^{1-1/\nu}}{(z - z_0)^{1-1/\nu}}, \tag{4.55}$$

where S is a complex constant, and L is a characteristic length scale of the geometry to which the corner belongs. What does the coefficient S represent? In our discussion of the corner flow, we found that S has units of velocity and can be written in the form,

$$S = \mathrm{i}\sigma n_0^{-1/\nu}, \tag{3.106}$$

in which the corner's orientation is described by the unit normal, n_0, bisecting the adjacent flow region, and the corner flow's strength and directionality are given by its signed intensity, σ, defined in (3.103).

Within the context of the conformal mapping from the unit circle, by combining the transformed velocity (4.38) with the behavior (A.162) near a corner, we immediately find that

$$S = \frac{\hat{w}(\zeta_0)}{L^{1-1/\nu}} \lim_{\zeta \to \zeta_0} \frac{\zeta - \zeta_0}{\nu \left(z(\zeta) - z(\zeta_0)\right)^{1/\nu}}. \tag{4.56}$$

In other words, S is proportional to the (well-behaved) velocity, $\hat{w}(\zeta_0)$, at the corner's pre-image on the unit circle. The purely geometric constant of proportionality, given by the limit above, is finite and non-zero; it describes the reorientation and magnification of the flow by the mapping.

In fact, we can carry this relationship between S and $\hat{w}(\zeta_0)$ a bit further. From the no-penetration condition, expressed in terms of the circle-plane velocity (4.39), we note that at any corner, we must have that $\mathrm{Re}\,(\zeta_0 \hat{w}(\zeta_0)) = 0$, since the right-hand side vanishes at such points. In other words, ***the normal component of the velocity in the circle plane is exactly zero at any pre-image of a corner***; this is necessary in order to reconcile the no-penetration condition on either side of this pre-image and still ensure a smooth velocity field at this point. The velocity must be tangent to the unit circle C_c at ζ_0, and thus, of the form $\hat{w}(\zeta_0) = (\mathrm{i}\zeta_0 \hat{u}_\tau(\zeta_0))^*$, where $\hat{u}_\tau$ is real-valued and denotes the tangent component of $\hat{w}$. We thus reach the important conclusion that the constant S is proportional to $\hat{u}_\tau(\zeta_0) \equiv \mathrm{i}\zeta_0 \hat{w}(\zeta_0)$, the tangent component of the fluid velocity in the circle plane at the corner's pre-image. If we also make use of the relationship (A.166) between the corner normal n_0 and the mapping, then it is straightforward to show that ***the signed intensity of the corner, σ, is proportional to the tangential velocity in the circle plane***:

$$\sigma = -\frac{\hat{u}_\tau(\zeta_0)}{L^{1-1/\nu}} \lim_{\zeta \to \zeta_0} \frac{|\zeta - \zeta_0|}{\nu \left|z(\zeta) - z(\zeta_0)\right|^{1/\nu}}. \tag{4.57}$$

This relationship is purely real and only pertains to the strength of the flow about the corner; all aspects of orientation have been stripped away. As in (4.56), the proportionality factor is purely geometric, describing only how the flow is magnified by the mapping. The minus sign simply accounts for the difference in sign convention between σ and $\hat{u}_\tau$. Using Eq. (A.161), the geometric factor can be rewritten slightly in a form more conducive to calculation:

$$\sigma = -\frac{\hat{u}_\tau(\zeta_0)}{(\nu L)^{1-1/\nu}} \lim_{\zeta \to \zeta_0} \frac{|\zeta - \zeta_0|^{1-1/\nu}}{|z'(\zeta)|^{1/\nu}}. \tag{4.58}$$

Result 4.4: Signed Intensity of a Corner in Schwarz–Christoffel Mapping

Suppose we have generated a polygonal shape with the Schwarz–Christoffel transformation (A.188), consisting of n vertices z_j, with associated turning angle factors β_j and pre-images ζ_j in the circle plane. The signed intensity of corner k is then given by

$$\sigma_k = -\hat{u}_\tau(\zeta_0) \left[(1+\beta_k)^{\beta_k} L^{\beta_k} |C| \prod_{\substack{j=1 \\ j\neq k}}^{n} |1 - \zeta_j/\zeta_k|^{\beta_j} \right]^{-1/(1+\beta_k)}, \tag{4.59}$$

where L is the characteristic length scale and $\hat{u}_\tau(\zeta_0) \equiv \mathrm{i}\zeta_0 \hat{w}(\zeta_0)$ is the tangent component of velocity in the circle plane. The proof of this follows immediately by substituting the Jacobian of the mapping into Eq. (4.58).

As a special case, consider the infinitely-thin flat plate of length c (which is the characteristic length, L). The mapping has $n = 2$ vertices, $\beta_1 = \beta_2 = 1$, pre-images at $\zeta_1 = 1$ and $\zeta_2 = -1$, and coefficient $C = c/4$. Then, using the result above,

$$\sigma_1 = -\hat{u}_\tau(1)/c, \quad \sigma_2 = -\hat{u}_\tau(-1)/c. \tag{4.60}$$

Note 4.4.6: Corner Intensity and the Bound Vortex Sheet Strength

In Eq. (3.105) we derived a relationship between the signed intensity of a corner flow, σ, and the difference in tangential velocity along the walls that form the corner. In particular, this relationship holds for any constituent corner of a body, as long as we interpret the radial velocities that appear in the reference frame that moves with the body. These radial velocities are clearly the components locally tangent to the body surface, and thus, we can alternatively write the relationship (3.105) in terms of the local bound vortex strength:

$$\sigma = -\frac{1}{2L^{1-1/\nu}} \lim_{s\to s_0} |s - s_0|^{1-1/\nu} (\gamma_1(s) + \gamma_2(s)), \tag{4.61}$$

where $\gamma_1(s)$ denotes the strength of the bound sheet just above the corner, $\gamma_2(s)$ the strength of the bound sheet just below it, and $s = s_0$ the edge's position in the arc length coordinate along the surface; Fig. 5.3 provides a useful illustration of the sides 1 and 2 of the corner.

In the case of an infinitely-thin plate of length c, each end point of the plate is a cusp (i.e., an infinitely sharp edge, with zero internal angle), for which $\nu = 2$. Then, the sum in bound vortex sheet strengths on either side of the edge is, by Note 3.6.4, the overall bound vortex sheet strength, γ, and the characteristic length scale L is c. Thus, Eq. (4.61) is easily adapted to the following:

$$\sigma = -\frac{1}{2c^{1/2}} \lim_{s \to s_0} |s - s_0|^{1/2} \gamma(s). \tag{4.62}$$

The consequences of this relationship are important: If the fluid velocity has the inverse square-root singular behavior (4.55), then the bound vortex sheet strength γ near this edge must also have the same singular behavior. Thus, the parameter σ represents the (well-behaved) factor that multiplies the sheet's singularity at that edge.

4.5 The Non-uniqueness of Two-Dimensional Potential Flow

In the previous section, the potential flow about a two-dimensional body was developed by conformally mapping a corresponding flow about a unit circle. This approach provides a natural opportunity to make an important general point about such flows. Suppose, to any flow in the circle plane, we add a point vortex centered at the origin,

$$\hat{F}_h(\zeta) = \frac{\Gamma_0}{2\pi i} \log \zeta, \tag{4.63}$$

where, as usual for a vortex, the strength Γ_0 is real valued. The flow produced by this vortex in the circle plane is clearly holomorphic in the region outside the circle. Furthermore, the velocity associated with it is purely circumferential, so it trivially satisfies the no-penetration condition on the circle C_c, regardless of the value of Γ_0. (Recall that such a flow was part of the solution when we applied the circle theorem to a point vortex outside a body in Note 4.4.3.) Since the mapping is conformal, then the corresponding flow in the physical plane also satisfies the no-penetration condition. Thus, we can add the flow mapped from this point vortex to any other flow and still satisfy the governing equation and the boundary conditions, regardless of the value of Γ_0. In other words, ***the potential flow about a two-dimensional body is not unique***.

Null Space of the Operator The origin of this non-uniqueness is that the two-dimensional version of the integral operator, $\mathcal{F}$, defined in Eq. (4.22), and its complex equivalent $\mathcal{F}$ in (4.28), have a null space. In other words, in two-dimensional problems, there is a non-trivial homogeneous part of the solution of the integral equations for vortex sheet strength developed in the previous sections. This homogeneous solution, or any multiple, can be added to a particular solution without affecting the result. The flow described by Eq. (4.63) constitutes the basis for this null space; the corresponding complex strength of the bound vortex sheet on the body is given by

$$g_h(z(\zeta)) = \frac{\Gamma_0}{2\pi \mathrm{i} z'(\zeta)\zeta}, \tag{4.64}$$

where $\zeta \in C_c$ and $z(\zeta)$ is holomorphic in the region exterior to the unit circle. Let us show that (4.64) is in the null space of the operator $\mathcal{F}$:

Proof The proof of this follows from the Cauchy integral formula and the application of the Plemelj formulae, which leads to Eq. (A.131). If we consider a complex velocity that is holomorphic and denoted by w in the region $\mathcal{V}_f$ (the – region in the discussion of Eq. (A.131)) and identically zero in $\mathcal{V}_b$ (the + region), then Eq. (A.131) leads to

$$\frac{1}{2}w(z) = -\frac{1}{2\pi \mathrm{i}} \int_{C_b} \frac{w(\lambda)}{\lambda - z}\,\mathrm{d}\lambda, \qquad z \in C_b. \tag{4.65}$$

Thus, any complex velocity defined in this manner—holomorphic in the exterior fluid region and zero inside the body—satisfies $\mathcal{F}[w](z) = 0$ for $z \in C_b$. However, this does not necessarily mean that the velocity satisfies the no-penetration condition on C_b. Let us transform the body contour to the unit circle, with the usual conformal transformation $z(\zeta)$ of the exterior of the circle to the region $\mathcal{V}_f$. A velocity field w satisfies the no-penetration condition on C_b if and only if $\hat{w}(\zeta) = w(z(\zeta))z'(\zeta)$ satisfies the condition on the unit circle (except at corners of C_b, where $z'(\zeta) = 0$). The velocity field must also vanish at infinity, which requires that $\hat{w}(\zeta)$ be expressible only in negative powers of ζ. The only velocity field that satisfies both of these conditions must have the form $C/(\mathrm{i}\zeta)$, where C is a real-valued constant. Thus, the corresponding velocity field in the z plane that satisfies the no-penetration condition on C_b and vanishes at infinity, and which also satisfies (4.65), is equal to

$$w_h(z) = \frac{C}{\mathrm{i}z'(\zeta(z))\zeta(z)}, \quad z \in \mathcal{V}_f, \tag{4.66}$$

and identically zero for all $z \in C_b$. After rewriting the real-valued constant C as $\Gamma_0/(2\pi)$, the corresponding complex vortex sheet strength on C_b must thus be given by (4.64). □

Making the Solution Unique Thus, in two-dimensional problems, we need an additional constraint in order to ensure a unique solution. We will address this only in part here, but return to it in Chap. 5. A constraint we use throughout this book is that ***the total circulation about the body and the fluid vorticity is zero***. Using Eq. (3.38) and applying the no-penetration condition, this can be expressed as

$$\Gamma_{\text{tot}} = \int_{\mathcal{V}_f} \omega\,\mathrm{d}V + \int_{S_b} \gamma\,\mathrm{d}S + \int_{S_b} (\boldsymbol{n} \times \boldsymbol{V}_b) \cdot \boldsymbol{e}_3\,\mathrm{d}S = 0, \tag{4.67}$$

or, in complex form from (3.39) and the surface vortex sheet strength (3.198),

$$\Gamma_{\text{tot}} = \sum_{J=1}^{N_v} \Gamma_J + \int_{C_b} g(z)\,\mathrm{d}z + \int_{C_b} W_b(z)\,\mathrm{d}z = 0. \tag{4.68}$$

(This complex form of the equation has also made use of the condition (2.43) that the body's volume is preserved.)

One can verify, in particular, that the circulation of the vortex sheet with strength given by (4.64) is indeed Γ_0, using (3.37) and the residue theorem. However, there is a more intuitive way to show this. Let us first develop the real-valued strength of the null-space vortex sheet from the complex form (4.64). We note that the unit tangent vector along the body contour C_b is given by

$$\tau = \frac{\mathrm{d}z}{\mathrm{d}s} = \frac{\mathrm{i}\zeta z'(\zeta)}{|z'(\zeta)|}, \quad \zeta \in C_c, \tag{4.69}$$

where s is the arc length parameter along C_b. This result is shown in Sect. A.2.3 in the Appendix. Then, combining this tangent vector with relation (3.199) between the complex and the conventional vortex sheet strengths—g and γ, respectively—we can see that the null-space vortex sheet strength is described by

$$\gamma_h(z) = \frac{\Gamma_0}{2\pi|z'(\zeta(z))|}. \tag{4.70}$$

But the factor in the denominator, $|z'(\zeta(z))|$, represents the differential change of arc length along the contour corresponding to a differential change in angle along the unit circle, as the discussion before Eq. (A.147) explains. So to compute the total circulation of the sheet, we can reduce the integral to

$$\int_{C_b} g_h(z)\,\mathrm{d}z = \int_{C_b} \gamma_h(z)\,\mathrm{d}s = \int_0^{2\pi} \frac{\Gamma_0}{2\pi}\,\mathrm{d}\theta = \Gamma_0. \tag{4.71}$$

This principle of zero total circulation makes sense on physical grounds. Any flow of interest to us in this work is generated by the motion of a body or by the flow past a stationary body. In either case, we can always imagine that motion to have been initiated from rest in a quiescent fluid. This circulation is obviously zero before the motion starts, and Kelvin's circulation theorem (3.40) guarantees that it remains zero for all subsequent time, as discussed in Sect. 3.1.5. Thus, by requiring that Γ_{tot} is zero, we insist that any circulation that enters the fluid must be balanced by equal and opposite bound circulation about the body.

In the classical approach of steady aerodynamics, this partnership between the fluid vorticity and the bound circulation is reconciled by imagining a 'starting vortex' that was generated at the airfoil (via a process to be discussed in detail in Sect. 5.1) an infinitely long time ago and has since convected infinitely far into the wake, thus having no effect on the flow field around the airfoil. The effect of the bound circulation it left behind is thus contained in the null-space solution (4.63).

In this work, we take a slightly different view of this steady state, since we must consider it within the context of a wide variety of unsteady scenarios. Here, we will imagine that any steady state is achieved only asymptotically rather than completely, so that the starting vortex generated at the initiation of motion has traveled a very large—but still finite—distance from the body. The benefit of this approach is that all fluid vorticity is contained within a region of finite extent, which will clarify our discussions about many topics, including multipole expansions at large distances (in Sect. 4.7) and momentum conservation (in Chap. 6).

The constraint of zero total circulation is not, in itself, sufficient to determine a unique solution. After all, we have not yet given any guidance on how the strength of the fluid vorticity is determined. All we know now is that, for any distribution of vorticity in the fluid, we can ensure zero total circulation without disrupting the no-penetration condition. Below, in Chap. 5, we will discuss the **Kutta condition** and generalizations of it; these will give us the criterion required to uniquely determine the strength of any new vortex introduced into the fluid, and thereby ensure a unique solution to the two-dimensional inviscid flow about a body.

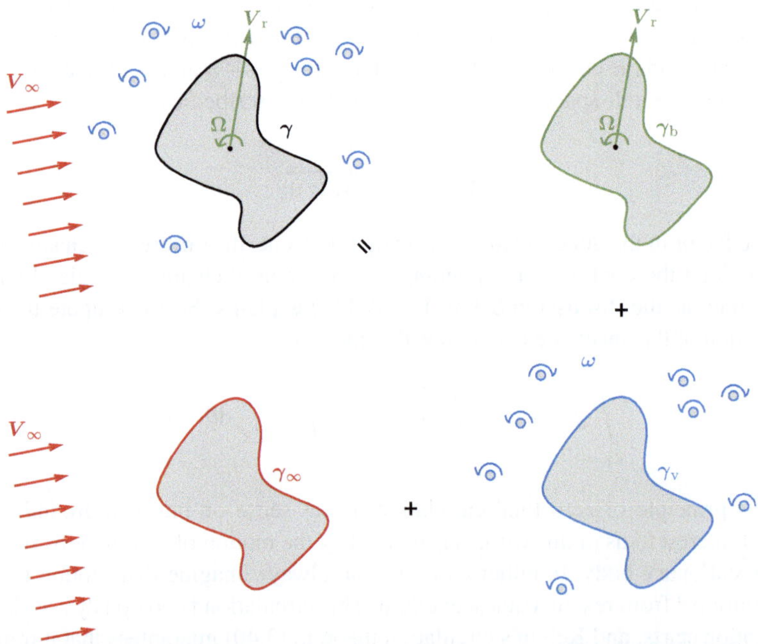

Fig. 4.7 Illustration of the decomposition of the overall flow into the basis flows from each flow contributor. The full flow field about the body (upper left) is the sum of the flows associated with uniform flow about the stationary body (lower left), the moving body in otherwise quiescent fluid (upper right), and fluid vorticity surrounding the stationary body (lower right)

4.6 Decomposition of the Flow into Basis Fields

In this section, we discuss an important aspect of the influence of a body on the fluid motion. Earlier in this chapter we introduced the concept of **flow contributors**: the body motion, the fluid vorticity, and, in a windtunnel reference frame, the background uniform flow. In each of the various equations developed for the vortex sheet in Sect. 4.3—Eqs. (4.23), (4.30) or (4.34)—the right-hand side depends linearly on these various flow contributors. Furthermore, the vortex sheet strength must, in turn, depend linearly on these contributors, since the integral operator $\mathcal{F}$ is a linear operator acting on this strength. This linear dependence was also clear in the conformal mapping solutions obtained individually for these contributors in Notes 4.4.2–4.4.4. Underlying all of these observations is the basic mathematical problem of potential flow—Laplace's equation for the vector or scalar potential—which the flow contributors enter linearly, either through the boundary conditions or the singular forcing function.

This linear dependence on the flow contributors will have substantial benefit for much of our analysis in the remainder of this book, because it means that we can linearly decompose many aspects of the flow field into the individual influences of these contributors. We will refer to these individual influences as **basis fields**, since they are loosely analogous to the basis vectors into which we decompose a vector. We will start by decomposing the unknown bound vortex sheet strength γ (or its complex version g) into the **basis vortex sheets** reacting individually to these flow contributors and solve for these basis sheets separately. Here, we use the term 'reacting' to convey that, for each flow contributor, the associated basis vortex sheet is an essential consequence of the no-penetration condition on the body surface. We should think of each of these basis sheets as an inseparable companion to the flow contributor with which it is associated. For a rigid body, many of these basis sheets, when normalized, are time invariant when expressed in the body coordinate system, making them intrinsic properties of the body.

The decomposition of this bound vortex sheet on the body surface will be used, in turn, to decompose other flow field quantities, as illustrated by Fig. 4.7. In general, for any linearly decomposable quantity f, we will use the notation

$$f = f_{\mathrm{v}} + f_{\mathrm{b}} + f_{\infty} \tag{4.72}$$

to denote, respectively, the contributions from fluid vorticity, body motion, and uniform flow. In particular, we can decompose the overall velocity field into the individual parts of flow contributors. This decomposition is clearly possible, since by (4.9) or its alternative forms, the velocity field in the fluid has a portion that depends linearly on γ. Thus, for each flow contributor, we will define a **basis velocity field**, defined as the part of the overall velocity field due solely to that contributor. Each of these basis velocity fields has associated basis potential fields (scalar potential, vector potential or streamfunction, and complex potential in two dimensions); all of these quantities' decompositions follows the same template (4.72).

In the case of uniform flow or fluid vorticity, the respective basis velocity fields are defined as the sum of the velocity field of the underlying flow contributor (in the absence of a body) and that induced by its basis vortex sheet. In other words, it is the contributor's velocity field, ***modified so that it also satisfies the no-penetration condition on the stationary body surface***. In contrast, the basis velocity field due to body motion is the velocity field induced in the otherwise ***quiescent fluid*** by this motion (i.e., in the absence of uniform flow or fluid vorticity). The overall velocity field is simply the linear superposition of each of these basis velocity fields. It should be noted that we include uniform flow in this list of flow contributors, as would be appropriate for an analysis in the windtunnel reference frame. The analysis can be easily adapted to the inertial frame by omitting the uniform flow.

For a rigid body, we can decompose the flow quantities even further, since the quantities depend linearly on ***each component*** of the translational $\boldsymbol{V}_{\mathrm{r}}$ and angular $\boldsymbol{\Omega}$ velocities of the body. For example, in three dimensions, there are six such basis fields, and this deeper decomposition of f_{b} is written for generic quantity f as follows:

$$f_{\mathrm{b}} = \sum_{k=1}^{3} \tilde{V}_{\mathrm{r},k} f_{\mathrm{bt}}^{(k)} + \sum_{k=1}^{3} \tilde{\Omega}_k f_{\mathrm{br}}^{(k)}. \tag{4.73}$$

It is important to note that we have used the body-fixed components of the motion vectors. This will have crucial benefits for ensuring the time-invariance of some of these basis fields. Thus, for example, $f_{\mathrm{bt}}^{(2)}$ denotes the basis f field induced by body translation—hence, the subscript bt—at unit velocity along its $\tilde{x}_2$ axis; $f_{\mathrm{br}}^{(1)}$ represents the basis field due to body rotation (subscript br) about the $\tilde{x}_1$ axis at unit angular velocity. The sums in (4.73) can be represented more succinctly if we use the matrix notation introduced in Chap. 2 to define $\tilde{\mathbf{f}}_{\mathrm{bt}}$ and $\tilde{\mathbf{f}}_{\mathrm{br}}$, whose components are $f_{\mathrm{bt}}^{(k)}$ and $f_{\mathrm{br}}^{(k)}$, respectively. Then

$$f_{\mathrm{b}} = \tilde{\mathbf{V}}_{\mathrm{r}}^{\mathsf{T}} \tilde{\mathbf{f}}_{\mathrm{bt}} + \tilde{\boldsymbol{\Omega}}^{\mathsf{T}} \tilde{\mathbf{f}}_{\mathrm{br}}. \tag{4.74}$$

In two dimensions, the six basis fields are reduced to three: two translational and one rotational; these are illustrated in Fig. 4.8.

Obviously, we can perform a similar component-wise decomposition of the uniform flow-induced motion:

$$f_{\infty} = \sum_{k=1}^{3} \tilde{V}_{\infty,k} f_{\infty}^{(k)} \equiv \tilde{\mathbf{V}}_{\infty}^{\mathsf{T}} \tilde{\mathbf{f}}_{\infty}. \tag{4.75}$$

Again, the body-fixed components of the uniform flow are used, to ensure invariance of the basis fields.

As we discussed in Sect. 3.3, it is generally possible to represent the fluid vorticity in an inviscid flow by a set of vortex filaments or point vortices. Thus, the vorticity-induced portion of a flow quantity can be decomposed into contributions from each such element. Let us denote the strength of the Jth element of a set of N_{v} filaments or point vortices by Γ_{J}. Then, we can write f_{v} as

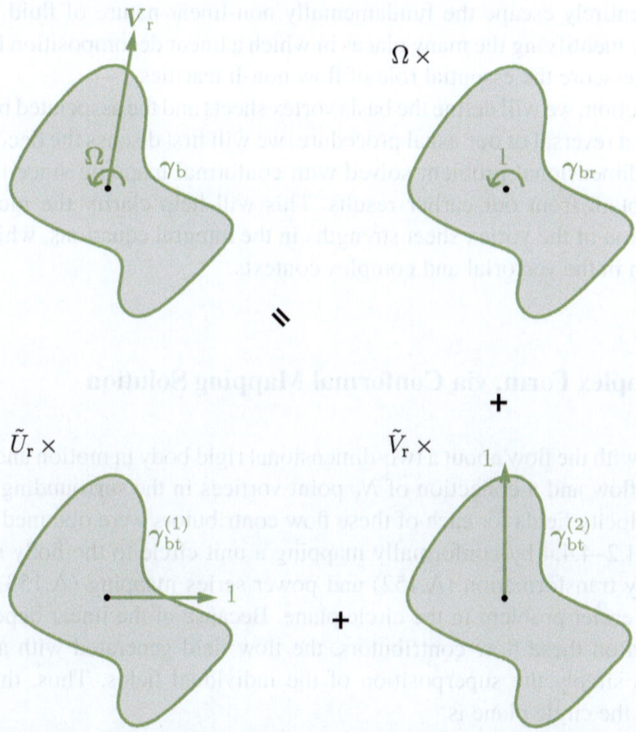

Fig. 4.8 The decomposition of the planar flow induced by body motion into its basis flows. The full body motion flow field (upper left) is the sum of the basis flow due to translation of the body reference point at unit velocity in the $\tilde{x}$ direction, multiplied by the corresponding component $\tilde{U}_r$ (lower left); the basis flow due to translation at unit velocity in the $\tilde{y}$ direction, multiplied by the corresponding component $\tilde{V}_r$ (lower right); and the basis flow due to rotation at unit angular velocity, multiplied by Ω (upper right). The three-dimensional case is analogous, with six such basis flow fields

$$f_v = \sum_{J=1}^{N_v} \Gamma_J f_J^v, \tag{4.76}$$

where f_J^v denotes the **unit vorticity-induced basis f field**, due to vortex element J with unit strength. Since each of these unit basis fields accounts for the no-penetration condition on the (stationary) body, the field depends on the configuration of the filament relative to the body and is therefore *not* time-invariant.

We will have various uses for the flow decomposition in the ensuing sections and chapters. Later in this chapter, we will decompose the kinetic energy, which will lead to the concept of added mass. In Chap. 6, based on the development of impulse in Sect. 6.2, we will use the basis flow fields to decompose the force and moment that the fluid applies to the body. It should be stressed that not all flow quantities are linearly decomposable; the pressure field, which depends on the square of the velocity field, is an obvious example. Despite the linearity of Laplace's equation, we cannot entirely escape the fundamentally non-linear nature of fluid dynamics! However, by identifying the many places in which a linear decomposition is possible, we also underscore the essential role of flow non-linearities.

In this section, we will define the basis vortex sheets and the associated basis velocity fields. In a reversal of our usual procedure, we will first discuss the decomposition in the two-dimensional problem solved with conformal mapping since it is almost trivial to obtain from our earlier results. This will help clarify the more general decomposition of the vortex sheet strengths in the integral equations, which we will discuss both in the vectorial and complex contexts.

4.6.1 Complex Form, via Conformal Mapping Solution

Let us start with the flow about a two-dimensional rigid body in motion and subjected to uniform flow and a collection of N_v point vortices in the surrounding fluid. The complex velocity fields for each of these flow contributors were obtained separately in Notes 4.4.2–4.4.4 by conformally mapping a unit circle to the body shape with a rigid-body transformation (A.152) and power series mapping (A.153), and then solving the easier problem in the circle plane. Because of the linear dependence of the problem on these flow contributors, the flow field generated with all of them in action is simply the superposition of the individual fields. Thus, the complex potential in the circle plane is

$$\hat{F}(\zeta) = \tilde{W}_\infty c_1 \zeta + \left(\tilde{W}_\infty^* - \tilde{W}_\mathrm{r}^*\right) \frac{c_1^*}{\zeta} + \tilde{W}_\mathrm{r} \sum_{k=1}^{\infty} \frac{c_{-k}}{\zeta^k} - \mathrm{i}\Omega \sum_{k=1}^{\infty} \frac{d_{-k}}{\zeta^k} + \sum_{\mathrm{J}=1}^{N_\mathrm{v}} \frac{\Gamma_\mathrm{J}}{2\pi\mathrm{i}} \left[\log(\zeta - \zeta_\mathrm{J}) - \log(\zeta - 1/\zeta_\mathrm{J}^*) + \log\zeta\right] + \frac{\Gamma_0}{2\pi\mathrm{i}} \log\zeta, \tag{4.77}$$

where the coefficients d_{-k} are defined in Eq. (A.157). Let us make a few observations before we proceed further.

- We have included in (4.77) an additional null-space solution, with strength Γ_0. This strength, in fact, represents the bound circulation of the body, since none of the other flow contributors has a net circulation about the body, as the reader can verify. As we discussed in Sect. 4.5, the only requirement we place on this bound circulation is that it should be equal and opposite to the circulation in the fluid, viz.

$$\Gamma_0 = -\sum_{\mathrm{J}=1}^{N_\mathrm{v}} \Gamma_\mathrm{J}, \tag{4.78}$$

so that the total circulation remains invariantly zero. Thus, the full potential (4.77) can be written simply as

$$\hat{F}(\zeta) = \tilde{W}_\infty c_1 \zeta + \left(\tilde{W}_\infty^* - \tilde{W}_r^*\right) \frac{c_1^*}{\zeta} + \tilde{W}_r \sum_{k=1}^{\infty} \frac{c_{-k}}{\zeta^k} - i\Omega \sum_{k=1}^{\infty} \frac{d_{-k}}{\zeta^k}$$
$$+ \sum_{J=1}^{N_v} \frac{\Gamma_J}{2\pi i} \left[\log(\zeta - \zeta_J) - \log(\zeta - 1/\zeta_J^*)\right]. \qquad (4.79)$$

- Though its total is constrained, we have freedom to partition this bound circulation Γ_0 among the various flow contributors. In classical treatments of unsteady aerodynamics, such as that of Theodorsen [68] and von Kármán and Sears [70], each contributor receives a piece of this bound circulation, as required for that contributor to enforce the Kutta condition (the topic of Chap. 5) at the trailing edge of the airfoil. The portion of the bound circulation associated with body motion (and uniform flow) is called the **quasi-steady circulation**. We will revisit the quasi-steady circulation when we discuss the classical results in detail in Sect. 8.5.
- For reasons that will be detailed in Sect. 5.3, in this book we take a different approach, and ***associate all of the bound circulation with the vorticity-induced field***.
- Strictly speaking, (4.79) is the complex potential in the windtunnel reference frame, since it includes a uniform flow at infinity. The potential in the inertial reference frame is recovered by simply setting $\tilde{W}_\infty$ to zero. Note that, in either reference frame, $\tilde{W}_r$ in this expression should be interpreted as the translational velocity of the body observed from that reference frame.

The complex velocity field associated with the potential (4.79) is easily found by taking the derivative with respect to z, which we can do by using the chain rule according to (4.38). This leads to the following:

Result 4.5: Two-Dimensional Velocity Field via Conformal Mapping

The velocity field in the fluid due to a two-dimensional rigid body, in motion with translational velocity $\tilde{W}_r = \tilde{U}_r - i\tilde{V}_r$ and angular velocity Ω, and subjected to a uniform flow of velocity $\tilde{W}_\infty = \tilde{U}_\infty - i\tilde{V}_\infty$, and a collection of N_v point vortices of strengths Γ_J and positions z_J, is

$$w(z) = \frac{1}{z'(\zeta)} \left[\tilde{W}_\infty c_1 - \left(\tilde{W}_\infty^* - \tilde{W}_r^*\right) \frac{c_1^*}{\zeta^2} - \tilde{W}_r \sum_{k=1}^{\infty} \frac{k c_{-k}}{\zeta^{k+1}} + i\Omega \sum_{k=1}^{\infty} \frac{k d_{-k}}{\zeta^{k+1}} \right.$$
$$\left. + \sum_{J=1}^{N_v} \frac{\Gamma_J}{2\pi i} \left(\frac{1}{\zeta - \zeta_J} - \frac{1}{\zeta - 1/\zeta_J^*} \right) \right], \qquad (4.80)$$

where z is any point on the body surface or in the fluid region exterior to it, corresponding to a point ζ on or exterior to the unit circle by the rigid transformation

$$z(\zeta) = Z_r + e^{i\alpha}\tilde{z}(\zeta), \tag{A.152}$$

and the conformal mapping

$$\tilde{z}(\zeta) = \sum_{k=-1}^{\infty} \frac{c_{-k}}{\zeta^k}. \tag{A.153}$$

The coefficients d_{-k} are related to these power-series coefficients by

$$d_{-k} = \begin{cases} \sum_{l=-1}^{\infty} c^*_{-l} c_{-l-k}, & k \geq 0, \\ \sum_{l=-1}^{\infty} c_{-l} c^*_{-l+k} = d^*_k, & k \leq -1. \end{cases} \tag{A.157}$$

The circle plane positions ζ_J in (4.80) correspond, via the mapping and rigid transformation above, to the vortex positions z_J in the physical plane.

Basis Velocity and Complex Potential Fields The velocity field (4.80) and the complex potential (4.79) clearly depend linearly on the uniform flow, the body's translational and angular velocities, and the strengths of the vortex elements in the fluid. Now, we write these fields in a form in which the basis velocity field associated with each flow contributor is explicitly identified, following the generic templates (4.72)–(4.75) described above:

$$w(z) = w_v(z) + \tilde{U}_\infty w^{(1)}_\infty(z) + \tilde{V}_\infty w^{(2)}_\infty(z) + \tilde{U}_r w^{(1)}_{bt}(z) + \tilde{V}_r w^{(2)}_{bt}(z) + \Omega w_{br}(z). \tag{4.81}$$

and

$$F(z) = F_v(z) + \tilde{U}_\infty F^{(1)}_\infty(z) + \tilde{V}_\infty F^{(2)}_\infty(z) + \tilde{U}_r F^{(1)}_{bt}(z) + \tilde{V}_r F^{(2)}_{bt}(z) + \Omega F_{br}(z). \tag{4.82}$$

Each of the basis velocity and complex potential fields appear in the expressions on the right-hand side of (4.81) and (4.82), respectively. These basis fields, expressed here in the inertial coordinate system, are defined as follows: The **vorticity-induced basis velocity field** is

$$w_v(z) = \sum_{J=1}^{N_v} \frac{\Gamma_J}{2\pi i z'(\zeta)} \left(\frac{1}{\zeta - \zeta_J} - \frac{1}{\zeta - 1/\zeta^*_J} \right). \tag{4.83}$$

This is the velocity field induced by the collection of point vortices, modified by the presence of the stationary body. Clearly, this can be further decomposed into contributions from each vortex. Indeed, let us write

$$w_{\mathrm{V}} = \sum_{\mathrm{J}=1}^{N_{\mathrm{V}}} \Gamma_{\mathrm{J}} w_{\mathrm{J}}^{\mathrm{V}}, \tag{4.84}$$

where we easily define the **unit vorticity-induced basis velocity field**—the velocity field due to vortex J, with unit strength, modified by the body's presence—as

$$w_{\mathrm{J}}^{\mathrm{V}}(z) := \frac{1}{2\pi \mathrm{i} z'(\zeta)} \left(\frac{1}{\zeta - \zeta_{\mathrm{J}}} - \frac{1}{\zeta - 1/\zeta_{\mathrm{J}}^*} \right). \tag{4.85}$$

Analogously, the **vorticity-induced basis complex potential field** is

$$F_{\mathrm{V}} = \sum_{\mathrm{J}=1}^{N_{\mathrm{V}}} \Gamma_{\mathrm{J}} F_{\mathrm{J}}^{\mathrm{V}}, \tag{4.86}$$

where the **unit vorticity-induced basis complex potential field**, modified by the presence of the body, is

$$F_{\mathrm{J}}^{\mathrm{V}}(z) := \frac{1}{2\pi \mathrm{i}} \left[\log(\zeta - \zeta_{\mathrm{J}}) - \log(\zeta - 1/\zeta_{\mathrm{J}}^*) \right]. \tag{4.87}$$

The two **uniform flow basis velocity fields** are

$$w_{\infty}^{(1)}(z) = \frac{1}{z'(\zeta)} \left(c_1 - \frac{c_1^*}{\zeta^2} \right), \quad w_{\infty}^{(2)}(z) = -\frac{\mathrm{i}}{z'(\zeta)} \left(c_1 + \frac{c_1^*}{\zeta^2} \right). \tag{4.88}$$

These represent the velocity fields about the stationary body when subjected to a uniform flow of unit speed in the $\tilde{x}$ (i.e., 1) and $\tilde{y}$ (i.e., 2) directions, respectively. The corresponding **uniform flow basis complex potential fields** are

$$F_{\infty}^{(1)}(z) = c_1 \zeta + \frac{c_1^*}{\zeta}, \quad F_{\infty}^{(2)}(z) = -\mathrm{i} \left(c_1 \zeta - \frac{c_1^*}{\zeta} \right). \tag{4.89}$$

The **basis velocity fields due to body translation** are

$$w_{\mathrm{bt}}^{(1)}(z) = \frac{1}{z'(\zeta)} \left(\frac{c_1^*}{\zeta^2} - \sum_{k=1}^{\infty} \frac{k c_{-k}}{\zeta^{k+1}} \right), \quad w_{\mathrm{bt}}^{(2)}(z) = \frac{\mathrm{i}}{z'(\zeta)} \left(\frac{c_1^*}{\zeta^2} + \sum_{k=1}^{\infty} \frac{k c_{-k}}{\zeta^{k+1}} \right). \tag{4.90}$$

These correspond to the velocity fields induced in the otherwise quiescent fluid due to translation of the body at unit speed in the $\tilde{x}$ and $\tilde{y}$ directions, respectively. It is easy to show that $w_{\infty}^{(1)}(z) = 1 - w_{\mathrm{bt}}^{(1)}(z)$ and $w_{\infty}^{(2)}(z) = -\mathrm{i} - w_{\mathrm{bt}}^{(2)}(z)$; these relations reflect the fact that these two basis fields differ by the uniform flow at infinity, but are otherwise equal and opposite. The **basis complex potential fields due to body translation** that accompany these are

$$F_{\mathrm{bt}}^{(1)}(z) = -\frac{c_1^*}{\zeta} + \sum_{k=1}^{\infty} \frac{c_{-k}}{\zeta^k}, \quad F_{\mathrm{bt}}^{(2)}(z) = -\mathrm{i} \left(\frac{c_1^*}{\zeta} + \sum_{k=1}^{\infty} \frac{c_{-k}}{\zeta^k} \right). \tag{4.91}$$

And finally, the **body rotation basis velocity field** is

$$w_{\mathrm{br}}(z) = \frac{\mathrm{i}}{z'(\zeta)} \sum_{k=1}^{\infty} \frac{k d_{-k}}{\zeta^{k+1}}, \tag{4.92}$$

and the **body rotation basis complex potential field** is

$$F_{\mathrm{br}}(z) = -\mathrm{i} \sum_{k=1}^{\infty} \frac{d_{-k}}{\zeta^{k}}; \tag{4.93}$$

these describe the flow induced by unit angular velocity of the body in quiescent fluid.

Note that each of the velocity fields obtained above has the form

$$w(z) = \frac{1}{z'(\zeta)} \hat{w}(\zeta), \tag{4.94}$$

where $\hat{w}$ was defined in (4.38) as the complex velocity field in the circle plane. Furthermore, we can readily transform any of these basis velocity fields into body-fixed coordinates, $\tilde{w}$, by using the rotation identity (2.38); equivalently,

$$\tilde{w}(\tilde{z}) = \frac{1}{\tilde{z}'(\zeta)} \hat{w}(\zeta). \tag{4.95}$$

Basis Vortex Sheets For future calculations, it is useful to identify the vortex sheet on the body surface that would produce the flow (4.80) when substituted into the integral form (4.13). This is found by evaluating the complex velocity at points on the surface of the body (or equivalently, the circle) and subtracting the body's own velocity at that point, as dictated by the definition of the complex vortex sheet strength (3.198). Analogous to the velocity, this vortex sheet can be decomposed into basis sheets due to the individual flow contributors. The overall vortex sheet strength can thus be written as a linear superposition of these basis sheets:

$$g(z) = g_{\mathrm{v}}(z) + \tilde{U}_{\infty} g_{\infty}^{(1)}(z) + \tilde{V}_{\infty} g_{\infty}^{(2)}(z) + \tilde{U}_{\mathrm{r}} g_{\mathrm{bt}}^{(1)}(z) + \tilde{V}_{\mathrm{r}} g_{\mathrm{bt}}^{(2)}(z) + \Omega g_{\mathrm{br}}(z), \quad z \in C_{\mathrm{b}}. \tag{4.96}$$

These basis vortex sheets can be obtained by evaluating the corresponding basis velocity field on the surface and subtracting the local body velocity, decomposed in analogous manner. For example, the strength of the basis vortex sheet due to point vortices in the fluid surrounding the stationary body can be shown to be

$$g_{\mathrm{v}}(z) = \sum_{\mathrm{J}=1}^{N_{\mathrm{v}}} \frac{\Gamma_{\mathrm{J}}}{2\pi \mathrm{i} \zeta z'(\zeta)} \mathrm{Re}\left(\frac{\zeta + \zeta_{\mathrm{J}}}{\zeta - \zeta_{\mathrm{J}}} \right), \tag{4.97}$$

where $\zeta \in C_c$, and the argument of function g_{v} is the corresponding point on the body in the physical plane, $z = z(\zeta) \in C_{\mathrm{b}}$. As with the velocity field, this sheet can obviously be decomposed further, into unit sheets reacting to each point vortex:

$$g_{\mathrm{v}} = \sum_{\mathrm{J}=1}^{N_{\mathrm{v}}} \Gamma_{\mathrm{J}} g_{\mathrm{J}}^{\mathrm{v}}, \tag{4.98}$$

where

$$g_{\mathrm{J}}^{\mathrm{v}}(z) := \frac{1}{2\pi \mathrm{i} \zeta z'(\zeta)} \mathrm{Re}\left(\frac{\zeta + \zeta_{\mathrm{J}}}{\zeta - \zeta_{\mathrm{J}}}\right). \tag{4.99}$$

The strengths of the basis vortex sheets due to uniform flow of unit speed in each direction about the stationary body are, respectively,

$$g_{\infty}^{(1)}(z) = -\frac{2}{\mathrm{i}\zeta z'(\zeta)} \mathrm{Im}(c_1 \zeta), \quad g_{\infty}^{(2)}(z) = \frac{2}{\mathrm{i}\zeta z'(\zeta)} \mathrm{Re}(c_1 \zeta), \tag{4.100}$$

in response to unit speed flow in the $\tilde{x}$ and $\tilde{y}$ directions. The basis vortex sheet strengths due to unit translational velocity in each body direction are simply the negative of the corresponding ones for uniform flow:

$$g_{\mathrm{bt}}^{(1)}(z) = \frac{2}{\mathrm{i}\zeta z'(\zeta)} \mathrm{Im}(c_1 \zeta), \quad g_{\mathrm{bt}}^{(2)}(z) = -\frac{2}{\mathrm{i}\zeta z'(\zeta)} \mathrm{Re}(c_1 \zeta). \tag{4.101}$$

This is because the overall vortex sheet strength only depends on the relative motion between the fluid and the body, and thus, only the difference between body translational velocity and the uniform flow velocity. Finally, with some effort, the strength of the basis vortex sheet due to unit angular velocity of the body is

$$g_{\mathrm{br}}(z) = \frac{2}{\mathrm{i}\zeta z'(\zeta)} \mathrm{Re}\left(\sum_{l=1}^{\infty} \zeta^{l} \sum_{k=-1}^{\infty} k c_{-k} c_{-k-l}^{*} - \frac{\mathcal{V}_{\mathrm{b}}}{2\pi}\right). \tag{4.102}$$

Using the integral over g in Eq. (4.68), it is straightforward to show that the last term in (4.102)—involving $\mathcal{V}_{\mathrm{b}}$, the area of the body—provides the circulation of the vortex sheet, equal to $-2\mathcal{V}_{\mathrm{b}}$. This circulation is equal and opposite to the circulation due to the body's unit angular velocity, ensuring that the total circulation of this basis flow field is zero.

We note that all basis sheet strengths defined above have the general form (4.40), rewritten here for reference:

$$g(z) = \frac{1}{\mathrm{i}\zeta z'(\zeta)} |z'(\zeta)| \gamma(z). \tag{4.103}$$

Thus, it is a simple matter to determine the real-valued vortex strength, γ, associated with each of these complex strengths. For example, the real-valued basis vortex sheet strength due to unit translation in the $\tilde{x}$ direction is

$$\gamma_{\mathrm{bt}}^{(1)}(z) = \frac{2}{|z'(\zeta)|} \mathrm{Im}(c_1 \zeta). \tag{4.104}$$

Time Dependence of the Basis Fields Thus far in this book, we have not had much need to investigate the time dependence of the flow fields we've obtained. The flow field construction has only required the solution of the equations of potential flow for a given instantaneous configuration of the body and the fluid vorticity; time dependence has been irrelevant to this solution. However, as we progress further, the time dependence of the flow field will play an increasingly important role. We use the opportunity of the decomposition of the flow into its component contributors in order to begin exploring this time dependence. Why? Because, as we will see, this decomposition has the particular benefit of isolating the time dependence into a few parameters, revealing several invariant aspects of the flow field. The results we obtain here will be very useful in later chapters.

Let us start by examining the time dependence of the basis velocity fields. To be clear, we will focus here on the time dependence when viewed from the *body-fixed* reference frame; the time dependence in the inertial frame can always be discerned later by accounting for the transformation between frames (2.23). Thus, we write any of the basis velocity fields in the body-fixed form $\tilde{w}(\tilde{z}, t)$ and examine its time dependence at a fixed point $\tilde{z}$.

We noted above that each of these basis fields can be separated into the Jacobian of the mapping and the velocity field in the circle plane, as guided by the template form (4.95). For a rigid body, a point $\tilde{z}$ that is stationary in the body-fixed reference frame of the physical plane is equivalent to a fixed point ζ in the circle plane; that is, the mapping $\tilde{z}(\zeta)$ and its Jacobian $\tilde{z}'(\zeta)$ are time invariant. This is true for any conformal mapping of a rigid body. For the particular case of the power series mapping (A.153), as was used here, the coefficients c_{-k} (and the derived coefficients d_{-k}) are constant. Thus, the circle-plane velocity fields due to body motion or uniform flow are also invariant. All that remains is the time dependence of the properties of the flow contributors, such as $\tilde{U}_\infty(t)$ or $\zeta_J(t)$, and we can now write the decomposed field $\tilde{w}(\tilde{z}, t)$ with the time dependencies explicitly acknowledged:

$$\tilde{w}(\tilde{z}, t) = \sum_{J=1}^{N_v} \Gamma_J(t)\tilde{w}_J^v(\tilde{z}, t) + \tilde{U}_\infty(t)\tilde{w}_\infty^{(1)}(\tilde{z}) + \tilde{V}_\infty(t)\tilde{w}_\infty^{(2)}(\tilde{z})$$
$$+ \tilde{U}_r(t)\tilde{w}_{bt}^{(1)}(\tilde{z}) + \tilde{V}_r(t)\tilde{w}_{bt}^{(2)}(\tilde{z}) + \Omega(t)\tilde{w}_{br}(\tilde{z}), \tag{4.105}$$

where

$$\tilde{w}_J^v(\tilde{z}, t) := \frac{1}{2\pi i \tilde{z}'(\zeta)}\left(\frac{1}{\zeta - \zeta_J(t)} - \frac{1}{\zeta - 1/\zeta_J^*(t)}\right). \tag{4.106}$$

It is straightforward, then, to compute the time derivative of this field, should it be necessary to do so.

It is interesting to note that each of the basis velocity fields due to body translation and rotation or due to uniform flow are frozen when viewed from the body's perspective. This is not true of the vorticity-induced basis velocity field in (4.105), which varies as the point vortices advect or change their strength. However, observe that ***only motion of the vortex relative to the body gives rise to changes in the unit vorticity-induced basis field***.

As with the basis velocity fields for uniform flow and body motion, the basis vortex sheet strengths associated with these flow contributors are time invariant when expressed in the body-fixed frame of reference. This invariance is most obvious in the real-valued vortex sheet strength, γ; the complex sheet strength, g, is proportional to the conjugate of the tangent vector, τ, and thus has additional time dependence due to rotation of the body. We have already noted above in (4.103) the clear relationship between the complex and real-valued strengths in each of the basis sheets. Note further that $|z'(\zeta)| = |\tilde{z}'(\zeta)|$, and thus the magnitude of the Jacobian does not change with time (for a rigid body) at a location $\tilde{z}$ fixed relative to the body axes (and thus, fixed ζ). It follows from inspection of (4.104) that, for example, $\gamma_{\text{bt}}^{(1)}$ is invariant at fixed $\tilde{z}$; the invariance of the other basis sheet strengths follows in kind. Only the vorticity-induced vortex sheet, γ_{v}, has additional time dependence through the vortex positions ζ_{J} and strengths Γ_{J}. We can immediately write the decomposed real-valued vortex sheet strength, as a function of body-fixed coordinates and time, $\gamma(\tilde{z}, t)$, with the time dependencies explicitly noted:

$$\gamma(\tilde{z}, t) = \sum_{\text{J}=1}^{N_{\text{v}}} \Gamma_{\text{J}}(t)\gamma_{\text{J}}^{\text{v}}(\tilde{z}, t) + \tilde{U}_{\infty}(t)\gamma_{\infty}^{(1)}(\tilde{z}) + \tilde{V}_{\infty}(t)\gamma_{\infty}^{(2)}(\tilde{z})$$
$$+ \tilde{U}_{\text{r}}(t)\gamma_{\text{bt}}^{(1)}(\tilde{z}) + \tilde{V}_{\text{r}}(t)\gamma_{\text{bt}}^{(2)}(\tilde{z}) + \Omega(t)\gamma_{\text{br}}(\tilde{z}), \tag{4.107}$$

where

$$\gamma_{\text{bt}}^{(1)}(\tilde{z}) = -\gamma_{\infty}^{(1)}(\tilde{z}) = \frac{2}{|\tilde{z}'(\zeta)|}\text{Im}(c_1\zeta), \quad \gamma_{\text{bt}}^{(2)}(\tilde{z}) = -\gamma_{\infty}^{(2)}(\tilde{z}) = -\frac{2}{|\tilde{z}'(\zeta)|}\text{Re}(c_1\zeta), \tag{4.108}$$

$$\gamma_{\text{br}}(\tilde{z}) = -\frac{2}{|\tilde{z}'(\zeta)|}\text{Re}\left(\sum_{l=1}^{\infty}\zeta^l \sum_{k=-1}^{\infty} kc_{-k}c_{-k-l}^{*} - \frac{\mathcal{V}_{\text{b}}}{2\pi}\right), \tag{4.109}$$

and

$$\gamma_{\text{J}}^{\text{v}}(\tilde{z}, t) = \frac{1}{|\tilde{z}'(\zeta)|}\frac{1}{2\pi}\text{Re}\left(\frac{\zeta + \zeta_{\text{J}}(t)}{\zeta - \zeta_{\text{J}}(t)}\right). \tag{4.110}$$

As for the other fields, this unit vorticity-induced basis vortex sheet only varies if the vortex moves relative to the body.

4.6.2 Vector Form

In the previous section, we found that the velocity field and the strength of the bound vortex sheet on the body surface could be decomposed into basis fields associated with each of the flow contributors. This decomposition was facilitated by the fact that the solution for the overall flow field was constructed by superposing the individual contributors' fields, each obtained by conformal mapping from the circle plane. In this section, we extend this decomposition to the vectorial analysis of the problem, in which the underlying solution is obtained from the integral equations presented in Sect. 4.3. We will rely on the linearity of the integral operator, $\mathcal{F}$, defined in Eq. (4.22). By extending in this manner, the decomposition will become available to us for any problem, in either two or three dimensions.

We start by decomposing the vectorial vortex sheet strength in a natural extension of the decomposition of the complex sheet strength (4.96). As before, there will be clear advantages in expressing the components of motion and uniform flow in body-fixed coordinates:

$$\boldsymbol{\gamma} = \boldsymbol{\gamma}_{\mathrm{v}} + \sum_{j=1}^{3} \tilde{V}_{\infty,j} \boldsymbol{\gamma}_{\infty}^{(j)} + \sum_{j=1}^{3} \tilde{V}_{\mathrm{r},j} \boldsymbol{\gamma}_{\mathrm{bt}}^{(j)} + \sum_{j=1}^{3} \tilde{\Omega}_{j} \boldsymbol{\gamma}_{\mathrm{br}}^{(j)}, \tag{4.111}$$

where $\boldsymbol{\gamma}_{\mathrm{v}}$ represents the basis vortex sheet strength reacting to fluid vorticity; $\boldsymbol{\gamma}_{\infty}^{(j)}$ is the sheet strength associated with uniform flow of unit speed in the jth body-fixed coordinate direction, $\tilde{\boldsymbol{e}}_j$; $\boldsymbol{\gamma}_{\mathrm{bt}}^{(j)}$ is the sheet strength due to body translation at unit speed in the jth body-fixed coordinate direction; and $\boldsymbol{\gamma}_{\mathrm{br}}^{(j)}$ is the sheet strength due to unit angular velocity of the body about the jth body-fixed axis. Note that each of these basis sheets, like $\boldsymbol{\gamma}$, is a vector field that lies in the body surface, $\mathcal{S}_{\mathrm{b}}$. The time dependence of each will be most clear when we express these vector fields in the body-fixed basis, $\{\tilde{\boldsymbol{e}}_j\}$. Furthermore, they are each the solution of an integral equation obtained by simplifying the right-hand side of (4.23) with the corresponding flow contributor. In two-dimensional problems, we also decompose the constraint on total circulation (4.67) to individually constrain the bound circulation to ensure a unique solution for each of these integral equations. In this constraint's decomposition, we will apportion all of the bound circulation to the vorticity-induced field; this is already implied in the complex solution given by (4.83) and (4.97), in which each image vortex has equal and opposite circulation to its real partner.

With this decomposed vortex sheet strength, the fluid velocity field can be decomposed in kind,

$$\boldsymbol{v} = \boldsymbol{v}_{\mathrm{v}} + \sum_{j=1}^{3} \tilde{V}_{\infty,j} \boldsymbol{v}_{\infty}^{(j)} + \sum_{j=1}^{3} \tilde{V}_{\mathrm{r},j} \boldsymbol{v}_{\mathrm{bt}}^{(j)} + \sum_{j=1}^{3} \tilde{\Omega}_{j} \boldsymbol{v}_{\mathrm{br}}^{(j)}, \tag{4.112}$$

where each of the basis velocity fields accounts for the corresponding no-penetration condition on the body surface. It is useful to note that, aside from the first of these (due to vorticity), each basis velocity field is irrotational everywhere in the fluid, and can thus be written as the gradient of a continuous scalar potential field. In this spirit, we can align our approach here with the Helmholtz decomposition (3.1) of the velocity field. Remember that, because we have restricted the rotational part of the flow field to singular vortex elements, the flow is irrotational almost everywhere, and we are free to associate *any* flow contributor with either the gradient of a scalar potential or the curl of a vector potential. Thus, the division we make here is of our own choosing. On that note, let us suppose that the vorticity-induced velocity field,

$\boldsymbol{v}_V$, is derived from the curl of a vector potential field, $\boldsymbol{\Psi}_V$, and group the rest of the flow contributors into a scalar potential field,[2] viz.

$$\boldsymbol{v} = \nabla \times \boldsymbol{\Psi}_V + \nabla\varphi. \tag{4.113}$$

As with the other flow field quantities, the underlying problem that the scalar potential satisfies is linearly dependent upon the flow contributors, so this potential is naturally decomposable into basis potentials,

$$\varphi = \sum_{j=1}^{3} \tilde{V}_{\infty,j}\left(\tilde{x}_j + \varphi_{\infty}^{(j)}\right) + \sum_{j=1}^{3} \tilde{V}_{r,j}\varphi_{bt}^{(j)} + \sum_{j=1}^{3} \tilde{\Omega}_j\varphi_{br}^{(j)}. \tag{4.114}$$

Note that we have explicitly identified the direct contribution of the uniform flow, whose scalar potential field is $\boldsymbol{V}_\infty \cdot \boldsymbol{x}$—and thus, whose contribution to the jth basis potential is $\tilde{x}_j$—so that $\varphi_{\infty}^{(j)}$ represents just the body's reaction to the jth component. These basis scalar potentials serve as an alternate means for expressing their associated flow field. We will describe the specific problem for each contributor below.

Vorticity-Induced Basis Field The vorticity-induced basis vortex sheet has a strength $\boldsymbol{\gamma}_V$ determined by solving

$$\mathcal{F}[\boldsymbol{\gamma}_V](\boldsymbol{x}) = -\int_{\mathcal{V}_f} \boldsymbol{K}(\boldsymbol{x}-\boldsymbol{y}) \times \boldsymbol{\omega}(\boldsymbol{y})\,\mathrm{d}V(\boldsymbol{y}). \tag{4.115}$$

In two dimensions, this is constrained with

$$\int_{\mathcal{S}_b} \gamma_V\,\mathrm{d}S = -\int_{\mathcal{V}_f} \omega\,\mathrm{d}V, \tag{4.116}$$

where the vectors are understood to have only a single component, out of the plane.

Now, with this vorticity-induced vortex sheet, we can define the **vorticity-induced basis velocity field** in the fluid region. This field, defined at all $\boldsymbol{x} \in \mathcal{V}_f$, includes the original velocity field induced by the vorticity plus the field induced by the reactive vortex sheet that solves integral equation (4.115):

$$\boldsymbol{v}_V(\boldsymbol{x}) := \int_{\mathcal{V}_f} \boldsymbol{K}(\boldsymbol{x}-\boldsymbol{y}) \times \boldsymbol{\omega}(\boldsymbol{y})\,\mathrm{d}V(\boldsymbol{y}) + \int_{\mathcal{S}_b} \boldsymbol{K}(\boldsymbol{x}-\boldsymbol{y}) \times \boldsymbol{\gamma}_V(\boldsymbol{y})\,\mathrm{d}S(\boldsymbol{y}). \tag{4.117}$$

By construction, this velocity field satisfies the no-penetration condition on the (stationary) body surface. Furthermore, all terms in this velocity field are linearly related to the vorticity field.

[2] Note that the scalar potential field defined here does not include the irrotational flow field that arises in reaction to the fluid vorticity. That field is contained in $\boldsymbol{v}_V = \nabla \times \boldsymbol{\Psi}_V$, together with the flow field induced directly by vorticity.

It is important to re-emphasize that the solution of the potential flow problem implied by (4.115) is linearly dependent on the flow contributors on the right-hand side, so the vorticity-induced basis vortex sheet and velocity field (4.117) are each linearly dependent on the vorticity field in the fluid. Since the vorticity field in our analysis framework is composed of a set of vortex elements, this linear dependence allows us to further decompose the basis field quantities into portions attributable to each vortex element in the fluid. This deeper decomposition was immediately accessible in the planar results (4.83) and (4.97) obtained by conformal mapping for a discrete set of point vortices. In a general problem, for which we may lack an explicit analytical solution, we will only formally represent the decomposition. We have shown previously in Sect. 3.3.3 that it is sufficiently general to interpret the vorticity field as a collection of vortex filaments in a three-dimensional flow or point vortices in two dimensions. Free vortex sheets are limiting forms of such collections.

In that spirit, let us consider a set of N_v vortex filaments (in three dimensions) or point vortices (in two dimensions). The Jth member of this set has strength Γ_J and lies along a space curve C_J, described parametrically by $\boldsymbol{x}_J(s)$, where s is the arc length along the curve. In two dimensions, this simply reverts to a single vector position $\boldsymbol{x}_J$ in the plane. The vorticity-induced basis vortex sheet can then be decomposed into a sum of unit vorticity-induced sheets reacting to each filament in the set. In particular, we can write

$$\boldsymbol{\gamma}_V = \sum_{J=1}^{N_v} \Gamma_J \boldsymbol{\gamma}_J^V, \tag{4.118}$$

where $\boldsymbol{\gamma}_J^V$ is the bound vortex sheet on the surface $\mathcal{S}_b$ that solves the following potential flow problem:

$$\mathcal{F}\left[\boldsymbol{\gamma}_J^V\right](\boldsymbol{x}) = -\int_{C_J} \boldsymbol{K}(\boldsymbol{x} - \boldsymbol{x}_J(s)) \times \mathrm{d}\boldsymbol{l}(s). \tag{4.119}$$

That is, $\boldsymbol{\gamma}_J^V$ ***is the strength of the bound vortex sheet that cancels the normal velocity induced on*** $\mathcal{S}_b$ ***by the Jth vortex filament, when that filament is assigned unit strength***. For a filament, the velocity kernel $\boldsymbol{K}$ in (4.119) is the three-dimensional kernel defined in Eq. (3.112). In two dimensions, when the filament reverts to a point vortex, the two-dimensional kernel (3.71) is used, and the right-hand side of (4.119) is simply $-\boldsymbol{K}(\boldsymbol{x}-\boldsymbol{x}_J)\times\boldsymbol{e}_3$. In such a case, the integral equation (4.119) is constrained so that the total strength of the sheet $\boldsymbol{\gamma}_J^V$ is equal and opposite to that of the unit vortex:

$$\int_{\mathcal{S}_b} \boldsymbol{\gamma}_J^V \, \mathrm{d}S = -\boldsymbol{e}_3. \tag{4.120}$$

The vorticity-induced basis velocity field can be decomposed accordingly:

$$\boldsymbol{v}_V = \sum_{J=1}^{N_v} \Gamma_J \boldsymbol{v}_J^V, \tag{4.121}$$

where the unit basis field, $\boldsymbol{v}_{\mathrm{J}}^{\mathrm{v}}$, is defined at all $\boldsymbol{x} \in \mathcal{V}_f$ as

$$\boldsymbol{v}_{\mathrm{J}}^{\mathrm{v}}(\boldsymbol{x}) := \int_{C_{\mathrm{J}}} \boldsymbol{K}(\boldsymbol{x} - \boldsymbol{x}_{\mathrm{J}}(s)) \times \mathrm{d}\boldsymbol{l}(s) + \int_{S_{\mathrm{b}}} \boldsymbol{K}(\boldsymbol{x} - \boldsymbol{y}) \times \boldsymbol{\gamma}_{\mathrm{J}}^{\mathrm{v}}(\boldsymbol{y})\, \mathrm{d}S(\boldsymbol{y}); \qquad (4.122)$$

the usual modifications are made in a two-dimensional context.

Uniform Flow Basis Fields The vortex sheet due to each uniform flow component, $\boldsymbol{\gamma}_{\infty}^{(k)}$, is the solution of

$$\mathcal{F}\left[\boldsymbol{\gamma}_{\infty}^{(k)}\right](\boldsymbol{x}) = -\tilde{\boldsymbol{e}}_k, \qquad (4.123)$$

for $k = 1, \ldots, n_{\mathrm{d}}$. In two-dimensional problems, the associated constraint on each sheet is given by

$$\int_{S_{\mathrm{b}}} \boldsymbol{\gamma}_{\infty}^{(k)}\, \mathrm{d}S = 0. \qquad (4.124)$$

Each of these sheets, $\boldsymbol{\gamma}_{\infty}^{(k)}$, induces a velocity whose normal component on the surface of the body is equal and opposite to the kth component of the normal, thereby annihilating the penetration of the kth component of the uniform flow on the surface when the overall $\boldsymbol{\gamma}_{\infty}$ is reconstituted from these sheets. Indeed, the velocity field associated with the uniform flow, modified by the no-penetration condition on the body, is given by

$$\boldsymbol{v}_{\infty}(\boldsymbol{x}) = \sum_{k=1}^{3} \tilde{V}_{\infty,k} \boldsymbol{v}_{\infty}^{(k)}(\boldsymbol{x}), \qquad (4.125)$$

where the basis fields are

$$\boldsymbol{v}_{\infty}^{(k)}(\boldsymbol{x}) = \tilde{\boldsymbol{e}}_k + \int_{S_{\mathrm{b}}} \boldsymbol{K}(\boldsymbol{x} - \boldsymbol{y}) \times \boldsymbol{\gamma}_{\infty}^{(k)}(\boldsymbol{y})\, \mathrm{d}S(\boldsymbol{y}). \qquad (4.126)$$

The first term in this basis velocity field expresses the contribution from the uniform flow itself, and the second term represents the reacting velocity field that enforces no-penetration.

As described above, an alternative to solving for the basis vortex sheets is to solve instead for the basis scalar potential field that satisfies the appropriate boundary conditions. In the case of uniform flow, the integral term in (4.126) is equivalently expressed as the gradient of the kth basis scalar potential, $\varphi_{\infty}^{(k)}$. Each of these potentials, of course, satisfies Laplace's equation,

$$\nabla^2 \varphi_{\infty}^{(k)} = 0. \qquad (4.127)$$

Furthermore, each basis potential must decay to zero at infinity (since it only represents the reaction to the uniform flow, not the uniform flow itself), and satisfy a Neumann condition on the body surface obtained from applying the no-penetration condition (3.11) to the basis velocity field:

$$\frac{\partial \varphi_{\infty}^{(k)}}{\partial n} = -\boldsymbol{n} \cdot \tilde{\boldsymbol{e}}_k, \tag{4.128}$$

at all $\boldsymbol{x} \in \mathcal{S}_{\mathrm{b}}$.

The flow field described by each $\boldsymbol{v}_{\infty}^{(k)}$ can equivalently be described by a vector potential. Though it has not been our practice to use the vector potential for the irrotational part of the flow, this particular potential will become important in our later discussion of force. The potential is defined, as expected, so that

$$\boldsymbol{v}_{\infty}^{(k)} = \nabla \times \boldsymbol{\Psi}_{\infty}^{(k)}. \tag{4.129}$$

Each basis vector potential satisfies Laplace's equation,

$$\nabla^2 \boldsymbol{\Psi}_{\infty}^{(k)} = 0, \tag{4.130}$$

The boundary conditions on $\mathcal{S}_{\mathrm{b}}$ are placed on the tangential components of this potential, which, from our discussion surrounding equation (3.15), ensures that the no-penetration condition is enforced on this surface. We also place a condition at infinity to ensure that the velocity field approaches $\tilde{\boldsymbol{e}}_k$:

$$\boldsymbol{n} \times \boldsymbol{\Psi}_{\infty}^{(k)}(\boldsymbol{x}) = 0, \text{ for } \boldsymbol{x} \in \mathcal{S}_{\mathrm{b}}, \qquad \boldsymbol{\Psi}_{\infty}^{(k)} \rightarrow \frac{1}{n_{\mathrm{d}} - 1} \tilde{\boldsymbol{e}}_k \times (\boldsymbol{x} - \boldsymbol{X}_{\mathrm{r}}), \text{ as } |\boldsymbol{x} - \boldsymbol{X}_{\mathrm{r}}| \rightarrow \infty. \tag{4.131}$$

Note the use of $\boldsymbol{x} - \boldsymbol{X}_{\mathrm{r}}$ rather than $\boldsymbol{x}$ in the condition at infinity. Remember that either of these conditions need only be enforced to within an arbitrary constant vector field, which can be adjusted as necessary to accommodate this shift in origin.

Body Translation Basis Fields For a rigid body in three dimensions, the vortex sheet strength due to each component of body translational velocity is the solution of

$$\mathcal{F}\left[\boldsymbol{\gamma}_{\mathrm{bt}}^{(k)}\right](\boldsymbol{x}) = \tilde{\boldsymbol{e}}_k, \tag{4.132}$$

for $k = 1, \ldots, n_{\mathrm{d}}$. (Here, we have made use of (A.52) to evaluate the principal value of the surface integral over a uniform vector field.) The associated constraint in a two-dimensional problem is

$$\int_{\mathcal{S}_{\mathrm{b}}} \gamma_{\mathrm{bt}}^{(k)} \, \mathrm{d}S = 0. \tag{4.133}$$

Note that $\boldsymbol{\gamma}_{\mathrm{bt}}^{(k)} = -\boldsymbol{\gamma}_{\infty}^{(k)}$. This is expected, since the vortex sheet's strength only depends on the relative velocity between the fluid at infinity and the body's translation.

The modified velocity field in $\mathcal{V}_f$ associated with the body's translational motion is given by

$$\sum_{k=1}^{3} \tilde{V}_{\mathrm{r},k} \boldsymbol{v}_{\mathrm{bt}}^{(k)}(\boldsymbol{x}), \tag{4.134}$$

where each basis velocity field is

$$\boldsymbol{v}_{\mathrm{bt}}^{(k)}(\boldsymbol{x}) = \int_{\mathcal{S}_\mathrm{b}} \boldsymbol{K}(\boldsymbol{x}-\boldsymbol{y}) \times \boldsymbol{\gamma}_{\mathrm{bt}}^{(k)}(\boldsymbol{y})\,\mathrm{d}S(\boldsymbol{y}). \tag{4.135}$$

The normal component of this velocity field (4.134) is equal to $\boldsymbol{V}_\mathrm{r} \cdot \boldsymbol{n}$ on the body surface. Also, from the results thus far, note that $\boldsymbol{v}_{\infty}^{(k)}(\boldsymbol{x}) = \tilde{\boldsymbol{e}}_k - \boldsymbol{v}_{\mathrm{bt}}^{(k)}(\boldsymbol{x})$. This reflects the fact already noted that the relative effects of these flow contributors—uniform flow and body translation—are indistinguishable; the additional $\tilde{\boldsymbol{e}}_k$ accounts for the change in the reference frame between the inertial and windtunnel frames. We note that, in a two-dimensional problem, there would only be two basis sheets and velocity fields.

We can also express the basis velocity field in terms of its basis scalar potential; the kth such field satisfies

$$\nabla^2 \varphi_{\mathrm{bt}}^{(k)} = 0, \tag{4.136}$$

and must decay to zero at infinity and be subject to

$$\frac{\partial \varphi_{\mathrm{bt}}^{(k)}}{\partial n} = \boldsymbol{n} \cdot \tilde{\boldsymbol{e}}_k \tag{4.137}$$

on $\mathcal{S}_\mathrm{b}$. The basis scalar potentials for uniform flow and body translation obviously are equal and opposite, $\varphi_{\infty}^{(k)} = -\varphi_{\mathrm{bt}}^{(k)}$.

As we did for the case with uniform flow, we can also define an equivalent basis vector potential field to describe this flow. This potential, too, satisfies Laplace's equation; it also satisfies the boundary conditions,

$$\boldsymbol{n} \times \boldsymbol{\Psi}_{\mathrm{bt}}^{(k)}(\boldsymbol{x}) = \frac{1}{n_\mathrm{d}-1} \boldsymbol{n} \times (\tilde{\boldsymbol{e}}_k \times (\boldsymbol{x} - \boldsymbol{X}_\mathrm{r})), \text{ for } \boldsymbol{x} \in \mathcal{S}_\mathrm{b},$$
$$\boldsymbol{\Psi}_{\mathrm{bt}}^{(k)} \to 0, \text{ as } |\boldsymbol{x} - \boldsymbol{X}_\mathrm{r}| \to \infty. \tag{4.138}$$

The correspondence between the kth potential and that due to uniform flow is, at all $\boldsymbol{x} \in \mathcal{V}_f$,

$$\boldsymbol{\Psi}_{\infty}^{(k)}(\boldsymbol{x}) = \frac{1}{n_\mathrm{d}-1} \tilde{\boldsymbol{e}}_k \times (\boldsymbol{x} - \boldsymbol{X}_\mathrm{r}) - \boldsymbol{\Psi}_{\mathrm{bt}}^{(k)}(\boldsymbol{x}). \tag{4.139}$$

Body Rotation Basis Fields Similarly, the basis vortex sheet due to body rotation about the kth axis, $\boldsymbol{\gamma}_{\mathrm{br}}^{(k)}$, is given by the solution of

$$\mathcal{F}\left[\boldsymbol{\gamma}_{\mathrm{br}}^{(k)}\right](\boldsymbol{x}) = \frac{1}{2}\boldsymbol{x}_{\perp}^{(k)} - \fint_{\mathcal{S}_\mathrm{b}} \left[\boldsymbol{K}(\boldsymbol{x}-\boldsymbol{y}) \times \left(\boldsymbol{n}_y \times \boldsymbol{y}_{\perp}^{(k)}\right) - \boldsymbol{K}(\boldsymbol{x}-\boldsymbol{y}) \left(\boldsymbol{n}_y \cdot \boldsymbol{y}_{\perp}^{(k)}\right)\right] \mathrm{d}S(\boldsymbol{y}), \tag{4.140}$$

where, for short-hand, we have defined $\boldsymbol{x}_{\perp}^{(k)} := \tilde{\boldsymbol{e}}_k \times (\boldsymbol{x} - \boldsymbol{X}_r)$, the local rigid-body velocity at $\boldsymbol{x}$ due to unit angular velocity about the kth body-fixed axis. In a two-dimensional problem, there is only a single unit sheet, corresponding to $k = 3$. The constraint on this sheet requires that its circulation balance the circulation due to the unit angular velocity, so that there is no net circulation about the body:

$$\int_{\mathcal{S}_b} \boldsymbol{\gamma}_{br}^{(3)} \, dS = -2\mathcal{V}_b \boldsymbol{e}_3, \tag{4.141}$$

where $\mathcal{V}_b$ denotes the area (the volume per unit depth) of the body. The modified velocity field associated with the body's angular velocity and its reactive vortex sheet is given by

$$\sum_{k=1}^{3} \tilde{\Omega}_k \boldsymbol{v}_{br}^{(k)}(\boldsymbol{x}). \tag{4.142}$$

where each basis field is

$$\boldsymbol{v}_{br}^{(k)}(\boldsymbol{x}) = \int_{\mathcal{S}_b} \left[\boldsymbol{K}(\boldsymbol{x} - \boldsymbol{y}) \times \left(\boldsymbol{n}_y \times \boldsymbol{y}_{\perp}^{(k)} + \boldsymbol{\gamma}_{br}^{(k)}(\boldsymbol{y}) \right) - \boldsymbol{K}(\boldsymbol{x} - \boldsymbol{y}) \left(\boldsymbol{n}_y \cdot \boldsymbol{y}_{\perp}^{(k)} \right) \right] dS(\boldsymbol{y}). \tag{4.143}$$

A two-dimensional cylinder of radius R provides a simple example to understand this modified velocity field. In this geometry, centered at the origin, positions on the body surface are aligned with the unit normal, so $\boldsymbol{x} = R\boldsymbol{n}$ for all $\boldsymbol{x} \in \mathcal{S}_b$. Thus, the second of the two terms in the surface integrals in (4.140) and (4.143) vanishes. The unit vortex sheet strength that satisfies the integral equation (4.140) and the constraint (4.141) is given by $\boldsymbol{\gamma}_{br}^{(3)}(\boldsymbol{x}) = -\boldsymbol{n} \times (\boldsymbol{e}_3 \times \boldsymbol{x}) = -R\boldsymbol{e}_3$. This sheet exactly cancels the effect of the body's angular velocity on the velocity field in the surrounding fluid, leaving a quiescent fluid in this region. This is, of course, the expected result, since the rotation of a cylinder cannot induce motion in the fluid in the absence of viscosity.

Alternatively, the integral in (4.143) can be expressed as the gradient of the basis scalar potential, $\varphi_{br}^{(k)}$, which satisfies Laplace's equation,

$$\nabla^2 \varphi_{br}^{(k)} = 0, \tag{4.144}$$

decays to zero at infinity and obeys

$$\frac{\partial \varphi_{br}^{(k)}}{\partial n} = ((\boldsymbol{x} - \boldsymbol{X}_r) \times \boldsymbol{n}) \cdot \tilde{\boldsymbol{e}}_k \equiv \boldsymbol{n} \cdot \boldsymbol{x}_{\perp}^{(k)} \tag{4.145}$$

for all $\boldsymbol{x} \in \mathcal{S}_b$.

The corresponding kth basis vector potential that equivalently describes this flow also satisfies Laplace's equation, as well as the boundary condition

$$\boldsymbol{n} \times \boldsymbol{\Psi}^{(k)}_{\text{br}}(\boldsymbol{x}) = -\frac{1}{n_{\text{d}}} \boldsymbol{n} \times [(\boldsymbol{x} - \boldsymbol{X}_{\text{r}}) \times (\tilde{\boldsymbol{e}}_k \times (\boldsymbol{x} - \boldsymbol{X}_{\text{r}}))], \text{ for } \boldsymbol{x} \in \mathcal{S}_{\text{b}}, \tag{4.146}$$

and vanishes at infinity.

Analogous to the relationship between the basis flow fields due to uniform flow and to body translation, one can define a basis flow field corresponding to a stationary body subjected to a fluid with uniform angular velocity at infinity. In particular, we define the basis velocity field

$$\boldsymbol{v}^{(k)}_{\text{r}\infty}(\boldsymbol{x}) := \tilde{\boldsymbol{e}}_k \times (\boldsymbol{x} - \boldsymbol{X}_{\text{r}}) - \boldsymbol{v}^{(k)}_{\text{br}}(\boldsymbol{x}). \tag{4.147}$$

This equivalently represents the negative of the velocity field observed from the body frame of reference when the body's unit angular velocity is about the $\tilde{x}_k$ axis passing through $\boldsymbol{X}_{\text{r}}$; it has vanishing normal component on $\mathcal{S}_{\text{b}}$. We use the subscript r∞ to describe these basis fields. The corresponding basis vector potential is given by

$$\boldsymbol{\Psi}^{(k)}_{\text{r}\infty}(\boldsymbol{x}) := -\frac{1}{n_{\text{d}}}(\boldsymbol{x} - \boldsymbol{X}_{\text{r}}) \times (\tilde{\boldsymbol{e}}_k \times (\boldsymbol{x} - \boldsymbol{X}_{\text{r}})) - \boldsymbol{\Psi}^{(k)}_{\text{br}}(\boldsymbol{x}). \tag{4.148}$$

and this has the boundary conditions

$$\begin{aligned} &\boldsymbol{n} \times \boldsymbol{\Psi}^{(k)}_{\text{r}\infty}(\boldsymbol{x}) = 0, \text{ for } \boldsymbol{x} \in \mathcal{S}_{\text{b}}, \\ &\boldsymbol{\Psi}^{(k)}_{\text{r}\infty} \to -\frac{1}{n_{\text{d}}}(\boldsymbol{x} - \boldsymbol{X}_{\text{r}}) \times (\tilde{\boldsymbol{e}}_k \times (\boldsymbol{x} - \boldsymbol{X}_{\text{r}})), \text{ as } |\boldsymbol{x} - \boldsymbol{X}_{\text{r}}| \to \infty. \end{aligned} \tag{4.149}$$

Time Dependence of the Basis Sheets and Velocity Fields By the conformal mapping approach used in the previous section, we found that the basis vortex sheets and modified velocity fields associated with body motion and uniform flow were time invariant when viewed from the body-fixed frame of reference. In fact, this time invariance holds in general, in both two- and three-dimensions, for a single rigid body. For example, if we write the basis vortex sheet due to body translation in the kth coordinate direction in the body-fixed basis, and express it as a function of position, $\boldsymbol{x}$, on the body surface (and therefore fixed in the body frame of reference), then the time dependence is limited to the basis vectors themselves, which change through body rotation:

$$\boldsymbol{\gamma}^{(k)}_{\text{bt}}(\boldsymbol{x}, t) = \tilde{\gamma}^{(k)}_{\text{bt},j}(\boldsymbol{x})\tilde{\boldsymbol{e}}_j(t), \tag{4.150}$$

where summation over $j = 1, 2, 3$ is implied in three dimensions (or just $j = 3$ in two dimensions). In other words, each of the components of the vector-valued basis vortex sheet strengths associated with body translation, when expressed in the body-fixed coordinate system, depends only on the shape of the body, and is therefore invariant in time. Similarly, one finds invariance of $\tilde{\gamma}^{(k)}_{\infty,j}(\boldsymbol{x})$ and $\tilde{\gamma}^{(k)}_{\text{br},j}(\boldsymbol{x})$, the components of the respective basis sheets induced by uniform flow and body rotation, in the body-fixed coordinate system. The same invariance also holds for the body-fixed components of the basis velocity fields when evaluated at any point, $\tilde{\boldsymbol{x}}$, in the fluid that remains fixed relative to the body. In contrast, the components of the vorticity-induced sheet, $\boldsymbol{\gamma}_{\text{v}}$, and the associated basis velocity field, $\boldsymbol{v}_{\text{v}}$, are generally time dependent, due to the advection (and possibly, the time-varying strength) of the fluid vorticity.

Note 4.6.1: Useful Properties of the Basis Scalar Potential Fields

As evident in their governing equations, the basis scalar potentials induced by body motion, $\varphi_{\mathrm{bt}}^{(k)}$ and $\varphi_{\mathrm{br}}^{(k)}$—as well as their associated velocity fields, $\boldsymbol{v}_{\mathrm{bt}}^{(k)}$ and $\boldsymbol{v}_{\mathrm{br}}^{(k)}$—are ***intrinsic properties of the body***, defined only by its shape, and as such, remain invariant as long as the shape does not change. This aspect endows these potential fields with certain useful properties, some of which we will highlight here.

Let us define a potential field $\Upsilon_k := \tilde{x}_k - \varphi_{\mathrm{bt}}^{(k)}$. It is easy to confirm that this field is simply the scalar potential field about the body due to a uniform flow of unit speed in the kth body-fixed coordinate direction; its gradient is the basis velocity field in that direction, $\nabla\Upsilon_k = \tilde{\boldsymbol{e}}_k - \nabla\varphi_{\mathrm{bt}}^{(k)} \equiv \boldsymbol{v}_{\infty}^{(k)}$. This potential field clearly satisfies Laplace's equation. Furthermore, one of its key properties is that its normal derivative vanishes on the surface of the body:

$$\boldsymbol{n} \cdot \nabla\Upsilon_k = \boldsymbol{n} \cdot \nabla\tilde{x}_k - \frac{\partial\varphi_{\mathrm{bt}}^{(k)}}{\partial n} = \boldsymbol{n} \cdot \tilde{\boldsymbol{e}}_k - \boldsymbol{n} \cdot \tilde{\boldsymbol{e}}_k = 0. \tag{4.151}$$

At large distances from the body, where $\varphi_{\mathrm{bt}}^{(k)}$ becomes weak, $\Upsilon_k \to \tilde{x}_k$ and $\nabla\Upsilon_k \to \tilde{\boldsymbol{e}}_k$.

Let us also define a potential $\Xi_k := \varphi_{\mathrm{br}}^{(k)} + (\tilde{\boldsymbol{e}}_k \times \boldsymbol{X}_{\mathrm{r}})_j \varphi_{\mathrm{bt}}^{(j)}$, where summation is implied over the repeated index j. This, too, satisfies Laplace's equation; on the surface $\mathcal{S}_{\mathrm{b}}$, it can be confirmed that

$$\boldsymbol{n} \cdot \nabla\Xi_k = (\tilde{\boldsymbol{e}}_k \times \boldsymbol{x}) \cdot \boldsymbol{n}. \tag{4.152}$$

Thus, Ξ_k represents the scalar potential field created when the body has angular velocity of unit speed about an axis parallel to $\tilde{x}_k$, but passing through the origin of the inertial coordinate system rather than the reference point $\boldsymbol{X}_{\mathrm{r}}$. It vanishes at points infinitely far from the body. Note that the normal component of $\tilde{\boldsymbol{e}}_k \times \boldsymbol{x} - \nabla\Xi_k$ clearly vanishes on $\mathcal{S}_{\mathrm{b}}$. Indeed, this is the (negative of the) velocity field that would be observed in the fluid from the reference frame fixed to the body, when the body has the aforementioned angular motion about the axis through the origin.

It will be useful, in some of their applications, to calculate these fields' partial derivative with respect to time. They are invariant when viewed from the body-fixed reference frame; thus, their rates of change at a fixed point in inertial space can be computed by using Eq. (2.23). For Υ_k, this leads to

$$\frac{\partial \Upsilon_k}{\partial t} = -\boldsymbol{V}_{\mathrm{b}} \cdot \nabla \Upsilon_k, \tag{4.153}$$

where $\boldsymbol{V}_{\mathrm{b}} := \boldsymbol{V}_{\mathrm{r}} + \boldsymbol{\Omega} \times (\boldsymbol{x} - \boldsymbol{X}_{\mathrm{r}})$ is the rigid-body velocity at $\boldsymbol{x}$. The rate of change of Ξ_k is slightly more complicated; one can verify that

$$\frac{\partial \Xi_k}{\partial t} = -\boldsymbol{V}_{\mathrm{b}} \cdot \nabla \Xi_k + (\tilde{\boldsymbol{e}}_k \times (\boldsymbol{V}_{\mathrm{r}} - \boldsymbol{\Omega} \times \boldsymbol{X}_{\mathrm{r}}))_j \varphi_{\mathrm{bt}}^{(j)}, \tag{4.154}$$

again, with summation implied over j.

Note 4.6.2: Useful Property of the Basis Vector Potential Fields

The basis vector potential fields due to body motion defined earlier in this section also have useful properties that we will have need to exploit later in the book. Let us denote by the generic symbol $\boldsymbol{\Psi}^{(k)}$ the vector potential field induced by either unit body translation or body rotation, $\boldsymbol{\Psi}_{\mathrm{bt}}^{(k)}$ or $\boldsymbol{\Psi}_{\mathrm{br}}^{(k)}$, in the kth body-fixed coordinate direction. This field, like the other basis fields that describe the body-induced flows, is an intrinsic property of the body. This means, in particular, that the components of the vector potential field, when expressed in the body-fixed coordinate system, are invariant at any fixed point $\tilde{\boldsymbol{x}} := \boldsymbol{x} - \boldsymbol{X}_{\mathrm{r}}$ in that body frame. However, we must also account for the rotation of the unit vectors in this basis. We considered such a situation in developing equation (2.27). Applying this, for either $\boldsymbol{\Psi}_{\mathrm{bt}}^{(k)}$ or $\boldsymbol{\Psi}_{\mathrm{br}}^{(k)}$, we get

$$\frac{\partial \boldsymbol{\Psi}^{(k)}}{\partial t} = \nabla \times \left(\boldsymbol{V}_{\mathrm{b}} \times \boldsymbol{\Psi}^{(k)} \right), \tag{4.155}$$

where $\boldsymbol{V}_{\mathrm{b}} := \boldsymbol{V}_{\mathrm{r}} + \boldsymbol{\Omega} \times \tilde{\boldsymbol{x}}$.

4.6.3 Flow Decomposition and the Kinetic Energy: Added Mass

The decomposition of the flow field presented in the previous section can be substituted into the kinetic energy of the flow to achieve some results of later utility. To facilitate this, we will apply the decomposition in its potential form (4.113); the expression of the kinetic energy in terms of the scalar and vector potentials was already given in (3.30), rewritten here with a slight change in notation:

$$T_f = \frac{1}{2}\rho \int_{\mathcal{V}_f} \boldsymbol{\Psi}_\mathrm{v} \cdot \omega \, \mathrm{d}V - \frac{1}{2}\rho \int_{\mathcal{S}_\mathrm{b}} \boldsymbol{\Psi}_\mathrm{v} \cdot (\boldsymbol{n} \times \boldsymbol{v}_\mathrm{v}) \, \mathrm{d}S - \frac{1}{2}\rho \int_{\mathcal{S}_\mathrm{b}} \varphi \frac{\partial \varphi}{\partial n} \, \mathrm{d}S. \tag{4.156}$$

As we noted just below its original definition (3.22), the kinetic energy is defined in the inertial reference frame, in which the fluid is at rest at infinity. Thus, we exclude the contribution of uniform flow from φ.

The form (4.156) distinguishes the vortical (the first two terms, which we will denote by $T_{f,\mathrm{v}}$) from the irrotational (the final term, denoted by $T_{f,\mathrm{ir}}$) contribution to the energy. Let us go further now, and decompose $T_{f,\mathrm{ir}}$ into flow contributors. We could substitute the decomposition for scalar potential directly into the form given in (4.156). However, in order to reveal an important feature of this term, let us first recall that this surface integral was originally developed from the integral of $\rho|\nabla\varphi|^2/2$ over the fluid region. Into that volumetric form, we can substitute the velocity decomposition (4.112), minus $\boldsymbol{v}_\mathrm{v}$ (which is accounted for in $T_{f,\mathrm{v}}$) and without the uniform flow contribution. It is easy to see that this product of $\nabla\varphi$ with itself will lead to a sum of terms involving interactions between the various basis velocity fields. For example, from the interactions among the basis velocity fields due to body translation, we have the sum of n_d^2 terms (9 terms in three dimensions, or 4 terms in a planar context),

$$\frac{1}{2} \sum_{j,k=1}^{3} \tilde{V}_{\mathrm{r},j} \tilde{V}_{\mathrm{r},k} \int_{\mathcal{V}_f} \rho \boldsymbol{v}_\mathrm{bt}^{(j)} \cdot \boldsymbol{v}_\mathrm{bt}^{(k)} \, \mathrm{d}V. \tag{4.157}$$

The integral has units of mass (or mass per unit depth in a two dimensional context), since the basis velocity fields are dimensionless. Indeed, the n_d^2 integrals in this sum constitute the components (in the body-fixed coordinate system) of a rank-2 tensor, called the **translational added mass tensor**. Using index notation to abbreviate the double summation, we can write this set of terms as

$$\frac{1}{2} \tilde{M}_{jk}^{FV} \tilde{V}_{\mathrm{r},j} \tilde{V}_{\mathrm{r},k}, \tag{4.158}$$

where the added mass tensor's components are defined as

$$\tilde{M}_{jk}^{FV} := \rho \int_{\mathcal{V}_f} \boldsymbol{v}_\mathrm{bt}^{(j)} \cdot \boldsymbol{v}_\mathrm{bt}^{(k)} \, \mathrm{d}V = -\rho \int_{\mathcal{S}_\mathrm{b}} \varphi_\mathrm{bt}^{(j)} \frac{\partial \varphi_\mathrm{bt}^{(k)}}{\partial n} \, \mathrm{d}S; \tag{4.159}$$

the superscript notation on $\tilde{M}_{jk}^{FV}$ will be explained in our later discussion of forces and moments in Chap. 6. Based on the manner in which the basis velocities enter their definition, these added mass components are clearly symmetric with respect to the indices j and k. Another form for these components can be obtained by replacing the normal derivative of the basis scalar potential with its boundary condition on $\mathcal{S}_\mathrm{b}$ from (4.137). Then,

$$\tilde{M}_{jk}^{FV} = \left(-\rho \int_{S_b} \varphi_{bt}^{(j)} \boldsymbol{n} \, dS \right) \cdot \tilde{\boldsymbol{e}}_k. \tag{4.160}$$

Analogously, the terms in $T_{f,ir}$ involving only interactions between the body rotation-induced basis fields can be written collectively as

$$\frac{1}{2} \tilde{M}_{jk}^{M\Omega} \tilde{\Omega}_j \tilde{\Omega}_k, \tag{4.161}$$

where

$$\tilde{M}_{jk}^{M\Omega} := \rho \int_{\mathcal{V}_f} \boldsymbol{v}_{br}^{(j)} \cdot \boldsymbol{v}_{br}^{(k)} \, dV = -\rho \int_{S_b} \varphi_{br}^{(j)} \frac{\partial \varphi_{br}^{(k)}}{\partial n} \, dS. \tag{4.162}$$

These are the body-fixed components of the **rotational added mass tensor**, and have units of moment of inertia (per unit depth, in two dimensions); there are 9 such components in a three-dimensional problem, but only 1 in two dimensions. These components, too, are symmetric with respect to j and k. Again, we can derive another form for them by substituting the boundary condition (4.145) for the normal derivative of the scalar potential, taking care to explicitly acknowledge that the motion is expressed relative to the body's reference point, $\boldsymbol{X}_r$:

$$\tilde{M}_{jk}^{M\Omega} = \left(-\rho \int_{S_b} (\boldsymbol{x} - \boldsymbol{X}_r) \times \varphi_{br}^{(j)} \boldsymbol{n} \, dS \right) \cdot \tilde{\boldsymbol{e}}_k. \tag{4.163}$$

The interactions between the body translation and rotation basis velocity fields in $|\boldsymbol{v}|^2$ lead to 18 terms—twice the product of the translational and rotational components—or, in two dimensions, with two translational and one rotational component, 4 terms:

$$\frac{1}{2} \tilde{M}_{jk}^{MV} \tilde{\Omega}_j \tilde{V}_{r,k} + \frac{1}{2} \tilde{M}_{jk}^{F\Omega} \tilde{V}_{r,j} \tilde{\Omega}_k, \tag{4.164}$$

where the components of the **translational-rotational** and **rotational-translational added mass tensors** are defined, respectively, as

$$\tilde{M}_{jk}^{MV} := \rho \int_{\mathcal{V}_f} \boldsymbol{v}_{br}^{(j)} \cdot \boldsymbol{v}_{bt}^{(k)} \, dV = -\rho \int_{S_b} \varphi_{br}^{(j)} \frac{\partial \varphi_{bt}^{(k)}}{\partial n} \, dS, \tag{4.165}$$

$$\tilde{M}_{jk}^{F\Omega} := \rho \int_{\mathcal{V}_f} \boldsymbol{v}_{bt}^{(j)} \cdot \boldsymbol{v}_{br}^{(k)} \, dV = -\rho \int_{S_b} \varphi_{bt}^{(j)} \frac{\partial \varphi_{br}^{(k)}}{\partial n} \, dS. \tag{4.166}$$

From their forms in terms of basis velocity fields, it is easy to see that $\tilde{M}_{jk}^{MV} = \tilde{M}_{kj}^{F\Omega}$—***they are transposes of each other***; thus, there are only 9 independent coupling terms (2 in two dimensions), and these can be unified under either one of these two tensors.

Alternative forms for each set of added mass components are derived by substituting the boundary conditions for the basis scalar potentials, as above:

$$\tilde{M}^{MV}_{jk} = \left(-\rho \int_{S_b} \varphi^{(j)}_{br} \boldsymbol{n}\, dS \right) \cdot \tilde{\boldsymbol{e}}_k, \qquad \tilde{M}^{F\Omega}_{jk} = \left(-\rho \int_{S_b} (\boldsymbol{x} - \boldsymbol{X}_r) \times \varphi^{(j)}_{bt} \boldsymbol{n}\, dS \right) \cdot \tilde{\boldsymbol{e}}_k. \tag{4.167}$$

Thus, we obtain the following:

Result 4.6: Kinetic Energy of the Flow About a Rigid Body

The total kinetic energy of an incompressible flow about a single rigid body, in a frame of reference in which the fluid is at rest at infinity, can be written as

$$T_f = T_{f,v} + \frac{1}{2}\tilde{M}^{FV}_{jk} \tilde{V}_{r,j}\tilde{V}_{r,k} + \frac{1}{2}\tilde{M}^{M\Omega}_{jk}\tilde{\Omega}_j\tilde{\Omega}_k + \tilde{M}^{F\Omega}_{jk}\tilde{V}_{r,j}\tilde{\Omega}_k, \tag{4.168}$$

where $T_{f,v}$ is the contribution of fluid vorticity about the (stationary) body. The remaining terms, collectively denoted by $T_{f,ir}$, measure the influence of body motion relative to the fluid medium; their contributions are assessed through the added mass tensor components, $\tilde{M}^{FV}_{jk}$, $\tilde{M}^{M\Omega}_{jk}$ and $\tilde{M}^{F\Omega}_{jk}$, defined in (4.159), (4.162) and (4.166), respectively.

If we collect the added mass coefficients into matrices, and write the body motion components as column arrays, then we can also write the irrotational part of this kinetic energy as

$$T_{f,ir} \equiv \frac{1}{2} \begin{pmatrix} \tilde{\boldsymbol{\Omega}}^T & \tilde{\mathbf{V}}_r^T \end{pmatrix} \begin{bmatrix} \tilde{\mathbf{M}}^{M\Omega} & \tilde{\mathbf{M}}^{MV} \\ \tilde{\mathbf{M}}^{F\Omega} & \tilde{\mathbf{M}}^{FV} \end{bmatrix} \begin{pmatrix} \tilde{\boldsymbol{\Omega}} \\ \tilde{\mathbf{V}}_r \end{pmatrix}. \tag{4.169}$$

Perhaps the most important conclusion we can draw from Result 4.6 is that the flow kinetic energy due to body motion is directly dependent upon the components of that motion, and thus, must vanish instantaneously when all motion ceases. Any persistence of non-zero kinetic energy in such circumstances is due to the vorticity in the fluid.

It is informative to compare this expression of kinetic energy with the intrinsic kinetic energy, T_b, of a rigid body of uniform density whose mass is $\mathcal{M}$ and whose moment of inertia components in the body-fixed coordinate system, about the reference point, are $\tilde{I}_{r,jk}$. This kinetic energy is given by

$$T_b = \frac{1}{2}\mathcal{M}\tilde{V}_{r,j}\tilde{V}_{r,j} + \frac{1}{2}\tilde{I}_{r,jk}\tilde{\Omega}_j\tilde{\Omega}_k. \tag{4.170}$$

The most notable difference between (4.168) and (4.170) can be observed for a body in purely translational motion. For a general body shape, the kinetic energy of the resulting fluid motion depends on the orientation of the body with respect to this motion, whereas the body's kinetic energy is unaffected by this orientation. The fluid kinetic energy is only independent of the orientation if the body's shape is isotropic (i.e., a sphere, or a circle in two dimensions), which ensures that the translational added mass tensor is itself isotropic.

An intuitive explanation for this dependence on orientation can be found by considering the squared magnitudes of the n_d basis velocity fields, $|\boldsymbol{v}_{\mathrm{bt}}^{(j)}|^2$, the integrals of which inhabit the corresponding diagonal entries of the translational added mass tensor. The magnitude of this velocity field is inherently weaker when the translation occurs in a direction along which the body exhibits a smaller profile. This is trivially true, for example, for an infinitely thin flat plate moving tangent to itself. The basis velocity field corresponding to any body-fixed axis tangent to the plate is identically zero, as are the row and column of the translational added mass tensor associated with this axis. Thus, this motion would create exactly zero kinetic energy in the fluid in a strictly inviscid setting.

In fact, as an integral of $\rho|\nabla\varphi|^2/2$, we know that the irrotational part of the kinetic energy $T_{f,\mathrm{ir}}$ must be greater than or equal to zero, and that it is strictly greater than zero if the body occupies non-zero volume. Thus, we can conclude the following, by definition:

Note 4.6.3: Positive Semi-definiteness of Added Mass

The overall added mass tensor in (4.169) is positive semi-definite, and is positive definite for a body of non-zero volume (or non-zero area in two dimensions). Furthermore, the same is true for the translational and rotational added mass tensors individually. We can prove this latter statement by simply considering cases in which the angular velocity and translational velocity are set to zero, respectively.

We will provide more intuition for added mass in Chap. 6.

4.7 Multipole Expansion of the Flow Field

On various occasions in this book, we will have need to evaluate the velocity field (4.8) at distances that are large compared with the extent of the body and the vorticity field. At such distances, this field can be re-written in a form in which it appears as though all flow contributors, except for the uniform flow, are concentrated at the origin. This form is realized by Taylor expanding the kernel function, $\boldsymbol{K}(\boldsymbol{x}-\boldsymbol{y})$, about $\boldsymbol{y}=0$. Since this kernel's form depends on whether the flow is two-dimensional or three-dimensional, we treat the expansion of these cases separately.

Also, note that we rely heavily on Cartesian index notation, reviewed in Sect. A.1.1, in order to write the results as succinctly as possible.

4.7.1 Two-Dimensional Flow

The expansion of the pth component of $\boldsymbol{K}(\boldsymbol{x}-\boldsymbol{y})$, defined from (3.71), about $\boldsymbol{y}=0$ is easily shown to be

$$2\pi K_p(\boldsymbol{x}-\boldsymbol{y}) = -\frac{\partial}{\partial x_p}(\log r) + y_i\frac{\partial^2}{\partial x_i \partial x_p}(\log r) - \frac{1}{2!}y_i y_j \frac{\partial^3}{\partial x_i \partial x_j \partial x_p}(\log r) - \cdots, \tag{4.171}$$

where $r := |\boldsymbol{x}| = (x_1^2 + x_2^2)^{1/2}$. Each of the terms in the expansion after the first term contains one or more **directional derivatives** of $\log r$ in the direction of $\boldsymbol{y}$. It is useful to note that the nth directional derivative can be written vectorially as $(\boldsymbol{y}\cdot\nabla)^n \log r$.

When we substitute the kernel expansion (4.171) into (4.8), the integrations over the fluid region $\mathcal{V}_f$ and the body surface $\mathcal{S}_b$ are reduced to moments about the origin of the vorticity and the surface distributions. The kth component of the velocity at large distance can be written as

$$\begin{aligned} &2\pi v_k(\boldsymbol{x}) \\ &= -\epsilon_{p3k}\left[S^{\text{v}}\frac{\partial}{\partial x_p}(\log r) - S_i^{\text{v}}\frac{\partial^2}{\partial x_i \partial x_p}(\log r) + \frac{1}{2!}S_{ij}^{\text{v}}\frac{\partial^3}{\partial x_i \partial x_j \partial x_p}(\log r) - \cdots\right] \\ &\quad + Q^{\text{s}}\frac{\partial}{\partial x_k}(\log r) - Q_i^{\text{s}}\frac{\partial^2}{\partial x_i \partial x_k}(\log r) + \frac{1}{2!}Q_{ij}^{\text{s}}\frac{\partial^3}{\partial x_i \partial x_j \partial x_k}(\log r) - \cdots. \end{aligned} \tag{4.172}$$

In this expression, we have made use of the permutation symbol, ϵ_{ijk}, with one of the indices fixed at 3—corresponding to the out of plane direction—in order to represent the planar cross product. All other indices in this expression take values for the planar directions, either 1 or 2, and as usual in index notation, repeated indices in a term imply summation.

Equation (4.172) is called a **multipole expansion** of the two-dimensional velocity field. It expresses the velocity field at large distances as a convergent series of singularities (at the origin) of progressively increasing powers of $1/r$ and stronger angular dependence. There are two sets of tensorial coefficients in the expansion, $S^{\text{v}}_{i_1 i_2 \cdots i_n}$ and $Q^{\text{s}}_{i_1 i_2 \cdots i_n}$, for all integers $n \geq 0$. The two sets of tensorial coefficients appear in distinctly different parts of the expansion. The first part, with coefficients denoted by S, has the form of the curl of a vector potential (streamfunction in two dimensions), while the second part, denoted by Q, is apparently derived from the gradient of a scalar potential. This distinction into two parts is consistent with the Helmholtz decomposition (3.1), and is similarly apparent in the underlying velocity integral formula (4.8). Their superscripts denote that they represent the effects of either vorticity (v) or source (s) distributions in the flow and body surface. In particular, $S^{\mathrm{v}}_{i_1 i_2 \cdots i_n}$ contain moments of the fluid vorticity as well as the surface vorticity distribution:

$$S^{\mathrm{v}}_{i_1 i_2 \cdots i_n} = \int\limits_{\mathcal{V}_f} y_{i_1} y_{i_2} \cdots y_{i_n} \omega(\boldsymbol{y})\, \mathrm{d}V(\boldsymbol{y}) + \int\limits_{\mathcal{S}_b} y_{i_1} y_{i_2} \cdots y_{i_n} \left(\boldsymbol{n}_y \times \boldsymbol{v}(\boldsymbol{y})\right) \cdot \boldsymbol{e}_3\, \mathrm{d}S(\boldsymbol{y}); \tag{4.173}$$

and the coefficients $Q^{\mathrm{s}}_{i_1 i_2 \cdots i_n}$ consist of moments of the surface source distribution:

$$Q^{\mathrm{s}}_{i_1 i_2 \cdots i_n} = \int\limits_{\mathcal{S}_b} y_{i_1} y_{i_2} \cdots y_{i_n} \left(\boldsymbol{n}_y \cdot \boldsymbol{v}(\boldsymbol{y})\right) \mathrm{d}S(\boldsymbol{y}). \tag{4.174}$$

From hereon, in order to simplify the notation, we will replace the dummy integration variable, $\boldsymbol{y}$, in these expressions with $\boldsymbol{x}$, and suppress the spatial dependence of the integrand where it is obvious.

Note that the vector potential coefficient for $n = 0$ is

$$S^{\mathrm{v}} := \int\limits_{\mathcal{V}_f} \omega\, \mathrm{d}V + \int\limits_{\mathcal{S}_b} (\boldsymbol{n} \times \boldsymbol{v}) \cdot \boldsymbol{e}_3\, \mathrm{d}S. \tag{4.175}$$

This is simply the total circulation, Γ_{tot}, in the fluid and about the body, defined by (3.38). As we stated earlier in Sect. 4.5, we require throughout this work that this circulation is zero. The scalar potential coefficient for $n = 0$ is

$$Q^{\mathrm{s}} := \int\limits_{\mathcal{S}_b} \boldsymbol{n} \cdot \boldsymbol{v}\, \mathrm{d}S. \tag{4.176}$$

This, by the no-penetration condition (3.11), represents the net volume flow rate through the surface (if it is permeable) or the volumetric expansion or contraction of the body. As we discussed in Chap. 3, and stated mathematically with Eq. (2.29), we assume throughout this book that this coefficient is identically zero. Thus, the leading terms in both parts of the expansion (4.172)—representing, respectively, a point vortex and a monopole source—are zero.

As mentioned above, the velocity expansion (4.172) has two distinct parts, attributed, respectively, to curl of a vector potential and gradient of a scalar potential. But because the flow is, by assumption, irrotational at large distances, we must also be able to write the velocity entirely as *either* the curl of a vector potential *or* the gradient of a scalar potential; thus, we should be able to reformulate the expansion (4.172) accordingly. This means that each set of coefficients should be representable in an alternative form that unifies its corresponding part of the expansion with the other. Here, we will develop a scalar potential form for the first part in order to unify its form with the second part. This will have value later, when we need the form of this potential at large distances.

To find this form, we first need an identity for the moments. In general, any vector potential coefficient, $S_{i_1 i_2 \cdots i_n}$ (for $n > 0$), can be written in terms of corresponding scalar potential coefficients:

$$nS_{i_1 i_2 \cdots i_n} = \epsilon_{i_1 3m} Q_{i_2 \cdots i_n m} + \epsilon_{i_2 3m} Q_{i_1 i_3 \cdots i_n m} + \cdots + \epsilon_{i_n 3m} Q_{i_1 \cdots i_{n-1} m}. \tag{4.177}$$

This relationship holds regardless of whether the coefficients consist of moments of vorticity or source distributions (i.e., regardless of the superscript). Using index notation, it can be shown that the inverse of the relationship is

$$Q_{i_1 i_2 \cdots i_n} = \epsilon_{m3i_n} S_{i_1 \cdots i_{n-1} m} + \nu_{i_1 i_2 \cdots i_n}, \tag{4.178}$$

where

$$\nu_{i_1 i_2 \cdots i_n} := \frac{1}{n} \left(Q_{i_2 \cdots i_{n-1} qq} \delta_{i_1 i_n} + Q_{i_1 i_3 \cdots i_{n-1} qq} \delta_{i_2 i_n} + \cdots + Q_{i_1 \cdots i_{n-2} qq} \delta_{i_{n-1} i_n} \right) \tag{4.179}$$

and δ_{ij} is the Kronecker delta. This last part in the inverse relationship, $\nu_{i_1 i_2 \cdots i_n}$, makes no contribution to the expansion, since the Kronecker deltas in this part contribute only expansion terms involving the Laplacian of $\log r$, which vanishes. Thus, we can effectively define scalar potential coefficients from vector potential coefficients as

$$Q_{i_1 i_2 \cdots i_n} = \epsilon_{m3i_n} S_{i_1 \cdots i_{n-1} m}. \tag{4.180}$$

Thus, we define the scalar potential form of the vorticity coefficients as

$$Q^{\mathrm{v}}_{i_1 i_2 \cdots i_n} = \int_{\mathcal{V}_f} x_{i_1} \cdots x_{i_{n-1}} (\boldsymbol{x} \times \boldsymbol{\omega})_{i_n} \, \mathrm{d}V + \int_{\mathcal{S}_b} x_{i_1} \cdots x_{i_{n-1}} [\boldsymbol{x} \times (\boldsymbol{n} \times \boldsymbol{v})]_{i_n} \, \mathrm{d}S, \tag{4.181}$$

for all $n > 0$. As a result of this transformation of the vorticity coefficients, we can develop a unified set of multipole coefficients for the scalar potential, and we have the following:

Result 4.7: Multipole Expansion of the Two-Dimensional Flow Field

The velocity field at large distances from a body or set of bodies is the sum of a uniform flow (if present) and the gradient of a scalar potential, φ, whose form at large distances is

$$\varphi(\boldsymbol{x}) = \sum_{n=1}^{\infty} \frac{(-1)^n}{2\pi n!} Q_{i_1 i_2 \cdots i_n} \frac{\partial^n}{\partial x_{i_1} \partial x_{i_2} \cdots \partial x_{i_n}} (\log r), \tag{4.182}$$

where the scalar potential coefficients are

$$Q_{i_1 i_2 \cdots i_n} = \int\limits_{\mathcal{V}_f} x_{i_1} \cdots x_{i_{n-1}} (\boldsymbol{x} \times \boldsymbol{\omega})_{i_n} \, \mathrm{d}V + \int\limits_{\mathcal{S}_b} x_{i_1} \cdots x_{i_{n-1}} [\boldsymbol{x} \times (\boldsymbol{n} \times \boldsymbol{v}) + \boldsymbol{x} (\boldsymbol{n} \cdot \boldsymbol{v})]_{i_n} \, \mathrm{d}S. \tag{4.183}$$

The first few derivatives of $\log r$ are calculated in the appendix in Sect. A.1.2.

Note that we could also have, in similar fashion, transformed the scalar potential coefficients for the surface source distribution into vector potential coefficients in order to develop a unified expansion of the streamfunction. This is left as an exercise to the reader.

All of the expansions show that, at large distances, in addition to any uniform flow, ***the leading-order contribution to the flow field is a dipole*** (3.79), whose velocity field decays like $1/r^2$ and whose scalar potential decays like $1/r$. The vector-valued strength of this dipole is

$$\boldsymbol{Q} = \int\limits_{\mathcal{V}_f} \boldsymbol{x} \times \boldsymbol{\omega} \, \mathrm{d}V + \int\limits_{\mathcal{S}_b} \boldsymbol{x} \times (\boldsymbol{n} \times \boldsymbol{v}) \, \mathrm{d}S + \int\limits_{\mathcal{S}_b} \boldsymbol{x} (\boldsymbol{n} \cdot \boldsymbol{v}) \, \mathrm{d}S. \tag{4.184}$$

As we will see in Chap. 6, the first two terms of this vector will serve a crucial role in the calculation of force on the body. It is also important to note that, in either of the multipole expansions shown here, (4.172) or (4.182), any time dependence of the field is contained in the multipole coefficients, whereas the spatial dependence is entirely provided by the derivatives of $\log r$.

4.7.2 Three-Dimensional Flow

The derivation of the expansion in three-dimensional flow follows the two-dimensional case closely. We start with the pth component of the velocity kernel $\boldsymbol{K}(\boldsymbol{x}-\boldsymbol{y})$, defined from (3.112), which is expanded about $\boldsymbol{y} = 0$:

$$4\pi K_p(\boldsymbol{x} - \boldsymbol{y}) = \frac{\partial}{\partial x_p}\left(\frac{1}{r}\right) - y_i \frac{\partial^2}{\partial x_i \partial x_p}\left(\frac{1}{r}\right) + \frac{1}{2!} y_i y_j \frac{\partial^3}{\partial x_i \partial x_j \partial x_p}\left(\frac{1}{r}\right) - \cdots . \tag{4.185}$$

Each of the terms in this expansion contains a **spherical harmonic**—a higher-order solution of Laplace's equation—that provides the term's angular dependence about the origin. Upon substitution into the integrals in (4.8), the kth component of the velocity at large distance can be written as

$$4\pi v_k(\boldsymbol{x})$$
$$= \epsilon_{pqk}\left[S^{\mathrm{v}}_q \frac{\partial}{\partial x_p}\left(\frac{1}{r}\right) - S^{\mathrm{v}}_{iq}\frac{\partial^2}{\partial x_i \partial x_p}\left(\frac{1}{r}\right) + \frac{1}{2!}S^{\mathrm{v}}_{ijq}\frac{\partial^3}{\partial x_i \partial x_j \partial x_p}\left(\frac{1}{r}\right) - \cdots\right]$$
$$- Q^{\mathrm{s}}\frac{\partial}{\partial x_k}\left(\frac{1}{r}\right) + Q^{\mathrm{s}}_i \frac{\partial^2}{\partial x_i \partial x_k}\left(\frac{1}{r}\right) - \frac{1}{2!}Q^{\mathrm{s}}_{ij}\frac{\partial^3}{\partial x_i \partial x_j \partial x_k}\left(\frac{1}{r}\right) + \cdots, \tag{4.186}$$

where now, the moment coefficients $S^{\mathrm{v}}_{i_1 i_2 \cdots i_n q}$—the coefficients of the multipole expansion of the vector potential—have an additional index compared to the two-dimensional case to distinguish the components of vorticity:

$$S^{\mathrm{v}}_{i_1 i_2 \cdots i_n q} = \int_{\mathcal{V}_f} y_{i_1} y_{i_2} \cdots y_{i_n} \omega_q(\boldsymbol{y})\, \mathrm{d}V(\boldsymbol{y}) + \int_{\mathcal{S}_{\mathrm{b}}} y_{i_1} y_{i_2} \cdots y_{i_n} \left(\boldsymbol{n}_y \times \boldsymbol{v}(\boldsymbol{y})\right)_q \, \mathrm{d}S(\boldsymbol{y}). \tag{4.187}$$

The scalar potential coefficients of the surface source distribution, $Q^{\mathrm{s}}_{i_1 i_2 \cdots i_n}$, are given, as in two dimensions, by

$$Q^{\mathrm{s}}_{i_1 i_2 \cdots i_n} = \int_{\mathcal{S}_{\mathrm{b}}} y_{i_1} y_{i_2} \cdots y_{i_n} \left(\boldsymbol{n}_y \cdot \boldsymbol{v}(\boldsymbol{y})\right) \, \mathrm{d}S(\boldsymbol{y}). \tag{4.188}$$

Note that each index, $i_1, i_2, \ldots, i_n$, takes values 1, 2, or 3, as with any other indices in this case. As in the two-dimensional case, we will replace the dummy integration variable, $\boldsymbol{y}$, in these expressions with $\boldsymbol{x}$ in order to simplify the notation.

Here, we note that the coefficient of the first term in the vector potential expansion, S^{v}_q, is identically zero; that is,

$$\int_{\mathcal{V}_f} \boldsymbol{\omega}\, \mathrm{d}V + \int_{\mathcal{S}_{\mathrm{b}}} \boldsymbol{n} \times \boldsymbol{v}\, \mathrm{d}S = 0, \tag{4.189}$$

which expresses that the total vorticity in the problem—in the fluid, on the surface, and contained in the body motion—vanishes identically. This is proved with the use of the divergence theorem over $\mathcal{V}_f$, and specifically (A.20), in which the surface normal vector is of opposite sign to that on $\mathcal{S}_{\mathrm{b}}$. Additionally, for the same reasons as in the two-dimensional case, the first term of the scalar potential expansion of the source distribution, Q^{s}, is assumed to be zero, due to the invariant volume of the body (or zero net flow rate through the surface). Thus, the leading-order behavior in (4.186), a monopole source and a point vortex, is absent.

Similar to two dimensions, we can transform the vector potential expansion into the form of a scalar potential, using coefficients related to the original ones via

$$n S_{i_1 i_2 \cdots i_n q} = \epsilon_{i_1 q m} Q_{i_2 \cdots i_n m} + \epsilon_{i_2 q m} Q_{i_1 i_3 \cdots i_n m} + \cdots + \epsilon_{i_n q m} Q_{i_1 \cdots i_{n-1} m}, \tag{4.190}$$

and the inverse relationship

$$Q_{i_1 i_2 \cdots i_n} = \frac{n}{n+1} \epsilon_{mqi_n} S_{i_1 \cdots i_{n-1} mq} + \nu_{i_1 i_2 \cdots i_n}, \tag{4.191}$$

where the final term,

$$\nu_{i_1 i_2 \cdots i_n} := \frac{1}{n+1} \left(Q_{i_2 \cdots i_{n-1} qq} \delta_{i_1 i_n} + Q_{i_1 i_3 \cdots i_{n-1} qq} \delta_{i_2 i_n} + \cdots + Q_{i_1 \cdots i_{n-2} qq} \delta_{i_{n-1} i_n} \right), \tag{4.192}$$

makes null contribution to the expansion for the same reason as in two dimensions.

Once again, we can now obtain a unified set of multipole coefficients for the scalar potential, so that

Result 4.8: Multipole Expansion of the Three-Dimensional Flow Field

In three dimensions, the velocity at large distances from a body or set of bodies is the sum of a uniform flow (if present) and the gradient of a scalar potential, φ, whose form at large distances is

$$\varphi(\boldsymbol{x}) = -\sum_{n=1}^{\infty} \frac{(-1)^n}{4\pi n!} Q_{i_1 i_2 \cdots i_n} \frac{\partial^n}{\partial x_{i_1} \partial x_{i_2} \cdots \partial x_{i_n}} \left(\frac{1}{r} \right), \tag{4.193}$$

where the coefficients are defined as

$$\begin{aligned} Q_{i_1 i_2 \cdots i_n} = & \frac{n}{n+1} \int_{\mathcal{V}_f} x_{i_1} \cdots x_{i_{n-1}} \left(\boldsymbol{x} \times \boldsymbol{\omega} \right)_{i_n} \mathrm{d}V \\ & + \int_{S_b} x_{i_1} \cdots x_{i_{n-1}} \left[\frac{n}{n+1} \boldsymbol{x} \times (\boldsymbol{n} \times \boldsymbol{v}) + \boldsymbol{x} \left(\boldsymbol{n} \cdot \boldsymbol{v} \right) \right]_{i_n} \mathrm{d}S. \end{aligned} \tag{4.194}$$

The first few derivatives of $1/r$ are calculated in the appendix in Sect. A.1.2.

Thus, as in the two-dimensional case, ***the leading-order contribution to the flow field at large distance is a dipole***, in addition to any uniform flow; in this case, the velocity field decays like $1/r^3$ and the scalar potential like $1/r^2$. The vector-valued coefficient of this dipole has the form

$$\boldsymbol{Q} = \frac{1}{2} \int_{\mathcal{V}_f} \boldsymbol{x} \times \boldsymbol{\omega} \, \mathrm{d}V + \frac{1}{2} \int_{S_b} \boldsymbol{x} \times (\boldsymbol{n} \times \boldsymbol{v}) \, \mathrm{d}S + \int_{S_b} \boldsymbol{x} \left(\boldsymbol{n} \cdot \boldsymbol{v} \right) \mathrm{d}S. \tag{4.195}$$

Again, the first two terms of this coefficient have an important role in the force on the body, as we will see in Chap. 6.

The formulas for the multipole coefficients in this section and the last emphasize the roles of vorticity and of velocity at the surface of the body. However, in Note 4.7.1, we show an alternative form that underscores the more comprehensive role of vorticity in determining the far-field behavior of the flow.

Note 4.7.1: Multipole Coefficients in Terms of Body Vorticity

In either two- or three-dimensional flows, we can write the unified scalar potential coefficients of the expansion—(4.183) or (4.194), respectively—in a form that emphasizes the more general role of vorticity in determining the far-field behavior. To do so, let us recall once again the no-penetration condition (3.11) and its associated tangency form (3.195). These enable us to write the surface integral in these coefficients instead as a surface integral over the surface vortex sheet and, by the divergence theorem, a volume integral over the body velocity. This volume integral can be further manipulated into a form involving moments of the vorticity associated with the body motion, $\boldsymbol{\omega}_{\mathrm{b}} = \nabla \times \boldsymbol{V}_{\mathrm{b}}$, and additional moments that make identically zero contribution to the expansion. Thus, it can be shown that, modulo these null contributions, the scalar potential coefficients of the multipole expansion can be written in a general vorticity form as

$$Q_{i_1 i_2 \cdots i_n} = \frac{n}{n + n_{\mathrm{d}} - 2} \left(\int_{\mathcal{V}_f} x_{i_1} \cdots x_{i_{n-1}} (\boldsymbol{x} \times \boldsymbol{\omega})_{i_n} \, \mathrm{d}V + \int_{\mathcal{S}_{\mathrm{b}}} x_{i_1} \cdots x_{i_{n-1}} (\boldsymbol{x} \times \boldsymbol{\gamma})_{i_n} \, \mathrm{d}S + \int_{\mathcal{V}_{\mathrm{b}}} x_{i_1} \cdots x_{i_{n-1}} (\boldsymbol{x} \times \boldsymbol{\omega}_{\mathrm{b}})_{i_n} \, \mathrm{d}V \right). \quad (4.196)$$

Note that we have combined the two-dimensional and three-dimensional versions into a single equation here, reflected in the factor involving n_{d}, the number of spatial dimensions.

Note 4.7.2: Alternate Form of Dipole Coefficients

As described in the previous two sections, the leading-order behavior of the flow at large distances is described by a dipole. The vector-valued strength of this dipole, given by (4.184) and (4.195) in the respective dimensional spaces, can be slightly rewritten in a form that will be useful later when we calculate force. The reformulated version relies on the no-penetration condition, $\boldsymbol{n} \cdot \boldsymbol{v} = \boldsymbol{n} \cdot \boldsymbol{V}_{\mathrm{b}}$, to rewrite the last integral in terms of velocity interior to $\mathcal{S}_{\mathrm{b}}$. By the divergence theorem, and our general restriction that the interior velocity is divergence free, we can write the last integral as simply $\boldsymbol{V}_{\mathrm{c}}\mathcal{V}_{\mathrm{b}}$, where $\boldsymbol{V}_{\mathrm{c}}$ is the velocity of the centroid $\boldsymbol{X}_{\mathrm{c}}$ of region $\mathcal{V}_{\mathrm{b}}$. Note, as usual, our notational convention that $\mathcal{V}_{\mathrm{b}}$ represents an area in two dimensions. Thus, the vector-valued dipole coefficient can be written in general as

$$\boldsymbol{Q} = \frac{1}{n_{\mathrm{d}} - 1}\left(\int_{\mathcal{V}_f} \boldsymbol{x} \times \boldsymbol{\omega}\, \mathrm{d}V + \int_{\mathcal{S}_{\mathrm{b}}} \boldsymbol{x} \times (\boldsymbol{n} \times \boldsymbol{v})\, \mathrm{d}S\right) + \boldsymbol{V}_{\mathrm{c}}\mathcal{V}_{\mathrm{b}}. \qquad (4.197)$$

4.7.3 Two-Dimensional Expansion in Complex Form

Before we proceed from this discussion of multipole expansions, it is useful to express our vectorial two-dimensional expansions in complex form. In complex analysis, our starting point is the well-known Laurent series of a holomorphic function, in which only the negative powers of the series are non-zero because the domain of analyticity extends to infinity. We express this expansion for the complex potential:

$$2\pi F(z) = a_0 \log z + \frac{a_1}{z} + \frac{a_2}{z^2} + \cdots . \qquad (4.198)$$

We have left out any uniform flow, which contributes additively to all of the expansions. To obtain the values of the coefficients, a_n, in a form most amenable for comparison with our vector forms, we first differentiate the expansion (4.198) to form the expansion for the complex velocity:

$$2\pi w(z) = \frac{a_0}{z} - \frac{a_1}{z^2} - \frac{2a_2}{z^3} + \cdots . \qquad (4.199)$$

The coefficients, a_n, can be obtained from the classical formula

$$a_0 = -\mathrm{i} \int_C w(\lambda)\,\mathrm{d}\lambda, \qquad a_n = \frac{\mathrm{i}}{n} \int_C \lambda^n w(\lambda)\,\mathrm{d}\lambda, \quad n \geq 1, \tag{4.200}$$

where C is a contour that encloses all singularities in the flow. This positively oriented contour can be deformed into a contour that wraps tightly around the body and each vortex, with each vortex contour joined to the body contour C_b by two adjacent segments of opposite sense that cancel each other out. On each vortex contour, the velocity is dominated by the vortex's induced contribution. Thus, it can be shown that, for $n = 0$,

$$a_0 = -\mathrm{i} \sum_{J=1}^{N_v} \Gamma_J - \mathrm{i} \int_{C_b} w(z)\,\mathrm{d}z, \tag{4.201}$$

and for all $n > 0$,

$$a_n = \frac{\mathrm{i}}{n} \sum_{J=1}^{N_v} \Gamma_J z_J^n + \frac{\mathrm{i}}{n} \int_{C_b} z^n w(z)\,\mathrm{d}z, \tag{4.202}$$

where, in order to simplify notation, we have renamed the dummy variable from λ to z for integration on the body contour. It should not be surprising that the leading coefficient, a_0, contains the total circulation in the fluid, given by Eq. (3.39); its real part comprises the volume flux through C_b. From hereon, as we did in the vectorial case, we will assume that a_0 is zero, since we are restricting our attention to flows in which these are zero.

The multipole expansions expressed in vectorial and complex forms are necessarily equivalent. Indeed, by inspection, the coefficients a_n are equivalent in substance to the tensorial coefficients for the streamfunction or scalar potential, $S_{i_1 \cdots i_n}$ and $Q_{i_1 \cdots i_n}$ respectively, defined in (4.173) and (4.174), or in their unified form (4.183). For example, on the positively-oriented contour C_b, one can show the following equivalence:

$$[\boldsymbol{x} \times (\boldsymbol{n} \times \boldsymbol{v}) + \boldsymbol{x}\,(\boldsymbol{n} \cdot \boldsymbol{v})]\,\mathrm{d}S \Longleftrightarrow -\mathrm{i}zw\,\mathrm{d}z. \tag{4.203}$$

For a given value of n, many of those tensorial coefficients are identical, since they are invariant to permutations of the indices; the complex coefficient, in contrast, encapsulates all moments at that value of n, as can be seen by expanding $z^n = (x+\mathrm{i}y)^n$ with the binomial theorem. For the dipole coefficient, $n = 1$, we have already shown this correspondence in our discussion of the basic singularities, in Eq. (3.91), where $\boldsymbol{Q}$ is shown in (4.184). Furthermore, it is useful to note that $1/z^n$ is related to the nth derivative of $\log z$:

$$\frac{\mathrm{d}^n}{\mathrm{d}z^n} \log z = \frac{(-1)^{n-1}(n-1)!}{z^n}, \qquad n > 0. \tag{4.204}$$

The equivalence of the complex and vectorial expansions can then be observed with the help of the directional derivative expressed in complex form (A.119).

Note 4.7.3: Multipole Coefficients for Plates

We noted in Sect. 4.2.3 that when the body is infinitely thin—i.e., a plate—the fluid velocity in the surface integrals is replaced by the jump in velocity on either side of the plate. The same is true for the multipole coefficients in these expansions. In any of the definitions of coefficients in this section, the factor $\boldsymbol{n}_y \times \boldsymbol{v}(\boldsymbol{y})$ can be replaced by $\boldsymbol{\gamma}(\boldsymbol{y})$, the local circulation of the bound vortex sheet of the plate. The resulting form of the unified scalar potential coefficients, for example, would be given by (4.196), with the corresponding re-interpretation of $\boldsymbol{\gamma}$ in the surface integral as the jump in velocity across the plate, and the final integral over $\mathcal{V}_b$ identically zero. Analogously, in the complex form of the expansion, $w(\lambda)$ in the surface integral in (4.202) is replaced by $g(\lambda) := w^+(\lambda) - w^-(\lambda)$. In both cases, the surface integral is carried out over the finite area patch or finite length curve $\mathcal{S}$ constituting the plate. In the complex form, the sign of the integral must be switched in order to conform with our convention of integrating the sheet in the direction with the positive side to the left.

Note 4.7.4: Multipole Coefficients Under Conformal Mapping

As discussed in Sect. 4.4, the closed contour C_b can always be conformally mapped from the unit circle C_c in the ζ plane. With such a conformal mapping, the integral over C_b in the multipole coefficients (4.202) can be transformed to an integral over C_c. In this transformation, we note that $w(z)\,\mathrm{d}z \equiv F'(z)\,\mathrm{d}z = \hat{F}'(\zeta)\,\mathrm{d}\zeta = \hat{w}(\zeta)\,\mathrm{d}\zeta$. Thus, for later use, we express the multipole coefficients under this transformation:

$$a_0 = -\mathrm{i}\sum_{\mathrm{J}=1}^{N_\mathrm{v}} \Gamma_\mathrm{J} - \mathrm{i}\int_{C_b} \hat{w}(\zeta)\,\mathrm{d}\zeta, \tag{4.205}$$

and

$$a_n = \frac{\mathrm{i}}{n}\sum_{\mathrm{J}=1}^{N_\mathrm{v}} \Gamma_\mathrm{J} z_\mathrm{J}^n + \frac{\mathrm{i}}{n}\int_{C_c} [z(\zeta)]^n \hat{w}(\zeta)\,\mathrm{d}\zeta, \quad n > 0. \tag{4.206}$$

Chapter 5
Edge Conditions

Contents

5.1 The Kutta Condition 162
 5.1.1 Steady Flow 164
 5.1.2 Unsteady Flow 165
5.2 Application of the Kutta Condition 173
5.3 Traditional Enforcement of the Kutta Condition: A Contrast 176
5.4 Generalized Edge Conditions 178

In Sect. 4.5 we found that there are an infinite number of possible solutions to the potential flow about a two-dimensional body, since any choice of bound circulation will leave the no-penetration condition unaffected. We observed in that discussion that we could pin this bound circulation to a unique value by setting it equal and opposite to the circulation in the surrounding fluid, in order to ensure that the global circulation is constantly zero. But how did that fluid circulation get there in the first place? We avoided this question during the discussion in Sect. 4.5. However, the reader will we recall that in Chap. 1 we presented the broad outlines of a strategy for introducing vorticity into the fluid in an inviscid flow. We will fill in the details of that strategy now, and remind ourselves of its physical foundations.

Let us consider the classical aerodynamics problem of a steady flow around an airfoil at small angle of attack. It is reasonable to expect in this flow that only one of this infinity of solutions is physically admissible. For all but one solution, the flow involves a rapid turn around the sharp trailing edge of the airfoil, leading to a velocity field with a singular behavior at the edge. However, if we manage to create such a singular flow in a real fluid at some instant, viscosity causes the boundary layer to immediately separate from the trailing edge. This separation triggers a dynamical process that brings the flow to a 'regularized' state, in which the resulting flow smoothly enters the wake from the trailing edge, and the nonphysically large magnitude of the velocity is thereby removed.

J. D. Eldredge, *Mathematical Modeling of Unsteady Inviscid Flows*, Interdisciplinary Applied Mathematics 50, https://doi.org/10.1007/978-3-030-18319-6_5

Though viscosity clearly plays an essential role in triggering this process, we can generally dispense with its intrinsic importance once the flow has been regularized. Indeed, in potential flow analysis of simple geometries, it is generally possible to find a solution that annihilates the singular velocity at the trailing edge. Thus, it is common practice to represent this short-lived process of viscous regularization by imposing a constraint on the inviscid solution. This is generally known as the **Kutta condition**, or sometimes as the Kutta–Joukowski condition, or the Joukowski hypothesis. In this chapter, we will explore the implications of this condition, and in doing so, reveal several different mathematical statements that express the condition. We will also discuss a means of generalizing—or, more precisely, relaxing—the condition in order to mathematically model a broader range of scenarios for which the Kutta condition may be inadequate. We will refer to the overall class of these as **edge conditions**, as the title of this chapter suggests.

5.1 The Kutta Condition

The Kutta condition can be expressed in a variety of forms, but they are all based on the same principle:

Result 5.1: The Kutta Condition

Among all possible solutions to a potential flow analysis of a problem with a sharp edge, choose the one for which the velocity at the edge is finite and allows the flow to leave the edge smoothly.

It is important to make a few notes on this here:

- Though this chapter will often focus on its application in two-dimensional contexts, the condition can be applied in either two-dimensional or three-dimensional settings.
- As it was originally stated, the Kutta condition only applies to steady flow. However, it is now well accepted that the Kutta condition can be imposed instantaneously in unsteady flows [15]. Clearly this requires some careful interpretation, since the underlying physical process that leads to flow regularization requires time to resolve itself. Here, we will assume that the condition is appropriate whenever we apply it, and focus only on the mathematics of its application.
- Sometimes there are multiple sharp edges in a geometry. The Kutta condition can, at least in theory, be applied at any or all of them. However, there might be compelling physical reasons why not all edges should be so constrained.
- In a general problem, the Kutta condition is used as a determinant of circulation. In a steady flow, this determination is made, once and for all, of the bound circulation about the body that produces finite velocity at the edge. In an unsteady

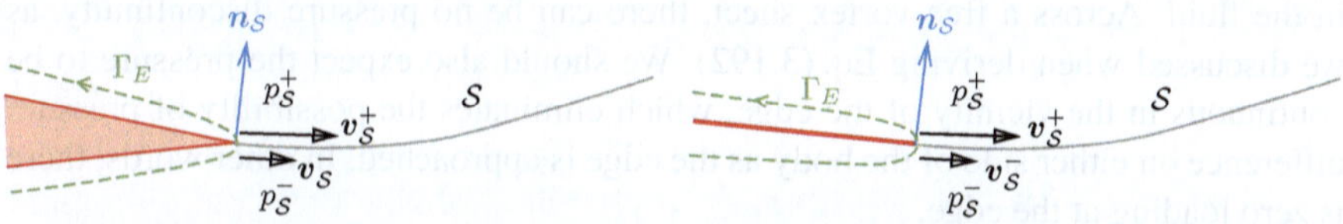

Fig. 5.1 Flow in the vicinity of a sharp edge with non-zero angle (left) and an edge with zero angle (right). In both cases, the edge is denoted by E. The free vortex sheet S is depicted in gray, and a portion of the contour defining circulation Γ_E is shown as the dashed green curve

flow, the condition (in conjunction with Kelvin's circulation theorem) specifies the instantaneous transfer of circulation from the body into the fluid required to maintain finite velocity at the edge.

There are several implications of the Kutta condition in different scenarios, and some of these implications are used (sometimes incorrectly) in place of the original condition. Let us discuss some of these in a two-dimensional setting. For example, in potential flow analysis of a body with a sharp trailing edge, in any solution that does not satisfy the Kutta condition there is, along with the singular velocity at the edge, a stagnation point that lies somewhere in the vicinity of the edge. Thus, it appears that the Kutta condition could be naively interpreted as a general requirement that this stagnation point exist at the edge itself. However, a stagnation point can only possibly arise at such an edge if the angle of the edge is non-zero, since the two streamlines along the walls that form the edge intersect at an angle, and even in that geometry, there is some subtlety, as we will see. If, instead, the edge is a cusp, or simply part of an infinitely-thin body, then the velocity remains non-zero (but finite) at this edge when the condition is enforced. (This is obviously true for the special case of a uniform flow past an infinitely-thin flat plate at zero angle of incidence, for which the Kutta condition is trivially satisfied with a velocity equal to that of the uniform flow everywhere.)

To discuss another set of implications, let us first state a corollary of the Kutta condition:

Result 5.2: The Kutta Condition, Alternatively Stated

The pressure in the fluid near any sharp edge must be finite and single valued.

That the pressure is finite follows from the same requirement on the velocity. On the other hand, the requirement that the pressure be single valued seems to have little relevance to the condition as originally stated. However, it eliminates solutions that satisfy the Kutta condition but for which there exists a discontinuity in the pressure in the fluid. Across a free vortex sheet, there can be no pressure discontinuity, as we discussed when deriving Eq. (3.192). We should also expect the pressure to be continuous in the vicinity of the edge, which eliminates the possibility of pressure difference on either side of the body as the edge is approached. In other words, there is zero loading at the edge.

Let us consider this in the contexts of the scenarios depicted in Fig. 5.1, in which we suppose that there is a vortex sheet emanating from each of the edges. Based on the requirement that $p_S^+ = p_S^-$, we can apply the relationship (3.192) obtained earlier in our discussion of vortex sheets, where here, our evaluation point is E, and we base our analysis in a reference frame moving with the edge. With a re-arrangement of the triple product, we obtain

$$\frac{\mathrm{d}\Gamma_E}{\mathrm{d}t} = \boldsymbol{n}_S \cdot \left[\frac{1}{2}(\boldsymbol{v}_S^+ + \boldsymbol{v}_S^-) \times \boldsymbol{\gamma}_S\right]. \tag{5.1}$$

It is important to emphasize that the circulation Γ_E in this expression is evaluated on a contour—a portion of which is depicted in Fig. 5.1—that encloses the body and any vorticity not topologically connected with the sheet emanating from edge E; that is, Γ_E is equal and opposite to the circulation of the sheet rooted at E. For example, if there is only one edge on the body from which vorticity enters the fluid, then this Γ_E is the bound circulation about the body. Thus, this equation expresses (the negative of) the flux of circulation into the wake. ***The edge itself is not a source of circulation, but merely a conduit for passing circulation from the body into the fluid.*** It is useful to note that this fact, along with the zero pressure difference on either side of the edge, implies that the velocity must also be continuous at the edge, as the reader can verify.

What is the meaning of the term in brackets? Let us separately consider the cases of steady and unsteady flow.

5.1.1 Steady Flow

In a steady flow, the flux of circulation is zero by definition, so this requires that

$$\frac{1}{2}(\boldsymbol{v}_S^+ + \boldsymbol{v}_S^-) \times \boldsymbol{\gamma}_S = 0. \tag{5.2}$$

This is trivially satisfied at an edge with non-zero angle, since as we discussed, such an edge will be a stagnation point. However, it must also hold at any other point along the supposed sheet, where generally the fluid velocity is non-zero. At an edge of zero angle, the fluid velocity does not vanish, as discussed earlier. Thus, for either type of edge, the vorticity in the wake must either be parallel to the mean fluid velocity (as in three-dimensional flow) or identically zero (as in two-dimensional flow). In other words, in steady two-dimensional flow, there can be *no* vortex sheet emanating from the edge; and in steady three-dimensional flow, the sheet can only have components of vorticity in the streamwise direction. This latter fact serves as an essential aspect of the lifting-line method of Prandtl, for example.

In passing, we note that, in steady conditions past an edge with non-zero angle, the flow must approach the edge from either side with equal tangential velocity to be consistent with the zero flux of circulation into the wake. This, in turn, implies that the stagnation streamline leaves the edge along the bisector of the exterior angle.

5.1.2 Unsteady Flow

When the conditions of the flow are changing in time, as they are soon after the initiation of motion or during unsteady kinematics, then new circulation must be continuously transferred into the fluid in order to preserve the finite velocity at the edge. In a potential flow analysis, this dictates that there must be a vortex sheet rooted at the edge in order to accept this flux of circulation, which is described by Eq. (5.1). We have to be a bit more careful in the interpretation than we were in the discussion following Eq. (3.192), since we are now considering the conditions at a junction between a free vortex sheet and the bound vortex sheet of the body.

Before we discuss the mathematical details of the Kutta condition in unsteady flow, we should remind ourselves once again that the actual flow in the immediate vicinity of the edge is viscous, just as it is upstream (in which the flow approaches in a viscous boundary layer) and downstream (where the near wake's behavior involves a mixing of two streams of unequal magnitude) of the edge. In fact, for flows that separate at the edge, one can view this as a problem of matched asymptotic expansions, with the viscous flow serving as the *inner* edge flow (with the relevant length scale defined by the small boundary layer and wake thicknesses) and the inviscid flow behavior as the *outer* edge flow (with the airfoil geometry imposing the pertinent length scale). Thus far, we have been content to ignore the inner part of this problem, but it exists nonetheless in any real flow. Indeed, it has been argued convincingly by Crighton [15] that, for a wide variety of flows, including those of interest in this book, the imposition of the Kutta condition on the outer flow is consistent—in the sense of matched asymptotic expansions—with the viscous inner behavior. This inner flow must itself reconcile the upstream boundary layer with the downstream near wake, and the overall matching of the various behaviors generally takes the form of a *triple deck* of intermediate regions [11, 9]. One of the consequences of this analysis is that the Kutta condition is only valid if the time scale of the unsteady motion is sufficiently slow (compared with the time required for viscous diffusion across the region near the edge).

Rather than expound on the details of the matching process, we merely note that it places the Kutta condition on firm theoretical ground and that this condition is, strictly speaking, placed on the outer flow. This latter fact requires that we consider the conditions of the flow in the *vicinity* of the edge rather than just at the edge itself. And, though we do not need to acknowledge the details of the inner viscous flow to make progress, it can be helpful to remember its connection with the inviscid flow.

For example, let us rewrite the circulation flux in (3.191) in a somewhat different form:

$$\frac{d\Gamma_E}{dt} = \frac{1}{2}\left(|\boldsymbol{v}_S^+|^2 - |\boldsymbol{v}_S^-|^2\right); \tag{5.3}$$

again, we emphasize that the velocities here are expressed relative to the edge. Although we arrived at it from the conditions of zero pressure jump across a free vortex sheet, Eq. (5.3) is the same as what we would obtain in a viscous flow by calculating the convective flux of vorticity from the boundary layers on either side of the edge, through a control surface perpendicular to the edge; in the viscous case, the velocities are evaluated at the outer limits of the respective boundary layers. This equation is an example of the consistency between the inner and outer flows.

To illustrate the manner in which (5.3) describes the flux of vorticity into the free vortex sheet, we depict two contours in Fig. 5.2 in the vicinity of the sharp edge of a flat plate at some time t; our discussion here applies to edges of non-zero angle, as well. The contours are assumed to be material loops, convecting with the local flow. One contour (depicted in blue) had passed just around the edge E at time $t - \delta t$, but has since moved slightly downstream at time t, as shown, intersecting the free vortex sheet at the blue dot, a material marker on this sheet. Due to Kelvin's circulation theorem, its circulation has not changed during this convection, but remains fixed at the value that it had in its earlier configuration about the edge, $\Gamma_E(t - \delta t)$. The other contour (in green) currently passes immediately around E and has the current circulation about this edge, $\Gamma_E(t)$. The distance, δs, between the edge and the blue material marker is given by the convective speed multiplied by the short time interval, δt:

$$\delta s = \frac{1}{2}\delta t(\boldsymbol{v}_S^+ + \boldsymbol{v}_S^-) \cdot \boldsymbol{\tau}_S. \tag{5.4}$$

By definition, the difference in circulations between the ends of this short segment of length δs is equal to the local vortex sheet strength multiplied by this length, viz.,

$$\Gamma_E(t - \delta t) - \Gamma_E(t) = \frac{1}{2}\gamma_S(\boldsymbol{x}_E)\delta t(\boldsymbol{v}_S^+ + \boldsymbol{v}_S^-) \cdot \boldsymbol{\tau}_S. \tag{5.5}$$

But, of course, by Result 3.5, $\gamma_S(\boldsymbol{x}_E)\boldsymbol{\tau}_S = -(\boldsymbol{v}_S^+ - \boldsymbol{v}_S^-)$, and thus, in the limit $\delta t \to 0$, we recover (5.3).

For further progress on the conditions at the edge, let us first discuss the case of zero edge angle before we move to the more complicated case of non-zero edge angle.

An Edge of Zero Angle For the case of an edge with zero angle, both the free and bound vortex sheet that meet at this edge represent jumps in tangential fluid velocity. We should expect that, when the flow is regularized at the edge, the velocity is not only finite but also continuous there. As we mentioned above, this velocity is not zero. In other words, ***the strength of the bound vortex sheet is equal to the strength of the free vortex sheet at an edge of zero angle, and their common value is generally non-zero***:

$$\gamma(\boldsymbol{x}_E) = \gamma_S(\boldsymbol{x}_E), \tag{5.6}$$

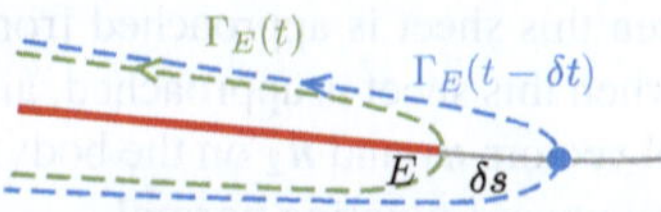

Fig. 5.2 Release of vorticity into a free vortex sheet at a sharp edge with zero angle. The contours used to define the edge circulations $\Gamma_E(t - \delta t)$ and $\Gamma_E(t)$ at two instants are depicted as dashed lines with the same color of the corresponding circulation label

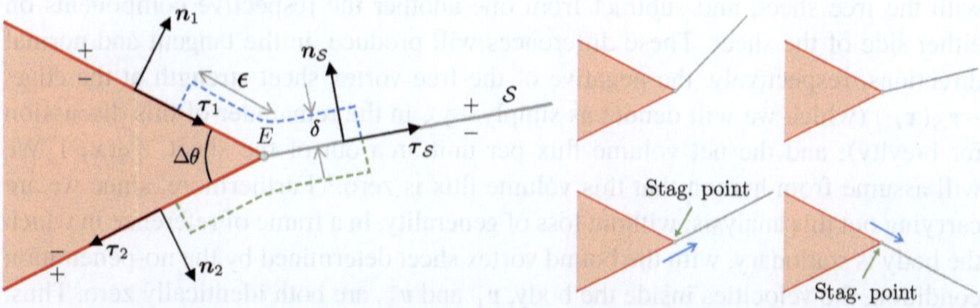

Fig. 5.3 (Left) The vicinity of a sharp edge E with non-zero interior angle $\Delta\theta$. The free vortex sheet S is depicted in gray, the interior of the body in light red, and the bound vortex sheet above (labeled with subscript 1) and below (labeled with 2) the edge are shown in dark red. Control volumes on either side of the edge are depicted with dashed lines. (Right) Inadmissible (top row) and admissible (bottom row) kinematic configurations for the free vortex sheet

at an edge at position $\boldsymbol{x}_E$. This continuity also means that the sheet must leave the edge along the body's tangent, for as we noted in the discussion of (3.192), the mean fluid velocity cannot have a relative component normal to the sheet due to conservation of mass. As a consequence of this, the vector expression in square brackets in Eq. (5.1) must either be parallel to $\boldsymbol{n}_S$, indicating a flux of circulation into the fluid, or be identically zero, such as when the flux is instantaneously absent. We will inspect the details of velocity continuity and the Kutta condition for an edge of zero angle in much deeper detail in Sect. 8.4.1.

An Edge of Non-zero Angle For an edge of non-zero angle there is some subtlety, because both the strength and the orientation of the wake vortex sheet are now in question. The analysis we present here is inspired from the work of [77]. Let us consider the general configuration posed near a two-dimensional edge of interior angle $\Delta\theta$ with a free vortex sheet rooted at the edge, as illustrated in Fig. 5.3. The two parts of the bound vortex sheet that meet at the edge are labeled with subscripts 1 and 2 and the free sheet by S. Each of these sheets has a + and − side, as shown; the + side of each bound vortex sheets is exterior to the body. We will denote by $\boldsymbol{v}_1^+$

the velocity on the + side of sheet 1 when this sheet is approached from that side, $\boldsymbol{v}_2^+$ the velocity on the + side of sheet 2 when this sheet is approached, and similarly for every other velocity limit. The normal vectors $\boldsymbol{n}_1$ and $\boldsymbol{n}_2$ on the body surface are consistent with our usual definition of an outward-pointing normal.

As edge E is approached from along the body or from along the free vortex sheet on the same side, we should expect that the same velocity is achieved. In particular, $\boldsymbol{v}_1^+ = \boldsymbol{v}_S^+$ and $\boldsymbol{v}_2^+ = \boldsymbol{v}_S^-$. Let us describe this continuity in the coordinates associated with the free sheet, and subtract from one another the respective components on either side of the sheet. These differences will produce, in the tangent and normal directions, respectively, the negative of the free vortex sheet strength at the edge, $-\gamma_S(\boldsymbol{x}_E)$ (which we will denote as simply $-\gamma_S$ in the remainder of this discussion for brevity); and the net volume flux per unit area out of the sheet, $q_S(\boldsymbol{x}_E)$. We will assume from hereon that this volume flux is zero.[1] Furthermore, since we are carrying out this analysis, without loss of generality, in a frame of reference in which the body is stationary, with the bound vortex sheet determined by the no-penetration condition, the velocities inside the body, $\boldsymbol{v}_1^-$ and $\boldsymbol{v}_2^-$, are both identically zero. Thus, the quantity $\boldsymbol{v}_1^+ \cdot \boldsymbol{\tau}_1$ is the negative of the local bound vortex sheet strength, denoted as $-\gamma_1$ for brevity; $\boldsymbol{v}_2^+ \cdot \boldsymbol{\tau}_2$ is the negative of the local sheet strength on this portion, $-\gamma_2$; and $\boldsymbol{v}_1^+ \cdot \boldsymbol{n}_1$ and $\boldsymbol{v}_2^+ \cdot \boldsymbol{n}_2$ are the local volume fluxes per unit area out of the body, which we will assume to be zero. Thus, the equations of continuous velocity can be written as

$$\gamma_S = \gamma_1 \boldsymbol{\tau}_1 \cdot \boldsymbol{\tau}_S - \gamma_2 \boldsymbol{\tau}_2 \cdot \boldsymbol{\tau}_S, \tag{5.7}$$

in which all sheet strengths are to be interpreted as local values at the edge.

In addition to the continuity of velocity, we also require that mass is conserved as the flow passes along either side of the body into the wake. Our objective is to enforce this conservation in the immediate vicinity of the edge. For this purpose, we use the control volumes depicted in Fig. 5.3, with the understanding that each control volume's dimensions are much smaller than the other scales of the flow. The dimensions δ and ϵ depicted in Fig. 5.3 can be varied independently, and in particular, if we allow ϵ to shrink to zero while δ remains small but non-zero, then the fluxes of mass through the segments of the boundary parallel to the sheets vanish, and we obtain that $\boldsymbol{v}_1^+ \cdot \boldsymbol{\tau}_1 = \boldsymbol{v}_S^+ \cdot \boldsymbol{\tau}_S$ and $\boldsymbol{v}_2^+ \cdot \boldsymbol{\tau}_2 = -\boldsymbol{v}_S^- \cdot \boldsymbol{\tau}_S$. If we add one to the other, then we arrive at the following simple condition:

Result 5.3: Continuity of Vortex Sheet Strengths

At an edge E of a body at which the flow has been regularized by the Kutta condition, the sum of the strengths of the bound vortex sheet approaching either side of the edge, denoted by γ_1 and γ_2, is equal to the strength of the free vortex sheet where it is rooted at the edge:

[1] Though this assumption of zero volume flux is consistent with our definition of a vortex sheet in this work, DeVoria and Mohseni [18] have recently extended the definition of a vortex sheet to allow for fluid entrainment. They argue that this vortex-entrainment sheet, rather than just a vortex sheet, is the more appropriate manifestation of the released vorticity at an edge.

$$\gamma_1 + \gamma_2 = \gamma_S. \tag{5.8}$$

If the edge has zero interior angle, then this sum of bound vortex strengths is simply the strength of the overall bound vortex sheet at the edge, as we described in Note 3.6.4, and the identity reduces to Eq. (5.6).

We can interpret this statement physically as a description of the origin of vorticity in the inviscid description of the wake. In the full viscous flow, vorticity enters the wake from the merging of the viscous boundary layers on either side of the body. In the inviscid (i.e., outer) problem, each boundary layer is replaced by the bound vortex sheet and its thickness is ignored. Thus, in the inviscid model, the origin of wake vorticity is not the edge, but rather, the bound vortex sheets on either side of the edge. Equation (5.8) can easily be generalized to three-dimensional edges, in which case the vector-valued sheet strengths would appear in the expression.

Let us now consider the combination of Eqs. (5.7)–(5.8). The unknowns in this set of equations are the strength of the wake adjacent to the edge, γ_S, and the wake's orientation at the edge, $\boldsymbol{\tau}_S$. The strengths γ_1 and γ_2 are determined, along with the rest of the distributed strength of the bound vortex sheet on the body, by the no-penetration condition. The elimination of γ_S from the equations reduces them to

$$\gamma_1 (1 - \boldsymbol{\tau}_1 \cdot \boldsymbol{\tau}_S) + \gamma_2 (1 + \boldsymbol{\tau}_2 \cdot \boldsymbol{\tau}_S) = 0. \tag{5.9}$$

This equation determines the orientation $\boldsymbol{\tau}_S$ of the free vortex sheet at the edge. One can show that the solution for this vector is a weighted average of the two body tangents:

$$\boldsymbol{\tau}_S = \frac{C\gamma_1}{\gamma_1 + \gamma_2}\boldsymbol{\tau}_1 - \frac{C\gamma_2}{\gamma_1 + \gamma_2}\boldsymbol{\tau}_2, \tag{5.10}$$

where $C = \left[1 - 2(1 + \boldsymbol{\tau}_1 \cdot \boldsymbol{\tau}_2)\gamma_1\gamma_2/(\gamma_1 + \gamma_2)^2\right]^{-1}$ is a normalization factor.

For an edge of non-zero angle, Eq. (5.10) makes it appear that the free vortex sheet can adopt a continuous range of orientations that depend only on γ_1 and γ_2. However, the choices are actually much more restricted than it appears. Let us consider some possibilities. In each, we will assume that γ_2 is positive and non-zero, corresponding to a flow directed along the lower side of the body toward the wake.

- Suppose that γ_1 is negative, which would correspond to flow along the upper side of the body directed toward the edge. Then the tangent $\boldsymbol{\tau}_S$ lies outside of the range between $\boldsymbol{\tau}_1$ and $\boldsymbol{\tau}_2$, as illustrated in the upper left panel of the quartet of configurations in Fig. 5.3. The free vortex sheet and the lower side of the body then form a convex edge, which, as we know from our discussion in Note 3.2.3, must lead to a singular velocity at the edge. Thus, this cannot be a regularized condition of the flow. Furthermore, there is ambiguity in the limiting orientation of the free vortex sheet as its strength $\gamma_1 + \gamma_2$ approaches zero from above or below; this orientation is directly upward and directly downward in these respective limits.

- However, the state $\gamma_1 = -\gamma_2 \leq 0$ corresponds to $\gamma_S = 0$ and is precisely the regularized steady-flow condition discussed above. There is no free sheet, and the stagnation streamline lies midway between $\boldsymbol{\tau}_1$ and $-\boldsymbol{\tau}_2$, forming concave regions on either side of this streamline. By Note 3.2.3, the flow necessarily stagnates at the edge on either side of the free vortex sheet, so strictly this requires that γ_1 and γ_2 are both zero at the edge. (They retain their equal and opposite disposition near the edge, however.)
- If γ_1 is also positive and non-zero, then $\boldsymbol{\tau}_S$ lies somewhere between horizontal and $-\boldsymbol{\tau}_2$, as in the upper right panel in Fig. 5.3. In this scenario, the regions on either side of the free vortex sheet form concave corners with the body, necessitating a stagnation point at the edge. This contradicts the supposition that γ_1 and γ_2 are non-zero. In fact, both must be zero at the edge, and therefore, so must be γ_S. The sheet's strength is zero at the edge and its orientation is ill-defined. (However, this configuration can be interpreted as a transient state, as we will discuss momentarily.)
- If γ_1 is zero, then $\gamma_S = \gamma_2 > 0$ and $\boldsymbol{\tau}_S = -\boldsymbol{\tau}_2$, as in the lower left panel of Fig. 5.3. In this configuration, the flow stagnates at the edge in the concave region between the free vortex sheet and the upper side of the body, but the lower region is neither concave nor convex, so the velocity at the edge there is finite and non-zero. This configuration is self-consistent and admissible. The analogous state, with $\gamma_2 = 0$, $\gamma_S = \gamma_1 < 0$ and $\boldsymbol{\tau}_S = \boldsymbol{\tau}_1$, shown in the lower right panel, is clearly also admissible.

This pair of admissible configurations of the free vortex sheet in regularized unsteady flow—tangent to one of the sides of the body that form the sharp edge—is sometimes referred to as the **Giesing–Maskell extension** of the Kutta condition, after the researchers who independently [25, 49] presented arguments in favor of the solution:

Result 5.4: Giesing–Maskell Extension of the Kutta Condition

At a body's edge of non-zero interior angle at which the flow has been regularized by the Kutta condition, the free vortex sheet at this edge can only adopt one of two possible configurations:

- For cases in which $\gamma_S < 0$, then it is tangent to the upper surface 1, and $\gamma_1 = \gamma_S$ and $\gamma_2 = 0$. The pressure is single valued, so that the flux of circulation into the sheet is then given by

$$\frac{\mathrm{d}\Gamma_E}{\mathrm{d}t} = \frac{1}{2}|\boldsymbol{v}_S^+|^2 = \frac{1}{2}\gamma_1^2. \tag{5.11}$$

- For cases in which $\gamma_S > 0$, then it is tangent to the lower surface 2, so that $\gamma_2 = \gamma_S$ and $\gamma_1 = 0$, and the flux is

$$\frac{d\Gamma_E}{dt} = -\frac{1}{2}|v_S^-|^2 = -\frac{1}{2}\gamma_2^2. \tag{5.12}$$

Both of the flux equations describe, as before, the transport of circulation away from the edge at a velocity equal to the average on either side of the sheet; here, the velocity on one side is zero. For an edge of zero interior angle, the body tangents $\boldsymbol{\tau}_1$ and $-\boldsymbol{\tau}_2$ are identical, and thus, the free vortex sheet's tangency to either is equivalent to leaving the edge parallel to the plate.

For non-zero angle, one should note two apparent weaknesses with this extended Kutta condition. First of all, there will be situations, particularly in oscillatory motions of the airfoil, in which the vorticity shed at the trailing edge changes sign, necessitating an awkward sudden switch from one configuration to the other. However, it should be borne in mind that the sheet's strength at the edge is zero during this transition and is small prior to it and just afterward, so the sudden switch does not actually disrupt the flow. As we noted above, all of the configurations between the two body angles serve as a reasonable intermediate state during this switch.[2] A second concern is that neither of these two body-tangent configurations agrees with the steady-state condition described above, so the limit of $d\Gamma_E/dt \to 0$ is different from the steady state. Though some investigators, such as Basu and Hancock [5] and Xia and Mohseni [77], have highlighted this issue and proposed reconciliations, the issue itself is arguably only of academic concern because the steady state is never actually achieved in finite time in any mathematical model of unsteady aerodynamics. Experimental studies have demonstrated that the Giesing–Maskell extension of the Kutta condition is sound in a variety of contexts [57].

On this last point, it is important to emphasize that the identities derived in this section—the continuity of vortex sheet strengths in Result 5.3 and the orientation of the free vortex sheet in Result 5.4—are not restatements of the Kutta condition, but are, instead, consequences of the regularized flow state. Hence, they can only be used in conjunction with, for example, the flux form of the Kutta condition (5.1). In the next section, we will offer another, constraint-based form of the Kutta condition. Collectively, they provide a local description of the release and immediate transport of new vorticity. Of course, this vorticity transport is already effected by tracking the trajectories of the vortex elements that comprise the sheet, to be discussed in Chap. 7. Thus, it should be emphasized that it is not strictly necessary to explicitly specify the continuity of the strengths or the orientation of the free vortex sheet, and instead, let these emerge from the natural kinematics of the vorticity. For practical numerical reasons, however, it can sometimes be helpful to augment the kinematics with these specifications since the flow in the vicinity of the regularized edge can be delicate.

---

[2] DeVoria and Mohseni [18] have shown that, by accounting for fluid entrainment into the free vortex sheet, all of the intermediate angles are admissible.

Before doing so, let us make two more observations. The first is this important note about the close relationship between the Kutta condition and Kelvin's circulation theorem:

Note 5.1.1: The Kutta Condition and Kelvin's Circulation Theorem

The Kutta condition, as we have expressed it here as a flux of circulation into the fluid from a body's edge, has a natural connection with Kelvin's circulation theorem in Result 3.3. Let us consider a closed material contour that, at some instant, envelops a body and passes through the edge E, as in Fig. 5.1. As time proceeds from this instant, the contour will deform away from the edge, enclosing any vortex elements that carry the fluxed circulation away. But since the total circulation inside the contour cannot vary with time, the flux of circulation into the fluid must come at the expense of an equal but opposite change in the bound circulation about the body. This is exactly the principle embodied by Eq. (5.1). In other words, ***by insisting that the pressure is single valued and finite at the edge, we have ensured that Kelvin's circulation theorem holds for any material contour that envelops the body***.

In this section, we have argued that the application of the Kutta condition at an edge in unsteady flow requires that a free vortex sheet be rooted at that edge. What would happen if, instead, we release a point vortex from this edge? This question has practical consequences, since point vortices are commonly used in inviscid models of unsteady flow.

Note 5.1.2: The Kutta Condition and the Release of Vortex Elements

Let us suppose that we enforce the Kutta condition at an edge, in the form dictated by Result 5.1, by releasing a point vortex—or, in three dimensions, a vortex filament—from the edge rather than rooting a free vortex sheet at this edge. The strength of this **nascent vortex element** is determined at each instant so that the velocity is made finite at the edge. Since the vortex convects with the flow throughout this process, its position relative to the edge continuously changes in time. Therefore, to maintain its role in canceling the singular velocity at the edge, ***the strength of the vortex necessarily varies with time***, as well. As we know from Note 3.3.1, this vortex of time-varying strength generates a discontinuity in pressure, $[p]_-^+ = \rho\dot{\Gamma}$, which we suppose to reside along a curve (or surface in three dimensions) extending from the vortex to the releasing edge, as discussed in Note 3.6.5.

This pressure discontinuity must be continuous as the edge is approached, so that ***the fluid loading at the edge is not zero***, as it is when a free vortex sheet is rooted at the edge. Obviously, with no such sheet, there is no discontinuity in tangential velocity in the fluid near the edge, so $\gamma_S = 0$. Furthermore, by referring to the general expression for pressure at the edge, given by Eq. (3.193), it is clear that ***the bound vortex sheet strength (or the sum of strengths, in the case of an edge of non-zero angle) must also be identically zero***. Thus, the continuity of vortex sheet strengths at the edge (5.8) remains valid even if the free vortex sheet is absent and a vortex element of time-varying strength is used in its place.

Note that the flux form of the Kutta condition (5.1) does not hold if such a discrete vortex element is used to enforce the Kutta condition. The pressure discontinuity between the edge and the element precludes a continuous flux of vorticity from one to the other. However, it is important to remember that Kelvin's circulation theorem still holds for a material contour that envelops the body and the convecting vortex element (and the discontinuity of pressure between them). Thus, in this approach, we have the loss of circulation from the bound vortex sheet and the equal gain of circulation by the element, but no physical connection between them to explicitly represent this transfer.

5.2 Application of the Kutta Condition

In Sect. 5.1, we discussed the Kutta condition and the various implications of regularizing the flow in the vicinity of a sharp edge of a body. However, we did not provide many practical details of the condition's enforcement. Here, we will discuss its implementation, focusing on a two-dimensional flow for definiteness. To get started, let us revisit some points we have previously discussed in the book, but not yet considered together:

- We have the Kutta condition itself, specifically in the form stated in Result 5.1, namely, that the velocity in the vicinity of a sharp edge must be finite and leave the edge smoothly. In the ensuing discussion of the condition in the context of unsteady flow, we indicated that the Kutta condition requires the transfer of circulation into the flow, most likely in the form of a vortex sheet.
- We also know the mathematical structure of the flow in the *absence* of regularization: the singular flow near a corner, as described in Note 4.4.5. In other words, we have specific understanding of the task to ensure that the flow is smooth and finite: Insert new circulation—e.g., into the vortex sheet rooted at the edge under consideration—as required to annihilate the singular part of the velocity field,

given by (4.55). We can achieve this ***by requiring that the signed intensity σ of the edge is identically zero.***

- The addition of new circulation into the fluid must be carried out in concert with the equal and opposite removal of bound circulation associated with the body, as required to maintain constant total circulation (4.67). This is a particular application of Kelvin's circulation theorem, Result 3.3.
- But how can we inject new circulation into the fluid, with equal and opposite bound circulation in the body, and still preserve the no-penetration condition on the body surface? Because, as discussed in Sect. 4.5, the flow about the body is not unique, and has a null-space solution (4.70) whose strength we can set as needed. Thus, for any distribution of vorticity in the fluid, the total circulation in the fluid can be maintained at zero, independently of the no-penetration condition.

Result 5.5: The Constraint Form of the Kutta Condition

Determine the strength of newly-created vorticity in the fluid so that the edge's signed intensity, σ, defined in (3.103), instantaneously satisfies

$$\sigma = 0. \tag{5.13}$$

By Note 4.4.6, this is equivalent to requiring that the strength of the bound vortex sheet at this edge—or, for an edge of non-zero angle, the sum of the sheet strengths on either side of the edge—is finite. When the body's geometry is obtained by a conformal mapping from the unit circle, then Note 4.4.5 shows that this condition is equivalent to requiring that the velocity tangent to the unit circle at the edge's pre-image is zero. (As Eq. (4.39) shows, the normal component of velocity is also zero at that pre-image, by virtue of the no-penetration condition and the vanishing Jacobian of the mapping.)

It is important to recognize that, in a general unsteady flow, the constraints dictated by the Kutta condition and Kelvin's circulation theorem must be considered simultaneously, since all of the constituents associated with a new vortex in the fluid—the vortex itself, the reactive bound vortex sheet on the body, and the null-space flow to balance the overall circulation—contribute to velocity at the edge. However, by decomposing the flow into contributors, as we did in Sect. 4.6, we have made this simultaneous consideration of constraints simple, since the vorticity-induced basis velocity field (4.83) (or, in vector form, (4.117)) contains all three of these elements. The Kutta condition is used to determine the strength required of this new vortex, such that its basis velocity field, in tandem with all other flow contributors, ensures $\sigma = 0$. This necessarily couples all of the flow contributors together. However, it is also useful to observe that the signed intensity, σ, depends linearly on each of the flow contributors, including the new vortex element. This is clear from the fact that σ is proportional to the tangent velocity component in the circle plane at the edge's pre-image.

In fact, we can use this linear dependence on flow contributors to devise a means of choosing the strength of a new vortex element to enforce the Kutta condition. Though we will frame this approach in a context in which we have represented fluid vorticity with discrete vortex elements, the approach is actually applicable to continuous free vortex sheets, as well. Let us suppose we already have N_v elements in the fluid, and their contributions to the signed intensity at an edge—along with those from other fluid vorticity and from body motion and uniform flow—are denoted by $\breve{\sigma}$. In general, this signed intensity is not zero, so the Kutta condition is violated. We seek to remedy this by introducing a new vortex element, labeled $N_v + 1$, whose strength is denoted by $\delta\Gamma_{N_v+1}$ and whose position by Z_{N_v+1}. In the case of a free vortex sheet, we can view this as a limiting process in which the new vortex element is a straight segment $Z_{N_v+1}(s)$ appended to the end of the sheet nearest the edge with circulation $\delta\Gamma_{N_v+1}$. Thus, in light of the linear dependence of σ on the vortex strength, we can write it as

$$\sigma = \delta\Gamma_{N_v+1}\sigma^v_{N_v+1} + \breve{\sigma}, \tag{5.14}$$

where $\sigma^v_{N_v+1}$ denotes the contribution of the new element, ***with strength equal to unity***, analogous to the unit basis fields we have defined earlier in this chapter. This unit contribution depends on Z_{N_v+1} and also accounts for the indirect influence of the body—the reactive bound vortex sheet and the null-space flow—the latter to ensure that Kelvin's theorem is automatically satisfied by this new addition. We know all of the contributions to $\breve{\sigma}$, and we can also assume that we know the element's position, Z_{N_v+1}; it remains to determine the newly-created strength, $\delta\Gamma_{N_v+1}$. By the constraint dictated in Result 5.5, this strength must clearly be

$$\delta\Gamma_{N_v+1} = -\frac{\breve{\sigma}}{\sigma^v_{N_v+1}}. \tag{5.15}$$

Equation (5.15) is only schematic, and implementational details will be provided in specific contexts later in the book, particularly in some of the example models in Chap. 9. Let us observe that, if there is more than one edge at which the flow must be regularized, then we can simultaneously enforce these constraints by adding new vortex elements at each edge. Their contributions are coupled, since each new element contributes to the value of σ at every edge. However, it is straightforward to extend the decomposition (5.14) accordingly and then solve the resulting linear equations for the unknown strengths of all new elements.

It should be clear that Eq. (5.14) is well suited for models in which vorticity is released into the fluid in a sequence of discrete vortex elements. If, instead, we model this release as an isolated point vortex of time-varying strength, denoted by index N_v, then it is more natural to express the constraint Result 5.5 in a time-derivative form, $\dot{\sigma} = 0$. This form leads to an equation that couples the rate of change of the point vortex's strength, $\dot{\Gamma}_{N_v}$, the vortex' velocity, $\dot{z}_{N_v}$, and the rate of change of other flow contributors. This equation would generally be solved simultaneously with the element transport equations.

In this book, we will present several examples in Chaps. 8 and 9 in which we have analytical access to the solution, enabling us to compute σ and thereby annihilate it exactly with a judicious choice of vortex strength via Eq. (5.15). But it is important to note that the approach we have described in Eq. (5.15) is not the only way to enforce the Kutta condition in an unsteady flow. To discuss other approaches, we need to take a more expansive view of our approach and contrast it with a more 'traditional' perspective.

5.3 Traditional Enforcement of the Kutta Condition: A Contrast

In a steady flow we have no explicit need to concern ourselves with new vorticity in the fluid or the preservation of total circulation. In such flow, as discussed in Sect. 4.5, we imagine a starting vortex to have traveled so far from the body that its influence on the flow near the body is negligible. However, its existence still requires the addition of the null-space flow, and only this flow contributes to the velocity at the edge. Thus, we directly determine the strength of the null-space flow—that is, we choose the *bound circulation*—to enforce $\sigma = 0$; we trivially enforce zero total circulation in the fluid by attributing equal and opposite circulation to the otherwise meaningless starting vortex.

In most classical aerodynamics studies of the subject, this is also the approach taken in *unsteady* flows, though the fluid vorticity is no longer meaningless. One sets the bound circulation to enforce the Kutta condition at the trailing edge, and then chooses the strength of new vorticity to enforce zero total circulation. In a numerical implementation, for example, this would consist of releasing a new control point of the discretized free vortex sheet from the regularized edge, with the point's invariant circulation equal to the local circulation at the edge. So before we leave this discussion of the Kutta condition, let us draw an important contrast between the strategy we have outlined in the previous section and this more traditional perspective.

When we only seek to enforce regularity at one edge (e.g., the trailing edge of an airfoil at low angle of attack), we note that the homogeneous solution provides a convenient free parameter—the bound circulation—that can be set as needed to ensure regularity of the field at the designated edge. In fact, this can be done individually in each basis field, including the motion-induced fields, so that the overall bound circulation is partitioned. As we noted in Sect. 4.6.1, this is the approach taken in much of the classical literature on steady and unsteady aerodynamics; see, e.g., von Kármán and Sears [70].

Of course, we still need to enforce the constraint of zero total circulation (4.67). In this traditional approach, the basis fields do not individually satisfy this constraint, so they must do so collectively. The strength of a newly-created vortex element in the fluid is chosen to balance the change in overall bound circulation resulting from all flow contributors. All of the flow contributors are coupled by this circulation constraint.

But the traditional approach suffers from one serious deficiency that obstructs the modeling of a wide class of flows: With only a single free parameter in the bound circulation, we can only enforce the Kutta condition at one edge. The approach we follow in this book is designed to overcome this limitation. We have at our disposal the strength of a new vortex element associated with *any* edge that we designate for regularity. So instead of letting the Kutta condition determine the bound circulation, we instead let this condition determine the strength(s) of newly-created vortex elements. This enables a more flexible modeling strategy, though at the cost of losing the ability to impose this regularity on the individual basis fields. Only in combination—by suitable choice of the strength of new vorticity—can they ensure finite velocity at a given edge. In contrast, we enforce the constraint of zero total circulation in each basis field. This is trivially done for the basis fields induced by body motion, leaving them entirely non-circulatory. All of the bound circulation about the body is partitioned into the unit vorticity-induced basis fields, (4.85) and (4.122), for example by endowing a vortex's image with equal and opposite strength.

It is important to stress that, in a case in which regularity is imposed only on one edge, the two approaches lead to the same strengths of vortex elements and the same overall bound circulation. ***This shows the close relationship between the Kutta condition and the constraint of zero total circulation: their roles in determining solution parameters might be interchanged, but, in concert, they achieve the same result.***

Let us contemplate the application of the traditional approach to our simple scenario of a steady airfoil flow. After the starting vortex has moved far from the airfoil, the Kutta condition is trivially enforced in the vorticity-induced basis field: the field is zero everywhere in the vicinity of the airfoil. But the body translation-induced basis field is not trivial and requires its own bound circulation to enforce regularity at the trailing edge. The starting vortex must have strength equal and opposite to this bound circulation. Traditionally, the bound circulation Γ_b associated with the irrotational basis fields is called the **quasi-steady circulation**, since, in an unsteady flow, it represents the bound circulation that would ultimately exist if the instantaneous motion were allowed to persist indefinitely.

It should be noted that there are other strategies for enforcing the Kutta condition that do not directly invoke the constraint form in Result 5.5. Instead, these rely on the flux form of the condition, expressed in Eq. (5.1) or (5.3). For example, Basu and Hancock [5] and Krasny [41] integrate this equation simultaneously with the transport equations in the free vortex sheet in order to determine the circulation to be endowed to new control points released from the edge into this sheet. Such an approach can be used separately at any edge of a body. The choice of the approach—whether to use the constraint form or the flux form—is often based on what is most conducive to the underlying numerical solution methodology. However, these flux-based forms of the conditions do not readily present a path to generalize the Kutta condition to more situations. For this generalization, we will rely on the constraint form, as we discuss in the next section.

5.4 Generalized Edge Conditions

In classical aerodynamics, in which the angle of attack is presumed small, the specification of flow regularity at the trailing edge—the Kutta condition—comprises an essential ingredient for obtaining a physically meaningful model of the flow. Our intention in this book is to extend the set of inviscid modeling tools to a much wider class of flows, such as those for which the angle of attack is no longer small and the flow is separated at the leading edge. But just as we needed the Kutta condition to specify smooth separation of flow from the trailing edge, we must endow the model with a mathematical condition that describes flow separation from any other edge. What condition should we impose at an edge from which the flow separates? The Kutta condition might be appropriate in some cases. Consider a flat plate translating normal to itself: the leading and trailing edges are indistinguishable from each other, and the Kutta condition is a good candidate to be applied at both. However, the Kutta condition imposed at a cusped edge produces a flow that emerges tangent to the edge, an unrealistic behavior at the leading edge of the same plate at small angle of attack. So we should concede that the Kutta condition is not satisfactory at the leading edge in many scenarios of separated flow.

This section is focused on postulating a ***generalized family of edge conditions***, of which the Kutta condition is a member. In Result 5.5 we expressed the Kutta condition at an edge as an annihilation of the intensity of the singular behavior, σ, associated with that edge. Though we will not prove it until Chap. 6, this signed intensity is directly related to the integrated negative pressure—the ***suction***—imposed by the flow on this edge, and as a consequence, we can describe σ as the edge's **suction parameter**. Thus, we can interpret Result 5.5 as a statement of the flow's inability to overcome the adverse pressure gradient as it circumnavigates the edge. Of course, because of its associated singular flow, any sharp edge imposes an infinitely adverse pressure gradient, so it would seem necessary to impose $\sigma = 0$—and therefore, to release vorticity—at any edge to remove this unphysical behavior. But we should recognize that our mathematical model of a flow may be based on an *approximation* of the shape of the real body, and an edge in our approximate geometry may not represent a true edge in the real geometry. Thin airfoil theory is a notable example of this: a real airfoil with a rounded leading edge is represented by a vortex sheet along the mid-camber line. While our mathematical model predicts an infinitely adverse pressure gradient around the leading edge of the geometry, the actual geometry at the same conditions imposes a more gentle pressure gradient in which the flow may remain attached. Thus, it seems reasonable to tolerate modest non-zero values of suction at an edge with no concomitant production of vorticity. This is the basic premise of the **critical leading-edge suction parameter**, proposed by Ramesh et al. [60] to govern the shedding of vortex elements near the leading edge of an airfoil.

The advantage of the parameter σ, as we defined it in (3.103), is that it contains information about both the magnitude and the directionality of the flow in the vicinity of the edge. The details of flow separation, such as the sign of generated vorticity, depend on the direction from which the edge is approached. Our edge condition based on σ will be the following:

Result 5.6: Critical Edge Suction Condition

At an edge on a body, the corresponding edge suction parameter σ, defined in (3.103), is required to remain in the range

$$\sigma_{\min} \leq \sigma \leq \sigma_{\max}. \tag{5.16}$$

If, at any instant, σ transgresses from these bounds, vorticity shall be generated near the edge with the strength just necessary to ensure that σ is brought back within the bounds.

For the definition of a leading edge suction parameter, Ramesh et al. [60] utilized the leading (scalar-valued) coefficient in the thin-airfoil Fourier–Chebyshev expansion, A_0. As we will see in Chap. 8, A_0 is identical to σ at the leading edge when the Kutta condition is enforced at the trailing edge. However, the definition (3.103) we provide in this book is general, so that Result 5.6 can be applied at any edge, independent of the enforcement of conditions at other edges. Note 4.4.6 provides another form for general calculation.

The bounds $\sigma_{\min}$ and $\sigma_{\max}$ are not specified here. In general, they will depend on the details of the body geometry, and perhaps on the type of motion. Indeed, because these bounds are meant to describe the behavior of boundary layer separation in a real viscous flow, they will also depend to some extent on the Reynolds number. Ramesh et al. [60] determined the critical upper bound $\sigma_{\max}$ for the leading edge of a two-dimensional airfoil by conducting a high-resolution numerical simulation of the viscous flow about the airfoil geometry undergoing a basic motion in which the flow separates, and a nominally identical inviscid discrete vortex simulation with a flat plate undergoing the same motion. The value of σ in the discrete vortex simulation at the instant at which the real flow was observed to begin shedding leading-edge vorticity was deemed to be the critical suction parameter. They found that, for the same geometry and Reynolds number, this critical value did not vary significantly for different types of motion.

We offer only the following general notes for now:

- If the Kutta condition is to be enforced at the edge, then $\sigma_{\min} = \sigma_{\max} = 0$. Thus, ***the critical edge suction condition is a generalization of the Kutta condition***.
- In general, $\sigma_{\min} \leq 0$ and $\sigma_{\max} \geq 0$. If the geometry is symmetric about the edge, then $\sigma_{\min} = -\sigma_{\max}$, but otherwise we expect their magnitudes to differ.
- The values of these critical parameters are most likely insensitive to the motion of the body. However, temporal changes in their values might be used as a surrogate for describing the influence of a flow feature that is not explicitly described with flow contributors. For example, if a disturbance is incident upon the body, then, rather than representing the disturbance with vortices or other potential flow elements, its effect can be represented by a time-varying value of critical suction parameters.

Fig. 5.4 Illustrations of a free vortex sheet $\mathcal{S}$ released from the edge of a body at which the Kutta condition has been enforced (left) or at which the suction parameter criterion in Result 5.6 has been enforced (right)

It is important to remember that σ represents the coefficient of the leading-order singular behavior of the flow about the edge. By forcing σ to remain in the range (5.16), we have capped the strength of this singular behavior, but the singularity still exists. This raises an important question regarding the release of nascent vortex elements from the edge. How can such elements' transport be well-behaved if they are released near a point at which the velocity is infinite? To answer this, it is helpful to consider the existence of stagnation points in the vicinity of the edge. Recall, from Note 4.4.5, that σ is proportional to the tangential velocity in the circle plane at the edge's pre-image. The Kutta condition forces this tangential velocity to be zero, pinning a stagnation point at this pre-image of the edge. In the physical plane, the Kutta condition does not make the edge itself a stagnation point, as we discussed in Sect. 5.1.2. However, it does render it a point of finite flow velocity that transports the released vorticity away, as illustrated in the left panel in Fig. 5.4, and furthermore, it ensures that there are no stagnation points in the immediate vicinity of the edge.

In contrast, by relaxing the constraint on σ so that it simply remains in a range around zero, we are no longer pinning the stagnation point to the pre-image of the edge, but instead, allowing this stagnation point to float along the unit circle in the neighborhood of this edge pre-image. But any stagnation point in the circle plane that does not lie at the edge translates into a stagnation point in the physical plane, since the Jacobian of the mapping does not vanish at such points. Thus, ***the suction parameter criterion in Result 5.6 causes a stagnation point to emerge in the vicinity of the edge***. Furthermore, since the resulting flow is only a perturbation from the behavior ensured by the Kutta condition, the streamlines that meet at this emergent stagnation point consist of two directed toward it along the body surface and another directed away from the body. Vorticity released from the edge itself would follow a trajectory similar to this (but, because the flow is unsteady, not identical to it): swept backward along the body from the edge toward the stagnation point and thence away from the body.

In a numerical implementation, in which the free vortex sheet is spatially discretized into a finite number of segments, it is not practically feasible to release vorticity exactly from the singular edge, so instead it is released from some nearby point where the velocity is finite. Thus, the vorticity's immediate trajectory would be slightly rounded, and we can expect a free vortex sheet associated with this edge to adopt a configuration akin to the one illustrated in the right panel in Fig. 5.4. It is interesting to note that, in this discrete setting, the body edge and the nearest segment of the sheet appear to collectively form an effective edge, convex and singular on one side and concave and well-behaved on the other, consistent with Note 3.2.3.

How do we determine the strength of a new vortex element released into the flow under this criterion on the suction parameter? Let us return again to the decomposition of σ in Eq. (5.14) that we introduced earlier in the context of the Kutta condition. It is easy to see how we can extend this to the new context: If the suction parameter due to existing contributors, $\check{\sigma}$, satisfies the criterion in Eq. (5.16), then we have no need to introduce a new element, so we set $\delta\Gamma_{N_v+1}$ to zero. However, if, for example, $\check{\sigma} > \sigma_{\max}$, then we set $\delta\Gamma_{N_v+1}$ so that the full suction parameter is brought just within the required bounds:

$$\delta\Gamma_{N_v+1} = \frac{\sigma_{\max} - \check{\sigma}}{\sigma^{v}_{N_v+1}}. \tag{5.17}$$

In other words, the strength of the new vortex element is directly proportional to the amount by which the maximum suction parameter is exceeded. A similar equation holds if $\check{\sigma}$ falls below the minimum threshold, $\sigma_{\min}$. This interpretation provides a useful intuition for the suction parameter constraint. ***The larger is the range of acceptable values of σ, the weaker is the strength of vorticity released into the fluid in order to correct violations of this range.*** The Kutta condition, with the narrowest allowable range, leads to the strongest released elements.

An example of the effect of the critical suction parameter, $\sigma_{\max}$, is shown in Fig. 5.5. Here, a flat plate has been accelerated from rest to a constant translational velocity at large angle of attack. We will describe the details of the numerical procedure used for this model later, in Sect. 9.4; here, we will only focus on the depicted results. These results only comprise the earliest times after the motion has initiated, so that the behavior at the edge shown is essentially independent of the other edge. Three different choices of the critical suction parameter are shown, including $\sigma_{\max} = 0$, the value corresponding to the Kutta condition. Strictly speaking, it is the minimum threshold, $\sigma_{\min}$, invoked in these results, since σ tends to be negative at this edge under these conditions, but we have set $\sigma_{\min} = -\sigma_{\max}$ due to the symmetry of the geometry, as discussed above. When we enforce the Kutta condition, the free vortex sheet adopts the expected configuration tangent to the plate. As the critical suction parameter is increased to $\sigma_{\max} = 0.2U$, this vortex sheet is swept slightly backward from the edge, toward the stagnation point that appears near this edge. The stagnation point moves further away from the edge when $\sigma_{\max} = 0.4U$, and the sheet is thus swept more significantly aft along the plate. As the upper-right panel in Fig. 5.5 shows, this progressive increase in $\sigma_{\max}$ leads to greater restriction on the growth of circulation in the sheet.

Let us make a few other notes before we move on:

- As we noted already in the discussion of the Kutta condition, we can simultaneously enforce edge conditions at multiple edges by solving the coupled system of constraint equations for the strengths of the released vortex elements at all edges.
- Although we demonstrated the application of the suction parameter criterion on the edge of a flat plate, the criterion can readily be used at edges of non-zero angle, as well.

- If an existing variable-strength point vortex, labeled N_v, is used instead of a new element to gather vorticity released from the edge, then, similar to the Kutta condition, constraint Result 5.6 can be applied in its time-derivative form. The resulting ordinary differential equation for $\dot{\Gamma}_{N_v}$ (which, as in the case of the Kutta condition, would couple with the vortex' transport, $\dot{z}_{N_v}$, would only be advanced when σ exceeds the acceptable range; otherwise its strength, Γ_{N_v}, would be held constant.

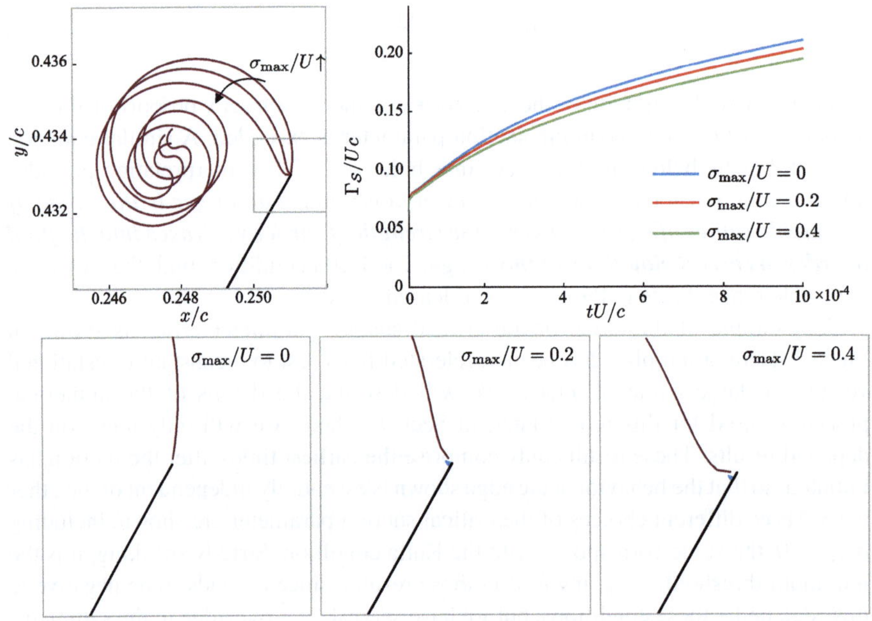

Fig. 5.5 Effect on the released vortex sheet of increasing σ_{max} in the suction criterion applied at the edge of a flat plate of length c. The upper left panel depicts the free vortex sheet generated by three different choices of σ_{max}/U (0, 0.2, 0.4) at time $tU/c = 10^{-3}$, after the plate has started impulsively translating from rest to the right at speed U and 60° angle of attack. The upper right panel shows the circulation of this vortex sheet versus time. The three lower panels depict the detailed view of the region around the edge (delimited by the square shown in the upper panel) for three different choices of σ_{max}. Note that these results exhibit the small gap between the edge and the root of the sheet due to the numerical implementation. The leftmost panel ($\sigma_{max} = 0$) corresponds to the Kutta condition. In the two other panels, the emergent stagnation point on the upper surface is depicted as a blue triangle

Chapter 6
Force and Moment on a Body

Contents

6.1 Force and Moment via Surface Traction 184
6.2 Force and Moment via Vorticity Impulse 187
6.2.1 Vectorial Forms 188
6.2.2 Complex Form for Planar Applications 196
6.3 Reconciliation of Force via Traction and via Impulse 199
6.3.1 The Force and Moment on a Region of Vorticity 200
6.3.2 The Spurious Force and Moment on Vorticity 204
6.3.3 Spurious Force and Moment on a Vortex in the Presence of a Body 207
6.3.4 Revisiting the Traction Force and Moment on the Body 211
6.4 Edge Suction 213
6.5 Decomposition of the Force into Contributors 216
6.5.1 Complex Form, via Conformal Mapping Solution 217
6.5.2 Decomposed Force and Moment, Generalized 220
6.6 The Contribution of Fluid Vorticity to Force and Moment 226
6.6.1 Basic Definitions of the Vorticity-Induced Impulses 227
6.6.2 The Rates of Change of the Vorticity-Induced Impulses 228
6.6.3 The Rate of Change of Impulse for Singular Vortex Elements 237
6.6.4 Alternative Derivation of the Force and Moment 239
6.7 The Mechanical Energy Equation, Revisited 243

In this chapter, we are interested in developing expressions to compute the force and moment on a body—either two- or three-dimensional, possibly in motion, and immersed in a fluid of infinite extent. We will develop two approaches to calculate these forces and moments. The first will be the most straightforward: we will compute the force and moment by integrating the traction on the surface of the body. The second, in contrast, will be based on the rate of change of **impulse** in the flow. One can think of this approach as accounting for the force and moment by its effect on the momentum in the surrounding fluid. We will discover that these impulse-based formulas are quite generally applicable, even to viscous flows, but that they also

J. D. Eldredge, *Mathematical Modeling of Unsteady Inviscid Flows*, Interdisciplinary Applied Mathematics 50, https://doi.org/10.1007/978-3-030-18319-6_6

rely on a few assumptions about the underlying problem. In order to clarify these assumptions, we will return to the surface traction form, and use this to reconcile the two approaches, under our usual restriction to inviscid flows. Consistent with our treatment thus far, we will neglect the effect of gravity, though it is easy to include it in any of the results we obtain.

After presenting the two different perspectives on force and moment, we will use our decomposition of the flow field in order to distinguish the contributions to the force and moment from the various basis fields of a general flow. This will naturally lead us again to added mass, first introduced in our decomposition of kinetic energy in Sect. 4.6.3, and culminate in the most useful result of the chapter, Result 6.14.

6.1 Force and Moment via Surface Traction

The force and moment exerted by a fluid on a body are due entirely to the integral of traction and its first moment over the surface of the body. In an inviscid flow, this traction is simply $-p\boldsymbol{n}$, where p is the pressure and $\boldsymbol{n}$ is the unit normal vector pointing into the fluid. Thus, the basic expressions for the force and the moment (about the origin of an inertial coordinate system) are, respectively,

$$\boldsymbol{f} = -\int_{\mathcal{S}_b} p\boldsymbol{n}\,\mathrm{d}S, \qquad \boldsymbol{m}_O = -\int_{\mathcal{S}_b} \boldsymbol{x}\times p\boldsymbol{n}\,\mathrm{d}S, \tag{6.1}$$

where $\boldsymbol{x}$ is a point on the surface, $\mathcal{S}_b$. Recall that the moment about any other point, P, can easily be computed from the transformation (2.48). Thus, we can focus our attention on the quantities defined in (6.1).

Let us assume henceforth in this section that the vorticity is concentrated in singular elements (i.e. point vortices, filaments or sheets), so that the fluid region $\mathcal{V}_f$ is otherwise irrotational. If we substitute for pressure in the force expression with the unsteady Bernoulli equation (3.60), we get straightforward expressions for force and moment in terms of velocity and scalar potential on the body surface:

Result 6.1: Surface Integral Form 1 of Force and Moment

The force and moment (about the origin) on a body in inviscid incompressible flow, irrotational almost everywhere, are given, respectively, by

$$\boldsymbol{f} = \rho\int_{\mathcal{S}_b}\left(\frac{1}{2}|\boldsymbol{v}|^2 + \frac{\partial\varphi}{\partial t}\right)\boldsymbol{n}\,\mathrm{d}S, \tag{6.2}$$

$$\boldsymbol{m}_O = \rho\int_{\mathcal{S}_b}\boldsymbol{x}\times\left(\frac{1}{2}|\boldsymbol{v}|^2 + \frac{\partial\varphi}{\partial t}\right)\boldsymbol{n}\,\mathrm{d}S, \tag{6.3}$$

where $\mathcal{S}_b$ is the surface of the body and $\boldsymbol{n}$ is the unit normal directed into the surrounding fluid.

Note that the Bernoulli constant (or any spatially uniform part of pressure) contributes nothing to the integrals over the whole surface $\mathcal{S}_b$. This can be shown by using results (A.22) and (A.24).

The forms given by (6.2) and (6.3) are not always convenient, since they require computing the rate of change of the scalar potential on the (possibly moving) surface at points that are stationary relative to the inertial frame of reference. We can make use of the results in Sect. A.1.7 of the Appendix for interchanging surface integrals with time derivatives to express the force in a different form, in which attention is shifted to the rate of change of the net contribution of φ on the body surface. We have a choice of a few different forms. For the first, we apply identities (A.92) and (A.94) to arrive at the following:

Result 6.2: Surface Integral Form 2 of Force and Moment

The force and moment on a body in inviscid incompressible flow, irrotational almost everywhere, are given, respectively, by

$$\boldsymbol{f} = \rho \int_{\mathcal{S}_b} \left[\left(\frac{1}{2}|\boldsymbol{v}|^2 - \boldsymbol{V}_b \cdot \boldsymbol{v} \right) \boldsymbol{n} + \varphi \nabla \boldsymbol{V}_b \cdot \boldsymbol{n} \right] \mathrm{d}S + \rho \frac{\mathrm{d}}{\mathrm{d}t} \int_{\mathcal{S}_b} \varphi \boldsymbol{n} \, \mathrm{d}S, \quad (6.4)$$

and

$$\boldsymbol{m}_O = \rho \int_{\mathcal{S}_b} \boldsymbol{x} \times \left[\left(\frac{1}{2}|\boldsymbol{v}|^2 - \boldsymbol{V}_b \cdot \boldsymbol{v} \right) \boldsymbol{n} + \varphi \nabla \boldsymbol{V}_b \cdot \boldsymbol{n} \right] \mathrm{d}S$$

$$- \rho \int_{\mathcal{S}_b} \boldsymbol{V}_b \times \varphi \boldsymbol{n} \, \mathrm{d}S + \rho \frac{\mathrm{d}}{\mathrm{d}t} \int_{\mathcal{S}_b} \boldsymbol{x} \times \varphi \boldsymbol{n} \, \mathrm{d}S, \quad (6.5)$$

where $\boldsymbol{V}_b$ is the local velocity of the surface $\mathcal{S}_b$, and we have used the fact that $\boldsymbol{v} = \nabla \varphi$. Note that the third term in the leading integral in both expressions requires that $\boldsymbol{V}_b$ be defined in a neighborhood of the surface in order to calculate the gradient. When the surface is in rigid motion, then $\nabla \boldsymbol{V}_b \cdot \boldsymbol{n} = -\boldsymbol{\Omega} \times \boldsymbol{n}$, where $\boldsymbol{\Omega}$ is the surface's angular velocity.

This form for force and moment may be adequate for many applications. It requires knowledge of the velocity and scalar potential fields of the fluid on the surface of the body, as well as the surface's motion. It is also important to emphasize that this form has been obtained only by manipulation of the integration over the body's surface.

Thus, for a system of bodies, it enables us to distinguish the force and moment on each of the constituent bodies. However, a drawback is that it contains terms that are quadratic in the fluid and body fields, which prevents us from performing a decomposition into independent contributions from the various flow contributors, as we did for the flow field in Sect. 4.6. We should, of course, ask the obvious question: is such a decomposition even possible? After all, our experience with inertially-dominated flows suggests that the force and moment *should*, at least in part, depend quadratically on the velocity. As we will see in our analysis in the next section, we can indeed obtain a form of force and moment that is linearly dependent upon the flow contributors. Then, in the subsequent section, we will reconcile these two forms, taking care to satisfy ourselves that the quadratic dependence hasn't entirely vanished.

Note that we have refrained in this section from providing an equivalent complex form for the force and moment in planar applications. We will provide a complex version of the impulse-based forms in the next section. Also, once we have manipulated our traction-based form of the force and moment a bit further in Sect. 6.3, we will also identify a complex version of this form.

Note 6.1.1: Force and Moment in the Windtunnel Frame

The force and moment measured in the windtunnel reference frame, in which there is a (possibly unsteady) uniform flow at infinite distances, can also be calculated with the formulas provided in this section. How do these compare with the forces and moments observed from the inertial reference frame? In Sect. 3.2.1, we found that the difference in pressure between these two frames, expressed in Eq. (3.65), was attributable to a hydrostatic-like term due to acceleration of the uniform flow. Thus, substituting this into the basic formulas (6.1) for force and moment, and using identities (A.23) and (A.25) to transform the resulting surface integrals of the difference in pressures, we get

$$\boldsymbol{f}^{\dagger} = \boldsymbol{f} + \rho \mathcal{V}_{\mathrm{b}} \dot{\boldsymbol{V}}_{\infty}, \qquad \boldsymbol{m}^{\dagger}_{\mathrm{O}^{\dagger}} = \boldsymbol{m}_{\mathrm{O}} - \boldsymbol{X}_{\mathrm{O}^{\dagger}} \times \boldsymbol{f} + \boldsymbol{X}^{\dagger}_{\mathrm{c}} \times \rho \mathcal{V}_{\mathrm{b}} \dot{\boldsymbol{V}}_{\infty}, \tag{6.6}$$

where $\boldsymbol{f}^{\dagger}$ and $\boldsymbol{m}^{\dagger}_{\mathrm{O}^{\dagger}}$ are the force and moment that result from integrating, over the body surface, the pressure, $p^{\dagger}$, observed in the windtunnel reference frame, with the moment taken about the origin $\mathrm{O}^{\dagger}$ of the windtunnel coordinate system. The first difference between the moments is attributable to the difference in reference points for the moments. The final terms in each expression represent a kind of 'buoyancy' due to the accelerating uniform flow. This buoyant force, acting through the centroid $\boldsymbol{X}^{\dagger}_{\mathrm{c}}$, takes the form of the inertia of the fluid displaced by the body. Note that, if the body is infinitely thin, there is no such buoyancy.

In most cases, it is more convenient to analyze a flow in the inertial reference frame. Equations (6.6) provide an easy means for adjusting the resulting force and moment obtained in that analysis so that they are correct in this windtunnel frame of reference. In these equations, we have kept the same reference point of the moment—the origin in the inertial reference frame—with the caveat that this point is moving with velocity $\boldsymbol{V}_\infty$ in the windtunnel frame. The moment can be shifted to a fixed reference point in the windtunnel frame by using the standard change of axis formula (2.48).

6.2 Force and Moment via Vorticity Impulse

The approach that we follow in this section is to account for the force and moment by the rate of change of momentum in the fluid. This approach is, of course, predicated on the assumption that momentum in the fluid is otherwise conserved (i.e., there are no other entities—real or spurious—with which the fluid can exchange momentum).

One complication of this approach is that the momentum of an infinite expanse of fluid,

$$\int_{\mathcal{V}_f} \rho \boldsymbol{v} \, \mathrm{d}V, \tag{6.7}$$

cannot generally be calculated. To see why this is true, let's suppose we integrate over a fluid region $\mathcal{V}_{R\backslash b}$, bounded internally by the surface $\mathcal{S}_b$ of the body and externally by a spherical surface $\mathcal{S}_R$ of radius R (or a circular surface in two dimensions), and then allow R to approach infinity. As we know from Sect. 4.7, the velocity field (with any uniform flow subtracted) decays like $1/r^2$ in two dimensions and $1/r^3$ in three dimensions—that is, it decays at a rate that just balances the rate at which the region $\mathcal{V}_{R\backslash b}$ grows, and therefore the integral is not absolutely convergent as we let $R \to \infty$. In fact, as we will see, there is a crucial contribution from the fluid beyond $\mathcal{S}_R$ that persists for any choice of arbitrarily large R. A similar aspect holds for the angular momentum. So our strategy will be to explicitly account for this 'external' contribution as we allow R to approach large values. We will assume that this outer surface $\mathcal{S}_R$ encloses all of the fluid vorticity, and that its unit normal $\boldsymbol{n}$ points outward. Let us first highlight a particular set of identities that will help us in this process:

Lemma 6.1: The Integrals of v and $x \times v$ in a Region

Consider a volume $\mathcal{V}$ enclosed by surface $\mathcal{S}$, on which the outward unit normal is $\boldsymbol{n}$. Then the following pair of identities hold for the velocity field $\boldsymbol{v}$:

$$\int_{\mathcal{V}} \boldsymbol{v} \, \mathrm{d}V = \frac{1}{n_\mathrm{d} - 1} \left(\int_{\mathcal{V}} \boldsymbol{x} \times \boldsymbol{\omega} \, \mathrm{d}V - \int_{\mathcal{S}} \boldsymbol{x} \times (\boldsymbol{n} \times \boldsymbol{v}) \, \mathrm{d}S \right). \tag{6.8}$$

and

$$\int_{\mathcal{V}} \boldsymbol{x} \times \boldsymbol{v} \, \mathrm{d}V = \frac{1}{n_\mathrm{d}} \left(\int_{\mathcal{V}} \boldsymbol{x} \times (\boldsymbol{x} \times \boldsymbol{\omega}) \, \mathrm{d}V - \int_{\mathcal{S}} \boldsymbol{x} \times (\boldsymbol{x} \times (\boldsymbol{n} \times \boldsymbol{v})) \, \mathrm{d}S \right), \tag{6.9}$$

where n_d is the number of spatial dimensions (2 or 3).

Proof Let us start with the vector identity that we previously used in Chap. 3:

$$\boldsymbol{v} = \frac{1}{n_\mathrm{d} - 1} \left[\boldsymbol{x} \times \boldsymbol{\omega} - \nabla(\boldsymbol{x} \cdot \boldsymbol{v}) + \nabla \cdot (\boldsymbol{x}\boldsymbol{v}) \right]. \tag{3.26}$$

Then, when we compute the integral of this identity over $\mathcal{V}$, the latter two terms on the right-hand side can be transformed into a surface integral over $\mathcal{S}$ with the divergence theorem (A.18). A similar process can be carried out for the first moment of this identity. □

These identities will allow us to write the momentum over some region with as the integral of a moment of vorticity over the same region, plus a contribution from a moment of tangential velocity over the surface(s) bounding the region.

One note to make before we proceed: we will carry out the analysis of this section in the inertial frame of reference, in which the fluid is at rest at infinity. In Note 6.1.1 we showed how to reconcile the force and moment between this frame and the windtunnel frame, so if we seek these quantities in the latter frame of reference, we can easily transform our results. We will also include a brief note outlining how the analysis of the present section changes when a uniform flow is present.

6.2.1 Vectorial Forms

Force Consider conservation of linear momentum in $\mathcal{V}_{R\backslash \mathrm{b}}$. In the derivation that follows, we will unify the treatment of the two- and three-dimensional problems, with the usual understanding that force and moment in two dimensions must be interpreted as 'per unit depth'. There are forces acting on this region from the body ($-\boldsymbol{f}$, the reaction of the force from the fluid on the body) and from the fluid beyond

$\mathcal{S}_R$; the latter we will denote by $\boldsymbol{f}_R$. The overall conservation statement is

$$-\boldsymbol{f} + \boldsymbol{f}_R = \frac{\mathrm{d}}{\mathrm{d}t} \int_{\mathcal{V}_{R\backslash \mathrm{b}}} \rho \boldsymbol{v} \, \mathrm{d}V + \int_{\mathcal{S}_R} \rho \boldsymbol{v} (\boldsymbol{v} \cdot \boldsymbol{n}) \, \mathrm{d}S. \tag{6.10}$$

The final integral in the expression makes vanishingly small contribution as $R \to \infty$, based on the decay of the velocity exhibited in the leading terms in the multipole expansions, as discussed in Sect. 4.7. To compute the integral over $\mathcal{V}_{R\backslash \mathrm{b}}$ (in which ρ is uniform, as always in this work), we will make use of the identity (6.8) in Lemma 6.1, applied over $\mathcal{V}_{R\backslash \mathrm{b}}$ and its bounding surfaces $\mathcal{S}_\mathrm{b}$ and $\mathcal{S}_R$. By construction, $\mathcal{V}_{R\backslash \mathrm{b}}$ includes all fluid vorticity, so the volume integral of $\boldsymbol{x} \times \boldsymbol{\omega}$ can be extended to the entire fluid, $\mathcal{V}_f$, and the velocity in the integral over $\mathcal{S}_R$ can be replaced by the gradient of the scalar potential, $\nabla \varphi$. We will collectively denote the first two terms on the right-hand side by the vector $\boldsymbol{P}$:

$$\boldsymbol{P} := \frac{1}{n_\mathrm{d} - 1} \left(\int_{\mathcal{V}_f} \boldsymbol{x} \times \boldsymbol{\omega} \, \mathrm{d}V + \int_{\mathcal{S}_\mathrm{b}} \boldsymbol{x} \times (\boldsymbol{n} \times \boldsymbol{v}) \, \mathrm{d}S \right). \tag{6.11}$$

Thus, in either two or three dimensions,

$$\int_{\mathcal{V}_{R\backslash \mathrm{b}}} \boldsymbol{v} \, \mathrm{d}V = \boldsymbol{P} - \frac{1}{n_\mathrm{d} - 1} \int_{\mathcal{S}_R} \boldsymbol{x} \times (\boldsymbol{n} \times \nabla \varphi) \, \mathrm{d}S. \tag{6.12}$$

The force $\boldsymbol{f}_R$ applied on the outer surface, $\mathcal{S}_R$, by the surrounding fluid can be obtained by integrating the pressure over that surface. By following the same steps as we used in obtaining Eq. (6.4), but applying them now to the fixed surface $\mathcal{S}_R$ rather than $\mathcal{S}_\mathrm{b}$, we find that

$$\boldsymbol{f}_R = \rho \frac{\mathrm{d}}{\mathrm{d}t} \int_{\mathcal{S}_R} \varphi \boldsymbol{n} \, \mathrm{d}S + \rho \int_{\mathcal{S}_R} \frac{1}{2} |\boldsymbol{v}|^2 \boldsymbol{n} \, \mathrm{d}S. \tag{6.13}$$

Again, the final integral is negligible as $R \to \infty$ due to the rate of decay of the velocity with distance. We substitute the remaining term for $\boldsymbol{f}_R$ in (6.10), and use the result (6.12) for the linear momentum in $\mathcal{V}_{R\backslash \mathrm{b}}$, to obtain

$$\boldsymbol{f} = -\rho \frac{\mathrm{d}\boldsymbol{P}}{\mathrm{d}t} + \rho \frac{\mathrm{d}}{\mathrm{d}t} \left[\frac{1}{n_\mathrm{d} - 1} \int_{\mathcal{S}_R} \boldsymbol{x} \times (\boldsymbol{n} \times \nabla \varphi) \, \mathrm{d}S + \int_{\mathcal{S}_R} \varphi \boldsymbol{n} \, \mathrm{d}S \right]. \tag{6.14}$$

The overall expression in square brackets vanishes, based on identity (A.66), proved using Stokes' theorem and applied here to a closed surface on which φ is continuous. Thus, we are left with the delightfully simple formula:

Result 6.3: Force as Rate of Change of Linear Impulse

The force on a body of constant volume in an unbounded fluid is given by

$$\boldsymbol{f} = -\rho \frac{\mathrm{d}\boldsymbol{P}}{\mathrm{d}t}, \tag{6.15}$$

where $\boldsymbol{P}$, defined (in either dimension, $n_\mathrm{d} = 2$ or 3) as

$$\boldsymbol{P} := \frac{1}{n_\mathrm{d} - 1} \left(\int_{\mathcal{V}_f} \boldsymbol{x} \times \boldsymbol{\omega} \,\mathrm{d}V + \int_{\mathcal{S}_\mathrm{b}} \boldsymbol{x} \times (\boldsymbol{n} \times \boldsymbol{v}) \,\mathrm{d}S \right), \tag{6.16}$$

is called the **linear impulse**. In two dimensions, the force is to be interpreted per unit depth.

Moment About the Origin We will follow similar steps to obtain an expression for the moment applied by the fluid on the body about the origin. Here, we start with the conservation of angular momentum in the region $\mathcal{V}_{R\setminus\mathrm{b}}$, upon which act moments from the body ($-\boldsymbol{m}_\mathrm{O}$) and from the fluid beyond $\mathcal{S}_R$ ($\boldsymbol{m}_{O,R}$):

$$-\boldsymbol{m}_\mathrm{O} + \boldsymbol{m}_{O,R} = \frac{\mathrm{d}}{\mathrm{d}t} \int_{\mathcal{V}_{R\setminus\mathrm{b}}} \boldsymbol{x} \times \rho \boldsymbol{v} \,\mathrm{d}V + \int_{\mathcal{S}_R} \boldsymbol{x} \times \rho \boldsymbol{v} (\boldsymbol{v} \cdot \boldsymbol{n}) \,\mathrm{d}S. \tag{6.17}$$

The surface integral is negligible as $R \to \infty$, since the decay of the velocity is sufficiently rapid. We transform the integral of $\boldsymbol{x} \times \boldsymbol{v}$ with identity (6.9) in Lemma 6.1, where once again, we can extend the volume integral of the vorticity moment to all of $\mathcal{V}_f$. We define the **angular impulse** about the origin as

$$\boldsymbol{\Pi}_\mathrm{O} = \frac{1}{n_\mathrm{d}} \left(\int_{\mathcal{V}_f} \boldsymbol{x} \times (\boldsymbol{x} \times \boldsymbol{\omega}) \,\mathrm{d}V + \int_{\mathcal{S}_\mathrm{b}} \boldsymbol{x} \times [\boldsymbol{x} \times (\boldsymbol{n} \times \boldsymbol{v})] \,\mathrm{d}S \right). \tag{6.18}$$

Using (6.5) on $\mathcal{S}_R$, it is straightforward to show that the moment applied on $\mathcal{S}_R$ is

$$\boldsymbol{m}_{O,R} = \rho \frac{\mathrm{d}}{\mathrm{d}t} \int_{\mathcal{S}_R} \boldsymbol{x} \times \varphi \boldsymbol{n} \,\mathrm{d}S + \rho \int_{\mathcal{S}_R} \boldsymbol{x} \times \frac{1}{2} |\boldsymbol{v}|^2 \boldsymbol{n} \,\mathrm{d}S; \tag{6.19}$$

again, the integral of $|\boldsymbol{v}|^2$ makes negligible contribution as $R \to \infty$. Then, when these results are collected in (6.17), the integrals over $\mathcal{S}_R$ can be shown to vanish with the help of identity (A.67):

$$\int_{S_R} \boldsymbol{x} \times \varphi \boldsymbol{n} \, \mathrm{d}S + \frac{1}{n_\mathrm{d}} \int_{S_R} \boldsymbol{x} \times [\boldsymbol{x} \times (\boldsymbol{n} \times \nabla\varphi)] \, \mathrm{d}S = 0. \tag{6.20}$$

Thus, we arrive at an expression for moment on the body about the origin, as simple as that for the force:

Result 6.4: Moment as Rate of Change of Angular Impulse

The moment about the origin on an isolated body of constant volume in an unbounded fluid is given by

$$\boldsymbol{m}_\mathrm{O} = -\rho \frac{\mathrm{d}\boldsymbol{\Pi}_\mathrm{O}}{\mathrm{d}t}, \tag{6.21}$$

where the **angular impulse** about the origin, $\boldsymbol{\Pi}_\mathrm{O}$, is defined in either dimension ($n_\mathrm{d} = 2$ or 3) as

$$\boldsymbol{\Pi}_\mathrm{O} = \frac{1}{n_\mathrm{d}} \left(\int_{V_f} \boldsymbol{x} \times (\boldsymbol{x} \times \boldsymbol{\omega}) \, \mathrm{d}V + \int_{S_\mathrm{b}} \boldsymbol{x} \times [\boldsymbol{x} \times (\boldsymbol{n} \times \boldsymbol{v})] \, \mathrm{d}S \right). \tag{6.22}$$

A Few Short Observations on Results 6.3 and 6.4

- In deriving the force (6.15) and moment (6.21), we never relied on any assumption that the flow is inviscid. Indeed, ***these formulas are valid whether the flow is inviscid or viscous.***
- If the flow is inviscid, we can use the no-penetration condition (3.195) to replace $\boldsymbol{n} \times \boldsymbol{v}$ with $\boldsymbol{\gamma} + \boldsymbol{n} \times \boldsymbol{V}_\mathrm{b}$ in both the linear and angular impulse. (If the flow is viscous, the surface vortex sheet strength, $\boldsymbol{\gamma}$, is zero.)
- We would have failed to achieve these results if we had neglected to account for $\boldsymbol{f}_R$ and $\boldsymbol{m}_{O,R}$.
- If there are multiple bodies immersed in the fluid, it is impossible to differentiate the contributions to their individual forces from these formulas. Only the collective force can be calculated. This is a drawback that will be addressed in Sect. 6.3.
- Recall that, in deriving these forms, we have assumed that there are no other forces or moments applied on the fluid that would change its momentum. This will be a key issue in our reconciliation with the form derived from surface traction in Sect. 6.3.
- The linear impulse can be written in terms of the dipole coefficient of the multipole expansion by making use of (4.197) in Note 4.7.2. From this, it is clear that $\boldsymbol{P} = \boldsymbol{Q} - \boldsymbol{V}_\mathrm{r}\mathcal{V}_\mathrm{b}$. Thus, there is a direct connection between the force between the fluid and body and the dipolar field that dominates at large distances.

There are other forms in which we can express the impulse. Some of these are covered in the following notes.

Note 6.2.1: A Vorticity Form of the Impulse

The linear and angular impulses, as defined in (6.16) and (6.22), respectively, were written in a form that involved only the fluid vorticity and the surface velocity. However, by extending the surface velocity field $\boldsymbol{V}_\mathrm{b}$ into the bodies, another form can be derived that emphasizes moments of vorticity. This can be found by noting that identities (6.8) and (6.9) can also be found for the body region, $\mathcal{V}_\mathrm{b}$, and the enclosing surface, $\mathcal{S}_\mathrm{b}$. With our usual definition for the normal vector (i.e., pointing outward from $\mathcal{V}_\mathrm{b}$), these identities are

$$\int_{\mathcal{V}_\mathrm{b}} \boldsymbol{V}_\mathrm{b}\,\mathrm{d}V = \frac{1}{n_\mathrm{d}-1}\left[\int_{\mathcal{V}_\mathrm{b}} \boldsymbol{x}\times\boldsymbol{\omega}_\mathrm{b}\,\mathrm{d}V - \int_{\mathcal{S}_\mathrm{b}} \boldsymbol{x}\times(\boldsymbol{n}\times\boldsymbol{V}_\mathrm{b})\,\mathrm{d}S\right] \tag{6.23}$$

and

$$\int_{\mathcal{V}_\mathrm{b}} \boldsymbol{x}\times\boldsymbol{V}_\mathrm{b}\,\mathrm{d}V = \frac{1}{n_\mathrm{d}}\left[\int_{\mathcal{V}_\mathrm{b}} \boldsymbol{x}\times(\boldsymbol{x}\times\boldsymbol{\omega}_\mathrm{b})\,\mathrm{d}V - \int_{\mathcal{S}_\mathrm{b}} \boldsymbol{x}\times(\boldsymbol{x}\times(\boldsymbol{n}\times\boldsymbol{V}_\mathrm{b}))\,\mathrm{d}S\right], \tag{6.24}$$

where $\boldsymbol{\omega}_\mathrm{b} := \nabla\times\boldsymbol{V}_\mathrm{b}$, the vorticity associated with the body's motion. When these identities are added to the respective impulse definitions in (6.16) and (6.22), the surface integrals contain only the jump in tangential velocity, $\boldsymbol{n}\times(\boldsymbol{v}-\boldsymbol{V}_\mathrm{b})$, which can simply be replaced by the surface vortex sheet strength, $\boldsymbol{\gamma}$. Thus, we obtain vorticity forms of the impulses,

$$\boldsymbol{P} = \frac{1}{n_\mathrm{d}-1}\left(\int_{\mathcal{V}_f} \boldsymbol{x}\times\boldsymbol{\omega}\,\mathrm{d}V + \int_{\mathcal{S}_\mathrm{b}} \boldsymbol{x}\times\boldsymbol{\gamma}\,\mathrm{d}S + \int_{\mathcal{V}_\mathrm{b}} \boldsymbol{x}\times\boldsymbol{\omega}_\mathrm{b}\,\mathrm{d}V\right) - \int_{\mathcal{V}_\mathrm{b}} \boldsymbol{V}_\mathrm{b}\,\mathrm{d}V \tag{6.25}$$

and

$$\boldsymbol{\Pi}_O = \frac{1}{n_d}\left(\int_{\mathcal{V}_f} \boldsymbol{x} \times (\boldsymbol{x} \times \boldsymbol{\omega})\, dV + \int_{S_b} \boldsymbol{x} \times (\boldsymbol{x} \times \boldsymbol{\gamma})\, dS + \int_{\mathcal{V}_b} \boldsymbol{x} \times (\boldsymbol{x} \times \boldsymbol{\omega}_b)\, dV \right) - \int_{\mathcal{V}_b} \boldsymbol{x} \times \boldsymbol{V}_b\, dV. \tag{6.26}$$

Note that these vorticity forms of the impulses each include a term that essentially removes the total momentum (at unit density) contained in the body's motion, since the combination of vorticity integrals represents the total momentum contained in the entire region enclosed by S_R, including inside the bodies.

Note 6.2.2: Another Form of the Moment's Relationship with Impulse

Lamb [43] (§152) and Wu [76] derived another form of the moment, using a vector identity slightly different from the one that led to (6.9):

$$\boldsymbol{x} \times \boldsymbol{v} = -\frac{1}{2}|\boldsymbol{x}|^2 \boldsymbol{\omega} + \frac{1}{2}\nabla \times (|\boldsymbol{x}|^2 \boldsymbol{v}). \tag{6.27}$$

As a result, it can be shown that

$$\boldsymbol{m}_O = \rho \frac{d}{dt}\left[\frac{1}{2}\int_{\mathcal{V}_f} |\boldsymbol{x}|^2 \boldsymbol{\omega}\, dV + \frac{1}{2}\int_{S_b} |\boldsymbol{x}|^2 (\boldsymbol{n} \times \boldsymbol{v})\, dS \right]. \tag{6.28}$$

Note that the expression has identical coefficients for both two- and three-dimensional problems. It is trivial to show that this form is equivalent to (6.21) in two dimensions; in three dimensions, some work is required, which the reader is invited to carry out.

Note 6.2.3: Impulse Defined About Points Other than the Origin

The linear and angular impulses defined in (6.16) and (6.22) are evaluated about the origin of the coordinate system. What is their relationship to definitions about some other (perhaps moving) point $\boldsymbol{x}_P$? Let us first consider the linear impulse, now written about $\boldsymbol{x}_P$ as

$$\boldsymbol{P}_P = \frac{1}{n_\mathrm{d}-1}\left(\int_{V_f}(\boldsymbol{x}-\boldsymbol{x}_P)\times\boldsymbol{\omega}\,\mathrm{d}V + \int_{S_\mathrm{b}}(\boldsymbol{x}-\boldsymbol{x}_P)\times(\boldsymbol{n}\times\boldsymbol{v})\,\mathrm{d}S\right). \quad (6.29)$$

But, since we have required in (4.67) that the total circulation is zero in two dimensions—and since, in three dimensions, the total vorticity (4.189) is identically zero—then it is easy to show that $\boldsymbol{P}_P = \boldsymbol{P}$. In other words, ***the linear impulse defined about any other point is identical with our original definition about the origin***.

By similar steps, it can be shown that the angular impulse defined about $\boldsymbol{x}_P$,

$$\boldsymbol{\Pi}_P := \frac{1}{n_\mathrm{d}}\left[\int_{V_f}(\boldsymbol{x}-\boldsymbol{x}_P)\times[(\boldsymbol{x}-\boldsymbol{x}_P)\times\boldsymbol{\omega}]\,\mathrm{d}V + \int_{S_\mathrm{b}}(\boldsymbol{x}-\boldsymbol{x}_P)\times[(\boldsymbol{x}-\boldsymbol{x}_P)\times(\boldsymbol{n}\times\boldsymbol{v})]\,\mathrm{d}S\right], \quad (6.30)$$

is related to the angular impulse about the origin by

$$\boldsymbol{\Pi}_P = \boldsymbol{\Pi}_O - \boldsymbol{x}_P\times\boldsymbol{P}, \quad (6.31)$$

analogous to the relationship (2.48) between moments evaluated about different points. Indeed, taking both of these relationships, it is clear that the moment about $\boldsymbol{x}_P$ is related to the impulse about this point through

$$\boldsymbol{m}_P = -\rho\frac{\mathrm{d}\boldsymbol{\Pi}_P}{\mathrm{d}t} - \rho\dot{\boldsymbol{x}}_P\times\boldsymbol{P}, \quad (6.32)$$

where the last term accounts for any motion of the reference point $\boldsymbol{x}_P$.

Note 6.2.4: Impulse Formulas in the Presence of a Uniform Flow

If a uniform flow is present, then we need to make a few modifications to the analysis of this section, principally in the arguments used to eliminate some of the integrals over $\mathcal{S}_R$. First, let's replace $\boldsymbol{v}$ with $\boldsymbol{v}^\dagger = \boldsymbol{v} + \boldsymbol{V}_\infty$, the velocity field in the windtunnel reference frame, as we defined in Sect. 2.3; the first portion, $\boldsymbol{v}$, decays to zero in the usual fashion. Furthermore, we should now replace $\boldsymbol{x}$ with $\boldsymbol{x}^\dagger = \boldsymbol{x} - \boldsymbol{X}_{\mathrm{O}^\dagger}$, the position relative to the origin $\mathrm{O}^\dagger$ of the windtunnel reference frame. When $\boldsymbol{v} + \boldsymbol{V}_\infty$ is substituted for $\boldsymbol{v}$ into the integral over $\mathcal{S}_R$ in (6.10), it leads to additional terms $\boldsymbol{V}_\infty(\boldsymbol{v} \cdot \boldsymbol{n}) + \boldsymbol{v}(\boldsymbol{V}_\infty \cdot \boldsymbol{n}) + \boldsymbol{V}_\infty(\boldsymbol{V}_\infty \cdot \boldsymbol{n})$. The first two of these terms decay sufficiently rapidly so that the surface integral tends to zero as $R \to \infty$; the third integral vanishes identically over a closed surface. Similarly, in (6.13), the integral over $\mathcal{S}_R$ involves additional terms $(\boldsymbol{V}_\infty \cdot \boldsymbol{v})\boldsymbol{n} + \frac{1}{2}|\boldsymbol{V}_\infty|^2\boldsymbol{n}$; again, the first of these terms decays faster than the surface area grows as R increases, and the second vanishes identically when integrated over $\mathcal{S}_R$. Thus, the relationship between force and linear impulse is unchanged from its form in the inertial reference frame:

$$\boldsymbol{f}^\dagger = -\rho \frac{\mathrm{d}\boldsymbol{P}^\dagger}{\mathrm{d}t}. \tag{6.33}$$

However, the impulse itself contains an additional term, due to $\boldsymbol{x} \times (\boldsymbol{n} \times \boldsymbol{V}_\infty)$ evaluated over $\mathcal{S}_\mathrm{b}$:

$$\boldsymbol{P}^\dagger = \boldsymbol{P} - \mathcal{V}_\mathrm{b}\boldsymbol{V}_\infty. \tag{6.34}$$

Equation (6.33), with this modified impulse, gives exactly the modified force in (6.6).

The steps are more subtle in the case of the moment. Here, the integral over $\mathcal{S}_R$ in (6.17) leads to three additional terms of the same form as for (6.10), but now in cross product with $\boldsymbol{x}^\dagger$. The third of these terms, involving only $\boldsymbol{V}_\infty$, vanishes identically over $\mathcal{S}_R$, but the first two no longer decay at a sufficient rate to allow the integral to vanish with increasing radius. Similar conclusions are drawn regarding the integral over $\mathcal{S}_R$ in (6.19): the integral of $\boldsymbol{x}^\dagger \times \frac{1}{2}|\boldsymbol{V}_\infty|^2\boldsymbol{n}$ vanishes identically, but the integral of $\boldsymbol{x}^\dagger \times (\boldsymbol{V}_\infty \cdot \boldsymbol{v})\boldsymbol{n}$ does not decay. Collecting these persistent terms, we have to deal with the additional surface integral

$$\rho \int_{\mathcal{S}_R} \boldsymbol{x}^\dagger \times [(\boldsymbol{V}_\infty \cdot \boldsymbol{v})\boldsymbol{n} - \boldsymbol{v}(\boldsymbol{V}_\infty \cdot \boldsymbol{n}) - \boldsymbol{V}_\infty(\boldsymbol{v} \cdot \boldsymbol{n})]\, \mathrm{d}S. \tag{6.35}$$

This integral can be manipulated further. We leave it as an exercise to the reader to show that it is equivalent to $\rho \boldsymbol{V}_\infty \times (\boldsymbol{P}^\dagger + \mathcal{V}_\mathrm{b}\boldsymbol{V}_\mathrm{c}^\dagger)$, where $\boldsymbol{V}_\mathrm{c}^\dagger$ is the velocity of the body centroid in the windtunnel reference frame. Thus, the relationship in the windtunnel reference frame between moment and angular impulse, each expressed about the origin of this coordinate system, is

$$\boldsymbol{m}^{\dagger}_{\mathrm{O}^{\dagger}} = -\rho \frac{\mathrm{d}\boldsymbol{\Pi}^{\dagger}_{\mathrm{O}^{\dagger}}}{\mathrm{d}t} + \rho \boldsymbol{V}_{\infty} \times (\boldsymbol{P}^{\dagger} + \mathcal{V}_{\mathrm{b}} \boldsymbol{V}^{\dagger}_{\mathrm{c}}); \tag{6.36}$$

the angular impulse in this expression differs from that in the inertial frame by the change in reference point from O to $\mathrm{O}^{\dagger}$ and by the uniform flow's contribution to the surface integral over $\mathcal{S}_{\mathrm{b}}$:

$$\boldsymbol{\Pi}^{\dagger}_{\mathrm{O}^{\dagger}} = \boldsymbol{\Pi}_{\mathrm{O}} - \boldsymbol{X}_{\mathrm{O}^{\dagger}} \times \boldsymbol{P} - \boldsymbol{X}^{\dagger}_{\mathrm{c}} \times \mathcal{V}_{\mathrm{b}} \boldsymbol{V}_{\infty}. \tag{6.37}$$

The expression (6.36) is similar to (6.32) in Note 6.2.3 (bearing in mind that the velocity of reference point $\boldsymbol{X}_{\mathrm{O}^{\dagger}}$ is $-\boldsymbol{V}_{\infty}$), but with an additional modification for the linear impulse of the uniformly moving fluid displaced by the body. It is easy to confirm that, when the impulse is substituted in (6.36), the result is exactly the relationship between moments in (6.6).

6.2.2 Complex Form for Planar Applications

As with other calculations in this book, it is useful to develop a complex form for the two-dimensional expressions for force and moment based on impulse. Here, we can rely on previous results to obtain the complex form directly. For example, by the direct relationship between linear impulse and the dipole coefficient of the multipole expansion, and the equivalence (3.91) of this dipole coefficient with $-a_1$, the leading complex coefficient, it is straightforward to show the following:

Result 6.5: Complex Form of Force as Rate of Change of Linear Impulse

For an impenetrable body in motion and/or deformation and surrounded by a collection of N_{v} point vortices at locations z_{J}, the complex form of the linear impulse, $P = P_x + \mathrm{i}P_y$, is given by

$$P = -\mathrm{i} \sum_{\mathrm{J}=1}^{N_{\mathrm{v}}} \Gamma_{\mathrm{J}} z_{\mathrm{J}} - \mathrm{i} \int_{C_{\mathrm{b}}} z w(z)\, \mathrm{d}z - \mathcal{V}_{\mathrm{b}} W^{*}_{\mathrm{r}}, \tag{6.38}$$

where $\mathcal{V}_{\mathrm{b}}$ and W_{r} denote, respectively, the body's area and the translational velocity of its reference point, and C_{b} is a counterclockwise contour enclosing the body. From this, the complex force (per unit depth) on the body, $f = f_x + \mathrm{i} f_y$, is given by

$$f = -\rho \frac{\mathrm{d}P}{\mathrm{d}t}. \tag{6.39}$$

For an infinitely-thin plate, the velocity w in the contour integral in (6.38) is replaced with the jump $g = w^+ - w^-$, as in Sect. 4.2.3, and the sign of the integral is changed to reflect the change of the direction of integration. The volumetric term also vanishes, resulting in

$$P = -\mathrm{i} \sum_{J=1}^{N_v} \Gamma_J z_J + \mathrm{i} \int_S z g(z)\, \mathrm{d}z. \tag{6.40}$$

To derive a complex expression for the moment, we recall the identification made in (4.203) between the vectorial and complex expressions of velocity moment on the body surface. We note that, in the angular impulse (6.22), we only require the integral of the cross product of $\boldsymbol{x}$ with this expression. Furthermore, the second moment of fluid vorticity in this impulse is evaluated over a set of discrete point vortices. Thus, our expression for the complex form of angular impulse about the origin is the following:

Result 6.6: Complex Form of Moment as Rate of Change of Angular Impulse

For a body in the same conditions as in Result 6.5, the angular impulse about the origin in two dimensional flow is given, in complex form, by

$$\Pi_O = -\frac{1}{2} \sum_{J=1}^{N_v} \Gamma_J |z_J|^2 - \frac{1}{2} \mathrm{Re}\left(\int_{C_b} |z|^2 w(z)\, \mathrm{d}z \right), \tag{6.41}$$

from which the moment about the origin is simply

$$m_O = -\rho \frac{\mathrm{d}\Pi_O}{\mathrm{d}t}. \tag{6.42}$$

For an infinitely-thin plate, the angular impulse is modified to

$$\Pi_O = -\frac{1}{2} \sum_{J=1}^{N_v} \Gamma_J |z_J|^2 + \frac{1}{2} \mathrm{Re}\left(\int_{S} |z|^2 g(z)\, \mathrm{d}z \right), \tag{6.43}$$

where $g = w^+ - w^-$ is the jump in complex velocity across the plate.

As in Note 6.2.3 for the vectorial case, the angular impulse about another point P can be obtained by a simple transformation, here given by

$$\Pi_P = \Pi_O - \mathrm{Im}(z_P^* P). \tag{6.44}$$

Similarly, the moment about that point is related to the moment about the origin by the complex equivalent of (2.48):

$$m_P = m_O - \mathrm{Im}(z_P^* f). \tag{6.45}$$

Note 6.2.5: Impulses Obtained Through Conformal Mapping

In two-dimensional problems in which the body is conformally mapped from the unit circle, the surface integrals in the linear and angular impulses, (6.38) and (6.41), can be reformulated as contour integrals over the unit circle. Here, in order to obtain the most compact expression, we will compute the angular impulse about the reference point Z_r, whence

$$P = -\mathrm{i} \sum_{J=1}^{N_v} \Gamma_J z_J - \mathrm{i} \int_{C_c} z(\zeta) \hat{w}(\zeta)\, \mathrm{d}\zeta - \mathcal{V}_b W_r^*, \tag{6.46}$$

and

$$\Pi_r = -\frac{1}{2} \sum_{J=1}^{N_v} \Gamma_J |z_J - Z_r|^2 - \frac{1}{2} \mathrm{Re} \left(\int_{C_c} |z(\zeta) - Z_r|^2 \hat{w}(\zeta)\, \mathrm{d}\zeta \right). \tag{6.47}$$

These integrals can often be evaluated exactly by using Cauchy's residue theorem. For example, let's demonstrate this for the transformation described by (A.152) and (A.153) in Sect. A.2.3 in the Appendix, in which a power series map describes the shape, and the shape is rigidly translated to Z_r and rotated by angle α relative to the inertial coordinate system. The velocity field about this body, in the presence of an arbitrary set of point vortices in the surrounding fluid, is given in the physical plane by (4.80). It can be verified that this leads to

$$\begin{aligned} P = \mathrm{e}^{\mathrm{i}\alpha} \Big[&(\tilde{W}_r^* - \tilde{W}_\infty^*) \left(2\pi |c_1|^2 - \mathcal{V}_b \right) - (\tilde{W}_r - \tilde{W}_\infty) 2\pi c_1 c_{-1} - \tilde{W}_\infty^* \mathcal{V}_b \\ &+ 2\pi \mathrm{i} \Omega c_1 d_{-1} - \mathrm{i} c_1 \sum_{J=1}^{N_v} \Gamma_J (\zeta_J - 1/\zeta_J^*) \Big] \end{aligned} \tag{6.48}$$

for the linear impulse, and

$$\Pi_r = \mathrm{Im}[2\pi(\tilde{W}_\infty - \tilde{W}_r)c_1 d_{-1}] + \mathrm{Im}[W_r(Z_c - Z_r)\mathcal{V}_b] + \pi\Omega \sum_{k=1}^{\infty} k|d_{-k}|^2$$

$$+ \frac{1}{2}\sum_{J=1}^{N_v} \Gamma_J \left[d_0 + 2\mathrm{Re}\left(\sum_{k=1}^{\infty} \frac{d_{-k}}{\zeta_J^k} \right) - |\tilde{z}(\zeta_J)|^2 \right] \quad (6.49)$$

for the angular impulse, where $\mathcal{V}_b$ denotes the area of the body, Z_c the position of its centroid in the inertial coordinate system, and the coefficients d_{-k} are defined in Eq. (A.157). A few notes should be made:

- Strictly speaking, the impulses (6.48) and (6.49) correspond to the wind-tunnel reference frame, since we have accounted for a uniform flow at infinity. However, they are easily simplified to the inertial reference frame by setting $\tilde{W}_\infty$ to zero.
- The expression in square brackets in (6.48) represents the linear impulse expressed in body-fixed components, $\tilde{P}$.
- It is irrelevant whether we express the angular impulse in body-fixed or inertial components, since its single component in the out-of-plane direction is identical in either.
- As shown in Eq. (A.155) in the Appendix, the area of a shape generated from this power series mapping is given by

$$\mathcal{V}_b = -\pi \sum_{k=-1}^{\infty} k|c_{-k}|^2. \quad (6.50)$$

It is also described in the passage following (A.155) that the coefficients of the mapping can always be chosen so that the reference point Z_r and the centroid Z_c coincide, ensuring that (6.49) represents the angular impulse about the centroid, and the second term in the expression is zero. Thus, ***we will generally assume that Z_r coincides with the centroid, unless explicitly stated.***

6.3 Reconciliation of Force via Traction and via Impulse

In this section, the two approaches used in the preceding sections to derive the fluid dynamic force and moment on a body—by integrating the surface tractions directly or by accounting for them through rates of change of impulse in the surrounding fluid—are reconciled with each other. In principle, there is no difference in their net effect: by Newton's second law, the body, through its reaction, must change the momentum in the surrounding fluid at exactly the rate given by the rate of change of the impulses. However, as we will see in this section, there are subtle aspects of this reconciliation that must be explicitly discussed in order to more deeply understand the effects of the assumptions we've made.

6.3.1 The Force and Moment on a Region of Vorticity

In preparation for our overall goals of this discussion, let's establish some basic results. We will start by addressing a general question. Suppose that there is some region $\mathcal{V}_v$ of arbitrary vorticity distribution ω in the fluid, and that this region can be completely enclosed by a surface $\mathcal{S}_v$. What are the force and moment that the exterior fluid exert on this volume? In this inviscid setting, this force and moment are due only to pressure exerted on the surface $\mathcal{S}_v$. By construction, we have ensured that the fluid velocity on this surface is entirely irrotational, and therefore derivable from the scalar potential field, $\boldsymbol{v} = \nabla\varphi$. Furthermore, let us suppose that this scalar potential field on the surface is continuous. This requires, for example, that the inevitable branch cut in the scalar potential field that accompanies point vortices in the plane must be completely contained inside $\mathcal{V}_v$. But this can always be ensured as long as such vortices are only included in equal and opposite pairs in $\mathcal{V}_v$, as depicted in the right panel in Fig. 6.1, consistent with our assumption throughout this book of zero total circulation. The surface $\mathcal{S}_v$ evolves so that it always encloses this vorticity field; we will denote the local velocity of this control surface by $\boldsymbol{v}_v$.

The force and moment on $\mathcal{V}_v$ are given by Result 6.1, with $\mathcal{S}_b$ replaced by $\mathcal{S}_v$. However, Eqs. (6.2) and (6.3) aren't very revealing. To move us toward a more useful form, let's interchange the integration and the time derivative of the scalar potential on $\mathcal{S}_v$. The general result for doing this interchange is so useful in this section and those that follow that we highlight it:

Lemma 6.2: A Reformulation of Integrals of $\boldsymbol{n}\partial\varphi/\partial t$ on a Surface

Consider a time-varying surface $\mathcal{S}$ with unit normal $\boldsymbol{n}$ and local velocity $\boldsymbol{u}$, and bounded by a closed contour C, whose directionality is defined relative to $\boldsymbol{n}$ in the same manner as in Fig. A.3. Note that in two dimensions, the contour C reduces to the two endpoints of the surface $\mathcal{S}$ in the plane. On $\mathcal{S}$ we suppose there is a continuous differentiable scalar field φ. Then

$$\int_{\mathcal{S}} \frac{\partial\varphi}{\partial t}\boldsymbol{n}\,\mathrm{d}S = -\frac{\mathrm{d}}{\mathrm{d}t}\left[\frac{1}{n_{\mathrm{d}}-1}\int_{\mathcal{S}} \boldsymbol{x}\times(\boldsymbol{n}\times\nabla\varphi)\,\mathrm{d}S\right] - \int_{\mathcal{S}} \nabla\varphi\boldsymbol{u}\cdot\boldsymbol{n}\,\mathrm{d}S$$
$$+\frac{\mathrm{d}}{\mathrm{d}t}\left[\frac{1}{n_{\mathrm{d}}-1}\int_{C} \varphi\boldsymbol{x}\times\mathrm{d}\boldsymbol{l}\right] - \int_{C} \varphi\boldsymbol{u}\times\mathrm{d}\boldsymbol{l} \qquad (6.51)$$

and

$$\int_S \boldsymbol{x} \times \frac{\partial \varphi}{\partial t} \boldsymbol{n} \, \mathrm{d}S = -\frac{\mathrm{d}}{\mathrm{d}t}\left[\frac{1}{n_\mathrm{d}} \int_S \boldsymbol{x} \times [\boldsymbol{x} \times (\boldsymbol{n} \times \nabla\varphi)] \, \mathrm{d}S\right] - \int_S \boldsymbol{x} \times \nabla\varphi \boldsymbol{u} \cdot \boldsymbol{n} \, \mathrm{d}S$$
$$+ \frac{\mathrm{d}}{\mathrm{d}t}\left[\frac{1}{n_\mathrm{d}} \int_C \boldsymbol{x} \times (\varphi \boldsymbol{x} \times \mathrm{d}\boldsymbol{l})\right] - \int_C \boldsymbol{x} \times (\varphi \boldsymbol{u} \times \mathrm{d}\boldsymbol{l}) \, . \tag{6.52}$$

Proof The first result (6.51) is easily obtained by combining the identities (A.93) and (A.66). Similarly, the second result (6.52) arises from (A.95) and (A.67). □

When we apply these identities, with surface velocity $\boldsymbol{u}$ set to $\boldsymbol{v}_\mathrm{V}$ and $\nabla\varphi$ replaced by $\boldsymbol{v}$, we obtain the following forms for the force and moment applied by the external fluid on the vortical region:

$$\boldsymbol{f}^\mathrm{V} = \rho \int_{S_\mathrm{V}} \frac{1}{2}|\boldsymbol{v}|^2 \boldsymbol{n} \, \mathrm{d}S - \rho \frac{\mathrm{d}}{\mathrm{d}t}\left[\frac{1}{n_\mathrm{d}-1} \int_{S_\mathrm{V}} \boldsymbol{x} \times (\boldsymbol{n} \times \boldsymbol{v}) \, \mathrm{d}S\right] - \rho \int_{S_\mathrm{V}} \boldsymbol{v}\boldsymbol{v}_\mathrm{V} \cdot \boldsymbol{n} \, \mathrm{d}S \tag{6.53}$$

and

$$\boldsymbol{m}_\mathrm{O}^\mathrm{V} = \rho \int_{S_\mathrm{V}} \boldsymbol{x} \times \frac{1}{2}|\boldsymbol{v}|^2 \boldsymbol{n} \, \mathrm{d}S - \rho \frac{\mathrm{d}}{\mathrm{d}t}\left[\frac{1}{n_\mathrm{d}} \int_{S_\mathrm{V}} \boldsymbol{x} \times [\boldsymbol{x} \times (\boldsymbol{n} \times \boldsymbol{v})] \, \mathrm{d}S\right] - \rho \int_{S_\mathrm{V}} \boldsymbol{x} \times \boldsymbol{v}\boldsymbol{v}_\mathrm{V} \cdot \boldsymbol{n} \, \mathrm{d}S. \tag{6.54}$$

Before proceeding further, let's establish another useful pair of identities, which relate the quantity $\boldsymbol{v} \times \boldsymbol{\omega}$ inside a region with the velocity on the enclosing surface:

Lemma 6.3: Relationships Between Volume Vorticity and Surface Velocity

Consider a volume $\mathcal{V}$ enclosed by a surface S, on which the outward unit normal is $\boldsymbol{n}$. Then, for incompressible velocity field $\boldsymbol{v}$ and the associated vorticity $\boldsymbol{\omega}$,

$$\int_S \left(\frac{1}{2}|\boldsymbol{v}|^2 \boldsymbol{n} - \boldsymbol{v}\boldsymbol{v} \cdot \boldsymbol{n}\right) \mathrm{d}S = \int_{\mathcal{V}} \boldsymbol{v} \times \boldsymbol{\omega} \, \mathrm{d}V \tag{6.55}$$

and

$$\int_{\mathcal{S}} \boldsymbol{x} \times \left(\frac{1}{2} |\boldsymbol{v}|^2 \boldsymbol{n} - \boldsymbol{v}\boldsymbol{v} \cdot \boldsymbol{n} \right) \mathrm{d}S = \int_{\mathcal{V}} \boldsymbol{x} \times (\boldsymbol{v} \times \boldsymbol{\omega})\, \mathrm{d}V. \tag{6.56}$$

These are easily proved with the help of the divergence theorem (A.18), the standard vector identity $\boldsymbol{v} \times \boldsymbol{\omega} = \frac{1}{2}\nabla|\boldsymbol{v}|^2 - \boldsymbol{v} \cdot \nabla \boldsymbol{v}$, and the zero divergence of $\boldsymbol{v}$.

It is useful to make a few notes on these identities, in anticipation of their later application:

- As for the divergence theorem itself, the surface $\mathcal{S}$ need not be contiguous, but might consist of multiple portions that bound $\mathcal{V}$ internally and externally.
- Suppose one of these portions is an externally bounding surface like the spherical surface $\mathcal{S}_R$, used previously in this chapter. If the flow is irrotational beyond $\mathcal{S}_R$, then the volume integrals can be extended to the entire surrounding fluid. Furthermore, as we can see from the multipole expansions of the flow field in Sect. 4.7, as the radius R increases, the velocity field at that distance decays at least as fast as $1/R^3$ for a three-dimensional flow and $1/R^2$ for a planar flow with no net circulation, and thus, the surface integral over $\mathcal{S}_R$ vanishes in both cases as $R \to \infty$. Thus, ***these identities also hold for an external region $\mathcal{V}$ extending to infinity, bounded internally by $\mathcal{S}$***. We must take care to remember that the unit normal $\boldsymbol{n}$ is directed away from $\mathcal{V}$ in the forms (6.55) and (6.56). Furthermore, ***if there is no such internally bounding contour, then the volume integrals in (6.55) and (6.56) are identically zero***.

With these identities, we can now obtain an enlightening form for the force and moment on a vortical region:

Result 6.7: The Force and Moment on a Vortex

The force and moment exerted by external fluid on an arbitrary region $\mathcal{V}_\mathrm{v}$ of vorticity, enclosed by surface $\mathcal{S}_\mathrm{v}$ with outward-directed unit normal $\boldsymbol{n}$ and which moves locally with velocity $\boldsymbol{v}_\mathrm{v}$, are given respectively by

$$\boldsymbol{f}^\mathrm{v} = \rho \int_{\mathcal{V}_\mathrm{v}} \boldsymbol{v} \times \boldsymbol{\omega}\, \mathrm{d}V - \rho \frac{\mathrm{d}}{\mathrm{d}t} \left[\frac{1}{n_\mathrm{d} - 1} \int_{\mathcal{S}_\mathrm{v}} \boldsymbol{x} \times (\boldsymbol{n} \times \boldsymbol{v})\, \mathrm{d}S \right] + \rho \int_{\mathcal{S}_\mathrm{v}} \boldsymbol{v}\, (\boldsymbol{v} - \boldsymbol{v}_\mathrm{v}) \cdot \boldsymbol{n}\, \mathrm{d}S \tag{6.57}$$

and

$$m_O^{\mathrm{v}} = \rho \int_{\mathcal{V}_{\mathrm{v}}} \boldsymbol{x} \times (\boldsymbol{v} \times \boldsymbol{\omega})\, \mathrm{d}V - \rho \frac{\mathrm{d}}{\mathrm{d}t}\left[\frac{1}{n_{\mathrm{d}}} \int_{\mathcal{S}_{\mathrm{v}}} \boldsymbol{x} \times [\boldsymbol{x} \times (\boldsymbol{n} \times \boldsymbol{v})]\, \mathrm{d}S\right] + \rho \int_{\mathcal{S}_{\mathrm{v}}} \boldsymbol{x} \times \boldsymbol{v}\, (\boldsymbol{v} - \boldsymbol{v}_{\mathrm{v}}) \cdot \boldsymbol{n}\, \mathrm{d}S, \qquad (6.58)$$

where n_{d} is the number of spatial dimensions.

Let's make a few observations on Result 6.7 before we move on.

- If the vorticity moves according to Helmholtz' theorems—in which case, we call it **free vorticity**—then $\boldsymbol{v}_{\mathrm{v}} \equiv \boldsymbol{v}$, thereby eliminating the last term in each expression. The volume $\mathcal{V}_{\mathrm{v}}$ can also be interpreted as an impenetrable body, $\mathcal{V}_{\mathrm{b}}$, in which case it constitutes a region of **bound vorticity**. In this context, $\boldsymbol{v}_{\mathrm{v}} = \boldsymbol{V}_{\mathrm{b}}$, and the final integrals in each expression vanish due to the no-penetration condition on $\mathcal{S}_{\mathrm{b}}$.
- In either case—free vorticity or bound vorticity—the results suggest that the force and moment exerted on $\mathcal{V}_{\mathrm{v}}$ apparently arise from two contributions. The first is from integrals of $\boldsymbol{v} \times \boldsymbol{\omega}$ over the vortical region. However, if $\mathcal{V}_{\mathrm{v}}$ is surrounded by an unbounded irrotational medium, then, as we observed in Lemma 6.3, the integrals are identically zero. That is, these integrals only arise in situations in which there is some vorticity elsewhere in the surrounding fluid—for example, the wake of a wing, if $\mathcal{V}_{\mathrm{v}}$ represents the wing itself. Traditionally, the quantity $\boldsymbol{v} \times \boldsymbol{\omega}$ is called the **vortex force**.
- The second contribution to the force and moment is from the time derivative of the expressions in square brackets. By comparison with Results 6.3 and 6.4, these bracketed expressions are the linear and angular impulse in the surrounding fluid, and as we will show later in this chapter, their contribution here represents, in part, the inertial reaction, or 'added mass', of the surrounding fluid. They only contribute if the flow is unsteady or if the region $\mathcal{V}_{\mathrm{v}}$ is accelerating through the fluid.
- Thus, in a steady flow past free or bound vorticity surrounded by an otherwise irrotational medium, the force and moment on a vortical region are identically zero. This is sometimes referred to as ***d'Alembert's paradox***.

Our primary purpose for deriving Result 6.7 will be revealed in the next few sections as we inspect spurious effects of fluid vorticity on the force and moment on a body. In the discussion that follows, then, $\mathcal{V}_{\mathrm{v}}$ will comprise a region of fluid vorticity. However, because the vortex region $\mathcal{V}_{\mathrm{v}}$ can also be interpreted as an impenetrable body, it behooves us to contrast Result 6.7 for directly computing this force and moment on the body with the impulse-based results we derived in Sect. 6.2. Those latter formulas, in Results 6.3 and 6.4, direct our focus toward the motions of vorticity in the fluid; Eqs. (6.57) and (6.58) measure this force and moment via the motion of the bound vorticity of the body. In a steady two-dimensional aerodynamic

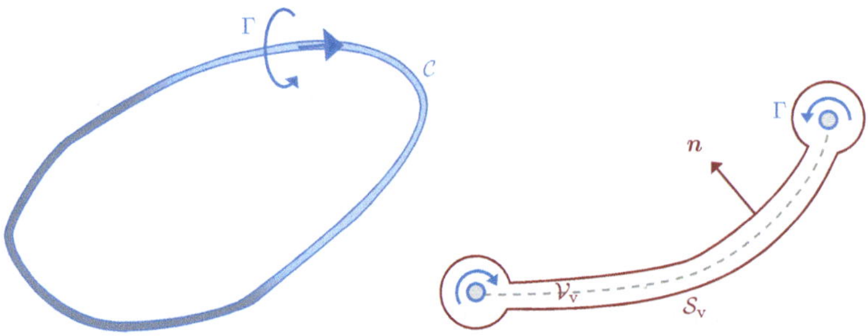

Fig. 6.1 (Left) A closed vortex filament C of strength Γ. (Right) A cross section of the filament, or a pair of equal and opposite strength point vortices in the plane. The discontinuity (or branch cut) in the scalar potential field is depicted as the gray dashed line. $\mathcal{S}_v$ is a control surface that envelops and convects with the filament (or pair), and encloses the control volume $\mathcal{V}_v$

flow past a wing section, this latter perspective is arguably the more intuitive, since it leads directly to the Kutta–Joukowski lift, $\rho \boldsymbol{U} \times \Gamma \boldsymbol{e}_3$. The impulse perspective is more awkward, since motion of fluid vorticity is limited to a starting vortex, infinitely far from the wing, but nonetheless whose continuous change of impulse accounts for the lift. In an unsteady flow, however, in which the surface impulse terms are operative in both forms, fluid vorticity must be accounted for even in the bound vorticity form in order to assess the fluid vorticity's contribution to these surface impulse terms. The impulse perspective of force and moment therefore provides a more self-consistent view of the force and moment. It also will have several advantages for decomposition into flow contributors, as we will discuss later in this chapter.

6.3.2 The Spurious Force and Moment on Vorticity

In the derivation that led to Result 6.7, we had no need to peek inside $\mathcal{V}_v$ to obtain our result, other than accepting that it contained vorticity. It wasn't necessary to know the nature of that vorticity, and particularly, whether it faithfully follows Helmholtz' theorems or is constrained in some fashion. In this section, we will analyze the behavior inside $\mathcal{V}_v$ by inspecting the conservation of momentum in this region. This will have several important benefits, particularly for achieving the main goal of our larger discussion of this section in reconciling the different accounts for force on a body. The analysis we pursue here will also be useful in Chap. 7 for describing the transport of these vortex elements.

Using the same notation as in the previous section, the conservation of linear momentum in the control volume $\mathcal{V}_v$ is described by

$$\frac{\mathrm{d}}{\mathrm{d}t} \int_{\mathcal{V}_v} \rho \boldsymbol{v} \, \mathrm{d}V + \int_{\mathcal{S}_v} \rho \boldsymbol{v}(\boldsymbol{v} - \boldsymbol{v}_v) \cdot \boldsymbol{n} \, \mathrm{d}S = - \int_{\mathcal{S}_v} p\boldsymbol{n} \, \mathrm{d}S + \boldsymbol{f}^{\mathrm{ext}}, \tag{6.59}$$

where $\boldsymbol{f}^{\text{ext}}$ is an external force applied on $\mathcal{V}_v$. If the vorticity is moving freely in accordance with Helmholtz' theorems, we anticipate that this external force will be zero. However, we seek the circumstances that will give rise to non-zero force, since these circumstances will then serve as a caveat on the basic assumptions we made in deriving the impulse-based formulas in Sect. 6.2. Similarly, the conservation of angular momentum in $\mathcal{V}_v$ may involve an externally-applied moment:

$$\frac{\mathrm{d}}{\mathrm{d}t}\int_{\mathcal{V}_v} \boldsymbol{x} \times \rho \boldsymbol{v}\, \mathrm{d}V + \int_{\mathcal{S}_v} \boldsymbol{x} \times \rho \boldsymbol{v}(\boldsymbol{v} - \boldsymbol{v}_v) \cdot \boldsymbol{n}\, \mathrm{d}S = -\int_{\mathcal{S}_v} \boldsymbol{x} \times p\boldsymbol{n}\, \mathrm{d}S + \boldsymbol{m}_O^{\text{ext}}. \tag{6.60}$$

The integrals of the pressure, p, constitute the force and moment applied by external fluid, $\boldsymbol{f}^{\text{v}}$ and $\boldsymbol{m}_O^{\text{v}}$, so we can directly substitute Result 6.7 for these. The flux integrals on $\mathcal{S}_v$ are cancelled by this substitution, and we get the following expressions for the external force and moment:

$$\boldsymbol{f}^{\text{ext}} = -\rho \int_{\mathcal{V}_v} \boldsymbol{v} \times \boldsymbol{\omega}\, \mathrm{d}V + \rho \frac{\mathrm{d}}{\mathrm{d}t}\left[\int_{\mathcal{V}_v} \boldsymbol{v}\, \mathrm{d}V + \frac{1}{n_d - 1}\int_{\mathcal{S}_v} \boldsymbol{x} \times (\boldsymbol{n} \times \boldsymbol{v})\, \mathrm{d}S\right] \tag{6.61}$$

and

$$\boldsymbol{m}_O^{\text{ext}} = -\rho \int_{\mathcal{V}_v} \boldsymbol{x} \times (\boldsymbol{v} \times \boldsymbol{\omega})\, \mathrm{d}V + \rho \frac{\mathrm{d}}{\mathrm{d}t}\left[\int_{\mathcal{V}_v} \boldsymbol{x} \times \boldsymbol{v}\, \mathrm{d}V + \frac{1}{n_d}\int_{\mathcal{S}_v} \boldsymbol{x} \times [\boldsymbol{x} \times (\boldsymbol{n} \times \boldsymbol{v})]\, \mathrm{d}S\right]. \tag{6.62}$$

But the terms in square brackets in these expressions can be rewritten in terms of moments of vorticity in $\mathcal{V}_v$ by Lemma 6.1. The result of these steps is an external force given by

$$\boldsymbol{f}^{\text{ext}} = -\rho \int_{\mathcal{V}_v} \boldsymbol{v} \times \boldsymbol{\omega}\, \mathrm{d}V + \frac{\rho}{n_d - 1}\frac{\mathrm{d}}{\mathrm{d}t}\int_{\mathcal{V}_v} \boldsymbol{x} \times \boldsymbol{\omega}\, \mathrm{d}V, \tag{6.63}$$

and an external moment about the origin given by

$$\boldsymbol{m}_O^{\text{ext}} = -\rho \int_{\mathcal{V}_v} \boldsymbol{x} \times (\boldsymbol{v} \times \boldsymbol{\omega})\, \mathrm{d}V + \frac{\rho}{n_d}\frac{\mathrm{d}}{\mathrm{d}t}\int_{\mathcal{V}_v} \boldsymbol{x} \times (\boldsymbol{x} \times \boldsymbol{\omega})\, \mathrm{d}V. \tag{6.64}$$

Thus, we have shown that an external force and moment must be applied whenever these terms differ from each other. When might this happen? To address this question, let's specialize the results (6.63) and (6.64) to a closed vortex filament C of strength Γ, as depicted in the left panel in Fig. 6.1. Using the notation of Sect. 3.3.2, in which $\boldsymbol{X}(s)$ denotes the configuration of the arc length-parameterized filament curve, the impulses become

$$\int_{\mathcal{V}_{\rm v}} \boldsymbol{x}\times\boldsymbol{\omega}\,{\rm d}V = \Gamma\int_C \boldsymbol{X}(s)\times{\rm d}\boldsymbol{l}(s), \quad \int_{\mathcal{V}_{\rm v}} \boldsymbol{x}\times(\boldsymbol{x}\times\boldsymbol{\omega})\,{\rm d}V = \Gamma\int_C \boldsymbol{X}(s)\times(\boldsymbol{X}(s)\times{\rm d}\boldsymbol{l}(s)). \tag{6.65}$$

The vortex force and moment of the filament are less obvious, because the fluid velocity on the filament axis includes the singular portion induced by the filament itself. Thus, the velocity that belongs here is some non-singular extraction of this field. We will have more to say about this velocity in Chap. 7; for now, we will just denote it generically as $\boldsymbol{v}_{-\rm v}$, where the subscript conveys that the vortex's own influence on this velocity is either removed or modified in some fashion. Thus,

$$\int_{\mathcal{V}_{\rm v}} \boldsymbol{v}\times\boldsymbol{\omega}\,{\rm d}V = \Gamma\int_C \boldsymbol{v}_{-\rm v}(s)\times{\rm d}\boldsymbol{l}(s), \quad \int_{\mathcal{V}_{\rm v}} \boldsymbol{x}\times(\boldsymbol{v}\times\boldsymbol{\omega})\,{\rm d}V = \Gamma\int_C \boldsymbol{X}(s)\times(\boldsymbol{v}_{-\rm v}(s)\times{\rm d}\boldsymbol{l}(s)). \tag{6.66}$$

Substituting these, and then applying the identities (A.98) and (A.100)) for the rates of change of the contour integrals (or their two-dimensional versions, (A.99) and (A.101)), we arrive at

$$\boldsymbol{f}^{\rm ext} = \frac{\rho}{n_{\rm d}-1}\dot{\Gamma}\int_C \boldsymbol{X}(s)\times{\rm d}\boldsymbol{l}(s) - \rho\Gamma\int_C \left(\boldsymbol{v}_{-\rm v}(s) - \dot{\boldsymbol{X}}(s)\right)\times{\rm d}\boldsymbol{l}(s), \tag{6.67}$$

and

$$\boldsymbol{m}_{\rm O}^{\rm ext} = \frac{\rho}{n_{\rm d}}\dot{\Gamma}\int_C \boldsymbol{X}(s)\times(\boldsymbol{X}(s)\times{\rm d}\boldsymbol{l}(s)) - \rho\Gamma\int_C \boldsymbol{X}(s)\times\left[\left(\boldsymbol{v}_{-\rm v}(s) - \dot{\boldsymbol{X}}(s)\right)\times{\rm d}\boldsymbol{l}(s)\right], \tag{6.68}$$

where $\dot{\boldsymbol{X}}(s)$ denotes the local velocity of the filament.

In two dimensions, for a vortex pair such as in the right panel in Fig. 6.1, these equations reduce to

$$\boldsymbol{f}^{\rm ext} = \rho\left[\dot{\Gamma}\boldsymbol{x}_{\rm v}^{+} - \Gamma\left(\boldsymbol{v}_{-\rm v}^{+} - \dot{\boldsymbol{x}}_{\rm v}^{+}\right)\right]\times\boldsymbol{e}_3 - \rho\left[\dot{\Gamma}\boldsymbol{x}_{\rm v}^{-} - \Gamma\left(\boldsymbol{v}_{-\rm v}^{-} - \dot{\boldsymbol{x}}_{\rm v}^{-}\right)\right]\times\boldsymbol{e}_3, \tag{6.69}$$

and

$$\begin{aligned}\boldsymbol{m}_{\rm O}^{\rm ext} = \rho\boldsymbol{x}_{\rm v}^{+}\times&\left[\left(\frac{1}{2}\dot{\Gamma}\boldsymbol{x}_{\rm v}^{+} - \Gamma\left(\boldsymbol{v}_{-\rm v}^{+} - \dot{\boldsymbol{x}}_{\rm v}^{+}\right)\right)\times\boldsymbol{e}_3\right]\\ &- \rho\boldsymbol{x}_{\rm v}^{-}\times\left[\left(\frac{1}{2}\dot{\Gamma}\boldsymbol{x}_{\rm v}^{-} - \Gamma\left(\boldsymbol{v}_{-\rm v}^{-} - \dot{\boldsymbol{x}}_{\rm v}^{-}\right)\right)\times\boldsymbol{e}_3\right],\end{aligned} \tag{6.70}$$

where the + and − superscripts distinguish the positions and velocities of the vortices of strength Γ and $-\Gamma$, respectively.

Equations (6.67) and (6.68), or their two-dimensional forms (6.69) and (6.70), show quite clearly that the external force and moment necessary to apply on a filament or vortex pair vanish if this vortex entity's strength remains constant in time and if it moves with velocity $\boldsymbol{v}_{-\rm v}$.[1] Though we have yet to precisely define $\boldsymbol{v}_{-\rm v}$, it

[1] Strictly speaking, the filament's motion can also include a velocity component along its axis without any effect on this force. This component does not modify its configuration in space.

represents the essential, non-singular part of the local fluid flow on the filament's (or point vortex's) axis. Thus, we have proved that, if the filament or vortex obeys Helmholtz' second and third theorems, it will be free of external force and moment—or equivalently, it will exert no spurious force or moment on the fluid.

In contrast, if we relax these theorems, then the equations indicate the effect that such a relaxation of the rules will have on the flow. Each infraction—the time-varying strength and the deviation of the motion from the de-singularized fluid velocity $\boldsymbol{v}_{-\mathrm{v}}$—contributes a spurious force and moment. The latter of these takes the form of a Kutta–Joukowski lift. It is useful to observe from the moment on the vortex pair (6.70) that the force from the deviation of velocity effectively acts at the location of the corresponding vortex; the force arising from time-varying strength, however, acts at the geometric center, $(\boldsymbol{x}_{\mathrm{v}}^{+} + \boldsymbol{x}_{\mathrm{v}}^{-})/2$, as can be verified by a simple rearrangement of the cross-products. This is simply because this latter force arises from the uniform discontinuity in pressure, $\rho\dot{\Gamma}$, along the branch cut that joins the two vortices. Thus, if we propose some strategy for canceling the net external force, we may nevertheless be left with an unbalanced moment.

6.3.3 Spurious Force and Moment on a Vortex in the Presence of a Body

Now let us adapt the results of the previous section to our objective of reconciling the impulse- and traction-based calculations of force and moment on a body. In the presence of a body, the vortex filament may intersect with the body surface, as illustrated in the right panel of Fig. 6.2. The intersection consists of exit and entry points, $\boldsymbol{x}^{i}$ and $\boldsymbol{x}^{f}$, joined by a contour, C_I, along which the scalar potential discontinuity meets the body surface. The union of C and $-C_I$—the contour C_I, reversed in direction—forms a closed contour, $\overline{C}$, traversed in the same direction as C. In a two-dimensional context, in which vortex axes are perpendicular to the flow plane, such entry and exit are impossible. However, one of the members of the point vortex pair is generally a bound vortex (e.g., an image vortex) inside the body, as shown in the left panel in Fig. 6.2, and the branch cut that joins them meets the body surface at a single point, $\boldsymbol{x}_I$.

In such cases of intersection with a body, we are only interested in the external force applied in the fluid. The derivation we provide here was originally given by Michelin and Llewelyn Smith [51] for a point vortex. In order to carry out our analysis, let's modify the surface $\mathcal{S}_{\mathrm{v}}$ so that it only lies in the fluid and does not completely enclose the vorticity, but rather, 'traps' it in a region $\mathcal{V}_{\mathrm{v}}$ between $\mathcal{S}_{\mathrm{v}}$ and the body surface $\mathcal{S}_{\mathrm{b}}$, as illustrated in the panels in Fig. 6.2. On the section $\mathcal{S}_{\mathrm{bv}}$ of the body surface that completes this enclosure of $\mathcal{V}_{\mathrm{v}}$, the adjacent flow need not be irrotational or have a continuous scalar potential. Surfaces $\mathcal{S}_{\mathrm{bv}}$ and $\mathcal{S}_{\mathrm{v}}$ are both enclosed by the same contour, C_{bv} (which reduces to two points in two dimensions), whose directionality is defined by the normal $\boldsymbol{n}$ of $\mathcal{S}_{\mathrm{v}}$, consistent with the generic illustrations in Fig. A.3. In general, the body will exert a force, $-\boldsymbol{f}^{\mathrm{bv}}$, and moment,

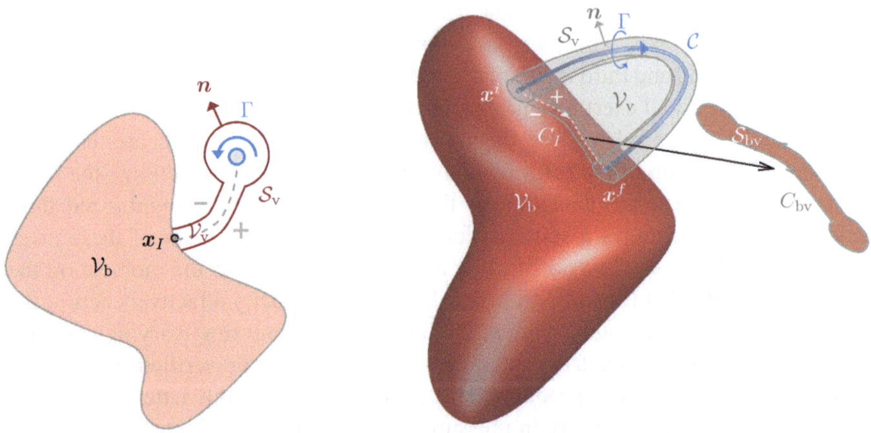

Fig. 6.2 Left: illustration of the control surface $\mathcal{S}_v$, with outward normal $\boldsymbol{n}$, surrounding a point vortex of strength Γ and its associated branch cut in scalar potential (the dashed line), in the vicinity of a planar body. The branch cut intersects the body surface at point $\boldsymbol{x}_I$. Right: illustration of the control surface $\mathcal{S}_v$ surrounding the fluid portion of a closed vortex filament C of strength Γ, and its enclosed surface of discontinuity of scalar potential, that intersect a three-dimensional body. The surface of discontinuity intersects the body's surface along contour C_I, and the filament exits and enters the body at points $\boldsymbol{x}^i$ and $\boldsymbol{x}^f$, respectively. The blown-up picture on the right depicts the section $\mathcal{S}_{bv}$ of the body surface that completes the enclosure of $\mathcal{V}_v$, and the section's bounding contour C_{bv}

$-\boldsymbol{m}_O^{bv}$, on the fluid over this section $\mathcal{S}_{bv}$. Thus, in the presence of a body, and with the Bernoulli equation used to determine the pressure on $\mathcal{S}_v$[2], the conservation of linear momentum in $\mathcal{V}_v$ becomes

$$\boldsymbol{f}^{ext} = -\rho \int_{\mathcal{S}_v} \left(\frac{1}{2}|\boldsymbol{v}|^2 \boldsymbol{n} - \boldsymbol{v}\boldsymbol{v}\cdot\boldsymbol{n} \right) \mathrm{d}S - \rho \int_{\mathcal{S}_v} \frac{\partial \varphi}{\partial t} \boldsymbol{n}\, \mathrm{d}S + \rho \frac{\mathrm{d}}{\mathrm{d}t} \int_{\mathcal{V}_v} \boldsymbol{v}\, \mathrm{d}V$$
$$- \rho \int_{\mathcal{S}_v} \boldsymbol{v}\boldsymbol{v}_v \cdot \boldsymbol{n}\, \mathrm{d}S + \boldsymbol{f}^{bv}, \tag{6.71}$$

and the conservation of angular momentum becomes

$$\boldsymbol{m}_O^{ext} = -\rho \int_{\mathcal{S}_v} \boldsymbol{x} \times \left(\frac{1}{2}|\boldsymbol{v}|^2 \boldsymbol{n} - \boldsymbol{v}\boldsymbol{v}\cdot\boldsymbol{n} \right) \mathrm{d}S - \rho \int_{\mathcal{S}_v} \boldsymbol{x} \times \frac{\partial \varphi}{\partial t} \boldsymbol{n}\, \mathrm{d}S + \rho \frac{\mathrm{d}}{\mathrm{d}t} \int_{\mathcal{V}_v} \boldsymbol{x} \times \boldsymbol{v}\, \mathrm{d}V$$
$$- \rho \int_{\mathcal{S}_v} \boldsymbol{x} \times \boldsymbol{v}\boldsymbol{v}_v \cdot \boldsymbol{n}\, \mathrm{d}S + \boldsymbol{m}_O^{bv}. \tag{6.72}$$

[2] In this section and the next, we are omitting non-zero contributions from the Bernoulli constant, B, on unclosed surfaces. However, these contributions ultimately cancel when the results are combined for our final results.

We can follow a similar set of steps that led to Result 6.7 for the force and moment on a vortex, taking care now to account for the edge C_{bv} of the surface $\mathcal{S}_v$ and the closure of $\mathcal{V}_v$ with $\mathcal{S}_{bv}$. The surface integrals of $\partial\varphi/\partial t$ over $\mathcal{S}_v$ can be reformulated with the identities from Lemma 6.2, and the integrals of $\boldsymbol{v}$ and $\boldsymbol{x} \times \boldsymbol{v}$ over $\mathcal{V}_v$ can be rewritten with the help of Lemma 6.1. These steps lead to

$$\begin{aligned}\boldsymbol{f}^{\text{ext}} = &-\rho \int_{\mathcal{S}_v} \left(\frac{1}{2}|\boldsymbol{v}|^2 \boldsymbol{n} - \boldsymbol{v}\boldsymbol{v} \cdot \boldsymbol{n} \right) \mathrm{d}S + \frac{\rho}{n_d - 1} \frac{\mathrm{d}}{\mathrm{d}t} \int_{\mathcal{V}_v} \boldsymbol{x} \times \boldsymbol{\omega} \, \mathrm{d}V + \boldsymbol{f}^{\text{bv}} \\ &+ \rho \frac{\mathrm{d}}{\mathrm{d}t} \left[\frac{1}{n_d - 1} \int_{\mathcal{S}_{bv}} \boldsymbol{x} \times (\boldsymbol{n} \times \boldsymbol{v}) \, \mathrm{d}S \right] - \rho \frac{\mathrm{d}}{\mathrm{d}t} \left[\frac{1}{n_d - 1} \int_{C_{bv}} \varphi \boldsymbol{x} \times \mathrm{d}\boldsymbol{l} \right] \\ &+ \rho \int_{C_{bv}} \varphi \boldsymbol{v}_{bv} \times \mathrm{d}\boldsymbol{l}\end{aligned} \tag{6.73}$$

and

$$\begin{aligned}\boldsymbol{m}_O^{\text{ext}} = &-\rho \int_{\mathcal{S}_v} \boldsymbol{x} \times \left(\frac{1}{2}|\boldsymbol{v}|^2 \boldsymbol{n} - \boldsymbol{v}\boldsymbol{v} \cdot \boldsymbol{n} \right) \mathrm{d}S + \frac{\rho}{n_d} \frac{\mathrm{d}}{\mathrm{d}t} \int_{\mathcal{V}_v} \boldsymbol{x} \times (\boldsymbol{x} \times \boldsymbol{\omega}) \, \mathrm{d}V + \boldsymbol{m}_O^{\text{bv}} \\ &+ \rho \frac{\mathrm{d}}{\mathrm{d}t} \left[\frac{1}{n_d} \int_{\mathcal{S}_{bv}} \boldsymbol{x} \times [\boldsymbol{x} \times (\boldsymbol{n} \times \boldsymbol{v})] \, \mathrm{d}S \right] - \rho \frac{\mathrm{d}}{\mathrm{d}t} \left[\frac{1}{n_d} \int_{C_{bv}} \boldsymbol{x} \times (\varphi \boldsymbol{x} \times \mathrm{d}\boldsymbol{l}) \right] \\ &+ \rho \int_{C_{bv}} \boldsymbol{x} \times (\varphi \boldsymbol{v}_{bv} \times \mathrm{d}\boldsymbol{l}) .\end{aligned} \tag{6.74}$$

Now, let's once again specialize the vorticity distribution inside $\mathcal{V}_v$ to the fluid-borne portion of a filament—or, in two dimensions, to a point vortex in the fluid—of strength Γ. In such a case, we can let the surface $\mathcal{S}_v$ wrap tightly around the vortex element and its associated surface of discontinuity in scalar potential. As a result, several of the integrals in (6.73) and (6.74) simplify.

First, the integrals over C_{bv} only receive contributions from the segments along C_I, on either side of the surface of discontinuity; the circular portions that surround the filament exit and entry make only a vanishingly small contribution. The segments along C_I run in opposite directions, and φ changes discontinuously by $-[\varphi] = \varphi^{-} - \varphi^{+}$ from the left to the right side as we traverse C_I. We showed in Note 3.3.1 that this jump is uniformly equal to the strength of the filament, $-[\varphi] = \Gamma$.

The second simplification concerns integrals over $\mathcal{S}_{bv}$. Those involving moments of $\boldsymbol{n} \times \boldsymbol{v}$ become vanishingly small as this surface section shrinks to zero. We can once again use the identities in Lemma 6.3 to replace the surface integral over $\mathcal{S}_v$ with integrals of the vorticity and fluid velocity inside $\mathcal{V}_v$, and thence write these integrals in their filament form, using $\boldsymbol{v}_{-v}$ for the unknown fluid velocity, as before.

This time, however, we must account for the fact $\mathcal{S}_{\mathrm{v}}$ is not closed, but must be combined with a contribution from $\mathcal{S}_{\mathrm{bv}}$. These integrals over $\mathcal{S}_{\mathrm{bv}}$ do not vanish as the section shrinks, but retain non-zero contributions from the neighborhoods of the exit and entry points of the filament. Similarly, the surface force and moment, $\boldsymbol{f}^{\mathrm{bv}}$ and $\boldsymbol{m}_{\mathrm{O}}^{\mathrm{bv}}$, remain non-zero in the limit, also due to contributions from the filament's exit and entry. As it turns out, these lingering contributions from the surface require a separate analysis that depends on the structure of the finite core of the filament. Indeed, this analysis, when carried out on arbitrary segments of the filament, can be used to determine the proper form of the de-singularized part of the fluid velocity $\boldsymbol{v}_{-\mathrm{v}}$ along the filament axis [54]. We will discuss this a bit further in Chap. 7. However, in this chapter, we need only satisfy ourselves that these effects from the filament's exit and entry regions have no net role in our current discussion once they have been used to settle the value of $\boldsymbol{v}_{-\mathrm{v}}$.

After performing these steps and then evaluating the time derivatives, we arrive at the following expressions for the external force and moment:

Result 6.8: Spurious Force and Moment on a Vortex Element Near a Body

For a vortex filament C of strength Γ, whose axis is described parametrically by $\boldsymbol{X}(s)$ and whose local transport is given by $\dot{\boldsymbol{X}}(s)$, the spurious force and moment that it exerts on the fluid are given, respectively, by

$$\boldsymbol{f}^{\mathrm{ext}} = \frac{\rho}{n_{\mathrm{d}}-1}\dot{\Gamma}\int_{\overline{C}} \boldsymbol{X}(s)\times \mathrm{d}\boldsymbol{l}(s) - \rho\Gamma\int_{C}\left(\boldsymbol{v}_{-\mathrm{v}}(s)-\dot{\boldsymbol{X}}(s)\right)\times \mathrm{d}\boldsymbol{l}(s), \quad (6.75)$$

and

$$\boldsymbol{m}_{\mathrm{O}}^{\mathrm{ext}} = \frac{\rho}{n_{\mathrm{d}}}\dot{\Gamma}\int_{\overline{C}} \boldsymbol{X}(s)\times(\boldsymbol{X}(s)\times \mathrm{d}\boldsymbol{l}(s)) - \rho\Gamma\int_{C}\boldsymbol{X}(s)\times\left[\left(\boldsymbol{v}_{-\mathrm{v}}(s)-\dot{\boldsymbol{X}}(s)\right)\times \mathrm{d}\boldsymbol{l}(s)\right], \quad (6.76)$$

where $\overline{C}$ represents the arbitrary closure of contour C in the body surface (if C intersects this surface), and $\boldsymbol{v}_{-\mathrm{v}}(s)$ is the local de-singularized fluid velocity on the filament axis.

The two-dimensional versions of these for a point vortex of strength Γ and position $\boldsymbol{x}_{\mathrm{v}}$ are, respectively,

$$\boldsymbol{f}^{\mathrm{ext}} = \rho\left[\dot{\Gamma}(\boldsymbol{x}_{\mathrm{v}}-\boldsymbol{x}_{I}) - \Gamma\left(\boldsymbol{v}_{-\mathrm{v}}-\dot{\boldsymbol{x}}_{\mathrm{v}}\right)\right]\times \boldsymbol{e}_3, \quad (6.77)$$

and

$$\boldsymbol{m}_{\mathrm{O}}^{\mathrm{ext}} = \rho\frac{1}{2}\dot{\Gamma}\left[\boldsymbol{x}_{\mathrm{v}}\times(\boldsymbol{x}_{\mathrm{v}}\times \boldsymbol{e}_3) - \boldsymbol{x}_{I}\times(\boldsymbol{x}_{I}\times \boldsymbol{e}_3)\right] - \rho\Gamma\boldsymbol{x}_{\mathrm{v}}\times\left[(\boldsymbol{v}_{-\mathrm{v}}-\dot{\boldsymbol{x}}_{\mathrm{v}})\times \boldsymbol{e}_3\right], \quad (6.78)$$

where $\boldsymbol{x}_I$ is the arbitrary point of intersection of the branch cut of scalar potential with the body surface.

If the vortex filament or point vortex has constant strength and moves with velocity $\boldsymbol{v}_{-\mathrm{v}}$, the spurious force and moment are identically zero.

Once again, we get separate contributions to this spurious force and moment from the time-varying strength of the vortex element and from deviation of the element's motion from the de-singularized local fluid velocity. The contribution from time-varying strength is particularly important to consider when, for example, the underlying model uses a point vortex or vortex filament to enforce the Kutta condition at an edge, as discussed in Note 5.1.2. The only difference from the results in the previous section is that, when the strength varies in time, the intersection contour C_I (or point $\boldsymbol{x}_I$) makes a contribution to the force. The result is a little surprising, since this intersection is arbitrarily chosen. However, in the most common situation in which the strength's time variation arises due to the enforcement of the Kutta condition at an edge of the body, we argued in Note 3.6.5 that this intersection should coincide with the edge. Similar to our observation of the results in the absence of a body, the force from the deviation in velocity of a point vortex effectively acts at the vortex, while the force from time-varying strength acts at the midpoint between the vortex and the intersection point on the body.

Finally, we note that, although the analysis of this section has been apparently specialized to a single vortex element, it is straightforward to show that the results also hold for any member of a collection of elements.

6.3.4 Revisiting the Traction Force and Moment on the Body

Now, let's tie everything together. We will return to the force and moment on the body that we derived in Sect. 6.1 from integrating the surface pressure, but now we will take care to avoid the section $\mathcal{S}_{\mathrm{bv}}$ designated for trapping the fluid vorticity in $\mathcal{V}_{\mathrm{v}}$. On the remaining portion of the body surface, $\mathcal{S}_{\mathrm{b\backslash v}} := \mathcal{S}_{\mathrm{b}} \setminus \mathcal{S}_{\mathrm{bv}}$, the force and moment exerted by the fluid are $\boldsymbol{f} - \boldsymbol{f}^{\mathrm{bv}}$ and $\boldsymbol{m}_{\mathrm{O}} - \boldsymbol{m}_{\mathrm{O}}^{\mathrm{bv}}$, and since by construction the flow is irrotational in the vicinity of this portion of the surface, we can obtain the pressure from the Bernoulli equation. The resulting equations for $\boldsymbol{f} - \boldsymbol{f}^{\mathrm{bv}}$ and $\boldsymbol{m}_{\mathrm{O}} - \boldsymbol{m}_{\mathrm{O}}^{\mathrm{bv}}$ are identical to those in Result 6.1, but with the integration carried out only over $\mathcal{S}_{\mathrm{b\backslash v}}$. We can replace the integration of $\partial\varphi/\partial t$ using the identities in Lemma 6.2, noting that the surface $\mathcal{S}_{\mathrm{b\backslash v}}$ is bounded by the contour $-C_{\mathrm{bv}}$, the reversal of the bounding contour of $\mathcal{S}_{\mathrm{v}}$. The resulting equations are

$$
\begin{aligned}
\boldsymbol{f} - \boldsymbol{f}_{\mathrm{bv}} = \rho \int\limits_{\mathcal{S}_{\mathrm{b}\setminus\mathrm{v}}} \left(\frac{1}{2}|\boldsymbol{v}|^2\boldsymbol{n} - \boldsymbol{v}\boldsymbol{v}\cdot\boldsymbol{n}\right) \mathrm{d}S - \rho\frac{\mathrm{d}}{\mathrm{d}t}\left[\frac{1}{n_{\mathrm{d}}-1}\int\limits_{\mathcal{S}_{\mathrm{b}\setminus\mathrm{v}}} \boldsymbol{x}\times(\boldsymbol{n}\times\boldsymbol{v})\,\mathrm{d}S\right] \\
- \rho\frac{\mathrm{d}}{\mathrm{d}t}\left[\frac{1}{n_{\mathrm{d}}-1}\int\limits_{C_{\mathrm{bv}}} \varphi\boldsymbol{x}\times\mathrm{d}\boldsymbol{l}\right] + \rho\int\limits_{C_{\mathrm{bv}}} \varphi\boldsymbol{v}_{\mathrm{bv}}\times\mathrm{d}\boldsymbol{l}
\end{aligned}
\tag{6.79}
$$

and

$$
\begin{aligned}
\boldsymbol{m}_{\mathrm{O}} - \boldsymbol{m}_{\mathrm{O}}^{\mathrm{bv}} = \rho \int\limits_{\mathcal{S}_{\mathrm{b}\setminus\mathrm{v}}} \boldsymbol{x}\times\left(\frac{1}{2}|\boldsymbol{v}|^2\boldsymbol{n} - \boldsymbol{v}\boldsymbol{v}\cdot\boldsymbol{n}\right) \mathrm{d}S - \rho\frac{\mathrm{d}}{\mathrm{d}t}\left[\frac{1}{n_{\mathrm{d}}}\int\limits_{\mathcal{S}_{\mathrm{b}\setminus\mathrm{v}}} \boldsymbol{x}\times[\boldsymbol{x}\times(\boldsymbol{n}\times\boldsymbol{v})]\,\mathrm{d}S\right] \\
- \rho\frac{\mathrm{d}}{\mathrm{d}t}\left[\frac{1}{n_{\mathrm{d}}}\int\limits_{C_{\mathrm{bv}}} \boldsymbol{x}\times(\varphi\boldsymbol{x}\times\mathrm{d}\boldsymbol{l})\right] + \rho\int\limits_{C_{\mathrm{bv}}} \boldsymbol{x}\times(\varphi\boldsymbol{v}_{\mathrm{bv}}\times\mathrm{d}\boldsymbol{l}),
\end{aligned}
\tag{6.80}
$$

using $\boldsymbol{v}\cdot\boldsymbol{n} = \boldsymbol{V}_{\mathrm{b}}\cdot\boldsymbol{n}$ on the body surface.

In order to obtain the principal results of this section, these equations can be combined with the results (6.73) and (6.74) that we obtained in our analysis of momentum conservation in $\mathcal{V}_{\mathrm{v}}$ in the previous section. The integrals over their common contour, C_{bv}, cancel. Furthermore, the leading integrals in each expression combine to form an integral of $\frac{1}{2}|\boldsymbol{v}|^2\boldsymbol{n} - \boldsymbol{v}\boldsymbol{v}\cdot\boldsymbol{n}$ and its first moment over a completely closed surface, $\mathcal{S}_{\mathrm{b}\setminus\mathrm{v}} \cup \mathcal{S}_{\mathrm{v}}$, exterior to which the flow is completely free of vorticity. Thus, Lemma 6.3 shows that this integral must be identically zero. The remaining surface integrals, of moments of $\boldsymbol{n}\times\boldsymbol{v}$, combine into a single integral over all of $\mathcal{S}_{\mathrm{b}}$. We are left with the following result, the primary objective of this section:

Result 6.9: Force and Moment on a Body via Traction over Its Surface

The force and moment exerted by a fluid on a body are given by

$$
\boldsymbol{f} = -\rho\frac{\mathrm{d}\boldsymbol{P}}{\mathrm{d}t} + \boldsymbol{f}^{\mathrm{ext}} \tag{6.81}
$$

and

$$
\boldsymbol{m}_{\mathrm{O}} = -\rho\frac{\mathrm{d}\boldsymbol{\Pi}_{\mathrm{O}}}{\mathrm{d}t} + \boldsymbol{m}_{\mathrm{O}}^{\mathrm{ext}}, \tag{6.82}
$$

where $\boldsymbol{P}$ and $\boldsymbol{\Pi}_{\mathrm{O}}$ are the linear and angular impulse, defined in Results 6.3 and 6.4, and $\boldsymbol{f}^{\mathrm{ext}}$ and $\boldsymbol{m}_{\mathrm{O}}^{\mathrm{ext}}$ are any spurious force and moment exerted on the fluid by vorticity, via the expressions given by Result 6.8.

Thus, in Result 6.9, we have shown that the force and moment derived in Sect. 6.2 by tracking the changes in impulse in the fluid are identical to those obtained by

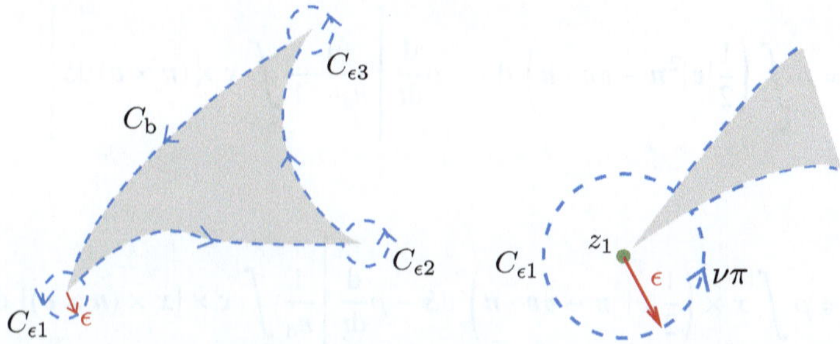

Fig. 6.3 Body contour on a body with edges. A detailed view of one edge in the right panel

integrating the traction over the body surface, provided that the spurious force and moment—defined in Result 6.8—are zero. In Chap. 7 we will discuss models for vortex transport in which we relax Helmholtz' third theorem and allow the strength of a vortex element to vary in time. Our analysis here has shed light on the consequences of this, and furthermore, has provided some notion of how to offset these consequences.

6.4 Edge Suction

A well-known observation from the inviscid aerodynamics of an infinitely-thin flat plate is that, while the exerted pressure can ostensibly only generate a force perpendicular to the plate, the classical Kutta–Joukowski theorem predicts a resultant force perpendicular instead to the uniform flow. In the classical steady problem, this apparent paradox is resolved by remembering that, in contrast to the trailing edge, where the Kutta condition is used to desingularize the flow, the velocity at the leading edge remains infinite. Thus, the pressure there is negatively infinite, and its integral over the vanishingly small region surrounding the edge results in a finite suction directed parallel to the plate. The sum of this so-called **leading-edge suction** and the normal force on the remainder of the plate is equal to the Kutta–Joukowski lift.

In the general context of unsteady inviscid flow, on any body with one or more sharp corners at which the velocity remains singular, there potentially will be a contribution to the integral of pressure from the vanishingly small region surrounding each such corner. Let us investigate this here. Equations (6.79) and (6.80) are the most useful forms of force and moment from which to start this investigation. For the present discussion, in which we are only concerned with general behaviors at corners, we can safely ignore the presence of the surface $\mathcal{S}_{bv}$ and its bounding contour, and thus utilize the following expressions of force and moment from integration of surface traction:

$$\boldsymbol{f} = \rho \int_{S_b} \left(\frac{1}{2} |\boldsymbol{v}|^2 \boldsymbol{n} - \boldsymbol{v}\boldsymbol{v} \cdot \boldsymbol{n} \right) \mathrm{d}S - \rho \frac{\mathrm{d}}{\mathrm{d}t} \left[\frac{1}{n_\mathrm{d} - 1} \int_{S_b} \boldsymbol{x} \times (\boldsymbol{n} \times \boldsymbol{v}) \, \mathrm{d}S \right] \tag{6.83}$$

and

$$\boldsymbol{m}_\mathrm{O} = \rho \int_{S_b} \boldsymbol{x} \times \left(\frac{1}{2} |\boldsymbol{v}|^2 \boldsymbol{n} - \boldsymbol{v}\boldsymbol{v} \cdot \boldsymbol{n} \right) \mathrm{d}S - \rho \frac{\mathrm{d}}{\mathrm{d}t} \left[\frac{1}{n_\mathrm{d}} \int_{S_b} \boldsymbol{x} \times [\boldsymbol{x} \times (\boldsymbol{n} \times \boldsymbol{v})] \, \mathrm{d}S \right]. \tag{6.84}$$

We will focus our attention for now on a two-dimensional body, as illustrated in Fig. 6.3, for which we can make use of complex forms of the force and moment. To develop these, we note the following equivalence, based on the results in Sect. A.2:

$$\left(\frac{1}{2} |\boldsymbol{v}|^2 \boldsymbol{n} - \boldsymbol{v}\boldsymbol{v} \cdot \boldsymbol{n} \right) \mathrm{d}S \Longleftrightarrow \left(\frac{\mathrm{i}}{2} w^2 \, \mathrm{d}z \right)^* . \tag{6.85}$$

We will utilize this equivalence to write the integrals over the contour enclosing the body, C_b. Our focus here is on the contribution from a portion of this contour in the vicinity of a corner. As depicted in the right panel of Fig. 6.3, we can carry out the integration near a corner over a circular arc of radius ϵ, which we will allow to shrink to zero. The final terms in (6.83) and (6.84) are essentially integrals of the scalar potential over the body surface. Since the scalar potential is finite everywhere, even at corners, we can safely omit those terms from our discussion.

Let us suppose that the corner, whose location we will denote generically as z_0, is of internal angle $(2 - \nu)\pi$, so the contour C_ϵ turns through angle $\nu\pi$. In Note 4.4.5, we identified the behavior of the fluid velocity near such a corner as

$$w(z) \approx \frac{SL^{1-1/\nu}}{(z - z_0)^{1-1/\nu}}, \tag{4.55}$$

where S is a constant parameter and L is a characteristic length scale of the geometry. Let's denote the contributions of C_ϵ to the overall force and moment as f_s and $m_{\mathrm{O},s}$, respectively. Thus, the contribution from the corner to force and moment are given by

$$f_s = \lim_{\epsilon \to 0} \left(\frac{\mathrm{i}\rho}{2} \int_{C_\epsilon} w^2(z) \, \mathrm{d}z \right)^* = \lim_{\epsilon \to 0} \left[\frac{\mathrm{i}\rho}{2} \int_{C_\epsilon} \left(\frac{SL^{1-1/\nu}}{(z - z_0)^{1-1/\nu}} \right)^2 \mathrm{d}z \right]^*, \tag{6.86}$$

$$m_{\mathrm{O},s} = \lim_{\epsilon \to 0} -\frac{1}{2} \rho \, \mathrm{Re} \int_{C_\epsilon} z w^2(z) \, \mathrm{d}z = \lim_{\epsilon \to 0} -\frac{1}{2} \rho \, \mathrm{Re} \int_{C_\epsilon} z \left(\frac{SL^{1-1/\nu}}{(z - z_0)^{1-1/\nu}} \right)^2 \mathrm{d}z. \tag{6.87}$$

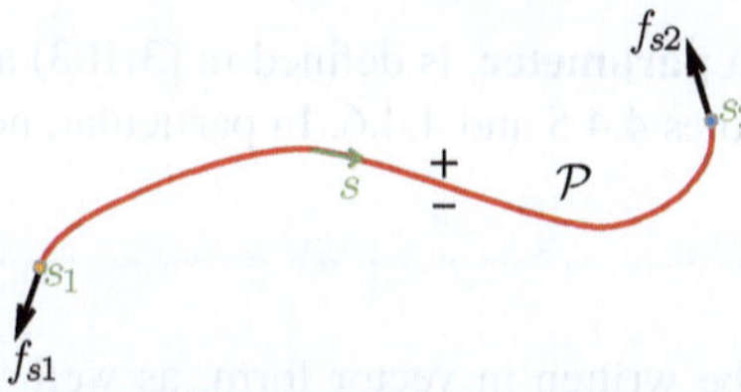

Fig. 6.4 Edge suction forces on an infinitely-thin plate

Positions on the contour C_ϵ are described by $z = z_0 + \epsilon e^{i\beta}$, where $0 \le \beta \le \nu\pi$. When this is substituted, and the integrals transformed into ones over β, then it is simple to show that only $\nu = 2$ will lead to a non-vanishing result for f_s and $m_{O,s}$; all smaller values of ν retain a factor of ϵ raised to a positive power, which causes the result to vanish in the limit. For the same reason, no values of ν will lead to a non-zero result for $m_{O,s}$. Thus, when $\nu = 2$—in other words, when the corner is a cusp—the results are

$$f_s = -\pi\rho L\,(S^*)^2\,, \qquad m_{O,s} = \pi\rho L \operatorname{Im}\left(z_0 S^2\right) \equiv -\operatorname{Im}\left(z_0 f_s^*\right). \qquad (6.88)$$

This contribution to force is called **edge suction**; it applies a corresponding moment about the origin.

How do we know that the force represents a suction? Equation (3.106) expresses the coefficient S in terms of the signed intensity of a corner, σ, and the corner's normal, n_0. When we substitute this into (6.88) for the case of a cusped edge, we obtain

$$f_s = \pi\rho\sigma^2 L n_0, \qquad (6.89)$$

The factor $\pi\rho\sigma^2 L$ is purely real. Thus, the force is clearly directed away from and parallel to the plate, as illustrated in Fig. 6.4: in other words, it is suction, and the strength of this suction is determined by σ. So henceforth, we will call σ the edge's **suction parameter**. Let's summarize this result:

Result 6.10: Edge Suction and the Suction Coefficient

On a cusp z_0 of a body, such as the edge of an infinitely-thin plate, with outward unit normal n_0, the contributions to force and moment (about the origin) on the body are given, in complex form, by

$$f_s = \pi\rho\sigma^2 L n_0, \qquad m_{O,s} = -\operatorname{Im}\left(z_0 f_s^*\right), \qquad (6.90)$$

where σ, the **suction parameter**, is defined in (3.103) and given by any of the expressions in Notes 4.4.5 and 4.4.6. In particular, note that σ has units of velocity.

Equation (6.90) can be written in vector form, as well. A few notes should be made on edge suction:

- ***There's no need to account explicitly for edge suction in the force derived from impulse.*** Edge suction only represents an additive contribution to the overall force when the force is calculated as an integral of pressure over the plate's surface, such as (6.2). When the force is calculated from the rate of change of impulse, as in (6.15) or (6.39), edge suction is automatically accounted for, so there is no need to consider it explicitly.
- ***Edge suction is intimately connected with the Kutta condition.*** As we discussed in Sect. 5.2, the Kutta condition can be expressed as a requirement that the signed intensity—now called the suction parameter—at an edge be set to zero. Result 6.10 indicates the important point that, if the flow is regularized at an edge by the Kutta condition, then the edge suction vanishes. In other words, edge suction is only present at edges with singular velocity behavior. The Kutta condition and edge suction are therefore related principles. We will return to this point later.
- ***Edge suction couples the flow contributors.*** In Sect. 4.6 we decomposed the vortex sheet strength on a body into basis fields from each flow contributor. However, since the sheet strength enters (6.89) in a squared form, the edge suction couples the influences from all flow contributors.

6.5 Decomposition of the Force into Contributors

As we did for the velocity field and bound vortex sheet strength in Sect. 4.6, we can decompose the impulse and force into basis fields associated with each flow contributor. Once again, we will demonstrate this first for the complex form of the impulse and force obtained from the power series conformal mapping in Note 6.2.5, in which the decomposition is straightforward and allows explicit expressions. This will illuminate the more general case, presented in vector form.

For simplicity, we will carry out the following decomposition on the force and moment observed in the inertial frame of reference, where uniform flow is absent. The results of the analysis below can be used to obtain an analogous decomposition of force $\boldsymbol{f}^{\dagger}$ and moment $\boldsymbol{m}_{\mathrm{r}}^{\dagger}$ in a windtunnel reference frame in which a uniform flow $\boldsymbol{V}_{\infty}$ is present. We would substitute the body translational velocity $\boldsymbol{V}_{\mathrm{r}}$ and its components with $\boldsymbol{V}_{\mathrm{r}}^{\dagger} - \boldsymbol{V}_{\infty}$, and make the modifications to impulse, force and moment outlined in Note 6.2.4. This is left as an exercise to the reader.

6.5.1 Complex Form, via Conformal Mapping Solution

The impulses described by Eqs. (6.48) and (6.49) in Note 6.2.5 are immediately decomposable into component contributions from each basis field. In order to demonstrate an important, more general point, we set aside the complex notation and instead write this decomposition in a form in which the kinematic contributors (i.e., body motion) enter through a matrix–vector product. Recall from Chap. 2 that the body-fixed components of the translational velocity of the body, measured at its reference point, are denoted by $\tilde{W}_r = \tilde{U}_r - i\tilde{V}_r$. Also, remember that, in this two-dimensional setting, the single components of angular velocity and angular impulse are unaffected by the choice of coordinate system. The resulting equation can be written as follows:

$$\rho\begin{pmatrix}\Pi_r\\ \tilde{P}_x\\ \tilde{P}_y\end{pmatrix}=2\pi\rho\begin{bmatrix}\frac{1}{2}\sum_{k=1}^{\infty}k|d_{-k}|^2 & -\mathrm{Im}(c_1d_{-1}) & \mathrm{Re}(c_1d_{-1})\\ -\mathrm{Im}(c_1d_{-1}) & |c_1|^2-\mathrm{Re}(c_1c_{-1})-\frac{\mathcal{V}_b}{2\pi} & -\mathrm{Im}(c_1c_{-1})\\ \mathrm{Re}(c_1d_{-1}) & -\mathrm{Im}(c_1c_{-1}) & |c_1|^2+\mathrm{Re}(c_1c_{-1})-\frac{\mathcal{V}_b}{2\pi}\end{bmatrix}\begin{pmatrix}\Omega\\ \tilde{U}_r\\ \tilde{V}_r\end{pmatrix}$$
$$+\rho\sum_{J=1}^{N_v}\Gamma_J\begin{pmatrix}\mathrm{Re}\left(\frac{1}{2}d_0+\sum_{k=1}^{\infty}d_{-k}/\zeta_J^k\right)-\frac{1}{2}|\tilde{z}(\zeta_J)|^2\\ \mathrm{Im}\left[c_1(\zeta_J-1/\zeta_J^*)\right]\\ -\mathrm{Re}\left[c_1(\zeta_J-1/\zeta_J^*)\right]\end{pmatrix}. \quad (6.91)$$

The force and moment are obtained by differentiating the impulses, as dictated by Eqs. (6.39) and (6.42). Note that these equations require that all vectors be expressed in inertial components and the moment and angular impulse to be taken about the (fixed) origin of the inertial coordinate system. Equation (6.91), in contrast, is expressed in body-fixed components, and the angular impulse is taken about the reference point of the body, $Z_r = X_r + iY_r$. We therefore need to relate these with the rotation operator, $e^{i\alpha}$, and the translation formula (6.44). It is also useful, for consistency, to express the moment about the body reference point, via Eq. (6.45). Writing these in matrix–vector form, we get

$$\begin{pmatrix}m_r\\ f_x\\ f_y\end{pmatrix}=-\rho\begin{bmatrix}1 & Y_r & -X_r\\ 0 & 1 & 0\\ 0 & 0 & 1\end{bmatrix}\frac{d}{dt}\underbrace{\left(\begin{bmatrix}1 & -Y_r & X_r\\ 0 & 1 & 0\\ 0 & 0 & 1\end{bmatrix}\underbrace{\begin{bmatrix}1 & 0 & 0\\ 0 & \cos\alpha & -\sin\alpha\\ 0 & \sin\alpha & \cos\alpha\end{bmatrix}\begin{pmatrix}\Pi_r\\ \tilde{P}_x\\ \tilde{P}_y\end{pmatrix}}_{(\Pi_r\ P_x\ P_y)^T}\right)}_{(\Pi_O\ P_x\ P_y)^T}, \quad (6.92)$$

where the impulses are obtained from the expression (6.91). The reader is reminded that the force and moment are to be interpreted as per unit depth in this planar context.

From the structure of these sets of equations, we can observe a number of important points that we will generalize later:

- The components of the matrix multiplying the vector $(\Omega\ \tilde{U}_\mathrm{r}\ \tilde{V}_\mathrm{r})^\mathrm{T}$ in (6.91) express the contributions to impulse from unit velocities of the body. When substituted into (6.92), these components contain the individual influences to force and moment from unit accelerations of the body in an otherwise quiescent fluid. In other words, they embody the **inertial reaction** of the fluid in which the body is immersed. For this reason, the matrix is commonly called the added mass matrix (or added inertia, virtual mass, or virtual inertia matrix). More precisely, these are the components of the **added mass tensor**, expressed in body-fixed coordinates. We first encountered this tensor and its components in the discussion of flow kinetic energy in Sect. 4.6.3. In this two-dimensional context, the components have the dimension of inertia per unit depth. Using the matrix notation introduced in Result 4.6 for these added mass components, we can write them in this two-dimensional context as

$$\tilde{\mathbf{M}}^{FV} = 2\pi\rho \begin{bmatrix} |c_1|^2 - \mathrm{Re}(c_1 c_{-1}) - \dfrac{\mathcal{V}_\mathrm{b}}{2\pi} & -\mathrm{Im}(c_1 c_{-1}) \\ -\mathrm{Im}(c_1 c_{-1}) & |c_1|^2 + \mathrm{Re}(c_1 c_{-1}) - \dfrac{\mathcal{V}_\mathrm{b}}{2\pi} \end{bmatrix}, \tag{6.93}$$

$$\tilde{\mathbf{M}}^{M\Omega} = \pi\rho \sum_{k=1}^{\infty} k|d_{-k}|^2, \quad \tilde{\mathbf{M}}^{MV} = (\tilde{\mathbf{M}}^{F\Omega})^\mathrm{T} = 2\pi\rho \left[-\mathrm{Im}(c_1 d_{-1})\ \mathrm{Re}(c_1 d_{-1}) \right]. \tag{6.94}$$

These are sub-matrices of the overall added mass matrix, as we will discuss further in Note 6.5.1.
- As we already found in Sect. 4.6.3, the added mass matrix is symmetric. This means, for example, that the force in the $\tilde{x}$ direction due to unit angular acceleration is equal to the moment exerted by the fluid about the body's centroid due to unit translational acceleration in the $\tilde{x}$ direction.
- These body-fixed added mass components depend only on the coefficients of the conformal mapping, and thus, for a rigid body, are time invariant. In other words, the added mass tensor is an intrinsic property of the body, determined only by its shape.
- The vector multiplied by Γ_J in (6.91) contains the angular and linear impulses (in body-fixed components) due to the Jth vortex with unit strength. In Chap. 4, we identified a partitioning of the vorticity-induced basis flow field into contributions from individual vortex elements; let us develop a similar decomposition for the impulses

$$\tilde{P}_\mathrm{v} = \sum_{\mathrm{J}=1}^{N_\mathrm{v}} \Gamma_\mathrm{J} \tilde{P}_\mathrm{J}^\mathrm{v}, \qquad \Pi_{\mathrm{r,v}} = \sum_{\mathrm{J}=1}^{N_\mathrm{v}} \Gamma_\mathrm{J} \Pi_{\mathrm{r,J}}^\mathrm{v}, \tag{6.95}$$

where the **unit vorticity-induced impulses** of vortex J (with the planar components of the linear impulse expressed in the body-fixed system) are

$$\tilde{P}^{\mathrm{v}}_{\mathrm{J}} = -\mathrm{i}c_1(\zeta_{\mathrm{J}} - 1/\zeta^*_{\mathrm{J}}), \qquad \Pi^{\mathrm{v}}_{\mathrm{r,J}} = \frac{1}{2}\left[d_0 + 2\mathrm{Re}\left(\sum_{k=1}^{\infty}\frac{d_{-k}}{\zeta^k_{\mathrm{J}}}\right) - |\tilde{z}(\zeta_{\mathrm{J}})|^2\right]. \quad (6.96)$$

- It is important to remember that these unit vorticity-induced impulses account for both the vortex's own impulse as well as the indirect contribution due to the presence of the body (represented here by the image vortex). Let us write Eq. (6.96) again, but this time, distinguish the direct and indirect contributions:

$$\tilde{P}^{\mathrm{v}}_{\mathrm{J}} = -\mathrm{i}\tilde{z}_{\mathrm{J}} + \tilde{P}^{\mathrm{bv}}_{\mathrm{J}}, \quad (6.97)$$

where we have denoted the indirect contribution by $\tilde{P}^{\mathrm{bv}}_{\mathrm{J}}$; in term of the power-series mapping, this indirect contribution is

$$\tilde{P}^{\mathrm{bv}}_{\mathrm{J}} = \mathrm{i}\left(\frac{c_1}{\zeta^*_{\mathrm{J}}} + \sum_{k=0}^{\infty}\frac{c_{-k}}{\zeta^k_{\mathrm{J}}}\right). \quad (6.98)$$

Similarly, the angular impulse can be written as

$$\Pi^{\mathrm{v}}_{\mathrm{r,J}} = -\frac{1}{2}|\tilde{z}_{\mathrm{J}}|^2 + \Pi^{\mathrm{bv}}_{\mathrm{r,J}}, \quad (6.99)$$

where the indirect contribution is denoted by $\Pi^{\mathrm{bv}}_{\mathrm{r,J}}$, taking the form

$$\Pi^{\mathrm{bv}}_{\mathrm{r,J}} = \frac{1}{2}\left[d_0 + 2\mathrm{Re}\left(\sum_{k=1}^{\infty}\frac{d_{-k}}{\zeta^k_{\mathrm{J}}}\right)\right] \quad (6.100)$$

for the power-series mapping.

- Note that these unit impulses are *not* time invariant, but change as the vortex advects. Indeed, the rates of change of these impulses will be useful when we compute the force, so it is helpful to compute them now, at least for illustrative purposes,

$$\frac{\mathrm{d}\tilde{P}^{\mathrm{v}}_{\mathrm{J}}}{\mathrm{d}t} = -\mathrm{i}\frac{\mathrm{d}\tilde{z}_{\mathrm{J}}}{\mathrm{d}t} + \mathrm{Re}\left(2\frac{\partial\mathrm{Re}(\tilde{P}^{\mathrm{bv}}_{\mathrm{J}})}{\partial\tilde{z}_{\mathrm{J}}}\frac{\mathrm{d}\tilde{z}_{\mathrm{J}}}{\mathrm{d}t}\right) + \mathrm{iRe}\left(2\frac{\partial\mathrm{Im}(\tilde{P}^{\mathrm{bv}}_{\mathrm{J}})}{\partial\tilde{z}_{\mathrm{J}}}\frac{\mathrm{d}\tilde{z}_{\mathrm{J}}}{\mathrm{d}t}\right), \quad (6.101)$$

$$\frac{\mathrm{d}\Pi^{\mathrm{v}}_{\mathrm{r,J}}}{\mathrm{d}t} = -\mathrm{Re}\left(\tilde{z}^*_{\mathrm{J}}\frac{\mathrm{d}\tilde{z}_{\mathrm{J}}}{\mathrm{d}t}\right) + \mathrm{Re}\left(2\frac{\partial\Pi^{\mathrm{bv}}_{\mathrm{r,J}}}{\partial\tilde{z}_{\mathrm{J}}}\frac{\mathrm{d}\tilde{z}_{\mathrm{J}}}{\mathrm{d}t}\right). \quad (6.102)$$

The terms involving the indirect impulse in each expression result simply from the complex form of the chain rule, accounting for the changes in each component of the vortex position, $\tilde{z}_{\mathrm{J}}$. They involve gradients of the indirect impulses with respect to vortex position. We could obtain explicit expressions for these gradients by applying the chain rule once again, using Eqs. (6.98) and (6.100) and the mapping from the circle plane. This is left as an exercise for the reader. However, we will have more to say about these rates of change in Sect. 6.6.

- Equations (6.101) and (6.102) give us an opportunity to make an important general point about a limitation of our decomposition into flow contributors. The straightforward manner in which we have decomposed the impulses into independent contributions from each vortex seems to suggest that the force and moment are similarly decomposable, i.e., that we can unambiguously distinguish the contribution of one vortex element to the force on a body. However, the force and moment depend on the rate of change of the impulses, and, as the equations above show, this rate of change depends on the motions of the vortex elements. As we will discuss in detail in Chap. 7, the motion of each point vortex depends on the entire velocity field, composed from all of the flow contributors (except itself). Thus, ***the influences of all flow contributors are present in each vortex element's contribution to force and moment***.
  One way to understand this point is to consider what happens if we remove a single vortex element from the flow field: We obviously lose its contribution to each of the decomposed flow quantities. However, we also lose its influence on the motion of every other vortex element. Thus, the remaining vortex elements all move differently, and this affects their own contributions to the force and moment. This is the essential non-linearity of fluid dynamics.

A Note on Historical Decompositions of Force and Moment In the next section, we will generalize the ideas that we presented here to the more general context, in either two or three dimensions. But first, let us note that there are other ways that we could choose to decompose force and moment. As we have already discussed in Sect. 5.3 regarding the flow field and its decomposition, some classical treatments of aerodynamics (e.g., von Kármán and Sears [70]) have used an approach in which the bound circulation—rather than being apportioned entirely to the vorticity-induced basis field, as it is here—is partitioned among the flow contributors to enforce the Kutta condition at the trailing edge in each. This approach could also be used to decompose the force and moment. Recall from our earlier discussion that the bound circulation assigned to the body-induced motion is called the quasi-steady circulation. As we will discuss in Sect. 8.5.2 of Chap. 8 in the context of classical analysis of a flat plate, we can isolate the effect of this quasi-steady circulation in the vorticity-induced contribution to force and moment. This contribution is often called the **quasi-steady force** and **moment**, and represents the force and moment that would be exerted on the airfoil if the airfoil's current motion were allowed to persist indefinitely.

6.5.2 Decomposed Force and Moment, Generalized

The results of the previous section, though obtained in a specific two-dimensional context, are easily generalizable to both two- and three-dimensional flows.

General Formulas Using the Plücker notation and the associated rigid-body transformations described in Chap. 2, we can write the general forms for the decomposed force and moment in inviscid flows. First, we note that the Plücker expression for the relationships between force and moment and their corresponding impulses (6.15) and (6.21) is

$$\begin{pmatrix} \mathbf{m}_O \\ \mathbf{f} \end{pmatrix} = -\rho \frac{d}{dt} \begin{pmatrix} \mathbf{\Pi}_O \\ \mathbf{P} \end{pmatrix}. \tag{6.103}$$

These relationships are necessarily expressed relative to a fixed point (the origin) in the inertial coordinate system. To express them relative to the (possibly moving) body reference point, $\boldsymbol{X}_r$, in the coordinate system of the (possibly rotating) body, we apply the composite transformation operator (2.52) to both the angular impulse and the moment:

$$\begin{pmatrix} \tilde{\mathbf{m}}_r \\ \tilde{\mathbf{f}} \end{pmatrix} = -\rho\, {}^{b}\mathcal{T}_i^{(f)} \frac{d}{dt} \left({}^{i}\mathcal{T}_b^{(f)} \begin{pmatrix} \tilde{\mathbf{\Pi}}_r \\ \tilde{\mathbf{P}} \end{pmatrix} \right). \tag{6.104}$$

It is easy to verify that, by applying the product rule and using the result (2.63), we obtain the following:

$$\begin{pmatrix} \tilde{\mathbf{m}}_r \\ \tilde{\mathbf{f}} \end{pmatrix} = -\rho \frac{d}{dt} \begin{pmatrix} \tilde{\mathbf{\Pi}}_r \\ \tilde{\mathbf{P}} \end{pmatrix} - \rho \begin{bmatrix} \tilde{\mathbf{\Omega}}^{\times} & \tilde{\mathbf{V}}_r^{\times} \\ 0 & \tilde{\mathbf{\Omega}}^{\times} \end{bmatrix} \begin{pmatrix} \tilde{\mathbf{\Pi}}_r \\ \tilde{\mathbf{P}} \end{pmatrix}. \tag{6.105}$$

The additional terms in this equation, compared to (6.103), are due to expression of the vectors with respect to moving axes. They describe the rates of transfer of impulse between body-fixed components due to the body's rotation at rate $\tilde{\mathbf{\Omega}}$, as well as the additional rate of change of angular impulse that arises as the body's moving reference point deviates from a fixed position at rate $\tilde{\mathbf{V}}_r$.

Now, we will write a general form for the relationship between the impulses and the flow contributors—body motion and fluid vorticity. We will start by recognizing the underlying form found in the two-dimensional case (6.91). This form is simply a specialization to planar bodies of a more general decomposed form for impulse expressed in body-fixed components. Though we will have much more to say about this decomposition in the rest of the chapter, we can immediately write this general form as

$$\rho \begin{pmatrix} \tilde{\mathbf{\Pi}}_r \\ \tilde{\mathbf{P}} \end{pmatrix} = \begin{bmatrix} \tilde{\mathbf{M}}^{M\Omega} & \tilde{\mathbf{M}}^{MV} \\ \tilde{\mathbf{M}}^{F\Omega} & \tilde{\mathbf{M}}^{FV} \end{bmatrix} \begin{pmatrix} \tilde{\mathbf{\Omega}} \\ \tilde{\mathbf{V}}_r \end{pmatrix} + \rho \begin{pmatrix} \tilde{\mathbf{\Pi}}_{r,v} \\ \tilde{\mathbf{P}}_v \end{pmatrix}. \tag{6.106}$$

The matrix operator in the first right-hand-side term, multiplying the components of body motion, contains the components of the added mass tensor expressed in the body-fixed coordinate system; these components were defined in our discussion of kinetic energy decomposition in Sect. 4.6.3. The Plücker operator contains four 3×3 blocks that individually measure the influence of a type of body motion (superscribed with Ω for angular motion, V for linear translation) on the moment (superscript M)

or force (superscript F). The final term in (6.106), subscripted with v, contains the contribution to the impulses from fluid vorticity.

Before providing detailed general expressions for the entries of the matrices and unit impulses, let us first put our results from (6.105) and (6.106) together to obtain a general set of expressions for the force and moment on a rigid body in inviscid flow. In developing these expressions, it is important to stress that ***the added mass matrices of a single rigid body, expressed in the body-fixed coordinate system, are time invariant***. We have not yet proved this in the general case, but we have seen that it is true in the two-dimensional case in the previous section.

Result 6.11: Decomposed Force and Moment on a Rigid Body

Consider an isolated rigid body moving with translational velocity $\boldsymbol{V}_\mathrm{r}$ and angular velocity $\boldsymbol{\Omega}$ in an inviscid incompressible fluid at rest at infinity and containing N_v vortex filaments (or point vortices in two dimensions) of strength Γ_J. The body-fixed components of the force and moment on the body can be written, respectively, as

$$\tilde{\mathbf{f}} = -\tilde{\mathbf{M}}^{FV}\dot{\tilde{\mathbf{V}}}_\mathrm{r} - \tilde{\mathbf{M}}^{F\Omega}\dot{\tilde{\boldsymbol{\Omega}}} - \rho\dot{\tilde{\mathbf{P}}}_\mathrm{v} - \rho\tilde{\boldsymbol{\Omega}}^{\times}\tilde{\mathbf{P}} \tag{6.107}$$

and

$$\tilde{\mathbf{m}}_\mathrm{r} = -\tilde{\mathbf{M}}^{M\Omega}\dot{\tilde{\boldsymbol{\Omega}}} - \tilde{\mathbf{M}}^{MV}\dot{\tilde{\mathbf{V}}}_\mathrm{r} - \rho\dot{\tilde{\boldsymbol{\Pi}}}_{\mathrm{r,v}} - \rho\tilde{\boldsymbol{\Omega}}^{\times}\tilde{\boldsymbol{\Pi}}_\mathrm{r} - \rho\tilde{\mathbf{V}}_\mathrm{r}^{\times}\tilde{\mathbf{P}}, \tag{6.108}$$

where the linear and angular impulses, in body-fixed components are, respectively,

$$\rho\tilde{\mathbf{P}} = \tilde{\mathbf{M}}^{FV}\tilde{\mathbf{V}}_\mathrm{r} + \tilde{\mathbf{M}}^{F\Omega}\tilde{\boldsymbol{\Omega}} + \rho\tilde{\mathbf{P}}_\mathrm{v} \tag{6.109}$$

and

$$\rho\tilde{\boldsymbol{\Pi}}_\mathrm{r} = \tilde{\mathbf{M}}^{M\Omega}\tilde{\boldsymbol{\Omega}} + \tilde{\mathbf{M}}^{MV}\tilde{\mathbf{V}}_\mathrm{r} + \rho\tilde{\boldsymbol{\Pi}}_{\mathrm{r,v}}, \tag{6.110}$$

and the added-mass components are defined in Sect. 4.6.3. Expressions for the vorticity-induced impulses, $\tilde{\mathbf{P}}_\mathrm{v}$ and $\tilde{\boldsymbol{\Pi}}_{\mathrm{r,v}}$, and their rates of change will be discussed in Sect. 6.6.

The components of the force and moment vectors along the axes of the inertial coordinate system can be obtained by the transformation (2.52); the rates of change of the body-fixed velocity components can be computed from their inertial-system counterparts with Eq. (2.66).

Note 6.5.1: Plücker Added Mass Matrices in Two Dimensions

In Note 2.5.1 we reconciled our use of 6×6 Pücker transforms with the two-dimensional context: only the third, fourth and fifth row and column are relevant. We can equivalently rationalize the use of a 6×6 added mass matrix in two dimensions. The relevant sub-matrix is outlined by the box below, in which the individual 3×3 added mass matrices have been colored to distinguish their entries:

$$\begin{bmatrix} \tilde{\mathbf{M}}^{M\Omega} & \tilde{\mathbf{M}}^{MV} \\ \tilde{\mathbf{M}}^{F\Omega} & \tilde{\mathbf{M}}^{FV} \end{bmatrix} = \begin{bmatrix} \bullet & \bullet & \bullet & \bullet & \bullet & \bullet \\ \bullet & \bullet & \bullet & \bullet & \bullet & \bullet \\ \bullet & \bullet & \boxed{\bullet} & \bullet & \bullet & \bullet \\ \bullet & \bullet & \bullet & \bullet & \bullet & \bullet \\ \bullet & \bullet & \bullet & \bullet & \bullet & \bullet \\ \bullet & \bullet & \bullet & \bullet & \bullet & \bullet \end{bmatrix}. \tag{6.111}$$

This boxed sub-matrix can be compared with, for example, the 3×3 matrix in Eq. (6.91) to identify the entries in each of the added mass matrices. Thus, in two dimensions, $\tilde{\mathbf{M}}^{FV}$ is restricted to the 2×2 set of entries for translational motion in the plane, $\tilde{\mathbf{M}}^{M\Omega}$ has only a single relevant entry corresponding to rotational motion about the out-of-plane axis, and the coupling matrices $\tilde{\mathbf{M}}^{MV}$ and $\tilde{\mathbf{M}}^{F\Omega}$, which are equal to each other's transpose, each contain only two entries relevant to the plane.

The Decomposed Impulses For a general problem involving a single rigid body, the linear and angular impulse are decomposable in the same fashion that we have decomposed, for example, the velocity field (4.112): into contributions from each flow contributor. We omit the contribution from uniform flow, since we are focusing our discussion here on analysis in the inertial frame of reference. Using analogous notation, these decompositions are, respectively,

$$\boldsymbol{P} = \boldsymbol{P}_{\mathrm{v}} + \sum_{j=1}^{3} \tilde{V}_{\mathrm{r},j} \boldsymbol{P}_{\mathrm{bt}}^{(j)} + \sum_{j=1}^{3} \tilde{\Omega}_j \boldsymbol{P}_{\mathrm{br}}^{(j)}, \tag{6.112}$$

and

$$\boldsymbol{\Pi}_{\mathrm{O}} = \boldsymbol{\Pi}_{\mathrm{O,v}} + \sum_{j=1}^{3} \tilde{V}_{\mathrm{r},j} \boldsymbol{\Pi}_{\mathrm{O,bt}}^{(j)} + \sum_{j=1}^{3} \tilde{\Omega}_j \boldsymbol{\Pi}_{\mathrm{O,br}}^{(j)}. \tag{6.113}$$

The meanings of these basis impulses follow naturally from our now-standard notation: $\boldsymbol{P}_{\mathrm{v}}$ denotes the linear impulse vector associated with fluid vorticity, modified by the non-penetration of the body; $\boldsymbol{\Pi}_{\mathrm{O,bt}}^{(j)}$ represents the angular impulse about the origin induced by body translation of unit speed along the jth body-fixed coordinate axis. The other definitions follow in straightforward fashion. We have chosen to demonstrate the decomposition here on the angular impulse about the origin, but clearly, a similar decomposition holds for any other choice of center, such as the body reference point. Furthermore, each of the basis angular impulses obeys the translation identity (6.31) with the corresponding basis linear impulse.

The form of each of these basis impulses is derived from the general definition, Eqs. (6.16) and (6.22), respectively. However, we can also develop alternative forms of the impulses that might be more convenient for certain applications or illuminate important connections. For obtaining these alternative forms, we recall our discussion of the decomposition of the velocity field in Sect. 4.6.2. There, we noted that all of the basis velocity fields except for the vorticity-induced field are entirely irrotational, and can therefore be written as the gradient of a scalar potential field free of discontinuities. The definitions of these basis scalar potential fields were given by the decomposition (4.114), and the underlying mathematical problem satisfied by each basis field was described in the ensuing discussion of each flow contributor.

Body Translation-Induced Basis Impulses The basis impulses due to body translation are given simply by

$$\boldsymbol{P}_{\mathrm{bt}}^{(j)} = \frac{1}{n_{\mathrm{d}}-1}\int_{\mathcal{S}_{\mathrm{b}}} \boldsymbol{x}\times(\boldsymbol{n}\times\boldsymbol{v}_{\mathrm{bt}}^{(j)})\,\mathrm{d}S \tag{6.114}$$

and

$$\boldsymbol{\Pi}_{\mathrm{O,bt}}^{(j)} = \frac{1}{n_{\mathrm{d}}}\int_{\mathcal{S}_{\mathrm{b}}} \boldsymbol{x}\times[\boldsymbol{x}\times(\boldsymbol{n}\times\boldsymbol{v}_{\mathrm{bt}}^{(j)})]\,\mathrm{d}S, \tag{6.115}$$

where the basis velocity field, $\boldsymbol{v}_{\mathrm{bt}}^{(j)}$, is described by Eq. (4.135). We can also write these in terms of the basis vortex sheet, since these are related by $\boldsymbol{n}\times\boldsymbol{v}_{\mathrm{bt}}^{(j)} = \boldsymbol{\gamma}_{\mathrm{bt}}^{(j)} + \boldsymbol{n}\times\tilde{\boldsymbol{e}}_j$ on $\mathcal{S}_{\mathrm{b}}$. This relationship is already embodied in the vorticity-based forms of the impulses, (6.25) and (6.26), derived in Note 6.2.1. We can apply these forms, with the body assigned unit translational velocity along the jth axis—so that $\boldsymbol{V}_{\mathrm{b}} = \tilde{\boldsymbol{e}}_j$ and $\boldsymbol{\omega}_{\mathrm{b}} = 0$—and end up with

$$\begin{aligned}
\boldsymbol{P}_{\mathrm{bt}}^{(j)} &= -\mathcal{V}_{\mathrm{b}}\tilde{\boldsymbol{e}}_j + \frac{1}{n_{\mathrm{d}}-1}\int_{\mathcal{S}_{\mathrm{b}}} \boldsymbol{x}\times\boldsymbol{\gamma}_{\mathrm{bt}}^{(j)}\,\mathrm{d}S,\\
\boldsymbol{\Pi}_{\mathrm{O,bt}}^{(j)} &= -\mathcal{V}_{\mathrm{b}}\boldsymbol{X}_{\mathrm{c}}\times\tilde{\boldsymbol{e}}_j + \frac{1}{n_{\mathrm{d}}}\int_{\mathcal{S}_{\mathrm{b}}} \boldsymbol{x}\times(\boldsymbol{x}\times\boldsymbol{\gamma}_{\mathrm{bt}}^{(j)})\,\mathrm{d}S,
\end{aligned} \tag{6.116}$$

where $\mathcal{V}_{\mathrm{b}}$ is the volume (or area in two dimensions) enclosed by the body and $\boldsymbol{X}_{\mathrm{c}}$ denotes the body centroid, defined in (2.2). Each basis vortex sheet $\boldsymbol{\gamma}_{\mathrm{bt}}^{(j)}$ is the solution of the integral equation (4.132).

Alternatively, we can develop a form in terms of the basis scalar potential, defined in (4.114), using identities (A.66) and (A.67) applied to the forms (6.114) and (6.115), respectively (recalling that $\boldsymbol{v}_{\mathrm{bt}}^{(j)} = \nabla\varphi_{\mathrm{bt}}^{(j)}$, and $\varphi_{\mathrm{bt}}^{(j)}$ is continuous on the body surface, $\mathcal{S}_{\mathrm{b}}$). It is easy to show that these lead to

$$P^{(j)}_{\rm bt} = -\int_{S_{\rm b}} \varphi^{(j)}_{\rm bt} \boldsymbol{n}\, {\rm d}S, \qquad \boldsymbol{\Pi}^{(j)}_{\rm O,bt} = -\int_{S_{\rm b}} \boldsymbol{x} \times \varphi^{(j)}_{\rm bt} \boldsymbol{n}\, {\rm d}S. \tag{6.117}$$

This last form allows us to connect our current discussion of basis impulses with the **added mass**, originally defined in Sect. 4.6.3 in the context of the decomposition of kinetic energy. To make this connection, let us note that that the integrals in (6.117) have also arisen in one expression of the added mass components given in that earlier discussion, namely, in (4.160) for the translational added mass, $\tilde{M}^{FV}_{jk}$, and in (4.167) for the rotational-translational added mass, $\tilde{M}^{F\Omega}_{jk}$. Thus, we can immediately connect some of the added mass components with these basis impulses:

$$\tilde{M}^{FV}_{jk} = \rho \boldsymbol{P}^{(j)}_{\rm bt} \cdot \tilde{\boldsymbol{e}}_k, \qquad \tilde{M}^{F\Omega}_{jk} = \rho \left(\boldsymbol{\Pi}^{(j)}_{\rm O,bt} - \boldsymbol{X}_{\rm r} \times \boldsymbol{P}^{(j)}_{\rm bt} \right) \cdot \tilde{\boldsymbol{e}}_k. \tag{6.118}$$

The expression in parentheses on the right-hand side of the second of these equations is simply the angular impulse about the body reference point, $\boldsymbol{X}_{\rm r}$. This form of the angular impulse arises because the body motion on which the decomposition is based is defined about this reference point.

Body Rotation-Induced Basis Impulses The basis impulses due to body rotation are given by

$$\boldsymbol{P}^{(j)}_{\rm br} = \frac{1}{n_{\rm d} - 1} \int_{S_{\rm b}} \boldsymbol{x} \times (\boldsymbol{n} \times \boldsymbol{v}^{(j)}_{\rm br})\, {\rm d}S \tag{6.119}$$

and

$$\boldsymbol{\Pi}^{(j)}_{\rm O,br} = \frac{1}{n_{\rm d}} \int_{S_{\rm b}} \boldsymbol{x} \times [\boldsymbol{x} \times (\boldsymbol{n} \times \boldsymbol{v}^{(j)}_{\rm br})]\, {\rm d}S, \tag{6.120}$$

where the basis velocity field, $\boldsymbol{v}^{(j)}_{\rm br}$, is described by Eq. (4.143). As for the translation-induced impulses, we can also write these in terms of the basis vortex sheet, using forms (6.25) and (6.26) in Note 6.2.1; for the jth rotational component of motion, the fluid vorticity is zero, and $\boldsymbol{V}_{\rm b} = \tilde{\boldsymbol{e}}_j \times (\boldsymbol{x} - \boldsymbol{X}_{\rm r})$ and $\omega_{\rm b} = 2\tilde{\boldsymbol{e}}_j$. Thus, we obtain the forms

$$\boldsymbol{P}^{(j)}_{\rm br} = \mathcal{V}_{\rm b} \frac{n_{\rm d} + 1}{n_{\rm d} - 1} (\boldsymbol{X}_{\rm c} - \boldsymbol{X}_{\rm r}) \times \tilde{\boldsymbol{e}}_j + \frac{1}{n_{\rm d} - 1} \int_{S_{\rm b}} (\boldsymbol{x} - \boldsymbol{X}_{\rm r}) \times \boldsymbol{\gamma}^{(j)}_{\rm br}\, {\rm d}S, \tag{6.121}$$

$$\boldsymbol{\Pi}^{(j)}_{\rm O,br} = -\frac{n_{\rm d} + 2}{n_{\rm d} \rho_{\rm b}} \boldsymbol{I}_{\rm O} \cdot \tilde{\boldsymbol{e}}_j + \mathcal{V}_{\rm b} \boldsymbol{X}_{\rm c} \times (\tilde{\boldsymbol{e}}_j \times \boldsymbol{X}_{\rm r}) + \frac{1}{n_{\rm d}} \int_{S_{\rm b}} \boldsymbol{x} \times (\boldsymbol{x} \times \boldsymbol{\gamma}^{(j)}_{\rm br})\, {\rm d}S, \tag{6.122}$$

where $\boldsymbol{I}_{\rm O}$ is the moment of inertia tensor for the body, defined in (2.3), and $\rho_{\rm b}$ is the body's uniform density. Each basis vortex sheet $\boldsymbol{\gamma}^{(j)}_{\rm bt}$ is the solution of the integral equation (4.132).

The rotation-induced basis impulses can also be written in terms of the basis scalar potential field, again using (A.66) and (A.67) to re-express the basic forms in terms of $\varphi^{(j)}_{\rm br} \boldsymbol{n}$. These are simply

$$P_{\rm br}^{(j)} = -\int\limits_{\mathcal{S}_{\rm b}} \varphi_{\rm br}^{(j)} \boldsymbol{n}\, {\rm d}S, \qquad \boldsymbol{\Pi}_{\rm O,br}^{(j)} = -\int\limits_{\mathcal{S}_{\rm b}} \boldsymbol{x} \times \varphi_{\rm br}^{(j)} \boldsymbol{n}\, {\rm d}S. \tag{6.123}$$

As we found for the translation-induced basis impulses, these forms of the impulses allow us to relate them to the added mass tensor. It is immediately apparent that

$$\tilde{M}_{jk}^{MV} = \rho \boldsymbol{P}_{\rm br}^{(j)} \cdot \tilde{\boldsymbol{e}}_k, \qquad \tilde{M}_{jk}^{M\Omega} = \rho \left(\boldsymbol{\Pi}_{\rm O,br}^{(j)} - \boldsymbol{X}_{\rm r} \times \boldsymbol{P}_{\rm br}^{(j)} \right) \cdot \tilde{\boldsymbol{e}}_k. \tag{6.124}$$

The jkth component of the translational-rotational added mass tensor is equivalent to the jth rotation-induced basis linear impulse along the kth body axis, and the jkth component of the rotational added mass tensor is the same as the jth rotation-induced basis angular impulse, taken about the body reference point, projected along the kth body axis.

Symmetry Properties of the Motion-Induced Impulses The symmetry properties of the added mass tensors can be used to immediately identify some identities for the basis impulses. For example, based on the symmetry of the translational and rotational added mass tensors, we can now write that

$$\boldsymbol{P}_{\rm bt}^{(j)} \cdot \tilde{\boldsymbol{e}}_k = \boldsymbol{P}_{\rm bt}^{(k)} \cdot \tilde{\boldsymbol{e}}_j, \qquad \boldsymbol{\Pi}_{\rm r,br}^{(j)} \cdot \tilde{\boldsymbol{e}}_k = \boldsymbol{\Pi}_{\rm r,br}^{(k)} \cdot \tilde{\boldsymbol{e}}_j. \tag{6.125}$$

This exhibits the remarkable fact that the kth component of impulse (and thus, force or moment) generated by a rigid body in motion along the jth direction (or rotating about the jth axis) is equivalent to the jth component generated by motion along (or rotating about) the kth axis. We will return to this point later. Furthermore, because of the relationship between the translational-rotational and rotational-translational added mass tensor components, there is also an interesting relationship between the linear impulse due to rotation and the angular impulse due to translation:

$$\boldsymbol{P}_{\rm br}^{(j)} \cdot \tilde{\boldsymbol{e}}_k = \boldsymbol{\Pi}_{\rm r,bt}^{(k)} \cdot \tilde{\boldsymbol{e}}_j. \tag{6.126}$$

It should be noted, though, that, for bodies possessing some degree of symmetry, these components are often zero, provided the body-fixed axes are aligned with the axes of symmetry.

The remaining basis impulses are due to fluid vorticity. These require enough detail that we set aside the next section for their analysis.

6.6 The Contribution of Fluid Vorticity to Force and Moment

For calculation of force and moment, we ultimately need practical expressions for the rates of change of the impulses. We have already found such expressions for the contributions from body motion in the form of added mass coefficients. However, we have not yet addressed the influences from fluid vorticity. First, let us give some general expressions for the vorticity-induced basis impulses.

6.6.1 Basic Definitions of the Vorticity-Induced Impulses

For a general distribution of vorticity, the vorticity-induced impulses follow in straightforward manner from the definitions (6.16) and (6.22), restricted to the vorticity-induced basis field:

$$\boldsymbol{P}_{\mathrm{v}} := \frac{1}{n_{\mathrm{d}}-1}\left(\int_{\mathcal{V}_f} \boldsymbol{x}\times\boldsymbol{\omega}\,\mathrm{d}V + \int_{\mathcal{S}_{\mathrm{b}}} \boldsymbol{x}\times(\boldsymbol{n}\times\boldsymbol{v}_{\mathrm{v}})\,\mathrm{d}S\right) \tag{6.127}$$

and

$$\boldsymbol{\Pi}_{\mathrm{O,v}} := \frac{1}{n_{\mathrm{d}}}\left(\int_{\mathcal{V}_f} \boldsymbol{x}\times(\boldsymbol{x}\times\boldsymbol{\omega})\,\mathrm{d}V + \int_{\mathcal{S}_{\mathrm{b}}} \boldsymbol{x}\times[\boldsymbol{x}\times(\boldsymbol{n}\times\boldsymbol{v}_{\mathrm{v}})]\,\mathrm{d}S\right). \tag{6.128}$$

The vorticity-induced velocity field, $\boldsymbol{v}_{\mathrm{v}}$, is defined in (4.117), and contains both the ***direct contribution*** of the fluid vorticity as well as the ***indirect contribution*** from the reactive field that enforces the no-penetration condition on $\mathcal{S}_{\mathrm{b}}$ (regarded as stationary for this flow contributor). For purposes of calculation, it is helpful to remember that, by (3.195), this velocity field's tangential component(s) on $\mathcal{S}_{\mathrm{b}}$ are simply described by the strength of the basis vortex sheet: $\boldsymbol{n}\times\boldsymbol{v}_{\mathrm{v}} = \boldsymbol{\gamma}_{\mathrm{v}}$. On some occasions, it will be useful to distinguish this indirect contribution; for such purposes, we will denote them with the label 'bv' (for *body–vortex*):

$$\boldsymbol{P}_{\mathrm{bv}} := \frac{1}{n_{\mathrm{d}}-1}\int_{\mathcal{S}_{\mathrm{b}}} \boldsymbol{x}\times\boldsymbol{\gamma}_{\mathrm{v}}\,\mathrm{d}S, \qquad \boldsymbol{\Pi}_{\mathrm{O,bv}} := \frac{1}{n_{\mathrm{d}}}\int_{\mathcal{S}_{\mathrm{b}}} \boldsymbol{x}\times(\boldsymbol{x}\times\boldsymbol{\gamma}_{\mathrm{v}})\,\mathrm{d}S. \tag{6.129}$$

Here, we have expressed the vorticity-induced angular impulse about the origin, but we could easily have expressed it about any other point, most notably the body reference point, $\boldsymbol{X}_{\mathrm{r}}$:

$$\boldsymbol{\Pi}_{\mathrm{r,v}} := \frac{1}{n_{\mathrm{d}}}\left(\int_{\mathcal{V}_f} \tilde{\boldsymbol{x}}\times(\tilde{\boldsymbol{x}}\times\boldsymbol{\omega})\,\mathrm{d}V + \int_{\mathcal{S}_{\mathrm{b}}} \tilde{\boldsymbol{x}}\times[\tilde{\boldsymbol{x}}\times(\boldsymbol{n}\times\boldsymbol{v}_{\mathrm{v}})]\,\mathrm{d}S\right), \tag{6.130}$$

where $\tilde{\boldsymbol{x}} := \boldsymbol{x} - \boldsymbol{X}_{\mathrm{r}}$, and the surface integral again represents the indirect influence of the body.

We are generally focused on discrete vortex elements (filaments or point vortices), and in Sect. 4.6, we decomposed the vorticity-induced basis field quantities into unit fields associated with each of the vortex elements. Thus, we can make use of the unit vorticity-induced basis vortex sheet strength, $\boldsymbol{\gamma}_{\mathrm{J}}^{\mathrm{v}}$, to develop expressions for the unit vorticity-induced impulses, viz.

$$\boldsymbol{P}_{\mathrm{v}} = \sum_{\mathrm{J}=1}^{N_{\mathrm{v}}} \Gamma_{\mathrm{J}} \boldsymbol{P}_{\mathrm{J}}^{\mathrm{v}}, \qquad \boldsymbol{\Pi}_{\mathrm{O,v}} = \sum_{\mathrm{J}=1}^{N_{\mathrm{v}}} \Gamma_{\mathrm{J}} \boldsymbol{\Pi}_{\mathrm{O,J}}^{\mathrm{v}}, \tag{6.131}$$

where the unit linear impulse is

$$\boldsymbol{P}_{\mathrm{J}}^{\mathrm{v}} = \frac{1}{n_{\mathrm{d}} - 1} \int_{C_{\mathrm{J}}} \boldsymbol{x} \times \mathrm{d}\boldsymbol{l} + \boldsymbol{P}_{\mathrm{J}}^{\mathrm{bv}}, \tag{6.132}$$

with the indirect contribution from the body, $\boldsymbol{P}_{\mathrm{J}}^{\mathrm{bv}}$, defined by

$$\boldsymbol{P}_{\mathrm{J}}^{\mathrm{bv}} := \frac{1}{n_{\mathrm{d}} - 1} \int_{S_{\mathrm{b}}} \boldsymbol{x} \times \boldsymbol{\gamma}_{\mathrm{J}}^{\mathrm{v}} \, \mathrm{d}S; \tag{6.133}$$

and the unit vorticity-induced angular impulse is

$$\boldsymbol{\Pi}_{\mathrm{O,J}}^{\mathrm{v}} = \frac{1}{n_{\mathrm{d}}} \int_{C_{\mathrm{J}}} \boldsymbol{x} \times (\boldsymbol{x} \times \mathrm{d}\boldsymbol{l}) + \boldsymbol{\Pi}_{\mathrm{O,J}}^{\mathrm{bv}}, \tag{6.134}$$

with indirect contribution from the body defined by

$$\boldsymbol{\Pi}_{\mathrm{O,J}}^{\mathrm{bv}} := \frac{1}{n_{\mathrm{d}}} \int_{S_{\mathrm{b}}} \boldsymbol{x} \times (\boldsymbol{x} \times \boldsymbol{\gamma}_{\mathrm{J}}^{\mathrm{v}}) \, \mathrm{d}S. \tag{6.135}$$

Note that we abstain from expressing these impulses in terms of the basis scalar potential, since that potential field is discontinuous and requires inclusion of awkward terms involving jumps in that field on the surface.

The expressions given thus far are appropriate for either three- or two-dimensional flow, provided that the filament integrals are interpreted as point vortices in the latter context. In that two-dimensional context, we also have the explicit forms of these impulses in Eq. (6.96) (with angular impulse taken about the body reference point) obtained by conformal mapping with the power series.

6.6.2 The Rates of Change of the Vorticity-Induced Impulses

The force and moment expressions described in Result 6.11 require knowledge of the rates of change of the body-fixed components of the vorticity-induced impulses, $\dot{\mathsf{P}}_{\mathrm{v}}$ and $\dot{\mathsf{\Pi}}_{\mathrm{r,v}}$. Written another way, we seek expressions for

$$\frac{\mathrm{d}}{\mathrm{d}t}(\boldsymbol{P}_{\mathrm{v}} \cdot \tilde{\boldsymbol{e}}_j), \quad \frac{\mathrm{d}}{\mathrm{d}t}(\boldsymbol{\Pi}_{\mathrm{r,v}} \cdot \tilde{\boldsymbol{e}}_j), \tag{6.136}$$

for $j = 1, 2, 3$. The forms of vorticity-induced impulse (6.127) and (6.128) provided thus far require knowledge of the bound vortex sheet that the vorticity induces on the body surface. In the two-dimensional case, described in Sect. 6.5.1, we had the luxury of replacing this bound vortex sheet's contribution with that of the image vortices; with the help of the transformation to the circle plane, it was straightforward to calculate each image's explicit contribution to the rate of change of the impulse. However, we have no obvious tools for the general case, calculating the bound vortex sheet's time variation. We need to address this deficiency.

In the analysis that follows we will dig deeper into these vorticity-induced impulses and their rates of change. But first, let us make some intuitive observations. We have already noted that we expect the unit vorticity-induced impulses associated with a vortex element to vary in time as the element moves. In fact, for a single rigid body, when the impulses are expressed in the body-fixed coordinate system, this is the *only* source of their components' time dependence. Furthermore, they should depend only on the position and orientation of the vortex element *relative to the body*. In particular this means that, if the element were to remain in the same position and orientation relative to the body, the impulse's body-fixed components should not change. Indeed, we have already recognized these dependencies in developing (6.101) and (6.102) for a point vortex in the plane.

For some insight that will serve us in developing these rates of change in more general circumstances, let us now derive some interesting results that relate the vorticity-induced impulses to the basis vector potentials that we defined in Sect. 4.6.2. Let us state the main result first:

Result 6.12: Vorticity-Induced Impulse and the Basis Vector Potentials

The jth components of the vorticity-induced impulses in the body-fixed coordinate system can be written as

$$\boldsymbol{P}_{\mathrm{v}} \cdot \tilde{\boldsymbol{e}}_j = \int_{\mathcal{V}_f} \boldsymbol{\omega} \cdot \boldsymbol{\Psi}_{\mathrm{r}\infty}^{(j)} \, \mathrm{d}V, \quad \boldsymbol{\Pi}_{\mathrm{r,v}} \cdot \tilde{\boldsymbol{e}}_j = \int_{\mathcal{V}_f} \boldsymbol{\omega} \cdot \boldsymbol{\Psi}_{\mathrm{r}\infty}^{(j)} \, \mathrm{d}V, \tag{6.137}$$

where the basis vector potentials due to uniform flow, $\boldsymbol{\Psi}_{\infty}^{(j)}$, and uniform rotation at infinity, $\boldsymbol{\Psi}_{\mathrm{r}\infty}^{(j)}$, are defined in (4.139) and (4.148), respectively.

Furthermore, the indirect contributions from the body to these vorticity-induced impulses can be written as

$$\boldsymbol{P}_{\mathrm{bv}} \cdot \tilde{\boldsymbol{e}}_j = -\int_{\mathcal{V}_f} \boldsymbol{\omega} \cdot \boldsymbol{\Psi}_{\mathrm{bt}}^{(j)} \, \mathrm{d}V, \quad \boldsymbol{\Pi}_{\mathrm{r,bv}} \cdot \tilde{\boldsymbol{e}}_j = -\int_{\mathcal{V}_f} \boldsymbol{\omega} \cdot \boldsymbol{\Psi}_{\mathrm{br}}^{(j)} \, \mathrm{d}V, \tag{6.138}$$

using the basis vector potentials due to unit body translation (4.138) and unit body rotation (4.146).

Proof Using the relationship (4.139), we can write

$$\int_{\mathcal{V}_f} \boldsymbol{\omega} \cdot \boldsymbol{\Psi}_{\infty}^{(j)} \, \mathrm{d}V = \frac{1}{n_\mathrm{d} - 1} \int_{\mathcal{V}_f} \left[\tilde{\boldsymbol{e}}_j \times (\boldsymbol{x} - \boldsymbol{X}_\mathrm{r})\right] \cdot \boldsymbol{\omega} \, \mathrm{d}V - \int_{\mathcal{V}_f} \boldsymbol{\omega} \cdot \boldsymbol{\Psi}_{\mathrm{bt}}^{(j)} \, \mathrm{d}V. \quad (6.139)$$

The triple product in the first integral on the right-hand side can be manipulated easily to obtain the direct part of the vorticity-induced impulse. (Recall from Note 6.2.3 that the linear impulse can be equivalently defined about any point, such as $\boldsymbol{X}_\mathrm{r}$.) Thus, our attention turns to the second integral on the right-hand side. For this, let us first note the vector identity

$$-\boldsymbol{\omega} \cdot \boldsymbol{\Psi}_{\mathrm{bt}}^{(j)} = -(\nabla \times \boldsymbol{v}_\mathrm{v}) \cdot \boldsymbol{\Psi}_{\mathrm{bt}}^{(j)} = -\nabla \cdot \left(\boldsymbol{v}_\mathrm{v} \times \boldsymbol{\Psi}_{\mathrm{bt}}^{(j)}\right) - \boldsymbol{v}_\mathrm{v} \cdot \left(\nabla \times \boldsymbol{\Psi}_{\mathrm{bt}}^{(j)}\right). \quad (6.140)$$

Note that we have been careful to distinguish the vorticity-induced part of the velocity, $\boldsymbol{v}_\mathrm{v}$, from the full velocity field. In particular, we must remember that this velocity field, by definition, has vanishing normal component on the body surface. Also, note that, by definition, $\nabla \times \boldsymbol{\Psi}_{\mathrm{bt}}^{(j)} = \boldsymbol{v}_{\mathrm{bt}}^{(j)}$. Now we can integrate the volume integral by parts, using the divergence theorem on the first term:

$$-\int_{\mathcal{V}_f} \boldsymbol{\omega} \cdot \boldsymbol{\Psi}_{\mathrm{bt}}^{(j)} \, \mathrm{d}V = \int_{\mathcal{S}_\mathrm{b}} \boldsymbol{n} \cdot \left(\boldsymbol{v}_\mathrm{v} \times \boldsymbol{\Psi}_{\mathrm{bt}}^{(j)}\right) \mathrm{d}S - \int_{\mathcal{V}_f} \boldsymbol{v}_\mathrm{v} \cdot \boldsymbol{v}_{\mathrm{bt}}^{(j)} \, \mathrm{d}V. \quad (6.141)$$

(We have omitted a step here in which we show that a corresponding integral over $\mathcal{S}_R$ vanishes since the integrand decays sufficiently fast.) The volume integral on the right-hand side involves an inner product between a velocity field, $\boldsymbol{v}_\mathrm{v}$, whose normal component vanishes on $\mathcal{S}_\mathrm{b}$, and another velocity field that is entirely derivable from the gradient of a scalar potential, $\boldsymbol{v}_{\mathrm{bt}}^{(j)} = \nabla \varphi_{\mathrm{bt}}^{(j)}$. In our derivation of flow kinetic energy in Eq. (3.24), we showed that such an integral is exactly zero.

The triple product in the surface integral over $\mathcal{S}_\mathrm{b}$ in (6.141) can be manipulated into a form that allows us to substitute the defining boundary condition (4.138) on $\boldsymbol{\Psi}_{\mathrm{bt}}^{(j)}$ on $\mathcal{S}_\mathrm{b}$. Regrouping the resulting triple product, we arrive at

$$-\int_{\mathcal{V}_f} \boldsymbol{\omega} \cdot \boldsymbol{\Psi}_{\mathrm{bt}}^{(j)} \, \mathrm{d}V = \tilde{\boldsymbol{e}}_{(j)} \cdot \left[\frac{1}{n_\mathrm{d} - 1} \int_{\mathcal{S}_\mathrm{b}} (\boldsymbol{x} - \boldsymbol{X}_\mathrm{r}) \times (\boldsymbol{n} \times \boldsymbol{v}_\mathrm{v})\right] \mathrm{d}S, \quad (6.142)$$

which completes the proof for the linear impulse identities in (6.137) and (6.138). The proofs for the angular impulse identities follow analogous steps and makes use of the identity (4.148) and boundary condition (4.146). □

Result 6.12 deserves some observations before we proceed further.

The Role of the Basis Vector Potentials It is curious that, by taking the inner product of the fluid vorticity with these basis vector potentials, we have been able to eliminate the explicit contribution of the bound vortex sheet to the impulses. So how can we interpret the role of these basis vector potentials in the vorticity-induced impulses? Suppose we break the vorticity field into its differential elements, $\omega\,\delta V$, as we have done before. Each of these elements contributes linearly to the impulse, both directly through the volume integral, as well as indirectly through the surface vortex sheet that it induces on the body, as evident in Eq. (4.117). Furthermore, each component of the vorticity of this element can be individually considered separately in these contributions, so that the jth body-fixed component of linear impulse associated with the element located at $\tilde{\boldsymbol{x}} := \boldsymbol{x} - \boldsymbol{X}_{\mathrm{r}}$ can be written as

$$\tilde{\omega}_k(\tilde{\boldsymbol{x}})\tilde{\mathsf{P}}^{\mathrm{v}}_{kj}(\tilde{\boldsymbol{x}})\,\delta V, \tag{6.143}$$

where $\tilde{\mathsf{P}}^{\mathrm{v}}_{kj}(\tilde{\boldsymbol{x}})$ is the component of a tensor, representing the jth body-fixed component of linear impulse due to a vortex of unit strength in the $\tilde{\boldsymbol{e}}_k$ direction at $\tilde{\boldsymbol{x}}$. Like the overall vorticity-induced impulse, this can be split into the direct and indirect parts,

$$\tilde{\mathsf{P}}^{\mathrm{v}}_{kj}(\tilde{\boldsymbol{x}}) = \frac{1}{n_{\mathrm{d}}-1}\,(\tilde{\boldsymbol{x}} \times \tilde{\boldsymbol{e}}_k)\cdot \tilde{\boldsymbol{e}}_j + \tilde{\mathsf{P}}^{\mathrm{bv}}_{kj}(\tilde{\boldsymbol{x}}), \tag{6.144}$$

where we have used $\tilde{\mathsf{P}}^{\mathrm{bv}}_{kj}$ to denote the indirect contribution from the body's presence. The overall vorticity-induced impulse from all such elements then follows easily from the integral over their individual contributions,

$$\boldsymbol{P}_{\mathrm{v}} \cdot \tilde{\boldsymbol{e}}_j = \int_{\mathcal{V}_f} \tilde{\omega}_k(\tilde{\boldsymbol{x}})\tilde{\mathsf{P}}^{\mathrm{v}}_{kj}(\tilde{\boldsymbol{x}})\,\mathrm{d}V; \tag{6.145}$$

the indirect contribution from the body is similarly assembled,

$$\boldsymbol{P}_{\mathrm{bv}} \cdot \tilde{\boldsymbol{e}}_j = \int_{\mathcal{V}_f} \tilde{\omega}_k(\tilde{\boldsymbol{x}})\tilde{\mathsf{P}}^{\mathrm{bv}}_{kj}(\tilde{\boldsymbol{x}})\,\mathrm{d}V. \tag{6.146}$$

But, by comparison with Result 6.12, it is clear that the body-induced vector potentials are precisely these element-wise contributions to impulse:

$$\Psi^{(j)}_{\infty,k}(\tilde{\boldsymbol{x}}) \equiv \tilde{\mathsf{P}}^{\mathrm{v}}_{kj}(\tilde{\boldsymbol{x}}), \quad \Psi^{(j)}_{\mathrm{bt},k}(\tilde{\boldsymbol{x}}) \equiv -\tilde{\mathsf{P}}^{\mathrm{bv}}_{kj}(\tilde{\boldsymbol{x}}). \tag{6.147}$$

Analogously, with the angular impulse, we can define $\tilde{\Pi}^{\mathrm{v}}_{\mathrm{r},kj}$ as the jth body-fixed component of angular impulse about the body reference point, due to a vortex of unit strength in the $\tilde{\boldsymbol{e}}_k$ direction at $\tilde{\boldsymbol{x}}$:

$$\tilde{\Pi}^{\mathrm{v}}_{\mathrm{r},kj}(\tilde{\boldsymbol{x}}) = -\frac{1}{n_{\mathrm{d}}}\,(\tilde{\boldsymbol{x}} \times (\tilde{\boldsymbol{e}}_k \times \tilde{\boldsymbol{x}}))\cdot \tilde{\boldsymbol{e}}_j + \tilde{\Pi}^{\mathrm{bv}}_{\mathrm{r},kj}(\tilde{\boldsymbol{x}}), \tag{6.148}$$

in which $\tilde{\Pi}^{\mathrm{bv}}_{\mathrm{r},kj}$ represents the indirect contribution from the body. It is again clear that these are equivalent to the body-induced vector potentials:

$$\Psi^{(j)}_{\mathrm{r}\infty,k}(\tilde{\boldsymbol{x}}) \equiv \tilde{\Pi}^{\mathrm{v}}_{\mathrm{r},kj}(\tilde{\boldsymbol{x}}), \quad \Psi^{(j)}_{\mathrm{br},k}(\tilde{\boldsymbol{x}}) \equiv -\tilde{\Pi}^{\mathrm{bv}}_{\mathrm{r},kj}(\tilde{\boldsymbol{x}}). \tag{6.149}$$

To summarize,

Note 6.6.1: The Elemental Contribution to Vorticity-Induced Impulses

The basis vector potentials due to uniform flow, $\Psi^{(j)}_{\infty,k}(\tilde{\boldsymbol{x}})$, and uniform rotation at infinity, $\Psi^{(j)}_{\mathrm{r}\infty,k}(\tilde{\boldsymbol{x}})$, represent, respectively, the contribution to body-fixed component j of vorticity-induced linear and angular impulse (about the body reference point, $\boldsymbol{X}_\mathrm{r}$) from a vortex of unit strength in the $\tilde{\boldsymbol{e}}_k$ direction at $\tilde{\boldsymbol{x}}$. Furthermore, the negative of the basis vector potentials due to body translation, $-\Psi^{(j)}_{\mathrm{bt},k}(\tilde{\boldsymbol{x}})$, and body rotation, $-\Psi^{(j)}_{\mathrm{br},k}(\tilde{\boldsymbol{x}})$, provide the indirect contribution from the body to these vorticity-induced impulse components.

Time Variation of the Impulses The time derivatives of the body-fixed impulse components in Result 6.12 will be used later in this section to develop new expressions for the force. Here we develop a generic expression that is essential for evaluating these time derivatives:

Lemma 6.4: The Time Rate of Change of $\omega \cdot \Psi^{(j)}$

Consider a fluid vorticity field $\boldsymbol{\omega}$ and a basis vector potential $\boldsymbol{\Psi}^{(j)}$ induced by rigid-body motion (translation or rotation) or by uniform flow or rotation at infinity past the stationary body. Then, the time rate of change of the inner product of these vector fields over the fluid region $\mathcal{V}_f$ is given by

$$\frac{\mathrm{d}}{\mathrm{d}t}\int\limits_{\mathcal{V}_f} \boldsymbol{\omega}\cdot\boldsymbol{\Psi}^{(j)}\,\mathrm{d}V = \int\limits_{\mathcal{V}_f} [(\boldsymbol{v}-\boldsymbol{V}_\mathrm{b})\times\boldsymbol{\omega}]\cdot\boldsymbol{v}^{(j)}\,\mathrm{d}V - \int\limits_{\mathcal{S}_\mathrm{b}} \boldsymbol{n}\cdot\boldsymbol{\omega}(\boldsymbol{v}-\boldsymbol{V}_\mathrm{b})\cdot\boldsymbol{\Psi}^{(j)}\,\mathrm{d}S, \tag{6.150}$$

where $\boldsymbol{v}^{(j)} = \nabla\times\boldsymbol{\Psi}^{(j)}$ is the basis velocity field corresponding to vector potential $\boldsymbol{\Psi}^{(j)}$ and $\boldsymbol{V}_\mathrm{b} := \boldsymbol{V}_\mathrm{r} + \boldsymbol{\Omega}\times(\boldsymbol{x}-\boldsymbol{X}_\mathrm{r})$ is the rigid motion of the body-fixed reference frame at any point $\boldsymbol{x}\in\mathcal{V}_f$.

Proof The time derivative can be applied to the volume integral with the help of Eq. (A.88), in which the deforming velocity $\boldsymbol{u}$ is simply the fluid velocity, $\boldsymbol{v}$. Thus, expanding the resulting integrand,

$$\frac{\partial}{\partial t}\left(\omega\cdot\Psi^{(j)}\right)+\nabla\cdot\left(v\omega\cdot\Psi^{(j)}\right)=$$
$$\left(\frac{\partial\omega}{\partial t}+\nabla\cdot(v\omega)\right)\cdot\Psi^{(j)}+\left(\frac{\partial\Psi^{(j)}}{\partial t}+v\cdot\nabla\Psi^{(j)}\right)\cdot\omega. \quad (6.151)$$

The first expression in parentheses on the right-hand side represents transport of the fluid vorticity; it can be substituted with the vorticity transport equation (3.21). The time derivative of $\Psi^{(j)}$ was developed in Note 4.6.2. Using some further vector identities, and the fact that ∇V_b is an anti-symmetric tensor (since it only contains rigid-body rotation), we obtain

$$\frac{\partial}{\partial t}\left(\omega\cdot\Psi^{(j)}\right)+\nabla\cdot\left(v\omega\cdot\Psi^{(j)}\right)=$$
$$[(v-V_b)\times\omega]\cdot(\nabla\times\Psi^{(j)})+\nabla\cdot\left(\omega(v-V_b)\cdot\Psi^{(j)}\right). \quad (6.152)$$

After introducing this expression for the integrand of the volume integral, and then applying the divergence theorem on the final term, our proof is completed. □

Now we are ready to state the time derivatives of the vorticity-induced impulses:

Result 6.13: Time Derivative of the Vorticity-Induced Impulses

The rate of change of the body-fixed components of the vorticity-induced impulse are given by

$$\frac{d}{dt}(P_v\cdot\tilde{e}_j)=\int_{\mathcal{V}_f}[(v-V_b)\times\omega]\cdot v_\infty^{(j)}\,dV,$$

$$\frac{d}{dt}(\Pi_{r,v}\cdot\tilde{e}_j)=\int_{\mathcal{V}_f}[(v-V_b)\times\omega]\cdot v_{r\infty}^{(j)}\,dV. \quad (6.153)$$

Proof These follow almost immediately from Result 6.12 and Lemma 6.4. However, we first have to dispense with the surface integral in Eq. (6.150). To do so, we note that $v-V_b$ is tangent to $\mathcal{S}_b$—equal to $n\times\gamma$ by Eq. (3.196))—and thus, only the tangent components of $\Psi_\infty^{(j)}$ or $\Psi_{r\infty}^{(j)}$ are involved in the surface integrand. But these tangent components are zero, as specified by the defining boundary conditions of these basis vector potentials, (4.131) and (4.149). Thus, the surface integral vanishes for both impulses. □

Result 6.13 is very useful. Remember that the vorticity-induced impulses contain the moments of both the fluid vorticity as well as its associated bound vortex sheet on the body surface. However, in Result 6.13 we now have a means of calculating the rates of change of these impulses—and thus, the force and moment—by only tracking the fluid vorticity; the bound vortex sheet is absent from the calculation. Of course, the body's presence must still be accounted for, since the velocity field $\boldsymbol{v}$ that appears in these integrals includes its modification by the body. But we have developed many tools for computing this velocity field, so it poses little challenge.

Let's split these impulses' rates of change into their direct and indirect contributions from the vorticity, as we have done previously. Once again, we can rely on Result 6.12 and Lemma 6.4. However, we now can no longer nullify the surface integral in Eq. (6.150), because the tangent components of the basis vector potentials no longer vanish on the surface. For example, the direct contribution from fluid vorticity to the rates of change of linear and angular impulse are, respectively,

$$\tilde{\boldsymbol{e}}_j \cdot \left[\int\limits_{\mathcal{V}_f} (\boldsymbol{v} - \boldsymbol{V}_\mathrm{b}) \times \boldsymbol{\omega} \, \mathrm{d}V - \frac{1}{n_\mathrm{d} - 1} \int\limits_{\mathcal{S}_\mathrm{b}} (\boldsymbol{x} - \boldsymbol{X}_\mathrm{r}) \times (\boldsymbol{v} - \boldsymbol{V}_\mathrm{b}) \boldsymbol{\omega} \cdot \boldsymbol{n} \, \mathrm{d}S \right] \tag{6.154}$$

and

$$\tilde{\boldsymbol{e}}_j \cdot \left[\int\limits_{\mathcal{V}_f} (\boldsymbol{x} - \boldsymbol{X}_\mathrm{r}) \times ((\boldsymbol{v} - \boldsymbol{V}_\mathrm{b}) \times \boldsymbol{\omega}) \, \mathrm{d}V \right.$$
$$\left. - \frac{1}{n_\mathrm{d}} \int\limits_{\mathcal{S}_\mathrm{b}} (\boldsymbol{x} - \boldsymbol{X}_\mathrm{r}) \times [(\boldsymbol{x} - \boldsymbol{X}_\mathrm{r}) \times (\boldsymbol{v} - \boldsymbol{V}_\mathrm{b})] \, \boldsymbol{\omega} \cdot \boldsymbol{n} \, \mathrm{d}S \right]. \tag{6.155}$$

These contains the straightforward volume integrals that arise from differentiating the vorticity moments over $\mathcal{V}_f$. However, they also each contain a surface integral over $\mathcal{S}_\mathrm{b}$ and involving the vorticity component that pierces the body surface. What does this term represent? In any situation in which vorticity does not pierce the surface, vortex stretching, tilting, and convection all contribute to force and moment in the same manner, through the transport of vorticity-bearing fluid elements described by the 'vortex force', $(\boldsymbol{v} - \boldsymbol{V}_\mathrm{b}) \times \boldsymbol{\omega}$. However, if vortex lines intersect with the surface, then some of the internal moments of vorticity transport over the fluid do not completely cancel; these surface integrals represent this incomplete cancelation. Though we might spend more effort pondering these surface integrals' significance, it is important to remember that they do no actually contribute to the force and moment on the body. Indeed, the indirect contribution from the body to each vorticity-induced impulse's rate of change contains an equal and opposite surface integral, so the overall effect is nullified. In fact, our main objective in splitting the effect of vorticity into direct and indirect parts is not affected by simply ignoring these terms. Thus, we will define the direct contributions as

$$\tilde{e}_j \cdot \int_{\mathcal{V}_f} (v - V_b) \times \omega \, dV, \qquad \tilde{e}_j \cdot \int_{\mathcal{V}_f} (x - X_r) \times ((v - V_b) \times \omega) \, dV, \tag{6.156}$$

and the indirect contributions from the presence of the body to these impulses as

$$\frac{d}{dt}(P_{bv} \cdot \tilde{e}_j) = -\int_{\mathcal{V}_f} [(v - V_b) \times \omega] \cdot v_{bt}^{(j)} \, dV,$$

$$\frac{d}{dt}(\Pi_{r,bv} \cdot \tilde{e}_j) = -\int_{\mathcal{V}_f} [(v - V_b) \times \omega] \cdot v_{br}^{(j)} \, dV. \tag{6.157}$$

From the results of this section, we can now immediately write more practical expressions for the force and moment due to fluid vorticity. These are collected here into the following result, likely the most important in this chapter:

Result 6.14: The Force and Moment on a Rigid Body

The force applied by the fluid on a rigid body moving with translational velocity V_r and angular velocity Ω in inviscid incompressible flow can be written, in its components along the body-fixed axes, as

$$\tilde{f}_j = -\tilde{M}_{jl}^{FV}\dot{\tilde{V}}_{r,l} - \tilde{M}_{jl}^{F\Omega}\dot{\tilde{\Omega}}_l - \rho \int_{\mathcal{V}_f} ((v - V_b) \times \omega) \cdot v_\infty^{(j)} \, dV - \rho\,(\Omega \times P) \cdot \tilde{e}_j, \tag{6.158}$$

and the components of the moment about X_r as

$$\tilde{m}_{r,j} = -\tilde{M}_{jl}^{MV}\dot{\tilde{V}}_{r,l} - \tilde{M}_{jl}^{M\Omega}\dot{\tilde{\Omega}}_l - \rho \int_{\mathcal{V}_f} ((v - V_b) \times \omega) \cdot v_{r\infty}^{(j)} \, dV - \rho\,(\Omega \times \Pi_r + V_r \times P) \cdot \tilde{e}_j, \tag{6.159}$$

where $V_b := V_r + \Omega \times (x - X_r)$; $v_\infty^{(j)}$ is the basis velocity field due to uniform flow of unit speed along the jth body-fixed axis, defined in Eq. (4.126); $v_{r\infty}^{(j)}$, defined in Eq. (4.147), is the negative of the velocity field observed in the reference frame of the body when it rotates at unit angular velocity about the jth axis; and P and Π_r are the linear and angular impulse (about X_r), respectively, given in decomposed form in (6.112) and (6.113).

The contributions from fluid vorticity to the jth component of force can be split into direct and indirect parts:

$$- \rho \tilde{\boldsymbol{e}}_j \cdot \int\limits_{\mathcal{V}_f} (\boldsymbol{v} - \boldsymbol{V}_\mathrm{b}) \times \boldsymbol{\omega} \, \mathrm{d}V + \rho \int\limits_{\mathcal{V}_f} [(\boldsymbol{v} - \boldsymbol{V}_\mathrm{b}) \times \boldsymbol{\omega}] \cdot \boldsymbol{v}_\mathrm{bt}^{(j)} \, \mathrm{d}V, \tag{6.160}$$

and, similarly, the jth component of moment

$$- \rho \tilde{\boldsymbol{e}}_j \cdot \int\limits_{\mathcal{V}_f} (\boldsymbol{x} - \boldsymbol{X}_\mathrm{r}) \times ((\boldsymbol{v} - \boldsymbol{V}_\mathrm{b}) \times \boldsymbol{\omega}) \, \mathrm{d}V + \rho \int\limits_{\mathcal{V}_f} [(\boldsymbol{v} - \boldsymbol{V}_\mathrm{b}) \times \boldsymbol{\omega}] \cdot \boldsymbol{v}_\mathrm{br}^{(j)} \, \mathrm{d}V, \tag{6.161}$$

where $\boldsymbol{v}_\mathrm{bt}^{(j)}$ and $\boldsymbol{v}_\mathrm{br}^{(j)}$ are the basis velocity fields due to body translation (4.135) and rotation (4.143), respectively.

How are we to physically interpret the form of these expressions for vorticity's contributions to force and moment?

The basis velocity $\boldsymbol{v}_\mathrm{bt}^{(j)}$ measures the change of a vortex element's indirect contribution to the jth linear impulse component due to an infinitesimal variation of its configuration relative to the body. This role is emphasized in Note 6.6.1: the basis vector potential $\boldsymbol{\Psi}_\mathrm{bt}^{(j)}$ measures a vortex element's contribution to the impulse, and $\boldsymbol{v}_\mathrm{bt}^{(j)}$ is the derivative (the curl) of this vector potential. Thus, with a slight manipulation, it is easy to show that the translation-induced basis velocity field $\boldsymbol{v}_\mathrm{bt}^{(j)}$ contains the gradient of the elemental linear impulse with respect to the element's change of position; it also implicitly describes the influence of its change of orientation, through stretching and tilting. We can make similar interpretations of $\boldsymbol{v}_\mathrm{br}^{(j)}$ and its role in the rate of change of angular impulse.

Why is this so? As we noted early in this section, if, during the body's motion, the vortex element remains stationary relative to the body, then, from the perspective of the body reference frame, neither the reacting sheet nor the vorticity-induced impulses changes. This explains why the integral over $\mathcal{V}_f$ involves only the fluid velocity ***relative to the body-fixed frame of reference***, $\boldsymbol{v} - \boldsymbol{V}_\mathrm{b}$: only if this is non-zero will the vortex element move in that frame and give rise to a change in the body-fixed components of either impulse. But in assessing this contribution to impulse, it does not matter whether it is the vortex or the body that has moved. The basis velocity $\boldsymbol{v}_\mathrm{bt}^{(j)}$ represents the influence of a body's incremental change of position at all points in the fluid. Analogously, the basis velocity $\boldsymbol{v}_\mathrm{br}^{(j)}$ assesses the sensitivity of the jth component of angular impulse to the element's motion relative to the body. As we discussed in Note 4.6.1, both of these basis velocities are intrinsic properties of the body's geometry.

Finally, it is important to remember that basis fields $\boldsymbol{v}_\mathrm{bt}^{(j)}$ and $\boldsymbol{v}_\mathrm{br}^{(j)}$ decay with distance from the body, so a vortex's contribution to force and moment is increasingly borne by the direct portion as it moves away from the body.

6.6.3 The Rate of Change of Impulse for Singular Vortex Elements

It is useful to consider how the vorticity-induced contributions to force and moment simplify when we consider a set of vortex filaments or planar point vortices. In such a case, the velocities in the integral are evaluated at the position of each vortex element, and the fluid velocity, in particular, is replaced by the element's local velocity, $\dot{\boldsymbol{x}}_{\mathrm{J}}$. Thus, the rate of change of the jth body-fixed component of vorticity-induced linear impulse is

$$\frac{\mathrm{d}}{\mathrm{d}t}(\boldsymbol{P}_{\mathrm{v}} \cdot \tilde{\boldsymbol{e}}_j) = \sum_{\mathrm{J}=1}^{N_{\mathrm{v}}} \Gamma_{\mathrm{J}} \int_{C_{\mathrm{J}}} \left[\boldsymbol{v}_{\infty}^{(j)}(\boldsymbol{x}_{\mathrm{J}}) \times (\dot{\boldsymbol{x}}_{\mathrm{J}} - \boldsymbol{V}_{\mathrm{b}}(\boldsymbol{x}_{\mathrm{J}}))\right] \cdot \mathrm{d}\boldsymbol{l}, \tag{6.162}$$

and the rate of change of the jth component of angular impulse about the reference point is

$$\frac{\mathrm{d}}{\mathrm{d}t}(\boldsymbol{\Pi}_{\mathrm{r,v}} \cdot \tilde{\boldsymbol{e}}_j) = \sum_{\mathrm{J}=1}^{N_{\mathrm{v}}} \Gamma_{\mathrm{J}} \int_{C_{\mathrm{J}}} \left[\boldsymbol{v}_{\mathrm{r}\infty}^{(j)}(\boldsymbol{x}_{\mathrm{J}}) \times (\dot{\boldsymbol{x}}_{\mathrm{J}} - \boldsymbol{V}_{\mathrm{b}}(\boldsymbol{x}_{\mathrm{J}}))\right] \cdot \mathrm{d}\boldsymbol{l}. \tag{6.163}$$

Parsing these into their direct and indirect contributions, we get

$$\frac{\mathrm{d}}{\mathrm{d}t}(\boldsymbol{P}_{\mathrm{v}} \cdot \tilde{\boldsymbol{e}}_j) = \tilde{\boldsymbol{e}}_j \cdot \sum_{\mathrm{J}=1}^{N_{\mathrm{v}}} \Gamma_{\mathrm{J}} \int_{C_{\mathrm{J}}} (\dot{\boldsymbol{x}}_{\mathrm{J}} - \boldsymbol{V}_{\mathrm{b}}(\boldsymbol{x}_{\mathrm{J}})) \times \mathrm{d}\boldsymbol{l} + \frac{\mathrm{d}}{\mathrm{d}t}(\boldsymbol{P}_{\mathrm{bv}} \cdot \tilde{\boldsymbol{e}}_j), \tag{6.164}$$

where the indirect contribution is

$$\frac{\mathrm{d}}{\mathrm{d}t}(\boldsymbol{P}_{\mathrm{bv}} \cdot \tilde{\boldsymbol{e}}_j) = -\sum_{\mathrm{J}=1}^{N_{\mathrm{v}}} \Gamma_{\mathrm{J}} \int_{C_{\mathrm{J}}} \left[\boldsymbol{v}_{\mathrm{bt}}^{(j)}(\boldsymbol{x}_{\mathrm{J}}) \times (\dot{\boldsymbol{x}}_{\mathrm{J}} - \boldsymbol{V}_{\mathrm{b}}(\boldsymbol{x}_{\mathrm{J}}))\right] \cdot \mathrm{d}\boldsymbol{l}; \tag{6.165}$$

and, for the angular impulse,

$$\frac{\mathrm{d}}{\mathrm{d}t}(\boldsymbol{\Pi}_{\mathrm{r,v}} \cdot \tilde{\boldsymbol{e}}_j) = \tilde{\boldsymbol{e}}_j \cdot \sum_{\mathrm{J}=1}^{N_{\mathrm{v}}} \Gamma_{\mathrm{J}} \int_{C_{\mathrm{J}}} (\boldsymbol{x} - \boldsymbol{X}_{\mathrm{r}}) \times ((\dot{\boldsymbol{x}}_{\mathrm{J}} - \boldsymbol{V}_{\mathrm{b}}(\boldsymbol{x}_{\mathrm{J}})) \times \mathrm{d}\boldsymbol{l}) + \frac{\mathrm{d}}{\mathrm{d}t}(\boldsymbol{\Pi}_{\mathrm{r,bv}} \cdot \tilde{\boldsymbol{e}}_j), \tag{6.166}$$

with its indirect contribution given by

$$\frac{\mathrm{d}}{\mathrm{d}t}(\boldsymbol{\Pi}_{\mathrm{r,bv}} \cdot \tilde{\boldsymbol{e}}_j) = -\sum_{\mathrm{J}=1}^{N_{\mathrm{v}}} \Gamma_{\mathrm{J}} \int_{C_{\mathrm{J}}} \left[\boldsymbol{v}_{\mathrm{br}}^{(j)}(\boldsymbol{x}_{\mathrm{J}}) \times (\dot{\boldsymbol{x}}_{\mathrm{J}} - \boldsymbol{V}_{\mathrm{b}}(\boldsymbol{x}_{\mathrm{J}}))\right] \cdot \mathrm{d}\boldsymbol{l}. \tag{6.167}$$

Complex Notation for Point Vortices With the usual simplification, the results above also hold for point vortices: the contour C_{J} is parallel to the out-of-plane direction, $\boldsymbol{e}_3$, and the contour integral over C_{J} is eliminated by the interpretation of the force and moment as per unit length in $\boldsymbol{e}_3$. However, we can also write these results in the usual complex notation. In doing so, we note that the velocity of the Jth point vortex relative to the body frame of reference, has the following equivalent representations, as can be confirmed from the identities in Chap. 2:

$$\dot{\boldsymbol{x}}_{\mathrm{J}} - \boldsymbol{V}_{\mathrm{b}}(\boldsymbol{x}_{\mathrm{J}}) \Longleftrightarrow \dot{z}_{\mathrm{J}} - W_{\mathrm{b}}^{*}(z_{\mathrm{J}}) = \dot{\tilde{z}}_{\mathrm{J}} e^{\mathrm{i}\alpha}, \tag{6.168}$$

where $\dot{\tilde{z}}_{\mathrm{J}} \equiv \mathrm{d}\tilde{z}_{\mathrm{J}}/\mathrm{d}t$ denotes the rate of change of the body-fixed coordinates of the Jth vortex. From this, and the complex analysis tools in Sect. A.2, the rate of change of the jth body-fixed component of linear impulse is given by

$$\sum_{\mathrm{J}=1}^{N_{\mathrm{v}}} \Gamma_{\mathrm{J}} \mathrm{Im}\left(\tilde{w}_{\infty}^{(j)}(\tilde{z}_{\mathrm{J}})\dot{\tilde{z}}_{\mathrm{J}}\right). \tag{6.169}$$

Parsed into direct and indirect contributions, and with the components assembled together, we get

$$-\mathrm{i}\sum_{\mathrm{J}=1}^{N_{\mathrm{v}}} \Gamma_{\mathrm{J}}\dot{\tilde{z}}_{\mathrm{J}} - \sum_{\mathrm{J}=1}^{N_{\mathrm{v}}} \Gamma_{\mathrm{J}} \left[\mathrm{Im}\left(\tilde{w}_{\mathrm{bt}}^{(1)}(\tilde{z}_{\mathrm{J}})\dot{\tilde{z}}_{\mathrm{J}}\right) + \mathrm{i}\mathrm{Im}\left(\tilde{w}_{\mathrm{bt}}^{(2)}(\tilde{z}_{\mathrm{J}})\dot{\tilde{z}}_{\mathrm{J}}\right)\right]. \tag{6.170}$$

The contribution to the rate of change of angular impulse is

$$-\sum_{\mathrm{J}=1}^{N_{\mathrm{v}}} \Gamma_{\mathrm{J}} \mathrm{Re}\left(\tilde{z}_{\mathrm{J}}^{*}\dot{\tilde{z}}_{\mathrm{J}}\right) - \sum_{\mathrm{J}=1}^{N_{\mathrm{v}}} \Gamma_{\mathrm{J}} \mathrm{Im}\left(\tilde{w}_{\mathrm{br}}(\tilde{z}_{\mathrm{J}})\dot{\tilde{z}}_{\mathrm{J}}\right). \tag{6.171}$$

Again, it should be emphasized that the basis field in the latter term in these expressions decays with distance from the body, so the vortex's direct contribution—the leading term—is dominant when it is far from the body. This renders these expressions consistent with our understanding of the steady state flow, in which a starting vortex is infinitely far into the wake of a body and traveling away from it, and the force reverts to the Kutta–Joukowski lift.

However, when the vortex is in the vicinity of the body, the indirect contribution is non-negligible and must be retained. Above, we observed that the basis velocities due to body translation and rotation measure the sensitivity of vorticity's indirect contribution to the impulse to the change of position of vorticity-bearing elements. In particular, in the context of point vortices, by comparing Eqs. (6.170)–(6.171) directly with Eqs. (6.101) and (6.102), it is easy to verify that

$$\tilde{w}_{\mathrm{bt}}^{(1)}(\tilde{z}_{\mathrm{J}}) = -2\mathrm{i}\frac{\partial \mathrm{Re}(\tilde{P}_{\mathrm{J}}^{\mathrm{bv}})}{\partial \tilde{z}_{\mathrm{J}}}, \quad \tilde{w}_{\mathrm{bt}}^{(2)}(\tilde{z}_{\mathrm{J}}) = -2\mathrm{i}\frac{\partial \mathrm{Im}(\tilde{P}_{\mathrm{J}}^{\mathrm{bv}})}{\partial \tilde{z}_{\mathrm{J}}}, \quad \tilde{w}_{\mathrm{br}}(\tilde{z}_{\mathrm{J}}) = -2\mathrm{i}\frac{\partial \Pi_{\mathrm{r,J}}^{\mathrm{bv}}}{\partial \tilde{z}_{\mathrm{J}}}. \tag{6.172}$$

These relations clearly show the role of the basis fields as measures of sensitivity. They are perhaps more recognizable if we write them in vector form:

$$\boldsymbol{v}_{\mathrm{bt}}^{(j)}(\boldsymbol{x}_{\mathrm{J}}) = \boldsymbol{e}_3 \times \nabla(\boldsymbol{P}_{\mathrm{J}}^{\mathrm{bv}} \cdot \tilde{\boldsymbol{e}}_j), \; j = 1, 2, \quad \boldsymbol{v}_{\mathrm{br}}(\boldsymbol{x}_{\mathrm{J}}) = \boldsymbol{e}_3 \times \nabla(\Pi_{\mathrm{r,J}}^{\mathrm{bv}}), \tag{6.173}$$

in which the gradients are taken with respect to the vortex position, $\boldsymbol{x}_{\mathrm{J}}$. Written in this form, the equations show that, in two-dimensional flow about a single rigid body, ***each component of the indirect part of the vorticity-induced impulse—interpreted as a function of vortex position—acts as a streamfunction for the translation- or rotation-induced basis velocity field***.

A Modification for Time-Varying Vortex Strengths Finally, let us stress that the contributions of fluid vorticity to force and moment derived in this section are predicated on the assumption that the vortex elements maintain constant strength, as required by Helmholtz' third theorem. If, for whatever reason, we relax this assumption, then we must include an additional term in the force and moment that accounts for the rate of change of the element's strength, i.e., the terms proportional to $\dot{\Gamma}_{\mathrm{J}}$ in (6.107) and (6.108). For example, the rate of change of the body-fixed components of vorticity-induced linear impulse, in the complex notation, would be

$$\frac{\mathrm{d}}{\mathrm{d}t}\sum_{\mathrm{J}=1}^{N_{\mathrm{v}}}\Gamma_{\mathrm{J}}\tilde{P}_{\mathrm{J}}^{\mathrm{v}} = \sum_{\mathrm{J}=1}^{N_{\mathrm{v}}}\left[\dot{\Gamma}_{\mathrm{J}}\tilde{P}_{\mathrm{J}}^{\mathrm{v}} + \Gamma_{\mathrm{J}}\mathrm{Im}\left(\tilde{w}_{\infty}^{(1)}(\tilde{z}_{\mathrm{J}})\dot{\tilde{z}}_{\mathrm{J}}\right) + \mathrm{i}\Gamma_{\mathrm{J}}\mathrm{Im}\left(\tilde{w}_{\infty}^{(2)}(\tilde{z}_{\mathrm{J}})\dot{\tilde{z}}_{\mathrm{J}}\right)\right]. \qquad (6.174)$$

As we have discussed previously, such an allowance for variable strength will introduce spurious forces and moments in the fluid, so this will not, in general, be equal to the force and moment obtained by integrating surface traction. The difference was described in Result 6.9.

6.6.4 Alternative Derivation of the Force and Moment

In this section we provide a slightly different derivation of the force and moment expressions in Result 6.14. This derivation, due originally to Howe [31, 32], makes use of the special properties of the motion-induced basis scalar potential fields, Υ_j and Ξ_j, as well as the associated velocities $\boldsymbol{v}_{\infty}^{(j)}$ and $\boldsymbol{v}_{\mathrm{r}\infty}^{(j)}$, discussed in Sect. 4.6.2 and Note 4.6.1. We will present only the derivation of the force (6.158); the proof for the moment (6.159) follows a similar set of steps, using $\varphi_{\mathrm{br}}^{j}$ and $\boldsymbol{v}_{\mathrm{r}\infty}^{(j)}$, which the reader is invited to carry out. Our derivation differs a bit from that of Howe [31] because our basis scalar potentials are defined with respect to the body-fixed axes, whereas Howe defines his potentials about the axes in the inertial coordinate system.

The derivation follows some of the same steps used in developing our impulse form in Sect. 6.2. In fact, just as that form was applicable to both inviscid and viscous flows, the derivation we will present is a specialization of the version originally presented by Howe for both types of flows. We will make use again of the restricted fluid region $\mathcal{V}_{R\backslash\mathrm{b}}$, bounded internally by the surface $\mathcal{S}_{\mathrm{b}}$ of a rigid body and externally by the distant surface $\mathcal{S}_R$, stationary in the inertial reference frame and enclosing all vorticity, that we will allow to approach infinity. To develop the formulation of force and moment on the body, let us first derive a few key results.

First, consider a generic differentiable scalar field f in $\mathcal{V}_{R\backslash \mathrm{b}}$. Let us integrate the dot product between the gradient of this field and the gradient of Υ_j (which, the reader will recall from Note 4.6.1, is simply the basis velocity field $\boldsymbol{v}_\infty^{(j)}$). Since this velocity field is divergence free, then we can write this dot product as $\nabla \cdot (f\boldsymbol{v}_\infty^{(j)})$. By applying the divergence theorem to this integral, it is transformed to surface integrals over $\mathcal{S}_\mathrm{b}$ and $\mathcal{S}_R$. However, the integral over $\mathcal{S}_\mathrm{b}$ vanishes due to the fact that the normal component of $\boldsymbol{v}_\infty^{(j)}$ vanishes everywhere on that surface. On $\mathcal{S}_R$, as it is pushed outward toward infinity, this normal component approaches $\tilde{n}_j$. Thus, we get the useful identity

$$\int_{\mathcal{V}_{R\backslash \mathrm{b}}} \nabla f \cdot \boldsymbol{v}_\infty^{(j)}\,\mathrm{d}V = \int_{\mathcal{S}_R} f\tilde{n}_j\,\mathrm{d}S. \tag{6.175}$$

In other words, this serves as a modification of the divergence theorem, since if we had applied the basic form of the divergence theorem to ∇f, we would also obtain a surface integral over $\mathcal{S}_\mathrm{b}$.

Now, let us consider the integral of the dot product of $\partial \boldsymbol{v}/\partial t$ with $\boldsymbol{v}_\infty^{(j)}$ over $\mathcal{V}_{R\backslash \mathrm{b}}$. To manipulate this, we will make use of the fact that the fluid velocity field, $\boldsymbol{v}$, is divergence free, and that the partial derivative of Υ_j with respect to time is given by Eq. (4.153). This latter derivative involves the rate of deviation between the body-fixed and inertial reference frames, $\boldsymbol{V}_\mathrm{r} + \boldsymbol{\Omega} \times (\boldsymbol{x} - \boldsymbol{X}_\mathrm{r})$. Since this velocity is simply an extension of the rigid body's own velocity into the surrounding fluid, we will denote it by the same symbol, $\boldsymbol{V}_\mathrm{b}$. Thus, using the divergence theorem, and taking the usual limit of the surface $\mathcal{S}_R$,

$$\int_{\mathcal{V}_{R\backslash \mathrm{b}}} \frac{\partial \boldsymbol{v}}{\partial t} \cdot \boldsymbol{v}_\infty^{(j)}\,\mathrm{d}V = \int_{\mathcal{V}_{R\backslash \mathrm{b}}} \frac{\partial}{\partial t}\left(\boldsymbol{v} \cdot \boldsymbol{v}_\infty^{(j)}\right)\mathrm{d}V - \int_{\mathcal{S}_\mathrm{b}} \boldsymbol{n} \cdot \boldsymbol{v}\boldsymbol{V}_\mathrm{b} \cdot \boldsymbol{v}_\infty^{(j)}\,\mathrm{d}S + \int_{\mathcal{S}_R} \boldsymbol{n} \cdot \boldsymbol{v}\boldsymbol{V}_\mathrm{b} \cdot \tilde{\boldsymbol{e}}_j\,\mathrm{d}S. \tag{6.176}$$

For the remaining volume integral on the right-hand side, we can interchange the integration and the time derivative by using (A.88). Keeping in mind that the surface $\mathcal{S}_R$ is stationary, then we arrive at

$$\begin{aligned}\int_{\mathcal{V}_{R\backslash \mathrm{b}}} \frac{\partial \boldsymbol{v}}{\partial t} \cdot \boldsymbol{v}_\infty^{(j)}\,\mathrm{d}V = {} & \frac{\mathrm{d}}{\mathrm{d}t}\int_{\mathcal{V}_{R\backslash \mathrm{b}}} \boldsymbol{v} \cdot \tilde{\boldsymbol{e}}_j\,\mathrm{d}V + \frac{\mathrm{d}}{\mathrm{d}t}\int_{\mathcal{S}_\mathrm{b}} \varphi_\mathrm{bt}^{(j)} \boldsymbol{V}_\mathrm{b} \cdot \boldsymbol{n}\,\mathrm{d}S - \frac{\mathrm{d}}{\mathrm{d}t}\int_{\mathcal{S}_R} \varphi_\mathrm{bt}^{(j)} \boldsymbol{v} \cdot \boldsymbol{n}\,\mathrm{d}S \\ & + \int_{\mathcal{S}_\mathrm{b}} (\boldsymbol{n} \cdot \boldsymbol{V}_\mathrm{b}\boldsymbol{v} - \boldsymbol{n} \cdot \boldsymbol{v}\boldsymbol{V}_\mathrm{b}) \cdot \boldsymbol{v}_\infty^{(j)}\,\mathrm{d}S + \int_{\mathcal{S}_R} \boldsymbol{n} \cdot \boldsymbol{v}\boldsymbol{V}_\mathrm{b} \cdot \tilde{\boldsymbol{e}}_j\,\mathrm{d}S.\end{aligned} \tag{6.177}$$

At large distances, the fluid velocity field decays like r^{-n_d} and the basis scalar potential $\varphi_\mathrm{br}^{(j)}$ like r^{1-n_d}, so the integral of their product over $\mathcal{S}_R$ tends to zero. Furthermore, the rate of change of the unit vector $\tilde{\boldsymbol{e}}_j$ is given by the rotation of the body-fixed frame (2.12). We thus have that

$$\int_{\mathcal{V}_{R\backslash b}} \frac{\partial \boldsymbol{v}}{\partial t} \cdot \boldsymbol{v}_\infty^{(j)} \, dV = \tilde{\boldsymbol{e}}_j \cdot \frac{d}{dt} \int_{\mathcal{V}_{R\backslash b}} \boldsymbol{v} \, dV + (\boldsymbol{\Omega} \times \tilde{\boldsymbol{e}}_j) \cdot \int_{\mathcal{V}_{R\backslash b}} \boldsymbol{v} \, dV + \frac{d}{dt} \int_{\mathcal{S}_b} \varphi_{bt}^{(j)} \boldsymbol{V}_b \cdot \boldsymbol{n} \, dS$$
$$+ \int_{\mathcal{S}_b} (\boldsymbol{n} \cdot \boldsymbol{V}_b \boldsymbol{v} - \boldsymbol{n} \cdot \boldsymbol{v} \boldsymbol{V}_b) \cdot \boldsymbol{v}_\infty^{(j)} \, dS + \int_{\mathcal{S}_R} \boldsymbol{n} \cdot \boldsymbol{v} \boldsymbol{V}_b \cdot \tilde{\boldsymbol{e}}_j \, dS. \tag{6.178}$$

Another useful result, essentially a generalization of Eq. (6.55), is

$$\int_{\mathcal{V}_{R\backslash b}} (\boldsymbol{V}_b \times \boldsymbol{\omega}) \cdot \boldsymbol{v}_\infty^{(j)} \, dV = \int_{\mathcal{S}_b} (\boldsymbol{n} \cdot \boldsymbol{V}_b \boldsymbol{v} - \boldsymbol{n} \cdot \boldsymbol{v} \boldsymbol{V}_b) \cdot \boldsymbol{v}_\infty^{(j)} \, dS$$
$$- \int_{\mathcal{S}_R} (\boldsymbol{n} \cdot \boldsymbol{V}_b \boldsymbol{v} - \boldsymbol{n} \cdot \boldsymbol{v} \boldsymbol{V}_b - \boldsymbol{V}_b \cdot \boldsymbol{v} \boldsymbol{n}) \cdot \tilde{\boldsymbol{e}}_j \, dS. \tag{6.179}$$

This identity is proved with the help of the divergence theorem; it also makes use of the fact that the normal component of $\boldsymbol{v}_\infty^{(j)}$ vanishes on $\mathcal{S}_b$, and that both $\boldsymbol{V}_b$ and $\boldsymbol{v}$ are divergence free in $\mathcal{V}_{R\backslash b}$. Combining these last two results by eliminating their common terms, we get

$$\int_{\mathcal{V}_{R\backslash b}} \frac{\partial \boldsymbol{v}}{\partial t} \cdot \boldsymbol{v}_\infty^{(j)} \, dV = \tilde{\boldsymbol{e}}_j \cdot \frac{d}{dt} \int_{\mathcal{V}_{R\backslash b}} \boldsymbol{v} \, dV + (\boldsymbol{\Omega} \times \tilde{\boldsymbol{e}}_j) \cdot \int_{\mathcal{V}_{R\backslash b}} \boldsymbol{v} \, dV + \frac{d}{dt} \int_{\mathcal{S}_b} \varphi_{bt}^{(j)} \boldsymbol{V}_b \cdot \boldsymbol{n} \, dS$$
$$+ \int_{\mathcal{V}_{R\backslash b}} (\boldsymbol{V}_b \times \boldsymbol{\omega}) \cdot \boldsymbol{v}_\infty^{(j)} \, dV - \int_{\mathcal{S}_R} [\boldsymbol{V}_b \times (\boldsymbol{n} \times \boldsymbol{v})] \cdot \tilde{\boldsymbol{e}}_j \, dS. \tag{6.180}$$

Let's make two observations regarding this equation. First, in the second term on the right-hand side, we note the integral of fluid velocity over the region $\mathcal{V}_{R\backslash b}$. This same integral arose in our derivation of the impulse form of the force, and we proved identity (6.12) for that task. We will use it again here. Second, in the final term on the right-hand side, we recall that the fluid velocity on $\mathcal{S}_R$ is entirely irrotational, $\boldsymbol{v} = \nabla\varphi$. The part of this integral involving the translational motion, $\boldsymbol{V}_r$, in $\boldsymbol{V}_b$ only involves the integral of $\boldsymbol{n} \times \nabla\varphi$ over the surface; this is easily shown to vanish identically due to Stokes theorem in form (A.63) and the fact that φ is continuous on the closed surface $\mathcal{S}_R$. The remaining part, due to the rigid-body rotation, can be manipulated with a form of Stokes' theorem on $\mathcal{S}_R$, proved in similar fashion to (A.66):

$$\int_{\mathcal{S}_R} (\boldsymbol{\Omega} \times \boldsymbol{x}) \times (\boldsymbol{n} \times \nabla\varphi) \, dS = -\boldsymbol{\Omega} \times \int_{\mathcal{S}_R} \varphi \boldsymbol{n} \, dS. \tag{6.181}$$

(The part due to $\boldsymbol{\Omega} \times \boldsymbol{X}_r$ vanishes for the same reason as that due to $\boldsymbol{V}_r$.) But this will cancel with the $\mathcal{S}_R$ integral arising from our use of (6.12) in the second term. Thus, we arrive at the entirely kinematic result

$$\int_{\mathcal{V}_{R\backslash b}} \frac{\partial \boldsymbol{v}}{\partial t} \cdot \boldsymbol{v}_{\infty}^{(j)} \, dV = \tilde{\boldsymbol{e}}_j \cdot \left(\frac{d}{dt} \int_{\mathcal{V}_{R\backslash b}} \boldsymbol{v} \, dV - \boldsymbol{\Omega} \times \boldsymbol{P} \right)$$
$$+ \frac{d}{dt} \int_{\mathcal{S}_b} \varphi_{bt}^{(j)} \boldsymbol{V}_b \cdot \boldsymbol{n} \, dS + \int_{\mathcal{V}_{R\backslash b}} (\boldsymbol{V}_b \times \boldsymbol{\omega}) \cdot \boldsymbol{v}_{\infty}^{(j)} \, dV. \quad (6.182)$$

Now, with identities (6.175) and (6.182), we are ready to obtain an expression for the force on the body. Let us start by multiplying (6.182) by the uniform fluid density. Using the integral form of conservation of momentum in $\mathcal{V}_{R\backslash b}$, Eq. (6.10), to replace the time derivative of the fluid momentum in that region, we get

$$\int_{\mathcal{V}_{R\backslash b}} \rho \frac{\partial \boldsymbol{v}}{\partial t} \cdot \boldsymbol{v}_{\infty}^{(j)} \, dV = \tilde{\boldsymbol{e}}_j \cdot (-\boldsymbol{f} + \boldsymbol{f}_R - \boldsymbol{\Omega} \times \rho \boldsymbol{P})$$
$$+ \frac{d}{dt} \int_{\mathcal{S}_b} \rho \varphi_{bt}^{(j)} \boldsymbol{V}_b \cdot \boldsymbol{n} \, dS + \rho \int_{\mathcal{V}_{R\backslash b}} (\boldsymbol{V}_b \times \boldsymbol{\omega}) \cdot \boldsymbol{v}_{\infty}^{(j)} \, dV, \quad (6.183)$$

where $\boldsymbol{f}$ is the force applied by the fluid on the body, and $\boldsymbol{f}_R$ is the force applied by the fluid outside of $\mathcal{S}_R$ on that surface. This latter force is due to the pressure acting on that surface. We will cancel its effect by making use of identity (6.175): In the volume integral on the left-hand side of Eq. (6.183), the partial derivative of fluid velocity can be substituted with the Euler equations in their form (3.17). The integral of the part of these equations involving the gradient can be transformed into a surface integral over $\mathcal{S}_R$ using identity (6.175). That will give rise to contributions from pressure and from $|\boldsymbol{v}|^2/2$ on that surface; the latter's contribution is vanishingly small, and the pressure's contribution is identically $\boldsymbol{f}_R$. We therefore arrive at

$$\boldsymbol{f} \cdot \tilde{\boldsymbol{e}}_j = -(\boldsymbol{\Omega} \times \rho \boldsymbol{P}) \cdot \tilde{\boldsymbol{e}}_j + \frac{d}{dt} \int_{\mathcal{S}_b} \rho \varphi_{bt}^{(j)} \boldsymbol{V}_b \cdot \boldsymbol{n} \, dS - \rho \int_{\mathcal{V}_{R\backslash b}} ((\boldsymbol{v} - \boldsymbol{V}_b) \times \boldsymbol{\omega}) \cdot \boldsymbol{v}_{\infty}^{(j)} \, dV. \quad (6.184)$$

The final integral, over $\mathcal{V}_{R\backslash b}$, can be extended to all of the fluid, since the vorticity is simply zero beyond $\mathcal{S}_R$. This form leaves us just one step short of our final result. We need only observe that the remaining surface integral over $\mathcal{S}_b$, once the rigid-body velocity $\boldsymbol{V}_r + \boldsymbol{\Omega} \times (\boldsymbol{x} - \boldsymbol{X}_r)$ is substituted for $\boldsymbol{V}_b$, gives rise to the integrals (4.160) and (4.167) that define the added mass components in body-fixed coordinates. Recalling that these components are time invariant, we obtain our desired result.

6.7 The Mechanical Energy Equation, Revisited

In Sect. 3.1.4, we derived a mechanical energy equation (3.33) for an incompressible flow, expressing that the rate of change of kinetic energy in the fluid surrounding a body (or bodies) is equal to the rate of work done on the fluid by the body motion. That equation was applicable to bodies with rigid or deformable surfaces. However, let us now specialize the result to a single rigid body. The first step is to write the body velocity on the surface in its rigid-body components, $\boldsymbol{V}_\mathrm{b} = \boldsymbol{V}_\mathrm{r} + \boldsymbol{\Omega} \times (\boldsymbol{x} - \boldsymbol{X}_\mathrm{r})$, from which the work term in (3.33) reveals integrals of pressure on the surface. These integrals are simply the force and moment (about the body's reference point), so the kinetic energy equation becomes

$$\frac{\mathrm{d}T_f}{\mathrm{d}t} = -\boldsymbol{V}_\mathrm{r} \cdot \boldsymbol{f} - \boldsymbol{\Omega} \cdot \boldsymbol{m}_\mathrm{r}. \tag{6.185}$$

Into this, we can substitute our expressions for the components of the force and moment in Result 6.14. The terms due to the moving body reference frame, such as $-\rho\boldsymbol{\Omega} \times \boldsymbol{P}$, cancel from the expression, and thus, have no effect on the kinetic energy in the fluid. The added mass terms can be further manipulated by exploiting the symmetry of this tensor and the time-invariance of the components in the body-fixed coordinate system. Furthermore, the contributions from the fluid vorticity to force and moment, measured by the basis velocity fields, $\boldsymbol{v}_\infty^{(j)}$ and $\boldsymbol{v}_{\mathrm{r}\infty}^{()}(j)$, combine naturally with the components of rigid-body velocity; one can easily check from the definitions of these basis fields and the general decomposition (4.112) that

$$\tilde{V}_{\mathrm{r},j}\boldsymbol{v}_\infty^{(j)} + \tilde{\Omega}_j \boldsymbol{v}_{\mathrm{r}\infty}^{(j)} = -(\boldsymbol{v} - \boldsymbol{V}_\mathrm{b} - \boldsymbol{v}_\mathrm{v}). \tag{6.186}$$

The resulting equation is

$$\frac{\mathrm{d}T_f}{\mathrm{d}t} = \frac{\mathrm{d}T_{f,\mathrm{ir}}}{\mathrm{d}t} + \rho \int_{\mathcal{V}_f} \boldsymbol{v}_\mathrm{v} \cdot ((\boldsymbol{v} - \boldsymbol{V}_\mathrm{b}) \times \boldsymbol{\omega})\,\mathrm{d}V, \tag{6.187}$$

where $T_{f,\mathrm{ir}}$, the flow kinetic energy due to body motion in an irrotational fluid, was defined in Result 4.6 as

$$T_{f,\mathrm{ir}} \equiv \frac{1}{2} \begin{pmatrix} \tilde{\boldsymbol{\Omega}}^\mathrm{T} & \tilde{\mathbf{V}}_\mathrm{r}^\mathrm{T} \end{pmatrix} \begin{bmatrix} \tilde{\mathbf{M}}^{M\Omega} & \tilde{\mathbf{M}}^{MV} \\ \tilde{\mathbf{M}}^{F\Omega} & \tilde{\mathbf{M}}^{FV} \end{bmatrix} \begin{pmatrix} \tilde{\boldsymbol{\Omega}} \\ \tilde{\mathbf{V}}_\mathrm{r} \end{pmatrix}. \tag{4.169}$$

By comparison with the overall kinetic energy in Eq. (4.168), the remaining term in (6.187) necessarily represents the rate of change of the vortical part of the kinetic energy, $T_{f,\mathrm{v}}$, described by the final two terms in (3.30). It is important to emphasize that, because the rates of the change of the total kinetic energy of the fluid as well as the irrotational part are both zero when the body motion stops, the kinetic energy attributable to vorticity must also remain invariant without body motion. We cannot escape the fact that, ***in a purely inviscid medium, changes in kinetic energy can only occur through work done by (or on) the moving body***; there is no dissipation.

If the body motion stops, the vortical motion in the fluid persists, but the energy of this motion is conserved.

One final note on $T_{f,\mathrm{v}}$ before we leave this topic. In our discussion following the expression (3.30) for the kinetic energy, we noted that, in the surface integral of $\boldsymbol{\Psi}_\mathrm{v} \cdot (\boldsymbol{n} \times \boldsymbol{v}_\mathrm{v})$ over $\mathcal{S}_\mathrm{b}$, the tangential components of the vector potential can be treated as equal to a uniform value inside the body. Let us denote that value by $\boldsymbol{\Psi}_\mathrm{v}^\mathrm{b}$. The remaining surface integral is equal and opposite to the integral of fluid vorticity over the surrounding fluid: in three dimensions because of identity (4.189) and in two dimensions because we assume throughout this book that the total circulation (3.38) is zero. Thus, we can write the kinetic energy due to fluid vorticity as

$$T_{f,\mathrm{v}} = \frac{1}{2}\rho \int_{\mathcal{V}_f} \left(\boldsymbol{\Psi}_\mathrm{v} - \boldsymbol{\Psi}_\mathrm{v}^\mathrm{b}\right) \cdot \boldsymbol{\omega} \, \mathrm{d}V. \tag{6.188}$$

Chapter 7
Transport of Vortex Elements

Contents

7.1 Planar Vortex Elements 246
7.1.1 Justification for Ignoring the Self-induction of a Point Vortex 248
7.1.2 Vortex Transport in the Conformal Mapping Plane: The Routh Correction 249
7.1.3 Vortex Clouds and Sheets 252
7.1.4 Variable-Strength Point Vortices 255
7.1.5 Blob Regularization 260
7.2 Vortex Filaments 263

Thus far in this book, we have focused our attention on two principal aspects of inviscid flow: first, on obtaining the flow field about a moving body amidst a general distribution of vorticity in the surrounding fluid; and second, on computing the resulting fluid dynamic force and moment exerted on the body in this scenario. We have thus far had little to say about the motion of this vorticity. However, our analysis of the problem is incomplete until we have addressed this motion, particularly because—as we found in our discussion of impulse-based calculations in Chap. 6—the force and moment depend on the time rate of change of the vorticity's distribution.

In this chapter we will explore the inviscid kinematics of vorticity, particularly stressing the transport of the singular elements to which we have devoted our attention in this book. When such elements obey Helmholtz' third theorem—namely, when they maintain time-invariant strength—then all such transport can be distilled into the basic kinematic statement of Helmholtz' second theorem: ***vorticity is transported by the flow***. We will have to be thoughtful about what is meant here by 'the flow', since the vortex element whose transport is in question contributes to this flow, and this contribution has unbounded magnitude as the element is approached. How do we interpret this? In a planar context, when the element is a point vortex, there is strong justification—which we will briefly review—for simply ignoring the point

J. D. Eldredge, *Mathematical Modeling of Unsteady Inviscid Flows*, Interdisciplinary Applied Mathematics 50, https://doi.org/10.1007/978-3-030-18319-6_7

vortex' contribution to its own motion. The de-singularized velocity that results from ignoring this self-contribution is often referred to as the **Kirchhoff velocity**, since it was first explicitly presented in lecture notes by Kirchhoff [40]. We might choose to focus on the transport in a mapped plane, rather than the physical plane, due to the relative ease of describing the velocity field. The Kirchhoff velocity must be corrected in this plane, however, and we will describe the rationale for this so-called **Routh correction** in Sect. 7.1.2. When the point vortices constitute a continuous sheet of vorticity, whose infinitesimal elements all ostensibly contribute to the motion of the sheet, the Kirchhoff velocity is interpreted as the Cauchy principal value of the integral over the sheet; this will be discussed in Sect. 7.1.3.

If we relax Helmholtz' third theorem and allow for a vortex element's strength to vary in time, then we must also reconsider this transport. This reconsideration leads to modified transport equations—most notably, the Brown–Michael equation—which augment the Kirchhoff velocity with an additional velocity that attempts to account for the consequences of time-varying strength. We will discuss the origins of these modified transport equations in Sect. 7.1.4.

Though the Kirchhoff velocity eliminates the obviously singular contribution of a vortex element on itself, this does not free of us all difficulties. If vortex elements of finite strength come within close proximity of one another, their mutual contributions to each other's motion are unbounded. This is a matter of significant practical importance, because such close encounters are prevalent in many flows of interest, and their pernicious effect is compounded by the finite step size of a numerical time integration method. This issue has led to a large body of literature on **regularization** techniques, which are designed to ensure that mutual interactions remain smooth and bounded as the distance between two elements approaches zero. We will summarize some of these regularization strategies in Sect. 7.1.5.

Dealing with the singular nature of planar point vortices is annoying, but tolerable. However, when we consider the transport of filaments in three dimensions, we are further flummoxed by the fact that, when the filament has non-zero curvature, its various parts induce motion on one another, and this introduces its own special challenges. Unfortunately, our regard of the filament as simply a line of zero cross-sectional area is too simplistic; the distribution of vorticity inside the small, but finite, cross section makes an essential contribution to the filament's self-induction. We will discuss this, and the related issue of regularization, in Sect. 7.2, our closing section of the chapter.

7.1 Planar Vortex Elements

We will start our discussion of the transport of planar vortex elements with the most basic element: the point vortex. The characteristics of the point vortex were presented in Sect. 3.3.1. Using the notation introduced in that section, we are concerned here with the motion of a point vortex—the Jth such vortex in a set of N_{v} vortices—whose strength and position are denoted, respectively, by $\Gamma_{\mathrm{J}}(t)$ and $\boldsymbol{x}_{\mathrm{J}}(t)$. For now, we will assume that the strength obeys Helmholtz' third theorem: $\Gamma_{\mathrm{J}}(t) = \Gamma_{\mathrm{J}}$.

Result 7.1: Transport of a Point Vortex in the Physical Plane

The transport equation for a point vortex in the plane can be written (in vector form) as

$$\frac{\mathrm{d}\boldsymbol{x}_{\mathrm{J}}}{\mathrm{d}t} = \boldsymbol{v}_{-\mathrm{J}}(t), \tag{7.1}$$

where we have used a notation $(\cdot)_{-\mathrm{J}}$—and will consistently do so throughout the book—to denote the Kirchhoff velocity of vortex J, i.e., the instantaneous local velocity in the fluid, evaluated at the position of the vortex, *minus its own influence*. Its mathematical definition can be written as

$$\boldsymbol{v}_{-\mathrm{J}}(t) := \lim_{\boldsymbol{x}\to\boldsymbol{x}_{\mathrm{J}}} \left[\boldsymbol{v}(\boldsymbol{x},t) - \boldsymbol{K}(\boldsymbol{x}-\boldsymbol{x}_{\mathrm{J}}(t))\times\Gamma_{\mathrm{J}}\boldsymbol{e}_3\right], \tag{7.2}$$

where $\boldsymbol{v}(\boldsymbol{x},t)$ is the full fluid velocity field, including the influence of every flow contributor, and $\boldsymbol{K}(\boldsymbol{x})$ is the two-dimensional velocity kernel (3.71). Note that the Kirchhoff velocity is not a spatial field, but rather, a function of time only; it is also defined separately for each vortex in a set.

In complex notation, the transport equation is written as

$$\frac{\mathrm{d}z_{\mathrm{J}}}{\mathrm{d}t} = w^{*}_{-\mathrm{J}}(t), \tag{7.3}$$

where w, as usual, denotes the complex velocity; the conjugate of this is thus required to match with the components of z_{J}. In this notation, the conjugate Kirchhoff velocity is defined as

$$w_{-\mathrm{J}}(t) := \lim_{z\to z_{\mathrm{J}}} \left[w(z,t) - \frac{\Gamma_{\mathrm{J}}}{2\pi\mathrm{i}(z-z_{\mathrm{J}})}\right], \tag{7.4}$$

where $w(z,t)$ is the full velocity field.

As usual, w in Eq. (7.4) is the derivative of the complex potential, F, and this potential field is also (logarithmically) singular at the vortex location. We can define a potential field, $F_{-\mathrm{J}}$, that omits the contribution of the vortex, and is therefore well-behaved in its vicinity:

$$F_{-\mathrm{J}}(z) := F(z) - \frac{\Gamma_{\mathrm{J}}}{2\pi\mathrm{i}}\log(z-z_{\mathrm{J}}), \tag{7.5}$$

in which the time dependence has been omitted for brevity. Then, the complex form of the Kirchhoff velocity can be written simply as

$$w_{-\mathrm{J}}(t) = \lim_{z\to z_{\mathrm{J}}} \frac{\mathrm{d}F_{-\mathrm{J}}}{\mathrm{d}z}. \tag{7.6}$$

7.1.1 Justification for Ignoring the Self-induction of a Point Vortex

Before we continue our discussion of point vortex transport in the plane, let us rigorously justify the omission of a vortex' contribution to its own transport velocity. In the process of doing so, we will also get some ideas for how to modify the transport if we relax some of Helmholtz' theorems. The analysis we apply here is adapted from a related discussion by Michelin and Llewelyn Smith [51].

In Sect. 6.3.2 we derived the external force (6.69) that must be applied on a vortex pair if the pair's evolution deviates from Helmholtz' theorems. However, when we were faced with the question of what to use for the local velocity when computing the vortex force on the pair, we dodged the question and left the velocities $\boldsymbol{v}^{\pm}_{-\mathrm{V}}$ in the formula unspecified, suggesting only that they represent some desingularized form of the local fluid velocity. So what is the result of this vortex force integral when applied to a vortex pair? Clearly, this is the most crucial step of our analysis. To get the result, it helps to take a step backward: let us use the identity in Lemma 6.3 to transform the vortex force back to a surface integral over $\mathcal{S}_{\mathrm{V}}$ surrounding the pair. We will evaluate this form of the integral, but take care to distinguish the singular from the non-singular part of the velocity field. Though we could carry out this integration in the vector form, it is really helpful to make use of our toolbox from complex analysis. In this context, we interpret $\mathcal{S}_{\mathrm{V}}$ as a counter-clockwise contour, and the reader is invited to show the following equivalence:

$$\left(\frac{1}{2}|\boldsymbol{v}|^2\boldsymbol{n} - \boldsymbol{v}\boldsymbol{v}\cdot\boldsymbol{n}\right)\,\mathrm{d}S \Longleftrightarrow \left(\frac{1}{2}\mathrm{i}w^2(z)\,\mathrm{d}z\right)^*, \tag{7.7}$$

where w is the complex fluid velocity. Before proceeding, let us note that the segments of $\mathcal{S}_{\mathrm{V}}$ along the branch cut must cancel each other out, since w is continuous and single-valued along those segments. So any contribution must come from the portions that envelop each vortex. Let us concentrate on the portion surrounding the positively-oriented vortex; the analysis on the portion surrounding the other vortex is easily adapted. Around this positive vortex, let us write the velocity as

$$w(z) = w(z) - \frac{\Gamma_{\mathrm{J}}}{2\pi\mathrm{i}(z - z_{\mathrm{J}}^{+})} + \frac{\Gamma_{\mathrm{J}}}{2\pi\mathrm{i}(z - z_{\mathrm{J}}^{+})}. \tag{7.8}$$

Taken together, the first two terms on the right-hand side are smooth and continuous throughout the neighborhood of the vortex; in fact, if we let the contour wrap tightly around the vortex center, these terms are well approximated on that contour by the Kirchhoff velocity (7.4) of the vortex:

$$w(z) \approx w^{+}_{-\mathrm{J}} + \frac{\Gamma_{\mathrm{J}}}{2\pi\mathrm{i}(z - z_{\mathrm{J}}^{+})}. \tag{7.9}$$

When this is substituted into (7.7) for the portion of the integral on the contour surrounding the vortex, this portion can be easily evaluated using the residue theorem. The result of this portion of the integral is

$$\rho i \Gamma_J (w^+_{-J})^*. \tag{7.10}$$

It is easy to see that the result for the portion about the negatively-oriented vortex would be $-\rho i(w^-_{-J})^*$. The equivalent vector form of the result of the full integral (7.7) is thus

$$-\rho\left(\boldsymbol{v}^+_{-J} - \boldsymbol{v}^-_{-J}\right) \times \Gamma_J \boldsymbol{e}_3. \tag{7.11}$$

Thus, in the external force on the pair (6.69), the velocities $\boldsymbol{v}^\pm_{-v}$ are simply the Kirchhoff velocities, and writing this expression in our current notation,

$$\boldsymbol{f}^{ext} = \rho\left[\dot{\Gamma}_J \boldsymbol{x}^+_J - \Gamma_J\left(\boldsymbol{v}^+_{-J} - \dot{\boldsymbol{x}}^+_J\right)\right] \times \boldsymbol{e}_3 - \rho\left[\dot{\Gamma}_J \boldsymbol{x}^-_J - \Gamma_J\left(\boldsymbol{v}^-_{-J} - \dot{\boldsymbol{x}}^-_J\right)\right] \times \boldsymbol{e}_3. \tag{7.12}$$

To carry out the same steps on the external moment on the pair (6.70), the reader is invited to show the equivalence

$$\boldsymbol{x} \times \left(\frac{1}{2}|\boldsymbol{v}|^2 \boldsymbol{n} - \boldsymbol{v}\boldsymbol{v}\cdot\boldsymbol{n}\right) \mathrm{d}S \Longleftrightarrow -\mathrm{Re}\left(\frac{1}{2} z w^2(z)\,\mathrm{d}z\right), \tag{7.13}$$

and thence, the external moment also contains the Kirchhoff velocities in place of $\boldsymbol{v}^\pm_{-v}$:

$$\begin{aligned}\boldsymbol{m}^{ext}_O = \rho \boldsymbol{x}^+_J \times &\left[\left(\frac{1}{2}\dot{\Gamma}_J \boldsymbol{x}^+_J - \Gamma_J\left(\boldsymbol{v}^+_{-J} - \dot{\boldsymbol{x}}^+_J\right)\right) \times \boldsymbol{e}_3\right] \\ -\rho \boldsymbol{x}^-_J \times &\left[\left(\frac{1}{2}\dot{\Gamma}_J \boldsymbol{x}^-_J - \Gamma_J\left(\boldsymbol{v}^-_{-J} - \dot{\boldsymbol{x}}^-_J\right)\right) \times \boldsymbol{e}_3\right].\end{aligned} \tag{7.14}$$

Results (7.12) and (7.14) allow us to be more specific about the conclusion we made at the end of Sect. 6.3.2. When $\dot{\Gamma}_J = 0$, ***a point vortex of constant strength, free of externally-applied force and moment, must move at the Kirchhoff velocity,*** thus confirming Result 7.1.

7.1.2 Vortex Transport in the Conformal Mapping Plane: The Routh Correction

Suppose we are interested in expressing the transport of a point vortex in a plane described by complex coordinate ζ, from which the physical plane has been conformally mapped by $z(\zeta)$. From our previous discussions, notably Sect. 4.4, we know how to relate the velocity fields in these two planes: the complex potentials at corresponding points are equal, $F(z(\zeta)) = \hat{F}(\zeta)$, and the complex velocity fields, $w(z(\zeta))$ and $\hat{w}(\zeta)$, are related by the derivative of this equality—Eq. (4.38), repeated here for emphasis:

$$w(z) := \frac{\mathrm{d}F}{\mathrm{d}z} = \frac{\mathrm{d}\hat{F}}{\mathrm{d}\zeta}\frac{\mathrm{d}\zeta}{\mathrm{d}z} = \frac{\hat{w}(\zeta)}{z'(\zeta)}. \tag{4.38}$$

Does this relationship also hold for the Kirchhoff velocity? In other words, is the de-singularized velocity in the mapped plane simply (7.4) multiplied by $z'(\zeta)$? As we will show, it is not quite so. Let us write the equality of the full complex potentials in the two planes a bit differently, this time explicitly isolating the contribution of the vortex to this potential in each plane:

$$F_{-\mathrm{J}}(z(\zeta)) + \frac{\Gamma_{\mathrm{J}}}{2\pi\mathrm{i}} \log(z(\zeta) - z_{\mathrm{J}}) = \hat{F}_{-\mathrm{J}}(\zeta) + \frac{\Gamma_{\mathrm{J}}}{2\pi\mathrm{i}} \log(\zeta - \zeta_{\mathrm{J}}), \tag{7.15}$$

where we've used the usual subscript to denote the part of the field that is missing the contribution of the Jth vortex. To prepare ourselves for deriving the correction, let us move the vortex' contribution in the physical plane to the right-hand side, combining it with its contribution in the mapped plane:

$$F_{-\mathrm{J}}(z(\zeta)) = \hat{F}_{-\mathrm{J}}(\zeta) - \frac{\Gamma_{\mathrm{J}}}{2\pi\mathrm{i}} \log\left(\frac{z(\zeta) - z_{\mathrm{J}}}{\zeta - \zeta_{\mathrm{J}}}\right). \tag{7.16}$$

By (7.6), the derivative of the left-hand side with respect to z provides, in the limit as the vortex is approached, the desired Kirchhoff velocity in the physical plane. Thus, our interest is in the expression on the right-hand side after the same derivative and limit. Since we're obviously interested in the behavior of these fields in the vicinity of the vortex, it is reasonable to Taylor expand the conformal mapping in the logarithm about the vortex' position:

$$z(\zeta) - z_{\mathrm{J}} = (\zeta - \zeta_{\mathrm{J}})z'(\zeta_{\mathrm{J}}) + \frac{1}{2}(\zeta - \zeta_{\mathrm{J}})^2 z''(\zeta_{\mathrm{J}}) + O((\zeta - \zeta_{\mathrm{J}})^3). \tag{7.17}$$

Substituting this into the argument of the logarithm in (7.16),

$$F_{-\mathrm{J}}(z(\zeta)) = \hat{F}_{-\mathrm{J}}(\zeta) - \frac{\Gamma_{\mathrm{J}}}{2\pi\mathrm{i}} \log\left(z'(\zeta_{\mathrm{J}}) + \frac{1}{2}(\zeta - \zeta_{\mathrm{J}})z''(\zeta_{\mathrm{J}}) + O((\zeta - \zeta_{\mathrm{J}})^2)\right). \tag{7.18}$$

Now, we're ready to take the derivative of both sides. We differentiate both sides with respect to ζ, using the chain rule on the left-hand side:

$$\frac{\mathrm{d}F_{-\mathrm{J}}}{\mathrm{d}z} z'(\zeta) = \frac{\mathrm{d}\hat{F}_{-\mathrm{J}}}{\mathrm{d}\zeta} - \frac{\Gamma_{\mathrm{J}}}{2\pi\mathrm{i}} \frac{\frac{1}{2}z''(\zeta_{\mathrm{J}}) + O(\zeta - \zeta_{\mathrm{J}})}{z'(\zeta_{\mathrm{J}}) + \frac{1}{2}(\zeta - \zeta_{\mathrm{J}})z''(\zeta_{\mathrm{J}}) + O((\zeta - \zeta_{\mathrm{J}})^2)}. \tag{7.19}$$

Finally, divide both sides by $z'(\zeta)$, and take the limit $\zeta \to \zeta_{\mathrm{J}}$, to obtain our final result:

Result 7.2: The Routh Correction

The Kirchhoff velocity in the physical plane, $w_{-\mathrm{J}}$, can be obtained from the solution in the conformally mapped plane as follows:

$$w_{-\mathrm{J}} = \frac{1}{z'(\zeta_{\mathrm{J}})}\left(\hat{w}_{-\mathrm{J}} - \frac{\Gamma_{\mathrm{J}}}{4\pi \mathrm{i}}\frac{z''(\zeta_{\mathrm{J}})}{z'(\zeta_{\mathrm{J}})}\right), \tag{7.20}$$

where we have defined $\hat{w}_{-\mathrm{J}}$ as

$$\hat{w}_{-\mathrm{J}} := \lim_{\zeta\to\zeta_{\mathrm{J}}}\left[\frac{\mathrm{d}\hat{F}}{\mathrm{d}\zeta} - \frac{\Gamma_{\mathrm{J}}}{2\pi\mathrm{i}(\zeta-\zeta_{\mathrm{J}})}\right]. \tag{7.21}$$

The final term inside parentheses in (7.20) is called the **Routh correction**, and is necessary whenever the vortex motion in the physical plane is obtained from the velocity field in a conformally mapped plane.

The Routh correction arises because, although the streamlines in the vicinity of a vortex in the physical plane are concentric circles, these circles are distorted in the mapped plane by the mapping's own curvature. One interpretation of this distortion is that a vortex in the mapped plane does make a contribution—the Routh correction—to its own motion, in order to reconcile it with the physical plane.

In Result 7.2, we defined a velocity $\hat{w}_{-\mathrm{J}}$ in the mapped plane that appears to be the analog of the Kirchhoff velocity in the physical plane. Suppose we wish to create a transport equation for the mapped position of the vortex, ζ_{J}, rather than for its physical position. One is tempted to just write this as $\mathrm{d}\zeta_{\mathrm{J}}/\mathrm{d}t = \hat{w}_{-\mathrm{J}}$. However, it is important to remember that the mapped plane coordinate system is fixed to the body. Thus, in deriving a transport equation for ζ_{J}, we must also account for the time rate of change of the transformation from this body-fixed system to the inertial system. This transformation, given in general by Eq. (A.152), is re-written here for the vortex:

$$z_{\mathrm{J}}(t) = Z_{\mathrm{r}}(t) + \mathrm{e}^{\mathrm{i}\alpha(t)}\tilde{z}(\zeta_{\mathrm{J}}(t)), \tag{7.22}$$

with the time dependence of all quantities explicitly noted. If we differentiate this transformation with respect to time, we obtain a relationship between the motion of the vortex in the physical plane and that of its pre-image in the mapped plane:

$$\frac{\mathrm{d}z_{\mathrm{J}}}{\mathrm{d}t} = W_{\mathrm{r}}^{*} + \mathrm{i}\Omega\mathrm{e}^{\mathrm{i}\alpha}\tilde{z}(\zeta_{\mathrm{J}}) + \mathrm{e}^{\mathrm{i}\alpha}\tilde{z}'(\zeta_{\mathrm{J}})\frac{\mathrm{d}\zeta_{\mathrm{J}}}{\mathrm{d}t}. \tag{7.23}$$

The first two terms on the right-hand side constitute $W_{\mathrm{b}}^{*}(z_{\mathrm{J}})$, the rigid-body motion of a point attached to the body-fixed frame. The remaining term measures the velocity of the vortex relative to this frame, which, by the chain rule, invokes the rate of change of ζ_{J}. From this relationship, we can solve for $\mathrm{d}\zeta_{\mathrm{J}}/\mathrm{d}t$, substituting the transport equation in the physical plane (7.3) for $\mathrm{d}z_{\mathrm{J}}/\mathrm{d}t$, to obtain

$$\frac{\mathrm{d}\zeta_{\mathrm{J}}}{\mathrm{d}t} = \frac{1}{\tilde{z}'(\zeta_{\mathrm{J}})}\left(\tilde{w}_{-\mathrm{J}}^{*} - \tilde{W}_{\mathrm{r}}^{*} - \mathrm{i}\Omega\tilde{z}(\zeta_{\mathrm{J}})\right). \tag{7.24}$$

In deriving this form of the relationship, we have removed a common factor $\mathrm{e}^{\mathrm{i}\alpha}$ from all terms in the numerator and denominator of the right-hand side; the velocities are consequently expressed in body-fixed coordinates. To complete our transport equation, we use the Routh-corrected Kirchhoff velocity in Result 7.2 to express the right-hand side entirely in terms of mapped plane information:

Result 7.3: Vortex Transport in the Mapped Plane

The time evolution of a point vortex's position ζ_J in a plane ζ from which the physical plane has been conformally mapped, $z(\zeta)$, is described by

$$\frac{d\zeta_J}{dt} = \frac{1}{\tilde{z}'(\zeta_J)}\left[\frac{1}{\tilde{z}'(\zeta_J)^*}\left(\hat{w}^*_{-J} + \frac{\Gamma_J}{4\pi i}\frac{\tilde{z}''(\zeta_J)^*}{\tilde{z}'(\zeta_J)^*}\right) - \tilde{W}^*_r - i\Omega\tilde{z}(\zeta_J)\right], \quad (7.25)$$

where $\hat{w}_{-J}$ is defined in Eq. (7.21).

7.1.3 Vortex Clouds and Sheets

As discussed in Result 7.1, the Kirchhoff velocity of a vortex is defined from the full velocity field in the fluid, with the contribution of the vortex itself omitted. The constitution of this full velocity field was discussed at length in Chap. 4, and, as exemplified by Sect. 4.6, includes separable contributions from body motion, uniform flow, and the fluid vorticity. The body motion and uniform flow create regular (i.e., non-singular) velocity fields[1] that enter the Kirchhoff velocity without modification, so there is no need to discuss them here. It is to be understood, of course, that these contributions, as appear in the generic template (4.72), must be added to the vorticity's contribution, which we have called the vorticity-induced basis velocity field, $\boldsymbol{v}_V$ (or w_V in complex coordinates). Remember that this basis field contains both the direct influence of the vorticity as well as its modification due to the presence of the body. It is this basis velocity field that suffers from the singular behavior we have sought to regularize with the Kirchhoff velocity, which suggests that we now seek a Kirchhoff form of the vorticity-induced basis velocity field, naturally denoted by $\boldsymbol{v}_{V,-J}$ or $w_{V,-J}$. The overall Kirchhoff velocity will then be given by the usual template:

$$\boldsymbol{v}_{-J} = \boldsymbol{v}_{V,-J} + \boldsymbol{v}_b + \boldsymbol{v}_\infty, \quad (7.26)$$

or

$$w_{-J} = w_{V,-J} + w_b + w_\infty. \quad (7.27)$$

Our focus of this section is on the two-dimensional context, and we will discuss the Kirchhoff velocity in two forms of singular vorticity distributions.

[1] Except at the corners of bodies, where we often will regularize the flow comprehensively according to the Kutta condition.

Vortex Clouds If the fluid vorticity is composed of a set of discrete point vortices, which we will refer to as a **vortex cloud**, then the vorticity-induced basis field comes as a summation over the set, which makes it simple to remove a vortex' own contribution. For example, in vector form, one can write the Kirchhoff version of Eq. (4.117) as

$$\boldsymbol{v}_{\mathrm{v},-\mathrm{J}} = \sum_{\substack{\mathrm{K}=1\\ \mathrm{K}\neq\mathrm{J}}}^{N_{\mathrm{v}}} \boldsymbol{K}(\boldsymbol{x}_{\mathrm{J}} - \boldsymbol{x}_{\mathrm{K}}) \times \Gamma_{\mathrm{K}} \boldsymbol{e}_3 + \int_{\mathcal{S}_{\mathrm{b}}} \boldsymbol{K}(\boldsymbol{x}_{\mathrm{J}} - \boldsymbol{y}) \times \boldsymbol{\gamma}_{\mathrm{v}}(\boldsymbol{y})\, \mathrm{d}S(\boldsymbol{y}), \tag{7.28}$$

where we recall that the integral of the bound vortex sheet strength $\boldsymbol{\gamma}_{\mathrm{v}}$ over $\mathcal{S}_{\mathrm{b}}$ represents the reaction from the impenetrable body to the fluid vorticity. There are two important notes to make here:

- The summation omits the self-contribution of vortex J, as dictated by the definition of the Kirchhoff velocity.
- The contribution from the bound vortex sheet $\boldsymbol{\gamma}_{\mathrm{v}}$ does *not* omit the self-contribution. In other words, the part of this sheet that is a reaction specifically to vortex J makes a contribution to the motion of this vortex. This contribution is not singular as long as the vortex itself is not on the body surface.

This distinction is more clear when the full velocity is obtained by conformal mapping from the circle plane, Eq. (4.83), whence

$$w_{\mathrm{v},-\mathrm{J}} = \frac{1}{z'(\zeta_{\mathrm{J}})} \left(\sum_{\substack{\mathrm{K}=1\\ \mathrm{K}\neq\mathrm{J}}}^{N_{\mathrm{v}}} \frac{\Gamma_{\mathrm{K}}}{2\pi\mathrm{i}(\zeta_{\mathrm{J}} - \zeta_{\mathrm{K}})} - \sum_{\mathrm{K}=1}^{N_{\mathrm{v}}} \frac{\Gamma_{\mathrm{K}}}{2\pi\mathrm{i}(\zeta_{\mathrm{J}} - 1/\zeta_{\mathrm{K}}^{*})} - \frac{\Gamma_{\mathrm{J}}}{4\pi\mathrm{i}} \frac{z''(\zeta_{\mathrm{J}})}{z'(\zeta_{\mathrm{J}})} \right). \tag{7.29}$$

Here, we have omitted the contribution of vortex J from the sum over the vortices (the first sum), but not from the sum over the image vortices (the second sum). In this vorticity-induced basis of the Kirchhoff velocity, we have also included the Routh correction, the final term, since it is naturally grouped with this flow contributor.

Note 7.1.1: Efficient Calculation of Vortex Cloud Transport

In the calculation of transport of elements in a vortex cloud, it is useful to remember that $\boldsymbol{K}$ is anti-symmetric, as we noted in Eq. (3.78). This means that, once we have calculated the velocity kernel $\boldsymbol{K}(\boldsymbol{x}_{\mathrm{J}} - \boldsymbol{x}_{\mathrm{K}})$ to compute the influence of vortex K on vortex J, we have also automatically obtained the reciprocal influence of J on K, $\boldsymbol{K}(\boldsymbol{x}_{\mathrm{K}} - \boldsymbol{x}_{\mathrm{J}}) = -\boldsymbol{K}(\boldsymbol{x}_{\mathrm{J}} - \boldsymbol{x}_{\mathrm{K}})$. This property can be easily exploited to reduce the number of calculations in a cloud's transport by nearly half.

Free Vortex Sheets In Sect. 3.6.1, we discussed the transport of a free vortex sheet. In particular, we showed that the only way to ensure that the pressure across the vortex sheet is continuous is to insist that each point on the sheet—described uniquely in two dimensions by a single material coordinate ξ—move with the mean of the local velocities in the fluid on either side:

$$\frac{\partial \boldsymbol{X}(\xi,t)}{\partial t} = \frac{1}{2}\left(\boldsymbol{v}^{+}(\xi,t) + \boldsymbol{v}^{-}(\xi,t)\right), \tag{3.190}$$

or its complex analog

$$\frac{\partial Z(\xi,t)}{\partial t} = \frac{1}{2}\left(w^{+}(\xi,t)^{*} + w^{-}(\xi,t)^{*}\right). \tag{7.30}$$

At the time we developed this transport equation for a sheet, we didn't yet have the specific forms of $\boldsymbol{v}^{\pm}$. However, those forms were developed in Sect. 4.3.3, and we make use of them now to write this transport equation in a more practical form. Suppose that the free vortex sheet is denoted by $\mathcal{S}$ and its strength distribution by γ. This is not to be confused with the bound vortex sheet on the body surface $\mathcal{S}_{\mathrm{b}}$ that reacts to the flow contributors; indeed, as with any fluid vorticity, the free vortex sheet has its own such reactive basis sheet, whose strength is denoted as usual by γ_{v}. Thus,

Result 7.4: Evolution Equation for a Free Vortex Sheet in Two Dimensions

The evolution of a free vortex sheet, $\mathcal{S}$—whose configuration and strength are described parametrically by $\boldsymbol{X}(\xi,t)$ and $\gamma(\xi)$, respectively—is given by the solution of

$$\frac{\partial \boldsymbol{X}(\xi,t)}{\partial t} = \int_{\mathcal{S}} \boldsymbol{K}\left(\boldsymbol{X}(\xi,t) - \boldsymbol{X}(\hat{\xi},t)\right)\gamma(\hat{\xi})\,\mathrm{d}s(\hat{\xi}) \times \boldsymbol{e}_3 + \boldsymbol{v}_{-\mathrm{v}}(\boldsymbol{X}(\xi,t)), \tag{7.31}$$

where $\boldsymbol{K}$ takes its two-dimensional form (3.71) and $\boldsymbol{v}_{-\mathrm{v}}$ denotes the velocity field *except* that induced directly by the free vortex sheet (i.e., due to the presence of a stationary or moving body, a uniform flow, and other fluid vorticity):

$$\boldsymbol{v}_{-\mathrm{v}}(\boldsymbol{X}(\xi,t)) := \int_{\mathcal{S}_{\mathrm{b}}} \boldsymbol{K}\left(\boldsymbol{X}(\xi,t) - \boldsymbol{y}\right)\gamma_{\mathrm{v}}(\boldsymbol{y})\,\mathrm{d}S(\boldsymbol{y}) \times \boldsymbol{e}_3 + \boldsymbol{v}_{\mathrm{b}}(\boldsymbol{X}(\xi,t)) + \boldsymbol{v}_{\infty}(\boldsymbol{X}(\xi,t)), \tag{7.32}$$

where γ_{v} denotes the strength of the basis vortex sheet on the body surface, $\mathcal{S}_{\mathrm{b}}$, arising in reaction to the free vortex sheet $\mathcal{S}$.

In Note 3.6.1 we showed that a two-dimensional vortex sheet can be conveniently parameterized by the local circulation, Γ, when it varies monotonically along the sheet (say, between 0 and Γ_S). This provides a useful alternative form of Eq. (7.31), attributed to Birkhoff [7] and Rott [62]:

Result 7.5: The Birkhoff–Rott Equation

The evolution of a two-dimensional free vortex sheet, S, described parametrically by $\boldsymbol{X}(\Gamma, t)$, where $\Gamma \in [0, \Gamma_S]$ is the local circulation, is given by the solution of the **Birkhoff–Rott equation**:

$$\frac{\partial \boldsymbol{X}(\Gamma,t)}{\partial t} = \int_0^{\Gamma_S} \boldsymbol{K}\left(\boldsymbol{X}(\Gamma,t) - \boldsymbol{X}(\hat{\Gamma},t)\right) \mathrm{d}\hat{\Gamma} \times \boldsymbol{e}_3 + \boldsymbol{v}_{-\mathrm{v}}(\boldsymbol{X}(\Gamma,t)), \quad (7.33)$$

where $\boldsymbol{v}_{-\mathrm{v}}$ has the same meaning as in Result 7.4, given by Eq. (7.32).

In complex notation, where the sheet's configuration is described by $Z(\Gamma, t)$, the evolution equation is given by

$$\frac{\partial Z(\Gamma,t)}{\partial t} = -\frac{1}{2\pi i}\int_0^{\Gamma_S} \frac{\mathrm{d}\hat{\Gamma}}{Z(\Gamma,t)^* - Z(\hat{\Gamma},t)^*} + w_{-\mathrm{v}}(Z(\Gamma,t))^*, \quad (7.34)$$

where $w_{-\mathrm{v}}$ denotes the complex equivalent of Eq. (7.32): the velocity field containing the influences of a body, uniform flow, and other fluid vorticity.

In particular, the Birkhoff–Rott form removes the need to keep track of the arc length along the sheet. This is particularly helpful in applications, in which the sheet is generally discretized by a finite number of material control points that are individually tracked.

7.1.4 Variable-Strength Point Vortices

Thus far in this chapter, we have remained steadfast in our enforcement of Helmholtz' third theorem—or, equivalently in two dimensions, Kelvin's circulation theorem (3.40)—by insisting that the strength of any point vortex or constituent element of a vortex sheet remain constant in time. However, at the end of our discussion in Sect. 7.1.1, we got a glimpse of how such theorems might be allowed to break: if we allow the strength of a point vortex to vary in time, then it should no longer move with the fluid (i.e., at the Kirchhoff velocity), but in some modified fashion. In this section, we will discuss a few choices for how we might modify the transport for these so-called **variable-strength vortices**. To organize the discussion, let us write the modified transport equation of vortex J as

$$\frac{\mathrm{d}\boldsymbol{x}_{\mathrm{J}}}{\mathrm{d}t} = \boldsymbol{v}_{-\mathrm{J}} + \Delta\boldsymbol{v}_{\mathrm{J}}, \tag{7.35}$$

where $\Delta\boldsymbol{v}_{\mathrm{J}}$ is an additional velocity of the vortex. In this section, we propose formulations for $\Delta\boldsymbol{v}_{\mathrm{J}}$.

Before we discuss this transport, let us give some motivation for proposing such vortices in the first place. In many flows in aerodynamics or hydrodynamics, vorticity in the fluid is naturally prone to roll up into distinct coherent structures. These structures tend to 'attract' vorticity-bearing elements toward them, condensing the fluid circulation into localized regions. When the vorticity is represented as a sheet, then the sheet stretches indefinitely as it rolls up into such a structure. Thus, from a modeling perspective, and to ensure computational efficiency, there is a compelling reason to substitute this indefinite roll-up process with a single element whose strength can vary in time to approximate this process. It will come as no surprise that, in both evolution equations we describe here, $\Delta\boldsymbol{v}_{\mathrm{J}}$ is related (linearly) to the rate of change of the vortex's strength, $\dot{\Gamma}_{\mathrm{J}}$, which we will often refer to as the **vorticity flux**.

The Brown–Michael Transport Equation The first modified form of transport for variable-strength point vortices was originally proposed in a model of the leading-edge vortices that develop over delta wing aircraft [22, 10]. As we saw in Sects. 6.3.2 and 6.3.3, a point vortex of variable strength gives rise to a spurious jump in pressure across the branch cut that joins a pair of vortices, or a single vortex to a body as in Result 6.8. The principle of the so-called Brown–Michael transport equation is that a vortex of variable strength should move in such a way that its modified motion cancels the force from this spurious pressure jump. This principle can be almost immediately enacted from (6.77), simply by setting the required external force on the vortex/branch cut to zero. As we showed in Sect. 7.1.1, the velocity $\boldsymbol{v}_{-\mathrm{v}}$ in (6.77) is the Kirchhoff velocity. Thus, substituting (7.35) into this expression for external force, we seek $\Delta\boldsymbol{v}_{\mathrm{J}}$ such that

$$\boldsymbol{f}^{\mathrm{ext}} = \rho\left[\dot{\Gamma}_{\mathrm{J}}(\boldsymbol{x}_{\mathrm{J}} - \boldsymbol{x}_{I}) + \Gamma_{\mathrm{J}}\Delta\boldsymbol{v}_{\mathrm{J}}\right] \times \boldsymbol{e}_3 = 0. \tag{7.36}$$

The only lingering issue is the choice of arbitrary intersection point, $\boldsymbol{x}_I$. For a general body, there isn't any single choice that is more sensible than another; the branch cut can intersect the body at any (possibly moving) point without affecting the force. However, for an infinitely thin plate, we reasoned in Note 3.6.5 that the branch cut should intersect the plate at one of its edges, as in the right panel in Fig. 3.14, in order to avoid awkward jumps in local circulation and pressure along the plate's length. But which edge? Typically, the choice will be natural, because, in the model of a flow, a variable-strength point vortex will be introduced near a particular edge, and this association will persist over time.

Thus, the Brown–Michael transport equation comprises an additional velocity

$$\Delta\boldsymbol{v}_{\mathrm{J}} = -\frac{\dot{\Gamma}_{\mathrm{J}}}{\Gamma_{\mathrm{J}}}(\boldsymbol{x}_{\mathrm{J}} - \boldsymbol{x}_{I}), \tag{7.37}$$

where $\boldsymbol{x}_I$ is an intersection point on the body, typically the edge of a plate from which the vortex emerges.

The transport of vortices by the Brown–Michael equation has the benefit of ensuring that the two means of calculating force on the body discussed in Chap. 6—by integrating the surface traction and by determining the rate of change of linear impulse in the fluid—are in agreement. But it is important to note that, although we have cancelled the spurious force on the vortex and branch cut by choosing the vortex's velocity by (7.37), we have not cancelled the spurious moment (6.78). By substituting the former equation into the latter, it is easy to show that the unbalanced external moment would be $\frac{1}{2}\rho\dot{\Gamma}_J|\boldsymbol{x}_J - \boldsymbol{x}_I|^2$.

Impulse Matching Transport Equation As we discussed above, variable-strength point vortices are generally most useful as a means of modeling the continuous flux of vorticity from a body into a nearby vortical region. In a real flow, this flux naturally stops once the vortical region has 'shed' from the body (e.g., by moving sufficiently far away), perhaps to be replaced by a new nascent vortical region, or, in the case of a wake, by the sufficient passage of time without new disturbances. One would expect the point vortex to model this behavior, wherein its type switches from one of variable strength to one of constant strength; we could specify some criterion that determines such a switch in behavior.

It is important to note that, for a general choice of criterion, the vorticity flux, $\dot{\Gamma}_J$—and therefore, $\Delta\boldsymbol{v}_J$—will suddenly jump to zero during such a switching event, giving rise to a sudden alteration of course for the vortex. One of the drawbacks of the Brown–Michael equation is that, unless our switching criterion is simply that $\dot{\Gamma}_J$ has naturally become zero, the force will also change discontinuously during a switching event.

In order to avoid this undesirable outcome, we propose another principle: choose $\Delta\boldsymbol{v}_J$ so that the vortex's modified motion cancels the effect of its variable strength on the force, rendering the force impervious to sudden changes in vorticity flux. By offsetting the contribution of $\dot{\Gamma}_J$ to force by modifying the vortex motion from its natural Kirchhoff velocity, the contribution that a vortex ultimately makes to force—as measured through the rate of change of impulse—is equivalent to that of a constant-strength point vortex with the same instantaneous position and strength: in other words, we are 'matching' the linear impulse and its rate of change with that of this imagined 'surrogate' point vortex of constant strength. For this reason, the principle is called **impulse matching** [72]. At any instant that we choose to freeze its strength, a variable-strength vortex naturally adopts the form of its surrogate, with a small change of course, but importantly, without producing a jump in force on the body.

Of course, this approach comes at the cost of a lingering spurious force on the vortex and its branch cut: the force computed from the integral of surface traction will not, in general, be equal to the force computed from rate of change of linear impulse. But, as we noted above, even the Brown–Michael equation cannot eliminate the spurious moment in the fluid, so the cost of failing to eliminate the spurious force is not as dear as it might seem. We will show below that the spurious force is relatively small in magnitude.

We will enact this principle of canceling the vorticity flux's influence on force by relying on the force's calculation from rate of change of linear impulse. To get the basic idea, let us consider the general expression of this force calculation, Eq. (6.107). The terms inside parentheses for each vortex element J comprise both the influence of variable strength $\dot{\Gamma}_{\mathrm{J}}$ as well as that of vortex motion, through $\dot{\tilde{\mathbf{P}}}^{\mathrm{v}}_{\mathrm{J}}$. We seek to modify this motion, via choice of $\Delta \boldsymbol{v}_{\mathrm{J}}$, to cancel the term proportional to $\dot{\Gamma}_{\mathrm{J}}$.

Our derivation of the rate of change of impulse in Sect. 6.6 is very helpful here, since it brings out the precise relationship between this rate of change and the vortex's transport. Specifically, for a point vortex, we can rely on Eq. (6.174), which incorporates all of the relevant parts of the rate of change of linear impulse. Let us first note that we can write the complex version of the modified transport equation (7.35) for point vortex J as

$$\frac{\mathrm{d}z_{\mathrm{J}}}{\mathrm{d}t} = w^{*}_{-\mathrm{J}} + \Delta w^{*}_{\mathrm{J}}, \tag{7.38}$$

where Δw_{J} is the complex conjugate modified velocity. We note that it is actually the velocity relative to the body frame, $\dot{\tilde{z}}_{\mathrm{J}} = (\dot{z}_{\mathrm{J}} - W^{*}_{\mathrm{b}}(z_{\mathrm{J}}))\mathrm{e}^{-\mathrm{i}\alpha}$, that enters the rate of change of impulse, but the modification Δw_{J} we seek affects the motion in either reference frame, modulo a change in coordinate system. For our present purposes, we consider each vortex's individual contribution to force, and only the parts due to $\dot{\Gamma}_{\mathrm{J}}$ and to the modification of the vortex's motion. Since we insist that these two parts offset each other, we set their sum to zero,

$$\dot{\Gamma}_{\mathrm{J}}\tilde{P}^{\mathrm{v}}_{\mathrm{J}} + \Gamma_{\mathrm{J}}\mathrm{Im}\left(\tilde{w}^{(1)}_{\infty}(\tilde{z}_{\mathrm{J}})\Delta\tilde{w}^{*}_{\mathrm{J}}\right) + \mathrm{i}\Gamma_{\mathrm{J}}\mathrm{Im}\left(\tilde{w}^{(2)}_{\infty}(\tilde{z}_{\mathrm{J}})\Delta\tilde{w}^{*}_{\mathrm{J}}\right) = 0, \tag{7.39}$$

and seek to solve this equation for $\Delta\tilde{w}^{*}_{\mathrm{J}}$. It can be confirmed that this solution is

$$\Delta\tilde{w}^{*}_{\mathrm{J}} = -\frac{\dot{\Gamma}_{\mathrm{J}}}{\Gamma_{\mathrm{J}}}\frac{\left(\tilde{w}^{(2)}_{\infty}(\tilde{z}_{\mathrm{J}})^{*}\,\mathrm{Re}(\tilde{P}^{\mathrm{v}}_{\mathrm{J}}) - \tilde{w}^{(1)}_{\infty}(\tilde{z}_{\mathrm{J}})^{*}\,\mathrm{Im}(\tilde{P}^{\mathrm{v}}_{\mathrm{J}})\right)}{\mathrm{Im}\left(\tilde{w}^{(1)}_{\infty}(\tilde{z}_{\mathrm{J}})\tilde{w}^{(2)}_{\infty}(\tilde{z}_{\mathrm{J}})^{*}\right)}. \tag{7.40}$$

For reference, and for additional insight, let us also write this modified velocity in vector form, with some help from the equivalences in Sect. A.2.1:

$$\Delta\boldsymbol{v}_{\mathrm{J}} = -\frac{\dot{\Gamma}_{\mathrm{J}}}{\Gamma_{\mathrm{J}}}\frac{\left(\boldsymbol{v}^{(2)}_{\infty}(\boldsymbol{x}_{\mathrm{J}})\tilde{\boldsymbol{e}}_1 - \boldsymbol{v}^{(1)}_{\infty}(\boldsymbol{x}_{\mathrm{J}})\tilde{\boldsymbol{e}}_2\right)\cdot\boldsymbol{P}^{\mathrm{v}}_{\mathrm{J}}}{\left(\boldsymbol{v}^{(1)}_{\infty}(\boldsymbol{x}_{\mathrm{J}})\times\boldsymbol{v}^{(2)}_{\infty}(\boldsymbol{x}_{\mathrm{J}})\right)\cdot\boldsymbol{e}_3}. \tag{7.41}$$

When the power-series mapping (A.153) has been used to obtain the solution in the physical plane, the right-hand side of (7.40) can be determined analytically, in terms of the circle-plane position of the vortex, ζ_{J}. Using the expressions (4.88) for the basis velocity fields and (6.96) for the vorticity-induced unit impulse, we arrive at

$$\Delta\tilde{w}^{*}_{\mathrm{J}} = -\frac{\dot{\Gamma}_{\mathrm{J}}}{\Gamma_{\mathrm{J}}}\tilde{z}'(\zeta_{\mathrm{J}})\frac{\zeta_{\mathrm{J}} - 1/\zeta^{*}_{\mathrm{J}}}{1 + 1/|\zeta_{\mathrm{J}}|^{2}}. \tag{7.42}$$

This form allows us to make a general observation about the directionality of the modified velocity. As a vector, the numerator, $\zeta_J - 1/\zeta_J^*$, is directed toward the vortex J from its image—along a radial line, normal to the unit circle, C_c; the denominator is real-valued and only modifies the length of the vector. As we discuss in Sect. A.2.3 in the Appendix, the Jacobian $\tilde{z}'$ just rotates the vector into the physical plane, preserving its orientation relative to the mapped coordinate lines. Thus, ***the modification to vortex transport described by the impulse matching principle is directed along the mapped radial coordinate lines***. The right panel in Fig. A.7 depicts an example of such mapped lines. The factor $-\dot{\Gamma}_J/\Gamma_J$ indicates that the motion tends toward the body when the vortex strength is increasing in magnitude, and away from the body when decreasing.

When the body is an infinitely-thin flat plate of length c—a geometry we will discuss at great length in Chap. 8—then the mapping is described by $\tilde{z}(\zeta) = \frac{1}{4}c(\zeta + 1/\zeta)$, and its inverse by $\zeta(\tilde{z}) = 2\tilde{z}/c + (2\tilde{z}/c + 1)^{1/2}(2\tilde{z}/c - 1)^{1/2}$. With some manipulation of this mapping, one can show [72] that the modified velocity (7.42), when finally expressed in vector form, becomes

$$\Delta \boldsymbol{v}_J = -\frac{\dot{\Gamma}_J}{\Gamma_J}\left(\frac{(\boldsymbol{x}_J - \boldsymbol{x}_T)|\boldsymbol{x}_J - \boldsymbol{x}_L| + (\boldsymbol{x}_J - \boldsymbol{x}_L)|\boldsymbol{x}_J - \boldsymbol{x}_T|}{|\boldsymbol{x}_J - \boldsymbol{x}_L| + |\boldsymbol{x}_J - \boldsymbol{x}_T|}\right), \tag{7.43}$$

where $\boldsymbol{x}_L$ and $\boldsymbol{x}_T$ are the positions of the leading and trailing edge of the plate, respectively. Thus, the vortex's modified velocity is effectively determined by a weighted average of its position relative to these edges. This should be contrasted with the modified velocity from the Brown–Michael principle (7.37), in which, for a plate, we choose $\boldsymbol{x}_I$ to be the edge from which the vortex emerges. The difference between these velocities is

$$\Delta \boldsymbol{v}_J^{\mathrm{IM}} - \Delta \boldsymbol{v}_J^{\mathrm{BM}} = \frac{\dot{\Gamma}_J}{\Gamma_J}\left(\frac{|\boldsymbol{x}_J - \boldsymbol{x}_L|}{|\boldsymbol{x}_J - \boldsymbol{x}_L| + |\boldsymbol{x}_J - \boldsymbol{x}_T|}\right) c\boldsymbol{\tau}, \tag{7.44}$$

where IM and BM designate the impulse matching and Brown–Michael forms, respectively, and $\boldsymbol{\tau} \equiv \tilde{\boldsymbol{e}}_1$ is the unit tangent along the plate, directed from the leading edge toward the trailing edge. The lingering spurious force on the branch cut and vortex is proportional (and orthogonal) to this difference in velocities, $\boldsymbol{f}^{\mathrm{ext}} = \rho\Gamma_J(\Delta \boldsymbol{v}_J^{\mathrm{IM}} - \Delta \boldsymbol{v}_J^{\mathrm{BM}}) \times \boldsymbol{e}_3$. The difference is negligibly small when the vortex is near its originating edge, $\boldsymbol{x}_J \to \boldsymbol{x}_L$. As it moves far from the plate, the difference becomes negligible by virtue of the tendency for Γ_J to become invariant under such circumstances.

Generalization of Impulse Matching to Vortex–Vortex Transfers Our derivation of the impulse matching principle above presumed that we were transferring circulation directly from the edge of the body into the point vortex. This process omits the free vortex sheet that would otherwise be rooted at the releasing edge. But some of this sheet's dynamics cannot be described by a single point vortex. Thus, rather than suppress the sheet's formation altogether, we might want to allow at least some portion of it to form, but still seek to avoid the complexity as it rolls up into coherent structures. In such circumstances, the circulation we wish to transfer into the point vortex is not bound to the body, but rather, is carried by the free vortex sheet that already resides in the fluid.

We can extend the impulse matching principle to this vortex–vortex transfer [17]. Let us return to the force contributed by point vortices in Eq. (6.174). Now, instead of focusing only on one vortex, we will consider two. One of these, which we will refer to as the ***target*** and label with a subscript t, is our intended point vortex. The other we will refer to as the ***source*** and label it with subscript s. This represents the origin of the circulation in the transfer process: the free end of the vortex sheet. The rate of change of these two elements' circulation is equal and opposite, $\dot{\Gamma}_t$ and $-\dot{\Gamma}_t$, respectively. Thus, both of these vortex elements are invoked in the contribution from time-varying circulation to force. But it is important to note that only the target's motion will be modified: the free end of the sheet is progressively annihilated as its circulation is transferred, and thus, there remains no need to track its motion. With these observations, it is easy to show that Eq. (6.174) results in a modified velocity for the target vortex that is only slightly different from Eq. (7.40):

$$\Delta\tilde{w}_t^* = -\frac{\dot{\Gamma}_t}{\Gamma_t}\frac{\left(\tilde{w}_\infty^{(2)}(\tilde{z}_t)^*\,\mathrm{Re}(\tilde{P}_t^{\mathrm{v}} - \tilde{P}_s^{\mathrm{v}}) - \tilde{w}_\infty^{(1)}(\tilde{z}_t)^*\,\mathrm{Im}(\tilde{P}_t^{\mathrm{v}} - \tilde{P}_s^{\mathrm{v}})\right)}{\mathrm{Im}\left(\tilde{w}_\infty^{(1)}(\tilde{z}_t)\tilde{w}_\infty^{(2)}(\tilde{z}_t)^*\right)}. \tag{7.45}$$

7.1.5 Blob Regularization

One of the challenges of conducting numerical simulations with convecting vortex elements is that they induce on each other a velocity that becomes unbounded as the distance between them tends to zero. If there were only two such vortices in an infinite medium free of boundaries, then their singular velocities would not be an issue because the distance between these vortices would never change. However, in practical flows with large sets of vortices, the vortex elements frequently do approach one another. This is particularly true in a free vortex sheet, whose numerical representation is a discrete set of point vortices. The constituent vortex elements in this sheet must maintain a tenuous balance and are prone to unphysical instability because of their interactions via the singular vortex kernel, $\boldsymbol{K}(\boldsymbol{x})$.

The general idea of regularization is to modify the velocity field induced by a point vortex to smoothly achieve a finite value as the center of the element is approached, but leave the behavior outside of a core region untouched. Another way to view this task is to replace the element's vorticity field—a Dirac delta function, δ, as in (3.120)—with a smooth (i.e., differentiable) representation, referred to as a **vortex blob** [12, 6], with an effective radius, ε. In other words, we replace Eq. (3.66) with

$$-\nabla^2 G_\varepsilon = \delta_\varepsilon(\boldsymbol{x}), \tag{7.46}$$

where δ_ε, the **blob function**, is a smooth function which tends toward δ as $\varepsilon \to 0$. The function's integral must equal unity, just like the Dirac delta function,

$$\int \delta_\varepsilon(\boldsymbol{x})\,\mathrm{d}V(\boldsymbol{x}) = 1. \tag{7.47}$$

One notable example of such a function is a Gaussian,

$$\delta_\varepsilon(\boldsymbol{x}) = \frac{1}{(\pi\varepsilon^2)^{n_\mathrm{d}/2}} \mathrm{e}^{-|\boldsymbol{x}|^2/\varepsilon^2}, \tag{7.48}$$

where n_d, as usual, is the number of spatial dimensions, 2 or 3.

The corresponding regularized Green's function, G_ε, and velocity kernel, $\boldsymbol{K}_\varepsilon = \nabla G_\varepsilon$, can be computed analytically in terms of δ_ε. For clarity of notation, we will focus only on two-dimensional regularization here. But first, let us make a few general observations: The regularized Green's function, like its singular counterpart, is radially symmetric in either dimension; in other words, it only depends on distance r from the center of the element. This means that, for a vortex centered at the origin—so that $r = |\boldsymbol{x}|$—we can write the relationship between G_ε and $\boldsymbol{K}_\varepsilon$ more specifically (with some help from index notation) as

$$K_{\varepsilon,i} = \frac{\partial G_\varepsilon}{\partial x_i} = \frac{\mathrm{d}G_\varepsilon}{\mathrm{d}r}\frac{\partial r}{\partial x_i} = \frac{x_i}{r}\frac{\mathrm{d}G_\varepsilon}{\mathrm{d}r}, \tag{7.49}$$

where we have used the fact that, in any dimension, $\partial r/\partial x_i = x_i/r$. In vector form,

$$\boldsymbol{K}_\varepsilon(\boldsymbol{x}) = \frac{\boldsymbol{x}}{r}\frac{\mathrm{d}G_\varepsilon}{\mathrm{d}r}. \tag{7.50}$$

Let us note that the blob function is an ε-scaled version of a certain dimensionless radially-symmetric template, $\tilde{\delta}(\rho)$, where $\rho := |\boldsymbol{x}|/\varepsilon$ is a measure of distance in units of blob radii:

$$\delta_\varepsilon(\boldsymbol{x}) = \frac{1}{\varepsilon^2}\tilde{\delta}(|\boldsymbol{x}|/\varepsilon). \tag{7.51}$$

For example, in the case of the Gaussian, $\tilde{\delta}(\rho) = \mathrm{e}^{-\rho^2}/\pi$. The integral constraint (7.47) can correspondingly be written as a polar integral of $\tilde{\delta}$ as

$$2\pi \int_0^\infty \rho' \tilde{\delta}(\rho')\,\mathrm{d}\rho' = 1. \tag{7.52}$$

The regularized Green's function is also an ε-scaled version of a radially-symmetric template, $\tilde{G}(\rho)$:

$$G_\varepsilon(\boldsymbol{x}) = \tilde{G}(|\boldsymbol{x}|/\varepsilon); \tag{7.53}$$

the Green's function in two dimensional flow has no units, so this relationship lacks the external factor involving ε. This template function satisfies the radially-symmetric Poisson equation

$$-\frac{1}{\rho}\frac{\mathrm{d}}{\mathrm{d}\rho}\left(\rho\frac{\mathrm{d}\tilde{G}}{\mathrm{d}\rho}\right) = \tilde{\delta}(\rho). \tag{7.54}$$

If we multiply both sides by ρ, integrate once, and then divide by ρ, we get

$$\frac{\mathrm{d}\tilde{G}}{\mathrm{d}\rho} = -\frac{1}{2\pi\rho}q(\rho), \tag{7.55}$$

where we have defined the function $q(\rho)$ as

$$q(\rho) := 2\pi\int_0^{\rho}\rho'\tilde{\delta}(\rho')\,\mathrm{d}\rho'. \tag{7.56}$$

In fact, this function q has the important role of regularizing the singular velocity kernel, since by (7.50), the regularized velocity kernel is proportional to the radial derivative of the regularized Green's function, and, after accounting for the rescaling $\rho \equiv r/\varepsilon = |\boldsymbol{x}|/\varepsilon$, it is easy to show that

$$\boldsymbol{K}_{\varepsilon}(\boldsymbol{x}) = -\frac{\boldsymbol{x}}{2\pi|\boldsymbol{x}|^2}q(|\boldsymbol{x}|/\varepsilon). \tag{7.57}$$

Similarly, in complex form, the regularized kernel follows by multiplying the singular velocity kernel $-1/(2\pi z)$ by $q(|z|/\varepsilon)$.

From the integral constraint (7.52) above, it is clear that $q(\rho) \to 1$ as $\rho \to \infty$. Thus, at distances farther than a few blob radii from the center of the singularity, this regularized kernel approaches the singular form (3.71). At the other extreme, a Taylor series expansion of (7.56) near $\rho = 0$ reveals that $q(\rho) = O(\rho^2)$, so the modification of the velocity kernel in (7.57) ensures that the velocity is smooth and finite as the center is approached.

The regularized Green's function then follows by integrating (7.55) once, upon which

$$\tilde{G}(\rho) = -\frac{1}{2\pi}\int\frac{q(\rho)}{\rho}\mathrm{d}\rho + C, \tag{7.58}$$

where the integration constant is chosen so that $\tilde{G}(\rho) \to -\frac{1}{2\pi}\log\rho$ as $\rho \to \infty$.

We've already given one example of the blob function: the Gaussian,

$$\tilde{\delta}(\rho) = \frac{1}{\pi}\mathrm{e}^{-\rho^2}. \tag{7.59}$$

This blob function is analogous to the Lamb–Oseen vortex, the vorticity distribution that evolves from an initially singular vortex in a viscous fluid. However, in this context of regularizing the singular vortex–vortex interactions, the radius of this blob function must generally be held constant rather than allowed to spread in time as a model for viscous diffusion [28, 61].

The corresponding velocity modifier, $q(\rho)$, of the Gaussian is easily obtained by evaluating the integral (7.56),

$$q(\rho) = 1 - e^{-\rho^2}. \tag{7.60}$$

Its regularized Green's function is

$$\tilde{G}(\rho) = -\frac{1}{2\pi}\left(\log\rho - \frac{1}{2}\mathrm{Ei}(-\rho^2)\right), \tag{7.61}$$

where Ei is the exponential integral. This tends toward a finite non-zero value as $\rho \to 0$, as desired.

Another common example is the algebraic form [41],

$$\tilde{\delta}(\rho) = \frac{1}{\pi(\rho^2+1)^2}, \tag{7.62}$$

whence

$$q(\rho) = \frac{\rho^2}{\rho^2+1}, \tag{7.63}$$

and

$$\tilde{G}(\rho) = -\frac{1}{2\pi}\log(\rho^2+1)^{1/2}. \tag{7.64}$$

Note that this blob function renders the velocity kernel into a form

$$\boldsymbol{K}_\varepsilon(\boldsymbol{x}) = -\frac{\boldsymbol{x}}{2\pi(|\boldsymbol{x}|^2+\varepsilon^2)}. \tag{7.65}$$

7.2 Vortex Filaments

We have discussed several types of three-dimensional vortex structures in this book: vortex filaments, vortex sheets, and point vortices. Each of these entities, as a material element, has a transport that follows from Helmholtz' theorems. In this section we will focus on the most subtle of these, the vortex filament. In Sect. 3.3.2, we presented the basic characteristics of a vortex filament: a vortex tube of narrow cross-section in which vorticity is concentrated. As such, Helmholtz' second theorem requires that the filament is a material structure, moving with the fluid. In many respects, a vortex filament may be regarded as infinitely narrow—effectively, a line vortex. But this interpretation is problematic when we consider the filament's own transport: portions of a curved filament induce velocity on one another in a manner that is logarithmically divergent. We only mentioned this caveat at the end of Sect. 3.3.2; here, let us provide a little more mathematical intuition for the root of the problem, and then review some of the ways in which it has been overcome.

Let us first recall the velocity field (3.122) induced by a filament, C, of strength Γ, described parametrically by a smooth space curve $\boldsymbol{X}(s)$, where s is the arc length. This velocity field is given at $\boldsymbol{x} \notin C$ by

$$\boldsymbol{v}(\boldsymbol{x}) = -\frac{\Gamma}{4\pi}\int_C \frac{(\boldsymbol{x}-\boldsymbol{X}(\sigma))}{|\boldsymbol{x}-\boldsymbol{X}(\sigma)|^3} \times \frac{\partial \boldsymbol{X}}{\partial \sigma}\,\mathrm{d}\sigma. \tag{7.66}$$

What happens when the evaluation point approaches a point on the axis of the idealized filament, $\boldsymbol{x} \to \boldsymbol{X}(s)$? To see this, let us split the integral over C into a portion near the evaluation point s and another portion over the remainder of the filament. We do not have to be precise about how we carry out this splitting. We need only observe that there is no challenge posed by the latter portion: its contribution to the velocity is bounded and $O(1)$. For the portion surrounding s, we can expand the integrand about this point, noting that nearby positions and tangents are described by

$$\boldsymbol{X}(\sigma) - \boldsymbol{X}(s) = (\sigma - s)\frac{\partial \boldsymbol{X}}{\partial s} + \frac{1}{2}(\sigma - s)^2\frac{\partial^2 \boldsymbol{X}}{\partial s^2} + O(\sigma - s)^3 \tag{7.67}$$

and

$$\frac{\partial \boldsymbol{X}}{\partial \sigma} = \frac{\partial \boldsymbol{X}}{\partial s} + (\sigma - s)\frac{\partial^2 \boldsymbol{X}}{\partial s^2} + O(\sigma - s)^2, \tag{7.68}$$

respectively. The leading terms of these expansions make no contribution when combined in the cross product, since they are both directed along the filament axis. And this would be the end of the story if the filament were straight. However, when the second-order terms—arising from local curvature, $\kappa = |\partial^2 \boldsymbol{X}/\partial s^2|$, of the filament—are included, there is a net contribution to the velocity field at $\boldsymbol{X}(s)$, given by

$$\boldsymbol{v}(\boldsymbol{X}(s)) = \frac{\Gamma}{8\pi}\frac{\partial \boldsymbol{X}}{\partial s} \times \frac{\partial^2 \boldsymbol{X}}{\partial s^2}\int \frac{\mathrm{d}\sigma}{|s-\sigma|} + O(1). \tag{7.69}$$

The integral in this expression results in a logarithmic divergence. Indeed, ***the self-induced velocity of a curved filament of infinitesimal cross-sectional area is unbounded, irrespective of the degree of curvature***.

At first glance, this behavior of the self-induced velocity is frustrating. But, on closer inspection, it becomes clear that we have missed an important detail: we have accounted for one length scale of the filament—the local radius of curvature, $1/\kappa$—but ignored its relationship with another length scale—the local radius of the filament core, a—that we have happily neglected thus far. For computing interactions between nearby elements of the curved filament, it should be no surprise that this relationship is essential. In short, we cannot calculate the motion of a filament without considering the internal structure of the filament cross-section.

So let us be a bit more careful. We will still restrict our attention to thin filaments, that is, $\kappa a \ll 1$, but now include the detailed influence of the core structure on the filament's self-induced motion. Moore and Saffman [54] performed a detailed analysis of a thin filament, determining the filament velocity that produced an equilibrium of the forces applied at the periphery of the filament core by the external fluid and from within the core. By retaining only the leading-order terms in κa, they found that the obtained velocity could be used to exactly reproduce some known results: the speed of self-propagation of a circular filament (i.e., a thin-cored vortex ring) with arbitrary vorticity distribution, and the dispersion relation for an imposed helical disturbance to a straight filament.

The filament velocity derived by Moore and Saffman [54] is tractable but complicated. Fortunately, they showed that there is a much simpler method of comparable accuracy for calculating the velocity. The idea, first used by Hama [29], is to cut off the integration over the idealized filamentary curve so that it excludes a portion $|\sigma - s| < \phi a$ surrounding the evaluation point. In the original proposition of this so-called **cutoff method** by Hama [29], the constant factor ϕ was left unspecified, simply interpreted to be of order 1. However, Crow [16] determined a suitable value for ϕ by matching the results obtained by the cutoff method with two of the classical results of Kelvin for a filament whose core is in rigid-body rotation. For both results, this value was the same: $\log 2\phi = 1/4$. This choice was later generalized in the work of Widnall et al. [74] and Moore and Saffman [54] to account for the effects of general core vorticity distribution and core flow along the filament's axis (such as is commonly found in aircraft trailing vortices). This general choice for cutoff (again, restricted to thin filaments) is

$$\log 2\phi = \frac{1}{2} - \frac{4\pi^2}{\Gamma^2}\int_0^a v_\theta^2(r')r'\,dr' + \frac{8\pi^2}{\Gamma^2}\int_0^a w^2(r')r'\,dr', \qquad (7.70)$$

where $v_\theta(r)$ and $w(r)$ are, respectively, the distributions of swirl and axial velocity within the core. Thus, we can now provide an equation for the evolution of a vortex filament:

Result 7.6: Evolution Equation for a Vortex Filament

For a thin vortex filament—that is, a filament for which the local radius of curvature is much larger than the radius, a, of the filament core—of strength Γ, the evolution of the curve C describing the filament's axis is described by

$$\frac{\partial \boldsymbol{X}(s,t)}{\partial t} = -\frac{\Gamma}{4\pi}\int_{C\backslash\phi a} \frac{(\boldsymbol{X}(s,t) - \boldsymbol{X}(\sigma,t))}{|\boldsymbol{X}(s,t) - \boldsymbol{X}(\sigma,t)|^3} \times d\boldsymbol{l}(\sigma) + \boldsymbol{v}_{-\mathrm{v}}(\boldsymbol{X}(s,t)) \qquad (7.71)$$

where s is the local arc length parameter, $d\boldsymbol{l}(s) := (\partial \boldsymbol{X}/\partial s)ds$ is a differential length vector along the filament, and $C \setminus \phi a$ denotes the filament without an interval of length $2\phi a$ surrounding the evaluation point at s, where the cutoff factor ϕ is provided for general core structure by Eq. (7.70). For a filament in rigid-body rotation and no axial velocity, $\log 2\phi = 1/4$.

It is important to remember that the vorticity transport equation (3.20) contains stretching and tilting terms in three dimensions. So it is natural to ask: Are we missing something in the filament transport described in Result 7.6? The answer is no—the local advection of the filament curve automatically accounts for the stretching and tilting. The modification of the vorticity itself by these processes is not of interest to us, since the vorticity's integrated value—the filament strength—remains invariant during these processes.

Discretization and Regularization of a Filament The discretization of a vortex filament is straightforward: we simply subdivide the curve into straight finite-length segments and track the end points of these segments [45]. In Sect. 3.3.2 we presented the analytical expression for the velocity induced by a straight segment.

The blob regularization that we used for point vortices can also be naturally adapted to filaments, for example, by modifying the velocity kernel in Eq. (7.71). This regularized kernel follows from the same principle, Eq. (7.46), as in the two-dimensional case, with the blob function similarly constrained to have unit integral. Analogous to Eq. (7.57) in the two-dimensional context, the regularized kernel takes a form that removes the singular behavior of the original:

$$\boldsymbol{K}_\varepsilon(\boldsymbol{x}) = -\frac{\boldsymbol{x}}{4\pi|\boldsymbol{x}|^3}q(|\boldsymbol{x}|/\varepsilon). \tag{7.72}$$

Here, in three dimensions, the function q is defined as

$$q(\rho) := 4\pi\int_0^\rho \rho'^2\tilde{\delta}(\rho')\,\mathrm{d}\rho', \tag{7.73}$$

where $\tilde{\delta}$ is the basic form of the blob function,

$$\delta_\varepsilon(\boldsymbol{x}) = \frac{1}{\varepsilon^3}\tilde{\delta}(|\boldsymbol{x}|/\varepsilon). \tag{7.74}$$

The regularized Green's function follows from integrating the Poisson equation, which in its three-dimensional axisymmetric form is

$$-\frac{1}{\rho^2}\frac{\mathrm{d}}{\mathrm{d}\rho}\left(\rho^2\frac{\mathrm{d}\tilde{G}}{\mathrm{d}\rho}\right) = \tilde{\delta}(\rho); \tag{7.75}$$

its ε-scaled version is

$$G_\varepsilon(\boldsymbol{x}) = \frac{1}{\varepsilon}\tilde{G}(|\boldsymbol{x}|/\varepsilon). \tag{7.76}$$

Some examples of three-dimensional blob functions and their companion kernel regularizations are the Gaussian,

$$\tilde{\delta}(\rho) = \frac{1}{\pi^{3/2}}\mathrm{e}^{-\rho^2},\quad \tilde{G}(\rho) = \frac{\operatorname{erf}\rho}{4\pi\rho},\quad q(\rho) = \operatorname{erf}\rho - \frac{2}{\pi^{1/2}}\rho\mathrm{e}^{-\rho^2}, \tag{7.77}$$

where erf is the error function; and the algebraic kernel,

$$\tilde{\delta}(\rho) = \frac{3}{4\pi(\rho^2+1)^{5/2}},\quad \tilde{G}(\rho) = \frac{1}{4\pi(\rho^2+1)^{1/2}},\quad q(\rho) = \frac{\rho^3}{(\rho^2+1)^{3/2}}. \tag{7.78}$$

This last example leads to a regularized velocity kernel of the form

$$K_\varepsilon(x) = -\frac{x}{4\pi(|x|^2+\varepsilon^2)^{3/2}}. \tag{7.79}$$

Several other examples are given by Winckelmans and Leonard [75]. For any choice of blob function, the regularized filament transport equation would be given by

$$\frac{\partial X(s,t)}{\partial t} = \Gamma \int_{C\backslash\phi a} K_\varepsilon(X(s,t) - X(\sigma,t)) \times \mathrm{d}l(\sigma) + v_{-\mathrm{v}}(X(s,t)). \tag{7.80}$$

The cutoff radius is still included in this integral, though its role is less essential with the addition of regularization. Generally, one would choose ε to be of similar order to ϕa.

Chapter 8
Flow About a Two-Dimensional Flat Plate

Contents

8.1 Basic Notation 270
8.2 Three Approaches to Solving for the Flow 273
8.2.1 Approach 1: Conformal Transformation from the Circle Plane 274
8.2.2 Approach 2: Inversion of the Cauchy Integral 280
8.2.3 Approach 3: Fourier–Chebyshev Expansion 287
8.3 Force and Moment 295
8.4 Application of the Kutta Condition 302
8.4.1 At the Trailing Edge: Thin Airfoil Theory 302
8.4.2 At Both Edges 309
8.5 Classical Results 309
8.5.1 Steady Flow at Fixed Small Angle of Attack 310
8.5.2 General Background on Classical Unsteady Results 313
8.5.3 Oscillatory Motion: Theodorsen 316
8.5.4 Impulsive Change of Motion: Wagner 321
8.5.5 Sinusoidal Gust Response 327
8.5.6 Sharp-Edge Gust 330
8.6 Generalized Edge Conditions and Their Interpretation 331
8.7 A Deforming Plate 334

In the previous chapters, we have presented a general set of tools with which to obtain the flow field and the force and moment due to inviscid flow about a rigid body. In this chapter, we consider an important special case of this general problem: the rigid-body motion of an infinitely-thin flat plate of length c in two dimensions. As usual, we assume that the flow is incompressible and irrotational, except for one or more isolated vortex elements, as depicted in Fig. 8.1. We will first build from the results of Chap. 4 to seek an expression for the velocity field for this flow; then we will apply the results from Chap. 6 to obtain the force and moment on the plate in Sect. 8.3. Then, in Sect. 8.4, we will demonstrate the application of the Kutta condition to regularize the flow at just the trailing edge, or at both edges, and discuss the consequences of both. With all of these tools specialized to the particular

J. D. Eldredge, *Mathematical Modeling of Unsteady Inviscid Flows*, Interdisciplinary Applied Mathematics 50, https://doi.org/10.1007/978-3-030-18319-6_8

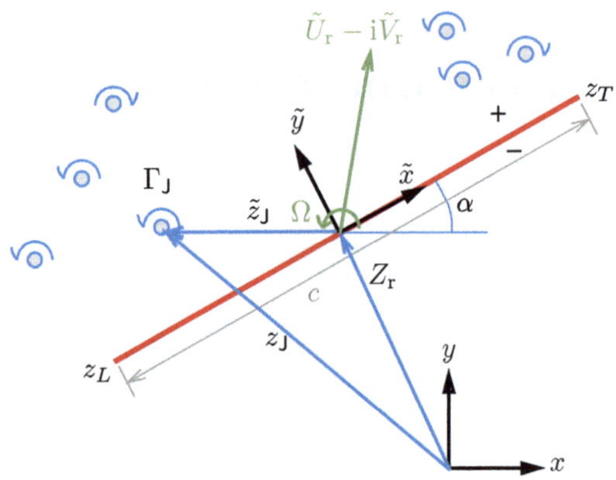

Fig. 8.1 Schematic of the two-dimensional flat plate, $\mathcal{P}$

geometry of a flat plate, we will find it straightforward in Sect. 8.5 to derive four of the classical results from unsteady aerodynamics, pertaining to small-amplitude perturbations of the plate's motion or to the fluid environment through which it is traveling.

In order to demonstrate the range of tools available to us, and to highlight their connections, we will pursue three different approaches in Sect. 8.2 to obtain the flow field. In the first approach, we will specialize our general solution based on conformal transformation from the circle plane. In the second, we will obtain the same result, but this time by first expressing the problem for the unknown vortex sheet strength in a Cauchy integral form. The solution of this problem can be written down immediately for the case of a flat plate. However, in order to demonstrate a general analytical technique, in the third approach we will expand the problem in Chebyshev polynomials and solve exactly for the unknown coefficients. This last approach is equivalent to the well-known Fourier series representation used in classical thin-airfoil theory.

8.1 Basic Notation

Before pursuing the approaches, let us review our basic notation. Primarily, our description of the flow will rely on the complex notation presented in Chap. 2. Here, we briefly review this notation, as well as some of the relevant fundamental concepts in Chap. 3.

Coordinates Let the inertial coordinate system be described by $z = x + iy$, and a system attached to the body by $\tilde{z} = \tilde{x} + i\tilde{y}$, as depicted in Fig. 8.1. These are related to each other through the rigid map

$$z = Z_r + \tilde{z}e^{i\alpha}, \tag{2.37}$$

using the position of the center of the plate, Z_r, and the rotation operator, $e^{i\alpha}$, based on the current angle of the plate, α, with respect to the x axis. In what follows, we will often liberally interchange the argument of a function between z and $\tilde{z}$. The meaning of each argument should be clear through the correspondence of these two descriptions of a given point. If other symbols are used to describe a spatial point, e.g., in dummy integration variables, we will always use the $\tilde{()}$ to denote body-fixed coordinates.

Note that $e^{i\alpha}$ rotates coordinates from the body-fixed system to the inertial system; the inverse rotation is provided by $e^{-i\alpha}$. The plate itself is described by points $z \in \mathcal{P}$ such that $\tilde{x} \in [-c/2, c/2]$ and $\tilde{y} = 0$. The + side of the plate is that into which $\tilde{y}$ is directed, and the − side is the opposite. Occasionally, we will make use of a scaled complex coordinate,

$$\xi = \frac{2\tilde{z}}{c} = \frac{2}{c}(z - Z_r)e^{-i\alpha}, \tag{8.1}$$

so that, along the plate, ξ takes values between −1 and 1. To be consistent with the bulk of the aerodynamics literature, we define the end of the plate at which $\xi = 1$ (and $\tilde{z} = c/2$) as the **trailing edge**, and the end at which $\xi = -1$ (and $\tilde{z} = -c/2$) as the **leading edge**. These points are labeled in the z coordinate system as z_T and z_L, respectively.

Correspondingly, the differential element $d\xi$ on the plate is related to the differential elements $d\tilde{z}$ and dz by

$$\frac{c}{2}d\xi = d\tilde{z} = e^{-i\alpha}dz. \tag{8.2}$$

Note that $e^{i\alpha}$ is the complex version of a unit tangent vector, τ, and dz is also tangent to the plate (with length $c\,|d\xi|/2$).

Complex Potential and Velocity In Sect. 3.2.2 we defined the complex potential (3.82) and its relation to the complex velocity (3.83). We also distinguished the velocity components expressed in the inertial coordinate system, w, from those expressed in the body-fixed system, $\tilde{w}$, related through the rotation operation (2.38), repeated here:

$$\tilde{w}(\tilde{z}) = w(z)e^{i\alpha}. \tag{8.3}$$

(We remind the reader that the appearance of $e^{i\alpha}$, rather than $e^{-i\alpha}$, to transform from inertial to body-fixed components is due to the fact that this is the complex velocity, $u - iv$, and thus requires the conjugate of the rotation operator.) Note that we have used different independent variables, z and $\tilde{z}$, for the velocity components in each coordinate system, but the correspondence of these variables is clear from (2.37).

Body Motion The rigid-body motion of the plate is described by the translational velocity of the reference point, $\dot{Z}_r$, and the angular velocity, Ω. The motion of any point $z \in \mathcal{P}$ is described by

$$W_b = W_r - i\Omega(z_b^* - Z_r^*), \tag{2.39}$$

where $W_r := \dot{Z}_r^*$, and expressed in the body-fixed coordinate system by

$$\tilde{W}_b \equiv \tilde{U}_b - i\tilde{V}_b = W_b e^{i\alpha} = \tilde{W}_r - i\Omega\tilde{z}^*, \tag{2.41}$$

where $\tilde{W}_r = \tilde{U}_r - i\tilde{V}_r \equiv W_r e^{i\alpha}$. Of course, for any point along the plate, $\tilde{z}$ is purely real, so its conjugate is identical to itself.

Uniform Flow A uniform flow at infinity is described in inertial coordinates by

$$W_\infty = U_\infty - iV_\infty, \tag{3.85}$$

and in body-fixed coordinates by

$$\tilde{W}_\infty = \tilde{U}_\infty - i\tilde{V}_\infty \equiv W_\infty e^{i\alpha}. \tag{3.87}$$

Vortices Point vortices in the plane were described in Sect. 3.3.1, and we summarize the important aspects here. Let there be N_v point vortices in the fluid. The Jth point vortex is described by its position z_J and strength Γ_J, both of which are time-varying, in general, though the latter will be time invariant if we enforce Helmholtz' theorems.

The velocity field induced collectively by all vortices is described by

$$w(z) = \frac{1}{2\pi i} \sum_{J=1}^{N_v} \frac{\Gamma_J}{z - z_J}. \tag{3.119}$$

For later use, we note that the form of this velocity is identical when rotated into the body-fixed coordinate system:

$$\tilde{w}(\tilde{z}) \equiv w(z) e^{i\alpha} = \frac{1}{2\pi i} \sum_{J=1}^{N_v} \frac{\Gamma_J}{\tilde{z} - \tilde{z}_J}. \tag{8.4}$$

It is important to add that we can easily adapt any of the results of this chapter to a scenario in which fluid vorticity takes the form of one or more free vortex sheets. We need only recall the steps that led to our derivation of the velocity field (3.153) induced by a vortex sheet, and use the template (3.170) (or its complex form, (3.171)) for replacing point vortex results with the corresponding result for a free vortex sheet. However, for clarity of presentation, it is preferable to focus most of our analysis on discrete point vortices, in order to easily distinguish fluid vorticity from the body's bound vortex sheet, which is heavily discussed in this chapter.

We will later have a need to associate each vortex with one of the two edges of the plate. For the purposes of this chapter, the edge to which a vortex is associated is completely arbitrary and there is no 'correct' choice. Let us define the sets $\mathcal{V}_L$ and $\mathcal{V}_T$ to represent the groups of vortices associated with the leading and trailing edges, respectively. Then we can naturally define the total circulations of each of these sets:

$$\Gamma_L := \sum_{J \in \mathcal{V}_L} \Gamma_J, \quad \Gamma_T := \sum_{J \in \mathcal{V}_T} \Gamma_J. \tag{8.5}$$

These summations are naturally replaced by integrals when the vorticity takes the form of free vortex sheets.

The local circulation, $\Gamma(z)$, of a general bound vortex sheet was related to the complex vortex sheet strength in Eq. (3.152). However, we will make a small change to this definition here in anticipation of its use later in our analysis. Now, we will define the circulation based on a contour $C(z)$, depicted in Fig. 8.2, that envelops the vortices in the set $\mathcal{V}_L$ and pierces the plate at z, enclosing the leading edge and a portion of the plate $\mathcal{P}(z)$. The reason for enclosing the leading edge was discussed in Sect. 3.6.2: each of the vortices has an associated branch cut, which we imagine to intersect the plate at the edge in order to ensure continuous local circulation along the plate. In terms of γ, the strength distribution of the bound vortex sheet, the local circulation is defined as

$$\Gamma(\tilde{z}) := \Gamma_L + \int_{\mathcal{P}(\tilde{z})} \gamma(\tilde{\lambda})\,\mathrm{d}\tilde{\lambda}. \tag{8.6}$$

Note that the local circulation at the trailing edge is

$$\Gamma(\tilde{z}_T) = \Gamma_L + \Gamma_{\mathcal{P}}, \tag{8.7}$$

where we have defined the total plate circulation, $\Gamma_{\mathcal{P}}$, as

$$\Gamma_{\mathcal{P}} := \int_{\mathcal{P}} \gamma(\lambda)\,\mathrm{d}\lambda. \tag{8.8}$$

Finally, the constraint of zero total circulation, Eq. (4.67), manifests itself for the plate as

$$\Gamma_L + \Gamma_{\mathcal{P}} + \Gamma_T = 0 \tag{8.9}$$

8.2 Three Approaches to Solving for the Flow

In this section, we will present three different approaches to solving for the flow about a flat plate. From the work in the preceding chapters, the approach that has been developed most thoroughly for solving for the flow in the plane is through the use of conformal mapping from the unit circle. Thus, we will present its application to the flat plate first, and use this as a basis of comparison for the other two approaches.

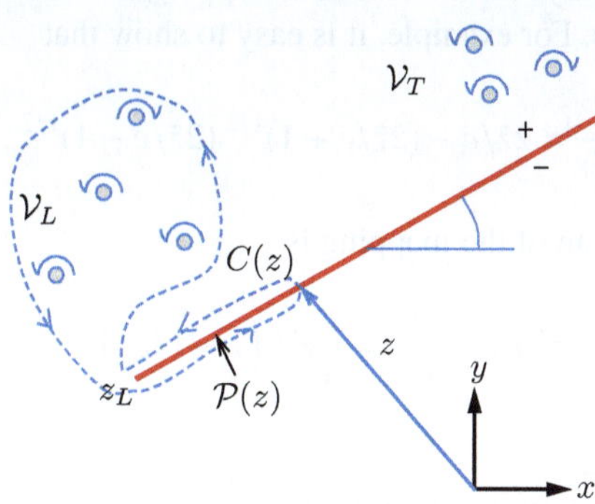

Fig. 8.2 Integration region $\mathcal{P}(z)$ for local circulation

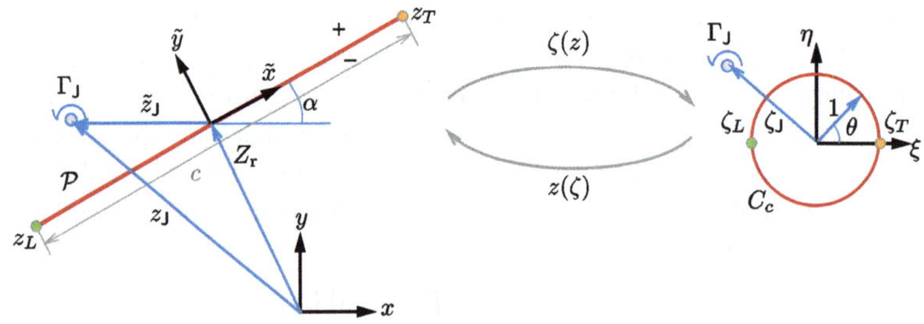

Fig. 8.3 Mapping of flat plate $\mathcal{P}$ to/from circular contour C_c

8.2.1 Approach 1: Conformal Transformation from the Circle Plane

Here, we specialize the general solution for a planar body obtained in Chap. 4 by conformally mapping the solution in the circle plane. That solution relied on a power series transformation (A.153), of which the mapping to a flat plate of length c is a special case: $c_1 = c_{-1} = c/4$, and all other coefficients are identically zero. This is the classical **Joukowski transformation**[1]:

$$\tilde{z}(\zeta) = \frac{c}{4}(\zeta + 1/\zeta), \tag{8.10}$$

which, when combined with the rigid map (A.152), provides the overall transformation depicted schematically in Fig. 8.3. The upper (+) side of the plate is mapped from the upper semi-circle, $C_c^+ := \left\{\zeta : \zeta = e^{i\theta}, \theta \in (0, \pi)\right\}$, and the lower (−) side from the lower semi-circle, $C_c^- := \left\{\zeta : \zeta = e^{i\theta}, \theta \in (\pi, 2\pi)\right\}$. Note that we exclude the edges from these definitions. The leading edge z_L is mapped from $\zeta_L := -1$ ($\theta = \pi$), and the trailing edge z_T from $\zeta_T := 1$ ($\theta = 0$).

All points $z \notin \mathcal{P}$ are mapped from the exterior region $|\zeta| > 1$. In particular, each vortex location z_J has a unique pre-image ζ_J. The inverse of the mapping (8.10) is

$$\zeta(\tilde{z}) = 2\tilde{z}/c + (2\tilde{z}/c + 1)^{1/2}(2\tilde{z}/c - 1)^{1/2}. \tag{8.11}$$

There are various auxiliary results from this mapping and its inverse that will serve us in future calculations. For example, it is easy to show that

$$\frac{1}{\zeta(\tilde{z})} = 2\tilde{z}/c - (2\tilde{z}/c + 1)^{1/2}(2\tilde{z}/c - 1)^{1/2}. \tag{8.12}$$

Furthermore, the Jacobian of the mapping is

$$z'(\zeta) := \frac{dz}{d\zeta} = \frac{c}{4}e^{i\alpha}\left(1 - 1/\zeta^2\right), \tag{8.13}$$

[1] In many texts, this transformation is defined not from a unit circle, but rather, from a circle of radius $c/4$, in which case $c_1 = 1$ and $c_{-1} = c^2/16$. There is no difference in the results in the physical plane, provided either choice of mapping is used consistently.

and the inverse of its magnitude, which enters frequently into the results that follow, is

$$\frac{1}{|z'(\zeta)|} = \frac{4\mathrm{i}}{c(\zeta - 1/\zeta)}. \tag{8.14}$$

We also note that the Jacobian is proportional to the roots of the distances from the respective edges of the plate:

$$z'(\zeta) = \mathrm{e}^{\mathrm{i}\alpha}\zeta^{-1}\left[\tilde{z}(\zeta) + c/2\right]^{1/2}\left[\tilde{z}(\zeta) - c/2\right]^{1/2}. \tag{8.15}$$

The second derivative of the mapping is also useful, particularly at the edges, so we note that it is given by

$$z''(\zeta) = \mathrm{e}^{\mathrm{i}\alpha}\frac{c}{2\zeta^3}, \tag{8.16}$$

which, at the edges, simplifies to $z''(\pm 1) = \pm c\mathrm{e}^{\mathrm{i}\alpha}/2$.

The Flow Field We can immediately write down the solution for the complex potential and the complex velocity fields from the results in Sect. 4.6.1. The coefficients d_{-k} that appear in the rotational results are defined by Eq. (A.157); for a flat plate, the only non-zero coefficients of relevance are $d_0 = c^2/8$ and $d_{-2} = c^2/16$. Then, the complex potential, $\hat{F}(\zeta)$, in the circle plane due to the rigid-body motion of the plate and the presence of uniform flow and vortices in the surrounding fluid follows from (4.79), and is

$$\hat{F}(\zeta) = \tilde{U}_\infty \frac{c}{4}(\zeta + 1/\zeta) - \mathrm{i}\tilde{V}_\infty \frac{c}{4}(\zeta - 1/\zeta) - \frac{\mathrm{i}}{2}\frac{\tilde{V}_{\mathrm{r}}c}{\zeta} - \frac{\mathrm{i}}{16}\frac{\Omega c^2}{\zeta^2}$$
$$+ \sum_{\mathrm{J}=1}^{N_{\mathrm{v}}} \frac{\Gamma_{\mathrm{J}}}{2\pi\mathrm{i}}\left[\log(\zeta - \zeta_{\mathrm{J}}) - \log(\zeta - 1/\zeta_{\mathrm{J}}^*)\right]. \tag{8.17}$$

Note that the first term in this potential is $\tilde{U}_\infty \tilde{z}$, and represents just an undisturbed uniform flow in the $\tilde{x}$ direction; the plate offers no disturbance to this component of the uniform flow. The point $1/\zeta_{\mathrm{J}}^*$, which lies inside C_c and on the same ray as ζ_{J}, represents the image of the vortex position ζ_{J} with respect to the circle.

Remember from Sect. 4.5 that, in writing the contribution from the point vortices in (8.17), we have made the choice to omit an image vortex at the origin in order to ensure that the circulation contributed by each vortex is exactly canceled by its image. Since the other parts of the potential contribute no circulation, this choice gives us a tidy and straightforward approach to enforce zero total circulation, Γ_{tot}, vortex by vortex. However, there are other choices that could be made to satisfy this constraint, and one of these will be discussed below in Note 8.5.1. It will be useful to see how this constraint manifests itself in the other solution approaches we follow in this chapter.

As required for a conformal transformation, the complex potentials in the physical ($\tilde{z}$) and circle planes are identical at corresponding points: $F(\tilde{z}) = \hat{F}(\zeta(\tilde{z}))$. Since the imaginary part of F comprises the streamfunction, we can plot the streamline patterns

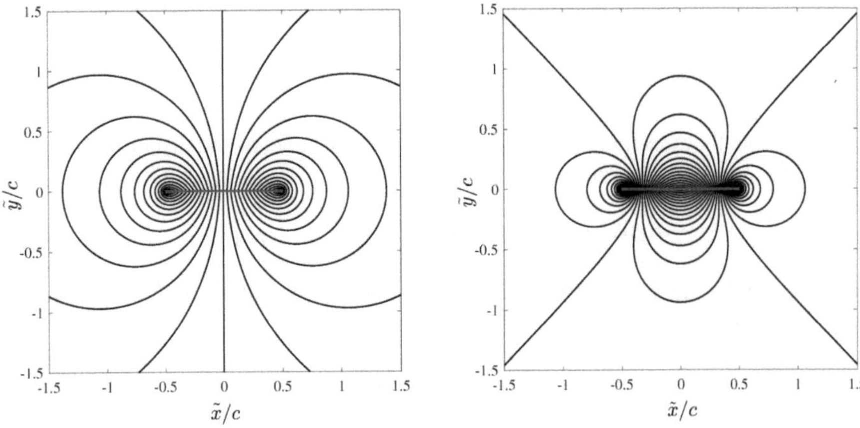

Fig. 8.4 Streamline patterns induced by unit translational (left) and angular (right) velocity of a flat plate

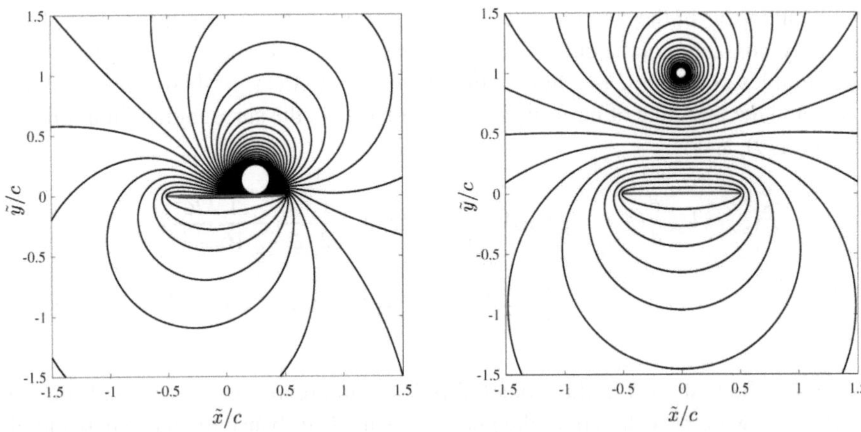

Fig. 8.5 Streamline patterns induced by point vortex at $(\tilde{x}/c, \tilde{y}/c) = (0.25, 0.05)$ (left) and $(\tilde{x}/c, \tilde{y}/c) = (0, 1)$ (right) in the vicinity of stationary flat plate

associated with each basis flow field, using a contour plot of this streamfunction field. These streamlines are depicted in Figs. 8.4 and 8.5.

Decomposed Flow The velocity field follows by differentiating the complex potential. As we discussed in Sect. 4.6, the linearity of the velocity field and the vortex sheet strength with respect to the flow contributors allows us to decompose these into basis velocity fields and basis sheets, exhibited in Eqs. (4.81) and (4.107) for the general planar body. Each of these basis sheets represents the reaction to a flow contributor considered in isolation, and the associated basis velocity field reflects the modification made by the presence of the body. For the flat plate, this decomposition is simplified by the fact that neither $\tilde{U}_r$ nor $\tilde{U}_\infty$ elicits a reactive sheet: the tangential component of motion of an infinitely-thin flat plate cannot create a disturbance in an inviscid flow, and thus, cannot influence the strength of the bound vortex sheet. If a

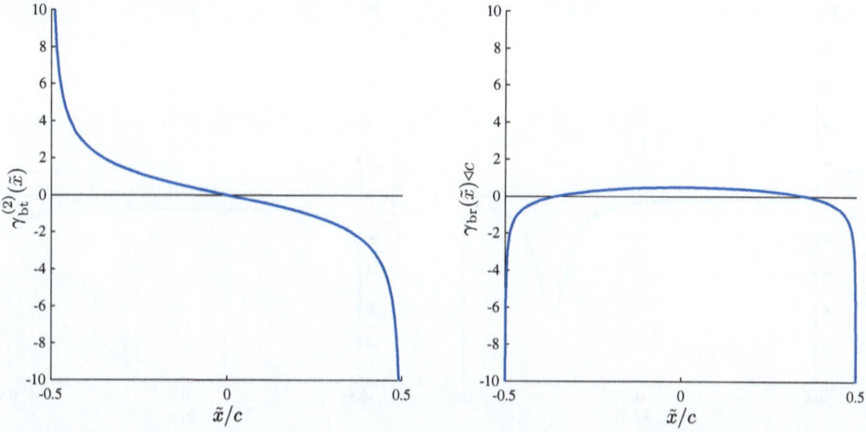

Fig. 8.6 Motion-induced vortex sheet strengths due to unit translational (left) and angular (right) velocity of the flat plate

uniform flow is present—that is, if we are analyzing the problem in the windtunnel frame—then its tangential component still contributes directly to the velocity field.

Thus, the decomposed velocity field, written in body-fixed coordinates, is

$$\tilde{w}(\tilde{z}) = \tilde{w}_{\text{v}}(\tilde{z}) + \tilde{U}_{\infty}\tilde{w}^{(1)}_{\infty}(\tilde{z}) + \tilde{V}_{\infty}\tilde{w}^{(2)}_{\infty}(\tilde{z}) + \tilde{V}_{\text{r}}\tilde{w}^{(2)}_{\text{bt}}(\tilde{z}) + \Omega\tilde{w}_{\text{br}}(\tilde{z}). \tag{8.18}$$

The basis velocity fields follow from Eqs. (4.83)–(4.92). Note that, in each of the following expressions, the explicit $\tilde{z}$ dependence has been suppressed, but is implicit in every instance of ζ via the inverse mapping. The vorticity-induced basis velocity field, decomposed into unit velocity fields from each vortex, is

$$\tilde{w}_{\text{v}}(\tilde{z}) = \sum_{\text{J}=1}^{N_{\text{v}}} \Gamma_{\text{J}}\tilde{w}^{\text{v}}_{\text{J}}(\tilde{z}), \quad \tilde{w}^{\text{v}}_{\text{J}}(\tilde{z}) = \frac{1}{2\pi\text{i}\tilde{z}'(\zeta)}\left(\frac{1}{\zeta - \zeta_{\text{J}}} - \frac{1}{\zeta - 1/\zeta^*_{\text{J}}}\right). \tag{8.19}$$

The uniform flow- and motion-induced basis velocity fields are

$$\tilde{w}^{(1)}_{\infty}(\tilde{z}) = 1, \quad \tilde{w}^{(2)}_{\infty}(\tilde{z}) = -\text{i}\frac{\zeta^2+1}{\zeta^2-1}, \quad \tilde{w}^{(2)}_{\text{bt}}(\tilde{z}) = \frac{2\text{i}}{\zeta^2-1}, \quad \tilde{w}_{\text{br}}(\tilde{z}) = \frac{\text{i}c}{2\zeta(\zeta^2-1)}. \tag{8.20}$$

Note that $\tilde{w}^{(2)}_{\infty} = -\tilde{w}^{(2)}_{\text{bt}} - \text{i}$. As noted in earlier chapters, only the difference between the translational velocity and the uniform flow is relevant, so their reacting velocity fields are equal and opposite; the additional term is due to the $\tilde{y}$ component of the undisturbed uniform flow at infinity.

The basis vortex sheet strengths in the decomposed form

$$\gamma = \gamma_{\text{v}} + \tilde{V}_{\infty}\gamma^{(2)}_{\infty} + \tilde{V}_{\text{r}}\gamma^{(2)}_{\text{bt}} + \Omega\gamma_{\text{br}} \tag{8.21}$$

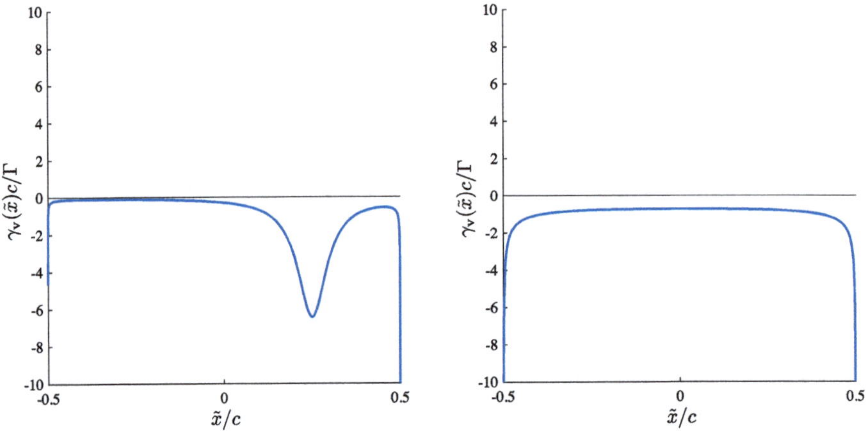

Fig. 8.7 Vorticity-induced vortex sheet strength due to a point vortex at $(\tilde{x}/c, \tilde{y}/c) = (0.25, 0.05)$ (left) and $(\tilde{x}/c, \tilde{y}/c) = (0, 1)$ (right)

follow from their general forms obtained through conformal mapping in Eqs. (4.97)–(4.102), specialized to the case of a flat plate and written in the form of the real-valued sheet strength, γ. Recall that γ_v is the strength of the vortex sheet—which we called the **vorticity-induced basis vortex sheet** in Sect. 4.6—that arises due to the plate's reaction to fluid vorticity; $\gamma_\infty^{(2)}$ is the strength of the **uniform flow-induced basis vortex sheet**, the sheet due to uniform flow of unit speed in the $\tilde{y}$ direction; $\gamma_{bt}^{(2)}$ is the strength of the vortex sheet due to unit normal translational velocity of the body; and γ_{br} is the strength due to unit angular velocity of the body. These latter two sheets are referred to as the **translation-** and **rotation-induced basis vortex sheet**, respectively. The sheets associated with uniform flow or translation components tangent to the plate, $\gamma_\infty^{(1)}$ and $\gamma_{bt}^{(1)}$, are identically zero and therefore omitted from further discussion.

As we discussed in Note 3.6.4, the vortex sheet on an infinitely-thin body such as a flat plate represents the jump in velocity across it, and can thus be interpreted as the combination of sheets on the positive and negative sides of the plate. Correspondingly, the overall complex sheet strength is the difference in the values of these positive and negative sheets, $g = g^+ - g^-$. A point and its companion on the other side of the plate are described, respectively, by $\zeta \in C_c^+$ and $\zeta^* = 1/\zeta \in C_c^-$ in the circle plane. The plate tangents at such points are equal and opposite; in the mapping from the circle plane, this is exemplified by the fact that

$$\tau(\zeta^*) = -\mathrm{i}\zeta^* z'(\zeta^*)/|z'(\zeta^*)| = \mathrm{i}\zeta z'(\zeta)/|z'(\zeta)| = -\tau(\zeta), \tag{8.22}$$

as can be easily proved using the results thus far. Thus, we can show that the basis vortex sheets are

$$\gamma_{bt}^{(2)}(\tilde{z}) = -\gamma_\infty^{(2)}(\tilde{z}) = -2\mathrm{i}\frac{\zeta^2+1}{\zeta^2-1}, \quad \gamma_{br}(\tilde{z}) = -\frac{\mathrm{i}c(\zeta^4+1)}{2\zeta(\zeta^2-1)}. \tag{8.23}$$

Also, the vorticity-induced basis sheet, decomposed further into unit contributions from each vortex, is

$$\gamma_v(\tilde{z}) = \sum_{J=1}^{N_v} \Gamma_J \gamma_J^v(\tilde{z}), \quad \gamma_J^v(\tilde{z}) = -\frac{4}{\pi c}\mathrm{Im}\left[(\zeta - 1/\zeta)^{-1}\left(\frac{\zeta}{\zeta - \zeta_J} + \frac{1}{1 - \zeta\zeta_J} - 1\right)\right]. \tag{8.24}$$

In each of these, the dependence of ζ on $\tilde{z}$ is omitted, but is implied by the inverse mapping.

All three vortex sheet strengths—Eqs. (8.23) and (8.24)—are plotted in Figs. 8.6 and 8.7. There are some important things to note about these sheet strengths:

1. ***Each vortex sheet has inverse square-root singularities at the ends of the plate.*** The basis velocity fields and vortex sheet strengths are singular at $\zeta = \pm 1$, the pre-images of the plate's edges. In this conformal mapping approach, this singularity has emerged via the inverse of the Jacobian of the mapping. As noted in (8.15), this Jacobian introduces square-root singularities with respect to distance from the edges. This singularity was anticipated based on our discussion in Note 4.4.5. This fact has many consequences. For example, we will sometimes use the Kutta condition (see Sect. 5.1) to eliminate one or more of the singularities by choosing other parameters (e.g., point vortex strengths) appropriately. The singularities that we do not remove will induce additional force—edge suction—on the plate.
2. ***The motion-induced sheets are time invariant.*** As we discussed in Sect. 4.6.1, and explicitly identified in (4.107), the translation- and rotation-induced sheet strengths (and thus, obviously, the uniform flow-induced sheet strength) depend only on the shape of the rigid plate and are therefore time invariant. In other words, the strength distributions shown in Fig. 8.6 never change, even if the motion of the body itself changes.
3. ***The vorticity-induced sheet is time varying.*** The sheet strength in (8.24) depends on the positions and strengths of vortices in the fluid. Thus, as these vortices evolve, the sheet itself must evolve with it.

Impulses The impulses were derived via the power-series conformal mapping for the general planar rigid body in Note 6.2.5 and decomposed into basis impulses in (6.91), re-written here for the flat plate (with the uniform flow's component included as a straightforward modification of the translational velocity):

$$\tilde{P} = \tilde{P}_v + (\tilde{V}_r - \tilde{V}_\infty)\tilde{P}_{bt}^{(2)} + \Omega\tilde{P}_{br} \tag{8.25}$$

and

$$\Pi_r = \Pi_{r,v} + (\tilde{V}_r - \tilde{V}_\infty)\Pi_{r,bt}^{(2)} + \Omega\Pi_{r,br}. \tag{8.26}$$

When we specialize the general forms to basis impulses of the flat plate (for which $\mathcal{V}_b = 0$), it is simple to verify that the body translation-induced impulses are

$$\tilde{P}_{bt}^{(2)} = \mathrm{i}\frac{\pi}{4}c^2, \qquad \Pi_{r,bt}^{(2)} = 0; \tag{8.27}$$

and rotation-induced impulses are

$$\tilde{P}_{\mathrm{br}} = 0, \qquad \Pi_{\mathrm{r,br}} = \frac{\pi}{128} c^4. \tag{8.28}$$

The vorticity-induced impulses are further decomposed into unit impulses,

$$\tilde{P}_{\mathrm{v}} = \sum_{\mathrm{J}=1}^{N_{\mathrm{v}}} \Gamma_{\mathrm{J}} \tilde{P}_{\mathrm{J}}^{\mathrm{v}}, \qquad \Pi_{\mathrm{r,v}} = \sum_{\mathrm{J}=1}^{N_{\mathrm{v}}} \Gamma_{\mathrm{J}} \Pi_{\mathrm{r,J}}^{\mathrm{v}}, \tag{8.29}$$

each of which takes the form

$$\tilde{P}_{\mathrm{J}}^{\mathrm{v}} = -\frac{\mathrm{i}c}{4}(\zeta_{\mathrm{J}} - 1/\zeta_{\mathrm{J}}^{*}), \tag{8.30}$$

$$\Pi_{\mathrm{r,J}}^{\mathrm{v}} = \frac{c^2}{32}\left[2 + 1/\zeta_{\mathrm{J}}^{2} + 1/\zeta_{\mathrm{J}}^{2*} - (\zeta_{\mathrm{J}} + 1/\zeta_{\mathrm{J}})(\zeta_{\mathrm{J}}^{*} + 1/\zeta_{\mathrm{J}}^{*})\right], \tag{8.31}$$

in terms of the position ζ_{J} of the Jth point vortex in the circle plane.

From this decomposition of impulse, we can develop expressions for the force and moment on the plate, using the general framework laid out in Sect. 6.5. In particular, the components of the added mass tensors for a flat plate expressed in body-fixed coordinates are

$$\tilde{\mathbf{M}}^{FV} = \begin{bmatrix} 0 & 0 \\ 0 & \rho\pi c^2/4 \end{bmatrix}, \quad \tilde{\mathbf{M}}^{M\Omega} = \rho\pi c^4/128, \quad \tilde{\mathbf{M}}^{MV} = (\tilde{\mathbf{M}}^{F\Omega})^{\mathrm{T}} = [0 \;\; 0]. \tag{8.32}$$

(Note that the shapes of these correspond to the distinctly colored regions inside the boxed sub-matrix of Note 6.5.1.)

This solution by conformal mapping was immediately accessible based on our more general discussion. However, there are other analytical approaches we can follow for this geometry that provide insight on the flow. The two we will discuss both rely on the Cauchy integral equation for the vortex sheet strength, so we will first set up this problem from the general integral equation derived in Sect. 4.3.

8.2.2 Approach 2: Inversion of the Cauchy Integral

From our discussion on vortex sheets and infinitely-thin plates in Sect. 4.2.3, the overall velocity field (due to the plate, the uniform flow, and the fluid vortices) at any point $z \notin \mathcal{P}$ is given by

$$w(z) = \frac{1}{2\pi\mathrm{i}} \int_{\mathcal{P}} \frac{g(\lambda)}{\lambda - z}\,\mathrm{d}\lambda + W_{\infty} + \frac{1}{2\pi\mathrm{i}} \sum_{\mathrm{J}=1}^{N_{\mathrm{v}}} \frac{\Gamma_{\mathrm{J}}}{z - z_{\mathrm{J}}}, \tag{4.15}$$

in which the velocity field due to the bound vortex sheet, with complex strength distribution g, takes the form of a Cauchy integral over the plate $\mathcal{P}$. The dummy integration variable λ is, of course, in the same inertial coordinate system as z. Recall that g represents the local difference in fluid velocity across the plate, $w^+ - w^-$. This distribution is, as yet, unknown, but will be determined by enforcing the no-penetration condition on the plate. This condition on g and its relationship to γ on the bound vortex sheet of an infinitely-thin (but not necessarily flat) plate are given in Result 3.5 by

$$\operatorname{Im}(g\tau) = 0, \qquad \gamma = -\operatorname{Re}(g\tau). \tag{3.136}$$

For a flat plate, $\tau = \mathrm{e}^{\mathrm{i}\alpha}$, so $\gamma(\tilde{z})\,\mathrm{d}\tilde{z} = -g(z)\,\mathrm{d}z$. Thus, $\gamma(\tilde{z})\,\mathrm{d}\tilde{z}$ is an equivalent representation of the local strength of the infinitesimal vortices comprising the bound sheet.

The velocity field in body-fixed coordinates at any point $\tilde{z} \notin \mathcal{P}$ can therefore be written in terms of γ as

$$\tilde{w}(\tilde{z}) = -\frac{1}{2\pi \mathrm{i}} \int_{\mathcal{P}} \frac{\gamma(\tilde{\lambda})}{\tilde{\lambda} - \tilde{z}}\,\mathrm{d}\tilde{\lambda} + \tilde{W}_\infty + \frac{1}{2\pi \mathrm{i}} \sum_{\mathrm{J}=1}^{N_\mathrm{v}} \frac{\Gamma_\mathrm{J}}{\tilde{z} - \tilde{z}_\mathrm{J}}. \tag{8.33}$$

For future reference, the complex potential is easily obtained from the integral of this velocity

$$F(\tilde{z}) = \int \tilde{w}(\tilde{z})\,\mathrm{d}\tilde{z} = \frac{1}{2\pi \mathrm{i}} \int_{\mathcal{P}} \gamma(\tilde{\lambda}) \log(\tilde{z} - \tilde{\lambda})\,\mathrm{d}\tilde{\lambda} + \tilde{W}_\infty \tilde{z} + \frac{1}{2\pi \mathrm{i}} \sum_{\mathrm{J}=1}^{N_\mathrm{v}} \Gamma_\mathrm{J} \log(\tilde{z} - \tilde{z}_\mathrm{J}); \tag{8.34}$$

the arbitrary integration constant is irrelevant and has been set to zero.

When Eq. (3.206) is applied to the plate via the Plemelj formulae, we obtain the integral equation (4.36) for a general infinitely-thin plate, which we will rewrite here in terms of the real-valued strength γ. First, we note that, on a flat plate, this is a condition on the imaginary part of $\tilde{w}$, since that imaginary part is the normal component of the velocity at all points on the plate. Furthermore, the factors $\tilde{\lambda} - \tilde{z}$ and $\mathrm{d}\tilde{\lambda}$ are parallel everywhere when $\tilde{z}$ is on the plate, so their ratio is real valued. Thus, the integral in (8.33), with its factor i, is purely imaginary. We can now write the integral equation for γ, and furthermore, find its solution using the techniques in [56]:

Result 8.1: Cauchy Integral Equation for a Rigid Flat Plate

The integral equation for the bound vortex sheet strength γ of a flat plate $\mathcal{P}$ in rigid-body motion in the presence of uniform flow and a set of point vortices of strengths Γ_J and positions z_J is

$$\frac{1}{2\pi} \underset{\mathcal{P}}{⨍} \frac{\gamma(\tilde{\lambda})}{\tilde{\lambda} - \tilde{z}} \,\mathrm{d}\tilde{\lambda} = \Delta V_{\perp}(\tilde{z}). \tag{8.35}$$

This constitutes a singular Fredholm integral equation of the first kind for γ. For brevity, we have written the right-hand side as

$$\Delta V_{\perp}(\tilde{z}) := -\mathrm{Im}\left(\Delta \tilde{W}(\tilde{z})\right), \tag{8.36}$$

where $\tilde{z} \in \mathcal{P}$ and

$$\Delta \tilde{W}(\tilde{z}) := -\tilde{W}_{\mathrm{r}} + \mathrm{i}\Omega\tilde{z} + \tilde{W}_{\infty} + \frac{1}{2\pi \mathrm{i}} \sum_{\mathrm{J}=1}^{N_{\mathrm{v}}} \frac{\Gamma_{\mathrm{J}}}{\tilde{z} - \tilde{z}_{\mathrm{J}}}. \tag{8.37}$$

The solution of the Cauchy integral equation (8.35) can be immediately written down (see equation (88.8) in §88 of [56]) as

$$\gamma(\tilde{z}) = \frac{1}{(c/2 - \tilde{z})^{1/2} (c/2 + \tilde{z})^{1/2}} \times \left[-\frac{2}{\pi} \underset{\mathcal{P}}{⨍} \frac{(c/2 - \tilde{\lambda})^{1/2} (c/2 + \tilde{\lambda})^{1/2} \Delta V_{\perp}(\tilde{\lambda})}{\tilde{\lambda} - \tilde{z}} \,\mathrm{d}\tilde{\lambda} + \frac{c B_0}{2} \right], \tag{8.38}$$

where B_0 is an arbitrary coefficient. Written in terms of the scaled plate variable, $\xi = 2\tilde{z}/c$, and the corresponding dummy variable $l = 2\tilde{\lambda}/c$, the solution is

$$\gamma(\xi) = \frac{1}{\left(1 - \xi^2\right)^{1/2}} \left[-\frac{2}{\pi} ⨍_{-1}^{1} \frac{(1 - l^2)^{1/2} \Delta V_{\perp}(\tilde{z}(l))}{l - \xi} \,\mathrm{d}l + B_0 \right]. \tag{8.39}$$

The fact that the constant B_0 in (8.38) is arbitrary signifies, once again, that the flow past a planar body is not uniquely determined by the no-penetration condition alone. This is consistent with our discussion in Sect. 4.5; the form of the term for which B_0 is the coefficient is identical to the general homogeneous solution (4.70). We addressed the relationship of this solution with total circulation in Sect. 4.5, and this will be discussed again below. Another important point on (8.38) is that it exhibits the expected inverse square root singularity at either edge of the plate, just as in the conformal mapping solution above.

Note that $\Delta V_{\perp}(\tilde{z})$ represents the normal component of the difference $\Delta \tilde{W}(\tilde{z})$ between the velocity in the uniform flow and induced by external vortices and the velocity of the plate itself; in the classical literature, $-\Delta V_{\perp}$ is typically called the **downwash**. The result (8.38) is the vortex sheet strength distribution that cancels this difference. We note that $\tilde{z}^* = \tilde{z}$ for all $\tilde{z} \in \mathcal{P}$. Thus, for purposes of computing the integrals, it is useful to write $\Delta V_{\perp}$ as

$$\Delta V_{\perp}(\tilde{z}) = -\frac{1}{2\mathrm{i}}\left(\Delta\tilde{W}(\tilde{z}) - \Delta\tilde{W}^*(\tilde{z})\right), \tag{8.40}$$

in which the complex-valued function $\Delta\tilde{W}^*(\tilde{z})$ signifies that the complex conjugate is first taken and then the function's argument is evaluated at $\tilde{z}$. We can thus write this normal velocity difference as

$$\Delta V_{\perp}(\tilde{z}) = \tilde{V}_{\infty} - \tilde{V}_{\mathrm{r}} - \Omega\tilde{z} - \frac{1}{2\mathrm{i}}\left(\frac{1}{2\pi\mathrm{i}}\sum_{\mathrm{J}=1}^{N_{\mathrm{v}}}\frac{\Gamma_{\mathrm{J}}}{\tilde{z}-\tilde{z}_{\mathrm{J}}} + \frac{1}{2\pi\mathrm{i}}\sum_{\mathrm{J}=1}^{N_{\mathrm{v}}}\frac{\Gamma_{\mathrm{J}}}{\tilde{z}-\tilde{z}_{\mathrm{J}}^*}\right). \tag{8.41}$$

As expected, only the normal component, $\tilde{V}_{\mathrm{r}}$, of the plate's translation figures in this problem, and only the difference between this translational velocity and the uniform flow is relevant to determining the vortex sheet strength.

For later use, let us denote the mean tangential component of fluid velocity along the plate as μ. Recall that, from the Plemelj formulae (A.125), the mean of a Cauchy integral obtains its principal value,

$$\begin{aligned}\mu(\tilde{z}) &:= \frac{1}{2}\mathrm{Re}\left(\tilde{w}^+(\tilde{z}) + \tilde{w}^-(\tilde{z})\right)\\ &= \mathrm{Re}\left(-\frac{1}{2\pi\mathrm{i}}\,\fint_{\mathcal{P}}\frac{\gamma(\tilde{\lambda})}{\tilde{\lambda}-\tilde{z}}\,\mathrm{d}\tilde{\lambda} + \tilde{W}_{\infty} + \frac{1}{2\pi\mathrm{i}}\sum_{\mathrm{J}=1}^{N_{\mathrm{v}}}\frac{\Gamma_{\mathrm{J}}}{\tilde{z}-\tilde{z}_{\mathrm{J}}}\right), \quad \tilde{z}\in\mathcal{P}. \end{aligned}\tag{8.42}$$

But since the integral is purely imaginary, as we just discussed, it makes no contribution to the mean tangential velocity:

$$\mu(\tilde{z}) = \mathrm{Re}\left(\tilde{W}_{\infty} + \frac{1}{2\pi\mathrm{i}}\sum_{\mathrm{J}=1}^{N_{\mathrm{v}}}\frac{\Gamma_{\mathrm{J}}}{\tilde{z}-\tilde{z}_{\mathrm{J}}}\right), \quad \tilde{z}\in\mathcal{P}. \tag{8.43}$$

It is often helpful to locate the instantaneous positions of any **stagnation points** on the surface of the plate. Such points are defined by vanishing tangential velocity in a reference frame moving with the plate; the normal velocity is obviously enforced to be zero in this reference frame. Since the tangential velocity is different on either side of the plate, we label it with a superscript indicating the side. Using the results obtained thus far, these relative tangential components on either side of the plate are

$$\tilde{u}^{\pm}(\tilde{z}) - \tilde{U}_{\mathrm{b}}(\tilde{z}) = \mp\frac{1}{2}\gamma(\tilde{z}) + \mu(\tilde{z}) - \tilde{U}_{\mathrm{r}}. \tag{8.44}$$

A stagnation point on the + or – side corresponds to a location at which this respective component vanishes. It is worth noting that the expressions (8.44) are, respectively, $-\gamma^+$ and γ^-, the strengths of the + and – 'side' sheets discussed in Note 3.6.4; the sign difference on the first of these is due to the fact that we have used the same tangent vector, $\tau = \mathrm{e}^{\mathrm{i}\alpha}$, in (8.44) to define the tangential velocity on both sides of the plate.

Decomposed Strength of Vortex Sheet When we substitute the normal velocity difference (8.41) into the integral expression (8.38) for the vortex sheet strength, the decomposition (8.21) of this sheet arises naturally. The motion-induced sheets can be obtained with the help of integral relations (A.248) and (A.249) in the Appendix, with $\tilde{z} \in \mathcal{P}$:

$$\gamma_{\mathrm{bt}}^{(2)}(\tilde{z}) = -\frac{2\tilde{z}}{(c/2 - \tilde{z})^{1/2}(c/2 + \tilde{z})^{1/2}}, \tag{8.45}$$

$$\gamma_{\mathrm{br}}(\tilde{z}) = -\frac{2(\tilde{z}^2 - c^2/8)}{(c/2 - \tilde{z})^{1/2}(c/2 + \tilde{z})^{1/2}}. \tag{8.46}$$

Again, the sheet due to uniform flow is simply the negative of the sheet due to translation, $\gamma_{\infty}^{(2)} = -\gamma_{\mathrm{bt}}^{(2)}$. The vorticity-induced basis sheet's strength is calculated with the help of relation (A.254) in the Appendix:

$$\gamma_{\mathrm{v}}(\tilde{z}) = -\frac{1}{(c/2 - \tilde{z})^{1/2}(c/2 + \tilde{z})^{1/2}} \left[\sum_{\mathrm{J}=1}^{N_{\mathrm{v}}} \frac{\Gamma_{\mathrm{J}}}{2\pi} \left(\frac{(\tilde{z}_{\mathrm{J}} - c/2)^{1/2}(\tilde{z}_{\mathrm{J}} + c/2)^{1/2}}{\tilde{z}_{\mathrm{J}} - \tilde{z}} + \frac{(\tilde{z}_{\mathrm{J}}^* - c/2)^{1/2}(\tilde{z}_{\mathrm{J}}^* + c/2)^{1/2}}{\tilde{z}_{\mathrm{J}}^* - \tilde{z}} - 2 \right) - \frac{c}{2} B_0 \right]. \tag{8.47}$$

In this equation, the first two terms inside the inner parentheses are a complex conjugate pair (since $\tilde{z} = \tilde{z}^*$ on the plate), so the parenthetical portion of the expression is real-valued. We have also included the arbitrary coefficient, B_0, in this sheet's strength. This can be related to the bound circulation $\Gamma_{\mathcal{P}}$. Using integral relations (A.245)–(A.247) in the Appendix, it is easy to show that $\gamma_{\mathrm{bt}}^{(2)}$ and γ_{br} contribute nothing to the sheet's circulation, and that the circulation of γ_{v} is directly attributed to the term with coefficient B_0,

$$\Gamma_{\mathcal{P}} = \int_{\mathcal{P}} \gamma_{\mathrm{v}}(\tilde{z})\,\mathrm{d}\tilde{z} = \frac{\pi c}{2} B_0. \tag{8.48}$$

This result confirms the arbitrariness of the bound circulation, unless we specify another constraint. The coefficient's value can be written in terms of the point vortex strengths with the constraint (4.68) that total circulation in the flow is zero, which for a flat plate takes the form

$$\Gamma_{\mathcal{P}} + \sum_{\mathrm{J}=1}^{N_{\mathrm{v}}} \Gamma_{\mathrm{J}} = 0. \tag{8.49}$$

We note in passing that this condition can be written in terms of the local circulation and the circulations of the separate groups $\mathcal{V}_T$ and $\mathcal{V}_L$ as

$$\Gamma(\tilde{z}_L) = \Gamma_T + \Gamma_{\mathcal{P}} = -\Gamma_L. \tag{8.50}$$

We also stress that, by Kelvin's circulation theorem, this constraint should hold at all instants, even if the vortex strengths are time-varying. Thus, we must set B_0 to enforce this condition:

$$B_0 = -\frac{2}{\pi c} \sum_{\mathrm{J}=1}^{N_{\mathrm{v}}} \Gamma_{\mathrm{J}}. \tag{8.51}$$

Using this constraint, we can now rewrite the strength of the vorticity-induced vortex sheet as the sum of the unit basis sheets reacting to each vortex with unit strength:

$$\gamma_{\mathrm{v}}(\tilde{z}) = \sum_{\mathrm{J}=1}^{N_{\mathrm{v}}} \Gamma_{\mathrm{J}} \gamma_{\mathrm{J}}^{\mathrm{v}}(\tilde{z}), \tag{8.52}$$

where

$$\gamma_{\mathrm{J}}^{\mathrm{v}}(\tilde{z}) = -\frac{1}{2\pi (c/2 - \tilde{z})^{1/2} (c/2 + \tilde{z})^{1/2}} \times \left(\frac{(\tilde{z}_{\mathrm{J}} - c/2)^{1/2} (\tilde{z}_{\mathrm{J}} + c/2)^{1/2}}{\tilde{z}_{\mathrm{J}} - \tilde{z}} + \frac{(\tilde{z}_{\mathrm{J}}^* - c/2)^{1/2} (\tilde{z}_{\mathrm{J}}^* + c/2)^{1/2}}{\tilde{z}_{\mathrm{J}}^* - \tilde{z}} \right), \tag{8.53}$$

for $\tilde{z} \in \mathcal{P}$. This result is purely real, as expected.

It can be shown that, as expected, these basis sheet strengths (8.45), (8.46), and (8.53) are identical to those obtained by conformal mapping: (8.23) and (8.24). This comparison, which requires substituting the mapping for the physical plane coordinates, is left as an exercise to the reader. One result that is helpful for showing this is

$$\tilde{z} - \tilde{z}_{\mathrm{J}} = -\frac{c}{4} \frac{(\zeta - \zeta_{\mathrm{J}})(1 - \zeta \zeta_{\mathrm{J}})}{\zeta \zeta_{\mathrm{J}}}, \tag{8.54}$$

which can be proved through manipulation of the Joukowski transformation.

Velocity Field Now, let us write the basis velocity field for each of the flow contributors, following decomposition (8.18). Each of these basis velocity fields is obtained by substituting the corresponding vortex sheet strength from (8.45), (8.46) and (8.53) into the Cauchy integral in Eq. (8.33); the uniform flow and vorticity-induced basis fields also contain the direct influences from those contributors. As can be readily checked using the integral results in Sect. A.3.3 in the Appendix,

$$\tilde{w}_{\mathrm{J}}^{\mathrm{v}}(\tilde{z}) = \frac{1}{2\pi \mathrm{i}(\tilde{z} - \tilde{z}_{\mathrm{J}})} + \frac{1}{4\pi \mathrm{i}} \left[\frac{1}{\tilde{z} - \tilde{z}_{\mathrm{J}}} \left(\frac{(\tilde{z}_{\mathrm{J}} - c/2)^{1/2} (\tilde{z}_{\mathrm{J}} + c/2)^{1/2}}{(\tilde{z} - c/2)^{1/2} (\tilde{z} + c/2)^{1/2}} - 1 \right) + \frac{1}{\tilde{z} - \tilde{z}_{\mathrm{J}}^*} \left(\frac{(\tilde{z}_{\mathrm{J}}^* - c/2)^{1/2} (\tilde{z}_{\mathrm{J}}^* + c/2)^{1/2}}{(\tilde{z} - c/2)^{1/2} (\tilde{z} + c/2)^{1/2}} - 1 \right) \right] \tag{8.55}$$

is the unit basis velocity field due to the Jth vortex in the fluid;

$$\tilde{w}_{\infty}^{(1)}(\tilde{z}) = 1, \quad \tilde{w}_{\infty}^{(2)}(\tilde{z}) = -\frac{\mathrm{i}\tilde{z}}{(\tilde{z} - c/2)^{1/2} (\tilde{z} + c/2)^{1/2}} \tag{8.56}$$

are the basis velocity fields due to uniform flow at unit speed in the $\tilde{x}$ and $\tilde{y}$ directions, respectively;

$$\tilde{w}_{\text{bt}}^{(2)}(\tilde{z}) = \mathrm{i}\left(\frac{\tilde{z}}{(\tilde{z}-c/2)^{1/2}(\tilde{z}+c/2)^{1/2}} - 1\right) \tag{8.57}$$

is the velocity field due to unit translational velocity of the plate in the $\tilde{y}$ direction; and

$$\tilde{w}_{\text{br}}(\tilde{z}) = \mathrm{i}\left(\frac{\tilde{z}^2 - c^2/8}{(\tilde{z}-c/2)^{1/2}(\tilde{z}+c/2)^{1/2}} - \tilde{z}\right) \tag{8.58}$$

is the velocity field due to unit angular velocity of the plate. Each of the basis velocity fields contains a factor $(\tilde{z}-c/2)^{1/2}(\tilde{z}+c/2)^{1/2}$ in the denominator, the expected square root singularity at each of the edges of the plate. We will return to this singular behavior later in this chapter.

Impulses To develop the basis impulses from the Cauchy integral solution, we will use the complex forms exhibited in Results 6.5 and 6.6 in Sect. 6.2. In Eqs. (6.40) and (6.43), the contour integrations can be readily carried out using the integral results in Sect. A.3.3. It can be confirmed that the motion-induced impulses (and those from the uniform flow) are identical to those derived in (8.27) and (8.28). The unit vorticity-induced impulses are

$$\tilde{P}_{\mathrm{J}}^{\mathrm{v}} = -\mathrm{i}\tilde{z}_{\mathrm{J}} - \mathrm{i}\,\mathrm{Re}\left((\tilde{z}_{\mathrm{J}}-c/2)^{1/2}(\tilde{z}_{\mathrm{J}}+c/2)^{1/2} - \tilde{z}_{\mathrm{J}}\right), \tag{8.59}$$

$$\Pi_{\mathrm{r,J}}^{\mathrm{v}} = -\frac{1}{2}|\tilde{z}_{\mathrm{J}}|^2 - \frac{1}{2}\mathrm{Re}\left[\tilde{z}_{\mathrm{J}}\left((\tilde{z}_{\mathrm{J}}-c/2)^{1/2}(\tilde{z}_{\mathrm{J}}+c/2)^{1/2} - \tilde{z}_{\mathrm{J}}\right)\right]; \tag{8.60}$$

these can be confirmed to match the results (8.31), once the conformal mapping is accounted for. Here, as in Eqs. (6.97) and (6.99), we have isolated a leading term in each expression that represents the 'direct' impulse of the point vortex; the remaining part of each expression comprises the vortex's indirect contribution, due to the reacting flow generated by the impenetrable plate. Using the notation we introduced earlier for this indirect contribution,

$$\tilde{P}_{\mathrm{J}}^{\mathrm{bv}} := -\mathrm{i}\,\mathrm{Re}\left((\tilde{z}_{\mathrm{J}}-c/2)^{1/2}(\tilde{z}_{\mathrm{J}}+c/2)^{1/2} - \tilde{z}_{\mathrm{J}}\right), \tag{8.61}$$

$$\Pi_{\mathrm{r,J}}^{\mathrm{bv}} := -\frac{1}{2}\mathrm{Re}\left[\tilde{z}_{\mathrm{J}}\left((\tilde{z}_{\mathrm{J}}-c/2)^{1/2}(\tilde{z}_{\mathrm{J}}+c/2)^{1/2} - \tilde{z}_{\mathrm{J}}\right)\right]. \tag{8.62}$$

The rates of change of the vorticity-induced impulses are crucial for computing the force and moment. In Eqs. (6.101) and (6.102) we already provided general expressions for these rates of change for any body. Let us develop expressions here for the specific case of the flat plate, in terms of the vortex velocity in the physical plane. These are

$$\frac{\mathrm{d}\tilde{P}_{\mathrm{J}}^{\mathrm{v}}}{\mathrm{d}t} = -\mathrm{i}\frac{\mathrm{d}\tilde{z}_{\mathrm{J}}}{\mathrm{d}t} - \mathrm{i}\,\mathrm{Re}\left[\left(\frac{\tilde{z}_{\mathrm{J}}}{(\tilde{z}_{\mathrm{J}}-c/2)^{1/2}(\tilde{z}_{\mathrm{J}}+c/2)^{1/2}} - 1\right)\frac{\mathrm{d}\tilde{z}_{\mathrm{J}}}{\mathrm{d}t}\right] \tag{8.63}$$

and

$$\frac{\mathrm{d}\Pi^{\mathrm{v}}_{\mathrm{r,J}}}{\mathrm{d}t} = -\mathrm{Re}\left(\tilde{z}^*_{\mathrm{J}}\frac{\mathrm{d}\tilde{z}_{\mathrm{J}}}{\mathrm{d}t}\right) - \mathrm{Re}\left[\left(\frac{\tilde{z}^2_{\mathrm{J}} - c^2/8}{(\tilde{z}_{\mathrm{J}} - c/2)^{1/2}(\tilde{z}_{\mathrm{J}} + c/2)^{1/2}} - \tilde{z}_{\mathrm{J}}\right)\frac{\mathrm{d}\tilde{z}_{\mathrm{J}}}{\mathrm{d}t}\right]. \tag{8.64}$$

Note that the factors multiplying $\mathrm{d}\tilde{z}_{\mathrm{J}}/dt$ in the latter terms in these two results closely resemble our expressions for $\tilde{w}^{(2)}_{\mathrm{bt}}$ and for $\tilde{w}_{\mathrm{br}}$, respectively. This is not a coincidence. In fact, these results are exactly of the form given by the formulation of force and moment given in Result 6.14, particularly in the complex forms (6.169) and (6.171). Thus, we have confirmed that the expressions given in Result 6.14 for the rate of change of the vorticity-induced impulse hold for the case of a two-dimensional flat plate. As a vortex moves away from the plate (so that $|\tilde{z}_{\mathrm{J}}|$ becomes large), the leading term in these expressions—the rate of change of the direct contribution of the vortex to the impulse—is dominant, while the remaining term—the indirect effect from the body's presence—vanishes.

8.2.3 Approach 3: Fourier–Chebyshev Expansion

Now, we will pursue an approach that starts from the same governing equation (8.35) used in the Cauchy integral approach. However, in this case, we will rely on an expansion in a judiciously-chosen set of basis functions. This approach will be of more general use in other contexts, so it is useful to demonstrate it here on a problem for which we have alternative tools.

Let us start by describing the body-fixed coordinate system in its scaled form ξ, defined in (8.1), which assumes the real-valued range $\xi \in [-1, 1]$ on the plate. This is the same range of values swept by a cosine function (in reverse) when acting over the interval $[0, \pi]$. Thus, the plate can be re-parametrized by a new coordinate, $\theta(\xi) := \arccos \xi$. From an inspection of our mapping from the unit circle in Sect. 8.2, this coordinate is simply the angle of the corresponding points on the upper semicircle, C^+_c. Note that differential elements in ξ and θ are related by

$$\mathrm{d}\xi = -\sin\theta\,\mathrm{d}\theta, \qquad \mathrm{d}\theta = -\frac{\mathrm{d}\xi}{(1-\xi^2)^{1/2}}. \tag{8.65}$$

The direct relationship between θ and the body-fixed coordinate of a point $\tilde{z} \in \mathcal{P}$ on the plate is $\tilde{z}(\theta) = (c/2)\cos\theta$; note that $\theta = \pi$ corresponds to the leading edge ($\tilde{z}_L := -c/2$) and $\theta = 0$ to the trailing edge ($\tilde{z}_T := c/2$).

Solution for the Vortex Sheet In the new coordinate, θ, we can expand the vortex sheet strength in a cosine series. However, we must also take care to acknowledge the known square root singularities in this strength distribution. When viewed as a function of ξ, this is equivalent to writing the vortex sheet strength in a **Fourier–Chebyshev expansion**,

$$\gamma(\tilde{z}(\xi)) = \frac{1}{(1-\xi^2)^{1/2}}\sum_{n=0}^{\infty} B_n T_n(\xi), \tag{8.66}$$

where T_n is the nth order **Chebyshev polynomial** of the first kind,

$$T_n(\xi) := \cos(n \arccos \xi). \tag{8.67}$$

Properties of the Chebyshev polynomials are summarized in Sect. A.3.2 of the Appendix. Let us introduce the vortex sheet expansion (8.66) into the integral equation (8.35) and evaluate the resulting integrals over the plate with Eq. (A.226), after rescaling the dummy variable with $\tilde{\lambda} = cl/2$, where $l \in [-1, 1]$. The result of this is

$$\frac{1}{2\pi} \sum_{n=0}^{\infty} B_n \int_{-1}^{1} \frac{T_n(l)}{(1-l^2)^{1/2}(l-\xi)} \, dl = \Delta V_\perp(\tilde{z}(\xi)) \longrightarrow \sum_{n=1}^{\infty} B_n U_{n-1}(\xi) = 2\Delta V_\perp(\tilde{z}(\xi)), \tag{8.68}$$

where $U_n(\xi)$ is the nth order Chebyshev polynomial of the second kind (A.210). It is important to note that B_0 has disappeared from the summation in this step. Why? Because the leading term in the expansion (8.66)—$B_0/(1-\xi^2)^{1/2}$—is the homogeneous solution of the integral equation, explicitly identified in the Cauchy solution (8.39). As in that approach, the value of B_0 can be related to the total plate circulation, and, by Kelvin's circulation theorem, to the sum of the strengths of the fluid vortices. This is easily verified with the help of (A.221) to be

$$B_0 = \frac{2\Gamma_{\mathcal{P}}}{\pi c} = -\frac{2}{\pi c} \sum_{\mathrm{J}=1}^{N_\mathrm{v}} \Gamma_\mathrm{J}. \tag{8.69}$$

The remaining coefficients, B_n, for $n > 0$, are determined from Eq. (8.68), which we invert with the orthogonality relations (A.222) for U_n. The results of this are

$$B_n = \frac{4}{\pi} \int_{-1}^{1} (1-\xi^2)^{1/2} U_{n-1}(\xi) \Delta V_\perp(\tilde{z}(\xi)) \, d\xi. \tag{8.70}$$

At this point, we could proceed to introduce the expression for $\Delta V_\perp$ and carry out the integration. However, in order to relate our approach here to a classical treatment—namely, Glauert's thin-airfoil theory—let us rewrite the integral in (8.70) in terms of Chebyshev polynomials of the first kind, using the recurrence relation (A.219). This leads to

$$B_n = \frac{2}{\pi} \int_{-1}^{1} \frac{\Delta V_\perp(\tilde{z}(\xi))}{(1-\xi^2)^{1/2}} \left(T_{n-1}(\xi) - T_{n+1}(\xi)\right) d\xi. \tag{8.71}$$

The function $\Delta V_\perp(\tilde{z})$ is smooth and bounded on the plate[2] and can be expanded in a series of Chebyshev polynomials of the first kind

$$\Delta V_\perp(\tilde{z}(\xi)) = -\sum_{n=0}^{\infty} A_n T_n(\xi). \tag{8.72}$$

The negative sign in (8.72) is included in anticipation of comparison with the classical results of Glauert's thin airfoil theory; that theory utilizes the downwash, defined earlier as the negative of $\Delta V_\perp$. It is then a simple matter of using the orthogonality

[2] If the fluid vorticity takes the form of a free vortex sheet, then $\Delta V_\perp(\tilde{z})$ has a logarithmic singularity at the edge where the sheet is rooted. We will discuss this in Sect. 8.4.1.

relations (A.221) to show that

$$A_n = -\frac{\sigma_n}{\pi}\int_{-1}^{1}\frac{\Delta V_\perp(\tilde{z}(\xi))}{(1-\xi^2)^{1/2}}T_n(\xi)\,\mathrm{d}\xi, \tag{8.73}$$

where we have defined the factor σ_n as

$$\sigma_n := \begin{cases} 1, & n = 0, \\ 2, & n > 0. \end{cases} \tag{8.74}$$

Thus, by simple comparison with (8.71), the set of coefficients for the vortex sheet can immediately be written in terms of the coefficients of the normal velocity induced on the plate:

$$B_1 = A_2 - 2A_0, \qquad B_n = A_{n+1} - A_{n-1}, \quad n > 1. \tag{8.75}$$

These relationships lead to the following form for the vortex sheet strength distribution:

$$\gamma(\xi) = \frac{1}{(1-\xi^2)^{1/2}}\left[-2A_0T_1(\xi) - A_1T_2(\xi) - \sum_{n=2}^{\infty}A_n\left(T_{n+1}(\xi) - T_{n-1}(\xi)\right) + B_0\right]. \tag{8.76}$$

Using the recurrence relation (A.219) again, we can rewrite the expression within the square brackets (mostly) in terms of Chebyshev polynomials of the second kind, allowing us to easily distinguish the behavior of the bound vortex sheet at its end points:

Result 8.2: Fourier–Chebyshev Form of the Bound Vortex Sheet Strength

The strength distribution of the bound vortex sheet on an infinitely-thin flat plate in two-dimensional flow, parameterized by $\xi \in [-1, 1]$, is given, in general, by

$$\gamma(\xi) = \frac{1}{(1-\xi^2)^{1/2}}\left(B_0 - A_1 - 2A_0T_1(\xi)\right) + 2\left(1-\xi^2\right)^{1/2}\sum_{n=1}^{\infty}A_nU_{n-1}(\xi), \tag{8.77}$$

where, for all $n \geq 0$, A_n are the coefficients of the expansion of the normal velocity difference (8.72), and the coefficient B_0 is related to the overall strength of the sheet (and, in turn, to the circulation in the fluid) by Eq. (8.69).

We can also write the vortex sheet strength in the general Fourier form in terms of $\theta = \arccos\xi$:

$$\gamma(\theta) = \frac{1}{\sin\theta}\left(B_0 - A_1 - 2A_0\cos\theta\right) + 2\sum_{n=1}^{\infty}A_n\sin(n\theta). \tag{8.78}$$

These forms in Result 8.2 have the important advantage that the singular behavior of γ is explicitly identified in the first three terms; the remaining terms, represented in the summation, are smooth over the entire plate. We will have need for the singular part later, for example, when we seek to apply the Kutta condition.

For later reference, let us derive an expression for the local bound circulation along the plate, defined in Eq. (8.6). Parameterizing this by the scaled coordinate, ξ, and using the integral results (A.223) and (A.224) and the relationship (8.69) between B_0 and the full bound circulation, we arrive at

$$\Gamma(\xi) = \Gamma_L + \Gamma_{\mathcal{P}} \left(1 - \frac{1}{\pi} \arccos \xi\right) + \frac{c}{2}(1-\xi^2)^{1/2} \left[2A_0 U_0(\xi) + \sum_{n=1}^{\infty} A_n \left(\frac{U_n(\xi)}{n+1} - \frac{U_{n-2}(\xi)}{n-1}\right)\right]. \tag{8.79}$$

(Note that we are relying on the interpretation $U_{-1}(\xi) := 0$ in this expression.) Evaluating this at the trailing edge, $\xi = 1$, clearly results in $\Gamma(1) = \Gamma_L + \Gamma_{\mathcal{P}}$, as expected. By the global conservation of circulation, Eq. (8.9), this is equal to the negative of the circulation of vorticity associated with the trailing edge, $-\Gamma_T$. It is interesting to note that, if we use this conservation to replace the plate's bound circulation, $\Gamma_{\mathcal{P}}$, with the sum of the circulation in the fluid vorticity, $-\Gamma_L - \Gamma_T$, then the first two terms in (8.79), written in terms of $\theta = \arccos \xi$, are $\Gamma_L \theta/\pi - \Gamma_T(1-\theta/\pi)$—a linear interpolation between the values of local circulation at the edges. The remaining terms in (8.79) modify this interpolation, but only away from the edges.

As we observed in Eq. (3.194) for any bound vortex sheet, the local pressure jump across the plate depends on the rate of change of this local circulation at each station ξ. This rate of change arises from the time variation of the circulations of fluid vorticity, Γ_L and Γ_T, and of the coefficients A_n, for all $n \geq 0$.

Decomposed Coefficients and the Basis Sheets The vortex sheet distribution in Result 8.2 depends on the coefficients A_n of the expanded normal velocity difference (8.72). We can take one of two approaches for computing these coefficients. The first approach is numerical: we recognize that the Chebyshev expansion is equivalent to a discrete cosine series in the independent variable $\theta = \arccos \xi$. When this series is truncated after N terms, the coefficients can be computed with a **discrete cosine transform (DCT)**. In the DCT we evaluate $\Delta V_\perp$ at points along the plate that are evenly spaced in $\theta \in [0, \pi]$. The transform itself is simply the explicit integral (8.73), discretized with a quadrature rule over these evaluation points.

However, we need not rely on a numerical technique, because the integral (8.73) can be evaluated analytically for each of the contributors in (8.41). Therefore, in this discussion, we will take this analytical approach. Note that if we decompose the normal velocity difference (8.41) into contributors in the usual manner, the coefficients A_n and B_n are decomposed analogously:

$$A_n = (\tilde{V}_r - \tilde{V}_\infty) A_{bt,n} + \Omega A_{br,n} + A_{v,n}, \quad n = 0, \ldots, \infty. \tag{8.80}$$

and

$$B_n = (\tilde{V}_r - \tilde{V}_\infty)B_{bt,n} + \Omega B_{br,n} + B_{v,n}, \quad n = 1, \ldots, \infty. \tag{8.81}$$

The coefficients in these respective decompositions are still related to each other through Eq. (8.75). Note that we have grouped the body translation and the uniform flow normal to the plate into a single term, since these have equal and opposite effect in the no-penetration condition. Also, observe that B_0 is omitted from this set of decomposed coefficients. As we have discussed already, this coefficient represents the plate's circulation and serves as our means of enforcing the constraint on global circulation, Eq. (8.69). In this book we group this coefficient entirely with the vorticity-induced contribution of the flow (i.e., $B_0 = B_{v,0}$), as we will show below. Later, in Note 8.5.1, we will discuss an alternative (and more traditional) decomposition that follows the approach used for the other coefficients.

First, let us calculate the coefficients due to body motion (and uniform flow). Using the orthogonality relations (A.221), we can immediately calculate the coefficients due to unit translational velocity normal to the plate (for which $\Delta V_\perp(\tilde{z})$ is set to -1),

$$A_{bt,n} = \begin{cases} 1, & n = 0, \\ 0, & n \neq 0, \end{cases} \longrightarrow B_{bt,n} = \begin{cases} -2, & n = 1, \\ 0, & n \neq 1. \end{cases} \tag{8.82}$$

Thus, the basis vortex sheets associated with this motion and with uniform flow normal to the plate are

$$\gamma_{bt}^{(2)}(\tilde{z}) = -\gamma_\infty^{(2)}(\tilde{z}) = -\frac{2T_1(2\tilde{z}/c)}{(1 - 4\tilde{z}^2/c^2)^{1/2}}, \tag{8.83}$$

identical to (8.45) and the first of Eq. (8.23), the results achieved by the other approaches.

For unit angular velocity (for which $\Delta V_\perp(\tilde{z})$ is set to $-\tilde{z} = -c\xi/2$), the coefficients are found to be

$$A_{br,n} = \begin{cases} c/2, & n = 1, \\ 0, & n \neq 1, \end{cases} \longrightarrow B_{br,n} = \begin{cases} -c/2, & n = 2, \\ 0, & n \neq 2, \end{cases} \tag{8.84}$$

so that

$$\gamma_{br}(\tilde{z}) = -\frac{cT_2(2\tilde{z}/c)/2}{(1 - 4\tilde{z}^2/c^2)^{1/2}}, \tag{8.85}$$

equal to (8.46) and the second of Eq. (8.23).

Note 8.2.1: Fourier–Chebyshev for Wavy Plate or Sinusoidal Gust

Though we have focused this chapter on a rigid plate, there will be occasions later in this book—and, indeed, even later in this chapter—in which we need to consider a plate undergoing a small sinusoidal deformation or subjected to a sinusoidal **gust** in the ambient fluid. In either case, we are faced with a normal velocity of the form $\Delta V_{\perp}(\tilde{z}) = \mathrm{Re}(v_0 \mathrm{e}^{-\mathrm{i}\kappa\xi})$, where v_0 is a complex, perhaps time-varying amplitude, and κ is a wavenumber. Here, we seek the Fourier–Chebyshev coefficients associated with the basic waveform, $\mathrm{e}^{-\mathrm{i}\kappa\xi}$. Substituting into (8.73), the coefficients are given by

$$A_{\kappa,n} = -\frac{\sigma_n}{\pi} \int_{-1}^{1} \frac{\mathrm{e}^{-\mathrm{i}\kappa\xi}}{(1-\xi^2)^{1/2}} T_n(\xi)\, \mathrm{d}\xi. \tag{8.86}$$

This integral can be evaluated analytically; its result is given in (A.225), so that, for all $n \geq 0$,

$$A_{\kappa,n} = -\sigma_n \mathrm{i}^{-n} J_n(\kappa), \tag{8.87}$$

where J_n is the nth-order Bessel function of the first kind. Using the recurrence relations (8.75), and a standard recurrence relation for Bessel functions, these correspond to the vortex sheet coefficients

$$B_{\kappa,n} = -\frac{4n\mathrm{i}^{-n-1}}{\kappa} J_n(\kappa), \tag{8.88}$$

for all $n > 0$.

The coefficients due to fluid vorticity, given by

$$A_{\mathrm{v},n} = \frac{\sigma_n}{2\pi} \sum_{\mathrm{J}=1}^{N_\mathrm{v}} \Gamma_\mathrm{J} \mathrm{Im}\left(\frac{1}{\pi \mathrm{i}} \int_{-1}^{1} \frac{T_n(\xi)\, \mathrm{d}\xi}{(1-\xi^2)^{1/2}(\tilde{z}(\xi) - \tilde{z}_\mathrm{J})} \right), \tag{8.89}$$

can be obtained from the integral relation (A.230), which invokes the **extended Chebyshev polynomials** of the second kind, $\overline{U}_n$, defined throughout the complex plane per (A.233). Since we are only interested in cases in which the vortex $\tilde{z}_\mathrm{J}$ is not on the plate, then it is straightforward to write the result:

$$\begin{aligned} A_{\mathrm{v},n} &= -\frac{\sigma_n}{\pi c} \sum_{\mathrm{J}=1}^{N_\mathrm{v}} \Gamma_\mathrm{J} \mathrm{Re}\left(\overline{U}_{n-1}(2\tilde{z}_\mathrm{J}/c) \right) \\ &= \frac{\sigma_n}{\pi c} \sum_{\mathrm{J}=1}^{N_\mathrm{v}} \Gamma_\mathrm{J} \mathrm{Re}\left(\frac{\left[2\tilde{z}_\mathrm{J}/c - (2\tilde{z}_\mathrm{J}/c - 1)^{1/2}(2\tilde{z}_\mathrm{J}/c + 1)^{1/2} \right]^n}{(2\tilde{z}_\mathrm{J}/c - 1)^{1/2}(2\tilde{z}_\mathrm{J}/c + 1)^{1/2}} \right), \end{aligned} \tag{8.90}$$

for all $n \geq 0$. It is useful to note that, based on the inverse Joukowski transform relations (8.11) and (8.12), these coefficients can also be written as

$$A_{\mathrm{v},n} = \frac{2\sigma_n}{\pi c} \sum_{\mathrm{J}=1}^{N_\mathrm{v}} \Gamma_\mathrm{J} \mathrm{Re}\left[(\zeta_\mathrm{J} - 1/\zeta_\mathrm{J})^{-1} \zeta_\mathrm{J}^{-n} \right]. \tag{8.91}$$

Using the relations between coefficients (8.75), it is straightforward to show that, in the forms of our three different approaches, the coefficients of the vortex sheet strength expansion are

$$B_{v,n} = -\frac{2\sigma_n}{\pi c}\sum_{J=1}^{N_v}\Gamma_J \mathrm{Re}(\zeta_J^{-n})$$

$$= -\frac{2\sigma_n}{\pi c}\sum_{J=1}^{N_v}\Gamma_J \mathrm{Re}\left[\left(2\tilde{z}_J/c - (2\tilde{z}_J/c-1)^{1/2}(2\tilde{z}_J/c+1)^{1/2}\right)^n\right] \tag{8.92}$$

$$= -\frac{2\sigma_n}{\pi c}\sum_{J=1}^{N_v}\Gamma_J \mathrm{Re}\left(\overline{T}_n(2\tilde{z}_J/c)\right), \tag{8.93}$$

for all $n \geq 0$. Note that we have added to this set the coefficient $B_{v,0} \equiv B_0$, corresponding to the homogeneous term, for which the right-hand side reduces to the form proportional to the total circulation of the fluid vortices (8.69). The vorticity-induced vortex sheet is thus

$$\gamma_v(\tilde{z}) = -\frac{2}{\pi c(1-4\tilde{z}^2/c^2)^{1/2}}\sum_{J=1}^{N_v}\Gamma_J\sum_{n=0}^{\infty}\sigma_n \mathrm{Re}\left(\overline{T}_n(2\tilde{z}_J/c)\right)T_n(2\tilde{z}/c). \tag{8.94}$$

At first appearance, it is not obvious that the infinite sum in this expression is convergent. However, by substituting the forms of the Chebyshev polynomials in circle plane coordinates (A.237), it is straightforward to show that the sum converges. One only needs to remember the standard expansion $1/(1-x) = \sum_{n=0}^{\infty} x^n$ for $|x| < 1$. With some manipulation, the resulting expression can be shown to be identical to (8.53) and (8.24).

Basis Velocity Fields The expansions of the vortex sheet distribution along the plate can be carried over to expansions of the fields in the surrounding fluid, particularly the complex velocity (8.33) fields. We simply substitute the expansion (8.66) for the vortex sheet strength and evaluate the resulting integrals. We make use of (A.230) to obtain

Result 8.3: Fourier–Chebyshev Form of the Velocity and Potential Fields

The complex velocity field (expressed in body-fixed components) about an infinitely-thin flat plate in two dimensions is given by

$$\tilde{w}(\tilde{z}) = \frac{\mathrm{i}}{2}\sum_{n=0}^{\infty}B_n\overline{U}_{n-1}(2\tilde{z}/c) + \tilde{W}_\infty + \frac{1}{2\pi\mathrm{i}}\sum_{J=1}^{N_v}\frac{\Gamma_J}{\tilde{z}-\tilde{z}_J}, \tag{8.95}$$

where $\overline{U}_n$ is the nth-order extended Chebyshev polynomial of the second kind (A.233), and B_n, $n \geq 0$, are the coefficients of the Fourier–Chebyshev expansion of the bound vortex sheet strength (8.66). These coefficients are related to those (A_n) of the normal velocity difference, $\Delta V_\perp$, on the plate by (8.75) for all $n > 0$; they are given in decomposed form in (8.81), where the expressions for the individual basis coefficients are provided in (8.82), (8.84) and (8.93). Among these, the coefficient B_0 is related to circulation about the plate and in the fluid by Eq. (8.69).

The complex potential field is formed from the integral of this velocity expression. We rely on the simple fact that the Chebyshev polynomials of the second kind are the derivatives of the first kind (A.215), with a caveat (A.240) at $n = 0$. This leads to

$$F(\tilde{z}) = -\frac{\mathrm{i}c}{4} B_0 \log\left(2\tilde{z}/c + (2\tilde{z}/c - 1)^{1/2}(2\tilde{z}/c + 1)^{1/2}\right)$$
$$+ \frac{\mathrm{i}c}{4}\sum_{n=1}^{\infty} \frac{1}{n} B_n \overline{T}_n(2\tilde{z}/c) + \tilde{W}_\infty \tilde{z} + \frac{1}{2\pi \mathrm{i}} \sum_{J=1}^{N_v} \Gamma_J \log(\tilde{z} - \tilde{z}_J), \quad (8.96)$$

where $\overline{T}_n$ is the nth-order extended Chebyshev polynomial of the first kind (A.232).

Let us decompose the velocity into contributors in the usual way. The velocity due to unit plate translational velocity in its normal direction and due to unit angular velocity are, respectively,

$$\tilde{w}_{\mathrm{bt}}^{(2)}(\tilde{z}) = -\mathrm{i}\overline{U}_0(2\tilde{z}/c), \qquad \tilde{w}_{\mathrm{br}}(\tilde{z}) = -\mathrm{i}\frac{c}{4}\overline{U}_1(2\tilde{z}/c). \quad (8.97)$$

The basis velocity fields due to uniform flow are given, as in the previous two approaches, by $\tilde{w}_\infty^{(1)}(\tilde{z}) = 1$ and $\tilde{w}_\infty^{(2)}(\tilde{z}) = -\tilde{w}_{\mathrm{bt}}^{(2)}(\tilde{z}) - \mathrm{i}$. And the vorticity-induced velocity is

$$\tilde{w}_{\mathrm{v}}(\tilde{z}) = \frac{1}{2\pi \mathrm{i}} \sum_{J=1}^{N_v} \Gamma_J \left[\frac{1}{\tilde{z} - \tilde{z}_J} + \frac{4}{c} \sum_{n=0}^{\infty} \mathrm{Re}\left(\overline{T}_n(2\tilde{z}_J/c)\right) \overline{U}_{n-1}(2\tilde{z}/c) \right]. \quad (8.98)$$

Using similar tricks as with (8.94), the infinite series in this expression can be shown to converge, and the full vorticity-induced velocity field is equivalent to (8.55) for each vortex element. For practical (numerical) calculation, that converged expression is clearly preferable in order to avoid unnecessary expense in calculating each term and the truncation of the infinite sum.

Impulses The linear and angular impulses can be easily computed in this Fourier–Chebyshev expansion approach. We rely on the expressions (6.40) and (6.43), just as we did for the Cauchy integral approach. When we substitute the expansion (8.66) and evaluate the resulting integrals with the orthogonality relations (A.221), it is easy to show that the results are

$$\tilde{P} = -\mathrm{i}\sum_{J=1}^{N_v}\Gamma_J \tilde{z}_J - \frac{\mathrm{i}\pi c^2}{8}B_1, \qquad \Pi_r = -\frac{1}{2}\sum_{J=1}^{N_v}\Gamma_J|\tilde{z}_J|^2 - \frac{\pi c^3}{32}B_0 - \frac{\pi c^3}{64}B_2. \quad (8.99)$$

It is useful to note that the homogeneous solution contributes to the angular impulse (as evident in the appearance of B_0), but not the linear impulse. These impulses can be decomposed into contributors in our usual manner; the results are identical to those reported in (8.27)–(8.31), as easily verified.

8.3 Force and Moment

In the previous section, we obtained the solution for the decomposed flow about a flat plate by three different approaches. In each approach, we were able to derive expressions for the impulses induced by the various flow contributors; their rates of change are obtained either directly, by differentiating, or by using the formulations derived in Eqs. (6.169) and (6.171). Let us assemble these here—guided by the general form for a rigid body in Result 6.14—into a full expression of the force and moment on the plate.

In body-fixed coordinates, the force on a flat plate of length c in rigid-body motion, amidst N_v point vortices of positions z_J and strengths Γ_J, is described by

$$\tilde{f}_x = \frac{\pi}{4}\rho c^2 \Omega \tilde{V}_r - \rho\frac{\mathrm{d}}{\mathrm{d}t}\sum_{J=1}^{N_v}\Gamma_J \mathrm{Im}\,(\tilde{z}_J) - \rho\Omega\sum_{J=1}^{N_v}\Gamma_J \mathrm{Re}\left((\tilde{z}_J - c/2)^{1/2}(\tilde{z}_J + c/2)^{1/2}\right) \quad (8.100)$$

$$\begin{aligned}\tilde{f}_y = &-\frac{\pi}{4}\rho c^2 \dot{\tilde{V}}_r \\ &+\rho\frac{\mathrm{d}}{\mathrm{d}t}\sum_{J=1}^{N_v}\Gamma_J \mathrm{Re}\,(\tilde{z}_J) + \rho\sum_{J=1}^{N_v}\Gamma_J \mathrm{Re}\left[\left(\frac{\tilde{z}_J}{(\tilde{z}_J - c/2)^{1/2}(\tilde{z}_J + c/2)^{1/2}} - 1\right)\frac{\mathrm{d}\tilde{z}_J}{\mathrm{d}t}\right] \\ &+\rho\sum_{J=1}^{N_v}\dot{\Gamma}_J \mathrm{Re}\left((\tilde{z}_J - c/2)^{1/2}(\tilde{z}_J + c/2)^{1/2} - \tilde{z}_J\right) - \rho\Omega\sum_{J=1}^{N_v}\Gamma_J \mathrm{Im}(\tilde{z}_J), \quad (8.101)\end{aligned}$$

and the moment about the reference point $\boldsymbol{X}_r$ is

$$\begin{aligned}
m_{\mathrm{r}} = &-\frac{\pi}{128}\rho c^4\dot{\Omega} \\
&+\frac{1}{2}\rho\frac{\mathrm{d}}{\mathrm{d}t}\sum_{\mathrm{J}=1}^{N_{\mathrm{v}}}\Gamma_{\mathrm{J}}|\tilde{z}_{\mathrm{J}}|^2+\rho\sum_{\mathrm{J}=1}^{N_{\mathrm{v}}}\Gamma_{\mathrm{J}}\mathrm{Re}\left[\left(\frac{\tilde{z}_{\mathrm{J}}^2-c^2/8}{(\tilde{z}_{\mathrm{J}}-c/2)^{1/2}(\tilde{z}_{\mathrm{J}}+c/2)^{1/2}}-\tilde{z}_{\mathrm{J}}\right)\frac{\mathrm{d}\tilde{z}_{\mathrm{J}}}{\mathrm{d}t}\right] \\
&+\frac{1}{2}\rho\sum_{\mathrm{J}=1}^{N_{\mathrm{v}}}\dot{\Gamma}_{\mathrm{J}}\mathrm{Re}\left(\tilde{z}_{\mathrm{J}}(\tilde{z}_{\mathrm{J}}-c/2)^{1/2}(\tilde{z}_{\mathrm{J}}+c/2)^{1/2}-\tilde{z}_{\mathrm{J}}^2\right) \qquad (8.102)\\
&-\frac{\pi}{4}\rho c^2\tilde{U}_{\mathrm{r}}\tilde{V}_{\mathrm{r}}+\rho\sum_{\mathrm{J}=1}^{N_{\mathrm{v}}}\Gamma_{\mathrm{J}}\left[\tilde{U}_{\mathrm{r}}\mathrm{Re}\left((\tilde{z}_{\mathrm{J}}-c/2)^{1/2}(\tilde{z}_{\mathrm{J}}+c/2)^{1/2}\right)+\tilde{V}_{\mathrm{r}}\mathrm{Im}(\tilde{z}_{\mathrm{J}})\right].
\end{aligned}$$

This form is quite general, but not easily dissected for deeper insight. Let us rewrite the expressions in vector form for some help with finding that insight:

Result 8.4: Force and Moment on a Flat Plate

For a flat plate of length c in general rigid-body motion described by $\boldsymbol{V}_{\mathrm{r}}$ and Ω, with N_{v} point vortices in the surrounding fluid, the force on the plate is given by

$$\begin{aligned}
\boldsymbol{f} = &-\frac{\pi}{4}\rho c^2\dot{\tilde{V}}_{\mathrm{r}}\boldsymbol{n}+\frac{\pi}{4}\rho c^2\Omega\tilde{V}_{\mathrm{r}}\boldsymbol{\tau}+\rho\frac{\mathrm{d}}{\mathrm{d}t}\sum_{\mathrm{J}=1}^{N_{\mathrm{v}}}\Gamma_{\mathrm{J}}\boldsymbol{e}_3\times\tilde{\boldsymbol{x}}_{\mathrm{J}} \\
&-\rho\sum_{\mathrm{J}=1}^{N_{\mathrm{v}}}\left[\Gamma_{\mathrm{J}}\boldsymbol{e}_3\cdot\left(\tilde{\boldsymbol{v}}_{\mathrm{J}}\times\boldsymbol{v}_{\mathrm{bt}}^{(2)}(\tilde{\boldsymbol{x}}_{\mathrm{J}})\right)+\dot{\Gamma}_{\mathrm{J}}\Delta P(\tilde{\boldsymbol{x}}_{\mathrm{J}})\right]\boldsymbol{n} \qquad (8.103)\\
&-\rho\Omega\sum_{\mathrm{J}=1}^{N_{\mathrm{v}}}\Gamma_{\mathrm{J}}\left[\tilde{\boldsymbol{x}}_{\mathrm{J}}\cdot\boldsymbol{v}_{\mathrm{bt}}^{(2)}(\tilde{\boldsymbol{x}}_{\mathrm{J}})\boldsymbol{n}-\Delta P(\tilde{\boldsymbol{x}}_{\mathrm{J}})\boldsymbol{\tau}\right]
\end{aligned}$$

and the moment about the plate's reference point is

$$\begin{aligned}
m_{\mathrm{r}} = &-\frac{\pi}{128}\rho c^4\dot{\Omega}-\frac{\pi}{4}\rho c^2\tilde{U}_{\mathrm{r}}\tilde{V}_{\mathrm{r}}+\frac{1}{2}\rho\frac{\mathrm{d}}{\mathrm{d}t}\sum_{\mathrm{J}=1}^{N_{\mathrm{v}}}\Gamma_{\mathrm{J}}|\tilde{\boldsymbol{x}}_{\mathrm{J}}|^2+\rho\sum_{\mathrm{J}=1}^{N_{\mathrm{v}}}\Gamma_{\mathrm{J}}\boldsymbol{V}_{\mathrm{r}}\cdot\tilde{\boldsymbol{x}}_{\mathrm{J}} \\
&-\rho\sum_{\mathrm{J}=1}^{N_{\mathrm{v}}}\left[\Gamma_{\mathrm{J}}\boldsymbol{e}_3\cdot(\tilde{\boldsymbol{v}}_{\mathrm{J}}\times\boldsymbol{v}_{\mathrm{br}}(\tilde{\boldsymbol{x}}_{\mathrm{J}})+\boldsymbol{V}_{\mathrm{r}}\times\boldsymbol{n}\Delta P(\tilde{\boldsymbol{x}}_{\mathrm{J}}))+\dot{\Gamma}_{\mathrm{J}}\Delta\Pi_{\mathrm{r}}(\tilde{\boldsymbol{x}}_{\mathrm{J}})\right] \\
&\qquad (8.104)\\
&-\rho\Omega\sum_{\mathrm{J}=1}^{N_{\mathrm{v}}}\Gamma_{\mathrm{J}}\tilde{\boldsymbol{x}}_{\mathrm{J}}\cdot\boldsymbol{v}_{\mathrm{br}}(\tilde{\boldsymbol{x}}_{\mathrm{J}}),
\end{aligned}$$

where, for brevity, we have defined relative vectors $\tilde{\boldsymbol{x}}_{\mathrm{J}} := \boldsymbol{x}_{\mathrm{J}} - \boldsymbol{X}_{\mathrm{r}}$ and $\tilde{\boldsymbol{v}}_{\mathrm{J}} := \mathrm{d}\tilde{\boldsymbol{x}}_{\mathrm{J}}/\mathrm{d}t = \dot{\boldsymbol{x}}_{\mathrm{J}} - \boldsymbol{V}_{\mathrm{r}}$. We have also denoted by $\Delta P(\tilde{\boldsymbol{x}}_{\mathrm{J}}) := \boldsymbol{P}_{\mathrm{J}}^{\mathrm{bv}}\cdot\tilde{\boldsymbol{e}}_2$ the sole component of the indirect part of the vortex-induced linear impulse (8.61); $\Delta\Pi_r(\tilde{x}_J) := \Pi^{bv}_{r,J}$ is the indirect part of the vorticity-induced angular impulse (8.62); and $\boldsymbol{v}^{(2)}_{bt}$ and $\boldsymbol{v}_{br}$ are, as usual, the vector forms of the basis velocity fields, $\tilde{w}^{(2)}_{bt}$ and $\tilde{w}_{br}$, given in (8.57) and (8.58), respectively. Recall, from Eq. (6.173), the indirect vorticity-induced impulses are streamfunctions of these basis velocity fields,

$$\boldsymbol{v}^{(2)}_{bt} = \boldsymbol{e}_3 \times \nabla(\Delta P), \qquad \boldsymbol{v}_{br} = \boldsymbol{e}_3 \times \nabla(\Delta\Pi_r). \tag{8.105}$$

The equations in Result 8.4 are still quite general, since they account for several possible behaviors. However, we have arranged the terms so that a few key aspects can be easily observed.

- The leading two terms in the force and the moment describe the inertial reaction (i.e., the added mass) response. The remaining terms on the first row describe the direct contribution of vortex motion to the force and moment, while the terms proportional to Γ_J on the second row represent the indirect contribution of the vortex, due to the body's presence.
- The terms proportional to $\dot{\Gamma}_J$—one in the second row and another in the first row that arises after applying the time derivative—clearly represent the effect of time-varying vortex strength. It was these terms in the force that we canceled by modifying the element velocity in the impulse matching principle described in Sect. 7.1.4.
- The last row of both equations accounts for the effect of plate rotation on the vortex contribution.
- For a given vortex in the set of elements, all of the terms in the second and third row of each equation vanish as the vortex moves far from the plate.

In particular, for a steadily-translating plate at small fixed angle of attack, in which only a single starting vortex emerges from the trailing edge of the plate, the vortex-related terms in the first row of each equation describe the steady-state force and moment, respectively. For example, the force is given by the direct contribution, $\rho\Gamma_J \boldsymbol{e}_3 \times \tilde{\boldsymbol{v}}_J$ for the single vortex, $J = 1$. This results in the well-known **Kutta–Joukowski lift theorem**, as will be discussed further in Sect. 8.5.1.

For more general plate motions that generate unsteady flows, the relative measures of a vortex element's direct and indirect contributions change along its trajectory. Its direct contribution provides easy intuition, and it is tempting to draw conclusions based solely on this part. Is this adequate? Its indirect contribution to force, for example, is proportional to $\tilde{\boldsymbol{v}}_J \times \boldsymbol{v}^{(2)}_{bt}(\tilde{x}_J)$. Thus, the force will be maximally affected by this part on vortex trajectories that are perpendicular to the $\boldsymbol{v}^{(2)}_{bt}$ streamlines—which, by (8.105), are also the contours of ΔP—and that lie in the high-$|\boldsymbol{v}^{(2)}_{bt}|$ region of the plate. This point is illustrated in Fig. 8.8, in which a flat plate has been accelerated from rest to constant speed at 45° angle of attack. Two typical trajectories of vortices originating from the leading and trailing edges are superimposed on a plot that

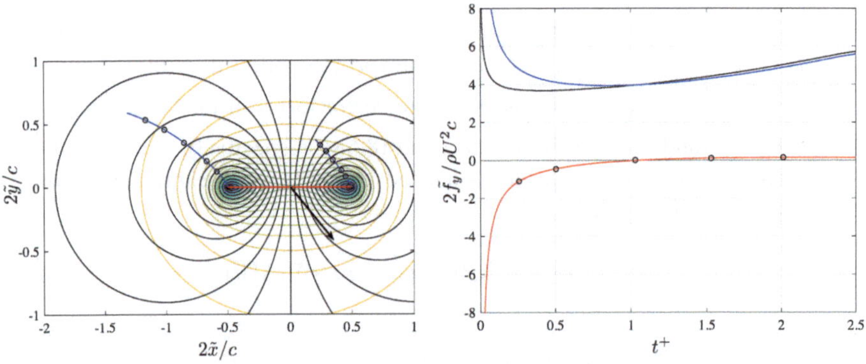

Fig. 8.8 A flat plate accelerated impulsively from rest to speed U at 45°, indicated by the arrow, generates trajectories of the initial leading- and trailing-edge vortices as shown in the left panel. Contours of the ΔP field, shown in black, represent the streamlines of the $\boldsymbol{v}_{\text{bt}}^{(2)}$ vector field, and contours of $|\boldsymbol{v}_{\text{bt}}^{(2)}|$ from 0.1 to 3 are shown in color, with darker (blue) colors indicating larger magnitude. (Right) Normal force coefficient on the plate versus convective time, $t^+ := tU/c$: direct vortex contribution (third term in Eq. (8.103)) (blue), indirect vortex contribution (fourth term in Eq. (8.103)) (red), total force (black). Circles on the indirect vortex force history correspond to circles along vortex trajectories on the left, with time progressing monotonically along trajectories from plate edges

contains both the streamlines and magnitude contours of the $\boldsymbol{v}_{\text{bt}}^{(2)}$ field of the plate. The resulting direct and indirect contributions made by these vortices to the plate's normal force—which, in this steady plate motion, constitute the entire force—are shown in the right panel of Fig. 8.8. The indirect contribution is significant at early times, but becomes negligibly weak after both vortices have exited the high-$|\boldsymbol{v}_{\text{bt}}^{(2)}|$ region. Since new vortices are continually developed at the salient edges of a plate at high angle of attack, this process tends to repeat itself. This example shows that in general unsteady flows ***we can never ignore the indirect contribution from the body's presence***.

There are a few other observations to make before we move on:

- The strengths and motions of the fluid vortex elements are left unspecified here. In Chap. 7 we discussed the vortex transport; in the next section, we will discuss the Kutta condition, which will allow us to relate these strengths to the motion parameters.
- In Note 2.5.2 we discussed how to transform the motion and force on a flat plate between the body-fixed coordinates and a coordinate system based at a certain location—e.g., a ***pivot point***—where the translational motion of the plate is particularly simple to describe. For example, we can utilize (2.70) and (2.71) to obtain the necessary body-fixed components of motion for use in Eqs. (8.103) and (8.104) from the underlying kinematics at this pivot point: velocity components $(U, \dot{h})$ along the two inertial axes, and angular velocity Ω about the out-of-plane axis. The force and moment can then be transformed to this pivot-point coordinate system with the transformation operator ${}^{\text{p}}\mathcal{T}_{\text{b}}^{(\text{f})}$:

$$\begin{pmatrix} m_{\mathrm{p}} \\ f_x \\ f_y \end{pmatrix} = \begin{bmatrix} 1 & 0 & -a \\ 0 & \cos\alpha & -\sin\alpha \\ 0 & \sin\alpha & \cos\alpha \end{bmatrix} \begin{pmatrix} m_{\mathrm{r}} \\ \tilde{f}_x \\ \tilde{f}_y \end{pmatrix}, \tag{8.106}$$

where the definitions of the angle α and the distance a are given in the diagram in Fig. 2.3.

- The results for force and moment, and those for the flow field in the prior sections of this chapter, are developed for a set of N_{v} point vortices in the fluid. However, they can easily be adapted when the fluid vorticity comprises a free vortex sheet (whose elements have time-invariant strengths), following the template for this adaptation outlined in Sect. 3.5. If we denote this vortex sheet by S and parameterize its configuration $\tilde{Z}(s)$ and real-valued strength distribution $\gamma_S(s)$ by arc length, s, along the sheet, then, for example, the normal component of force on the plate would be written as

$$\tilde{f}_y = -\frac{\pi}{4}\rho c^2 \dot{\tilde{V}}_{\mathrm{r}} + \rho \int_S \mathrm{Re}\left(\frac{\tilde{Z}(s)}{(\tilde{Z}(s)-c/2)^{1/2}(\tilde{Z}(s)+c/2)^{1/2}} \frac{\mathrm{d}\tilde{Z}}{\mathrm{d}t} \right) \gamma_S(s)\,\mathrm{d}s$$
$$- \rho\Omega \int_S \mathrm{Im}(\tilde{Z}(s))\gamma_S(s)\,\mathrm{d}s, \tag{8.107}$$

where $\mathrm{d}\tilde{Z}/\mathrm{d}t$ denotes the rate of change of the body-fixed coordinates of a point on the sheet—that is, the velocity of a material point on the sheet relative to the body-fixed reference frame. Similar adaptations of the fluid vorticity contributions can be made to free vortex sheets for all of the other results presented in this chapter.

We derived the force and moment in this section through the rate of change of the impulse in the fluid, our favored approach in this book. However, suppose we wish to know the pressure distribution on the plate surface. For this, we can rely on the expression for the pressure jump across a bound vortex sheet, Eq. (3.194). We can be more specific in this expression here, in which the relative tangential velocity between the fluid and the plate was discussed earlier in this chapter in Eq. (8.44).

Result 8.5: Local Pressure Jump Across a Flat Plate

The jump in pressure across a flat plate at point $\tilde{z}$ can be written as

$$[p]_-^+(\tilde{z}) = \rho\gamma(\tilde{z})\left(\mu(\tilde{z}) - \tilde{U}_{\mathrm{r}}\right) + \rho\frac{\mathrm{d}\Gamma}{\mathrm{d}t}(\tilde{z}), \tag{8.108}$$

where $\gamma(\tilde{z})$ is the local strength of the bound circulation, described in Result 8.2 in Fourier–Chebyshev form; $\mu(\tilde{z})$ is the average of the tangential components of fluid velocity on either side of the plate, given by (8.43); $\tilde{U}_{\mathrm{r}}$ is the tangential component of the plate's translational velocity; and $\Gamma(\tilde{z})$ is the local value of bound circulation, expressed in Eq. (8.79).

Of course, as we know from Sect. 6.4, this pressure distribution only describes the pressure along the side of the plate and misses an essential part of the contribution of pressure to force and moment: the edge suction. If we seek to compute the force and moment from this pressure, we must include these contributions:

$$f = -\mathrm{i} \int_{\mathcal{P}} [p]_-^+ \, \mathrm{d}z + \pi \rho c (\sigma^2(1) - \sigma^2(-1)) \mathrm{e}^{\mathrm{i}\alpha} \tag{8.109}$$

and

$$m_O = \mathrm{Re} \int_{\mathcal{P}} [p]_-^+ z^* \, \mathrm{d}z - \pi \rho c \mathrm{Im}(Z_r \mathrm{e}^{-\mathrm{i}\alpha})(\sigma^2(1) - \sigma^2(-1)), \tag{8.110}$$

where $\sigma(1)$ is the suction parameter at the trailing edge and $\sigma(-1)$ the value at the leading edge. The following note provides us with more specific values for these parameters, and prepares us for the next section in which we enforce the Kutta condition by annihilating this suction at one or both edges.

Note 8.3.1: Edge Suction on a Two-Dimensional Flat Plate

In Sect. 6.4, we presented general results for the edge suction on infinitely-thin plates, relating this suction force to the singular behavior of the bound vortex sheet at the edge. Here, let us explore the form of the edge suction in the different approaches we have taken to solve for the flow about the two-dimensional flat plate. We will also refer to the results in Notes 4.4.5 and 4.4.6, where we provided various forms of the edge suction parameter, σ. Before proceeding, we remind the reader that edge suction is automatically accounted for when the force is derived from impulse formulas; there is no need to explicitly acknowledge it in such calculations. In contrast, it must be accounted for when force is derived from integrals of traction (i.e., pressure). Furthermore, the consideration of edge suction has intrinsic value in its connection with other flow characteristics near an edge, including flow separation.

The form of σ given in terms of the tangent component of the circle plane velocity in Result 4.4 was shown to specialize for the flat plate of length c to

$$\sigma(\pm 1) = -\frac{1}{c} \hat{u}_\tau(\pm 1), \tag{8.111}$$

where the expression is evaluated at either the leading edge, $\zeta_L = -1$, or trailing edge, $\zeta_T = 1$. Substituting this into Result 6.10, with edge normals $n_O = -\mathrm{e}^{\mathrm{i}\alpha}$ and $\mathrm{e}^{\mathrm{i}\alpha}$ at the leading and trailing edge, respectively, we can then write the suction at either edge as

$$f_{s,\pm 1} = \pm \frac{\pi \rho}{c} \mathrm{e}^{\mathrm{i}\alpha} \hat{u}_{\tau}^{2}(\pm 1). \tag{8.112}$$

As we discussed in Note 4.4.5, the complex velocity in the circle plane at either edge's pre-image is purely imaginary, since it must be tangent to the circle's surface, and $\hat{u}_\tau(\pm 1) = \pm \mathrm{i}\hat{w}(\pm 1)$. With some manipulation of the derivative of the complex potential (8.17), we can show that

$$\hat{u}_{\tau}(\pm 1) = \pm \frac{(\tilde{V}_{\infty} - \tilde{V}_{\mathrm{r}})c}{2} - \frac{\Omega c^2}{8} - \sum_{\mathrm{J}=1}^{N_{\mathrm{v}}} \frac{\Gamma_{\mathrm{J}}}{2\pi} \mathrm{Re}\left(\frac{\zeta_{\mathrm{J}} \pm 1}{\zeta_{\mathrm{J}} \mp 1}\right). \tag{8.113}$$

It is useful to express the suction parameter entirely in terms of the physical plane coordinates, which we can do with some simple manipulation:

$$\sigma(\pm 1) = \pm \frac{1}{2}(\tilde{V}_{\mathrm{r}} - \tilde{V}_{\infty}) + \frac{\Omega c}{8} + \sum_{\mathrm{J}=1}^{N_{\mathrm{v}}} \frac{\Gamma_{\mathrm{J}}}{2\pi c} \mathrm{Re}\left(\frac{(\tilde{z}_{\mathrm{J}} \pm c/2)^{1/2}}{(\tilde{z}_{\mathrm{J}} \mp c/2)^{1/2}}\right). \tag{8.114}$$

An important conclusion that we can draw from this equation is that the edge suction parameter is directly proportional to the strengths of the various flow contributors—body translational and rotational velocity, uniform flow speed, and the strengths of the vortex elements in the fluid. By design, the suction parameter incorporates the direction of the local edge flow in its form; it is positive if the flow traverses the edge in a clockwise fashion. For example, if the plate translates in the positive $\tilde{y}$ direction, the flow will go around the leading edge $\tilde{z} = -c/2$ in the counter-clockwise direction and the trailing edge $\tilde{z} = c/2$ in the clockwise direction, as reflected in the signs in (8.114); positive (i.e., counter-clockwise) angular motion leads to clockwise flow relative to both edges.

This dependence on contributors is important for our interpretation of the critical suction parameters, introduced in Sect. 5.4, which we will revisit below in this context of the flat plate. Note that all contributors to the flow are coupled together in the edge suction (8.112), since σ enters its expression quadratically.

We can form another expression for the suction parameter on the plate edges, in terms of the Fourier–Chebyshev expansion coefficients, by starting with the relationship between σ and the bound vortex sheet strength, developed in Note 4.4.6. We note that, on a flat plate, the arc length of any point on the plate is equivalent to its body-fixed coordinate $\tilde{z}$, or equivalently, its scaled coordinate $c\xi/2$. Using the expression (8.77) for the flat plate's bound vortex sheet distribution, it is straightforward to show that

$$\sigma(\pm 1) = -\frac{1}{4}\left[(B_0 - A_1) \mp 2A_0\right]. \tag{8.115}$$

8.4 Application of the Kutta Condition

In the results obtained by three different strategies in Sect. 8.2, we repeatedly saw the evidence of the inverse square-root singularity of the velocity field at the edges of the flat plate. In this section, we will discuss approaches for annihilating this singularity at one or both edges, as dictated by the Kutta condition, discussed in Chap. 5. As we know from that latter discussion, the enforcement of the Kutta condition at an edge is equivalent to determining the flow parameters that set the edge suction parameter σ to zero. The results in Note 8.3.1 provide an easy framework for guiding this determination of parameters. To organize our discussion for the flat plate, we will first focus on the more classical task: to enforce the Kutta condition at the trailing edge, z_T, only. Then we will discuss the simultaneous enforcement at both edges. We will not yet provide any firm physical grounds for either choice, but rather, simply discuss the mathematics of the task.

8.4.1 At the Trailing Edge: Thin Airfoil Theory

In the classical aerodynamics problem of flow past a flat plate at low angle of attack, the solution is typically constrained with a Kutta condition applied at the trailing edge, z_T ($\xi = 1$ in scaled coordinates, or $\zeta_T = 1$ in the circle plane). This is the basis for classical **thin airfoil theory** [26], specialized here to a flat plate for demonstration purposes. Setting σ to zero at this edge leads, via Eq. (8.115), to a condition on the Fourier–Chebyshev coefficients,

$$B_0 - A_1 - 2A_0 = 0. \tag{8.116}$$

This constraint regularizes the singular term in the bound vortex sheet strength at this edge.

It is also useful to note that the suction parameter at the leading edge must therefore be $\sigma(-1) = -A_0$. In other words, when the Kutta condition is enforced at the trailing edge, ***the leading coefficient, A_0, of the Fourier–Chebyshev expansion of the normal velocity difference on the plate is equivalent to the leading-edge suction parameter.*** As we noted back in Sect. 5.4, Ramesh et al. [60] utilized this fact to define the leading-edge suction parameter as A_0 directly, and impose critical bounds on it in the manner of Result 5.6.

Using the definitions of A_0 and A_1 from (8.73), this condition on the Fourier–Chebyshev coefficients can be used to eliminate B_0 from the integral solution (8.39) and the constraint (8.69) on the total circulation. We can thus rewrite the solution for arbitrary distribution of normal velocity as

$$\gamma(\xi) = -\frac{2}{\pi}\left(\frac{1-\xi}{1+\xi}\right)^{1/2} \fint_{-1}^{1} \left(\frac{1+l}{1-l}\right)^{1/2} \frac{\Delta V_{\perp}(\tilde{z}(l))}{l-\xi}\,\mathrm{d}l, \tag{8.117}$$

subject to the constraint

$$\frac{2}{\pi}\int_{-1}^{1}\left(\frac{1+l}{1-l}\right)^{1/2}\Delta V_{\perp}(\tilde{z}(l))\,\mathrm{d}l = \frac{2}{\pi c}\sum_{\mathrm{J}=1}^{N_{\mathrm{v}}}\Gamma_{\mathrm{J}}. \tag{8.118}$$

This form of the bound vortex sheet is now unique and is singular only at the leading edge, $\xi = -1$.

However, these expressions (8.117) and (8.118) for the constrained solution are only a formality, because we already know the forms of the solution for each flow contributor, and can thus apply the constraint directly on these contributors. The constraint (8.116) imposed by the Kutta condition, along with the additional constraint of zero total circulation (8.69), imply a relationship:

$$\pi(\tilde{V}_{\infty} - \tilde{V}_{\mathrm{r}})c - \frac{\pi}{4}\Omega c^2 - \sum_{\mathrm{J}=1}^{N_{\mathrm{v}}}\Gamma_{\mathrm{J}}\mathrm{Re}\left(\frac{(\tilde{z}_{\mathrm{J}}+c/2)^{1/2}}{(\tilde{z}_{\mathrm{J}}-c/2)^{1/2}}\right) = 0. \tag{8.119}$$

Provided this relationship is satisfied at all times, then the overall solution—given in any of its three different forms in Sect. 8.2—is guaranteed to satisfy the Kutta condition at the trailing edge. The bound vortex sheet strength on the plate that results from applying the Kutta condition at the trailing edge can be obtained in one of two ways: we could evaluate the integral (8.117) with the form of $\Delta V_{\perp}$ given in (8.41), or—more easily—we could compose the unregularized vortex sheet strength from the basis sheets (8.45), (8.46) and (8.53), and then subtract the constraint (8.119) to develop the regularized sheet strength. Either way, we arrive at the following:

$$\gamma(\xi) = -2\left(\frac{1-\xi}{1+\xi}\right)^{1/2}(\tilde{V}_{\infty} - \tilde{V}_{\mathrm{r}}) + \Omega c(1-\xi^2)^{1/2}$$
$$+\frac{1}{\pi}\left(\frac{1-\xi}{1+\xi}\right)^{1/2}\mathrm{Re}\sum_{\mathrm{J}=1}^{N_{\mathrm{v}}}\frac{\Gamma_{\mathrm{J}}}{\tilde{z}_{\mathrm{J}} - c\xi/2}\left(\frac{\tilde{z}_{\mathrm{J}}+c/2}{\tilde{z}_{\mathrm{J}}-c/2}\right)^{1/2}. \tag{8.120}$$

Equation (8.119) is generally used to determine a vortex strength, as described in Sect. 5.2. That is, the singular behavior at the trailing edge is eliminated by appropriately choosing the strength of a vortex in the flow. It is important to remember that each vortex influences the flow at the trailing edge directly, as well as indirectly, via its image vortex (or, equivalently, its reacting bound vortex sheet on the plate surface). Equation (8.119) contains both effects.

Below, in Sect. 8.5, we will discuss the simplest application of this constraint: a steady uniform flow past a stationary plate at small fixed angle of attack. In this case, the constraint is used to determine the strength of the starting vortex at large distances aft of the flat plate. When the flow is unsteady, then condition (8.119) is used to determine the strength of a **nascent vortex**—that is, a developing vortex associated with the trailing edge. This nascent vortex might be an isolated vortex, but often is associated with a free vortex sheet, S, in which case (8.119) is replaced with its analog for a continuous distribution of strength γ_S along the sheet:

$$\fint_S \mathrm{Re}\left(\frac{(\tilde{Z}(s)+c/2)^{1/2}}{(\tilde{Z}(s)-c/2)^{1/2}}\right)\gamma_S(s)\,\mathrm{d}s = \pi(\tilde{V}_\infty - \tilde{V}_\mathrm{r})c - \frac{\pi}{4}\Omega c^2, \tag{8.121}$$

where only the principal value of the integral is needed. The bound vortex sheet strength associated with this free sheet in the fluid is readily obtained by replacing the last term in (8.120) with

$$\frac{1}{\pi}\left(\frac{1-\xi}{1+\xi}\right)^{1/2} \mathrm{Re}\int_S \left(\frac{\tilde{Z}(s)+c/2}{\tilde{Z}(s)-c/2}\right)^{1/2} \frac{\gamma_S(s)\,\mathrm{d}s}{\tilde{Z}(s)-c\xi/2}. \tag{8.122}$$

We will see some examples of this in our discussion of basic unsteady flows in Sect. 8.5.

Limiting Behavior of the Bound Vortex Sheet Strength at the Trailing Edge The forms of the bound vortex sheet strength given so far—the general forms in Result 8.2, and those regularized at the trailing edge in the vicinity of discrete vortices (8.120) or a free vortex sheet (8.122)—mask some important subtlety in the mathematical behavior of this strength as the trailing edge is approached. In particular, one is tempted to state that the bound sheet strength is zero at the regularized edge. After all, once the Kutta condition (8.116) is applied, all of the basis functions in the Fourier series in Result 8.2 are identically zero when evaluated at $\theta = 0$. Indeed, when the plate is surrounded by discrete vortices, disconnected from the edge, all of the terms in (8.120) continuously approach zero as $\xi \to 1$ (i.e., as $\theta \to 0$), including those associated with the set of vortices. This is consistent with our expectation that, once the edge's behavior is regularized, the jump in tangential velocity should be continuous between the plate and the fluid. By that same reasoning, if, in lieu of a discrete set of vortices, a free vortex sheet emerges from the trailing edge, we also expect that the strength of the bound sheet should match that of the free sheet at the trailing edge. We argued for this continuity in our original discussion of the Kutta condition in Sect. 5.1.2. But the strength of this free sheet is *non-zero* in general!

How do we reconcile this apparent paradox? The answer is that the bound sheet strength is *not* zero at the edge when a continuous free sheet is rooted there. Since there is an infinite variety of configurations that the free vortex sheet could take, from a similarly infinite variety of plate motions, it is difficult to see how we could prove this. However, with a little thought, it should be clear that the only crucial distinction between a free vortex sheet and a set of discrete vortices is that the former is continuously connected to the trailing edge. Thus, the behavior can be revealed by just examining a short segment of this sheet connected to the edge; the remainder of the free sheet induces zero strength in the bound vortex sheet at the edge, just as though it were a set of discrete point vortices. So, let us follow an approach taken by Alben [3] and construct a simple sheet segment of unspecified length L, emerging tangentially from the plate with uniform strength $\gamma_S(0)$. Since we know the other parts of the free sheet contribute nothing at the edge, we will evaluate the bound vortex sheet strength—as a function of the scaled coordinate $\xi = 2\tilde{z}/c$—based only on the term (8.122) due to this segment:

$$\gamma(\xi) = \frac{1}{\pi}\gamma_S(0)\left(\frac{1-\xi}{1+\xi}\right)^{1/2} \int_0^L \frac{(c+s)^{1/2}\mathrm{d}s}{s^{1/2}(s + c(1-\xi)/2)}. \tag{8.123}$$

This integral can be evaluated analytically:

$$\gamma(\xi) = \frac{2}{\pi}\gamma_S(0)\left[\arctan\left(\left(\frac{1+\xi}{1-\xi}\right)^{1/2}\frac{L^{1/2}}{(c+L)^{1/2}}\right) + 2\left(\frac{1-\xi}{1+\xi}\right)^{1/2}\log\left((L/c)^{1/2} + (1+L/c)^{1/2}\right)\right]. \tag{8.124}$$

The second term in this expression contributes a clear square-root dependence on distance from the edge, $(1-\xi)^{1/2}$, similar to the omitted terms involving plate motion and the remaining vortex elements in the free sheet in (8.120). The first term is different, however. Let us note that, when its argument z is large, $\arctan z \approx \pi/2 - 1/z$. Thus, the dominant behavior contributed by this term is finite and non-zero, and the overall behavior of the bound sheet—from all contributions—is

$$\gamma(\xi) \approx \gamma_S(0) + C(1-\xi)^{1/2}, \quad \xi \to 1. \tag{8.125}$$

where C is a constant that depends on the plate's motion and the free sheet configuration. Clearly $\gamma(\xi) \to \gamma_S(0)$ as $\xi \to 1$.

This analysis has revealed a deficiency in our Fourier–Chebyshev expansion of the normal velocity on the plate (8.72). If we return to the original form of this normal velocity (8.41), but written instead with a continuous vortex sheet in place of the set of discrete vortices, then it takes the form

$$\Delta V_\perp(\xi) = \tilde{V}_\infty - \tilde{V}_\mathrm{r} - \frac{\Omega c}{2}\xi + \frac{1}{c\pi}\mathrm{Re}\int_S \frac{\gamma_S(s)\,\mathrm{d}s}{\xi - 2\tilde{Z}(s)/c}. \tag{8.126}$$

Near the plate, $\gamma_S(s) \approx \gamma_S(0)$ and $\tilde{Z}(s) = c/2 + s\tilde{Z}'(0)$, where $\tilde{Z}'(0)$ is the tangent of the sheet at the edge. We will show in the next discussion that this sheet's tangent must be equal to the tangent of the plate, which, in complex notation and body-fixed coordinates, is simply equal to 1 at the trailing edge. This near-edge form of the integrand, $\gamma_S(0)/(\xi - 1 - 2s/c)$, contributes a logarithmic singularity in $\Delta V_\perp$ at the edge. Our Fourier–Chebyshev expansion (8.72) is not well suited for approximating this singular behavior, and the coefficients decay very slowly—that is, a very large number of terms are needed in the expansion. Furthermore, no matter how many terms we actually use, the series for $\gamma(\xi)$ still fails to converge uniformly: it converges toward a finite but non-zero limit at all $\xi < 1$ but is identically zero at $\xi = 1$.

We can overcome this deficiency by using a technique proposed by Jones [34] and Alben [3]. We isolate the logarithmically singular behavior by adding and subtracting the singular part to and from $\Delta V_\perp$:

$$\Delta V_\perp(\xi) = \Delta \check{V}_\perp(\xi) + \frac{\gamma_S(0)}{2\pi}\log(1-\xi), \tag{8.127}$$

where $\Delta \check{V}_\perp(\xi)$ denotes the well-behaved normal velocity with this logarithmic singularity removed. This de-singularized part of the normal velocity can now be expanded in Chebyshev polynomials in the manner of (8.72), but with the benefit that the coefficients, $\check{A}_n$, decay to zero much more rapidly without having to capture the singular behavior. The corresponding bound vortex sheet strength, $\check{\gamma}(\xi)$, has the same form as in Result 8.2, but with the coefficients from $\Delta \check{V}_\perp(\xi)$ rather than the full normal velocity. In particular, $\check{\gamma}(\xi)$ approaches zero uniformly as $\xi \to 1$.

Furthermore, the effect of the singular part of $\Delta V_\perp$ on the bound vortex sheet can be calculated explicitly, and the full bound vortex sheet strength can be written as

$$\gamma(\xi) = \check{\gamma}(\xi) + \frac{1}{\pi}\gamma_S(0)\left(\pi - \arccos\xi - \frac{1 + (\log 2)T_1(\xi)}{(1-\xi^2)^{1/2}}\right). \tag{8.128}$$

The first two terms proportional to $\gamma_S(0)$ cancel the logarithmic singularity at $\xi = 1$ induced by the free vortex sheet. However, they also contribute a net circulation about the plate and a uniform normal velocity proportional to log 2, both of which must be removed by subtracting the final term. This latter term needs to be accounted for in a modified statement of the Kutta condition at the trailing edge (8.116), regularizing all of the terms with square-root singularities in (8.128):

$$B_0 - \check{A}_1 - 2\check{A}_0 - \frac{\gamma_S(0)}{\pi}(1 + \log 2) = 0. \tag{8.129}$$

Provided this constraint is satisfied, then $\gamma(\xi)$ is finite at $\xi = 1$. Furthermore, (8.128) clearly shows that the vortex sheet strength is continuous at the trailing edge, $\gamma(1) = \gamma_S(0)$.

Limiting Form of the Regularized Velocity at the Edge Once the Kutta condition has been enforced at the trailing edge, what is the strength of the (now finite) velocity at that edge? To answer this, let us consider the form of the velocity obtained by the conformal transformation from the circle plane:

$$\tilde{w}(\tilde{z}) = \frac{\hat{w}(\zeta)}{\tilde{z}'(\zeta)}. \tag{8.130}$$

In the regularized flow, both the numerator and denominator of this expression vanish at the trailing edge $\tilde{z}_T = c/2$ ($\zeta_T = 1$). Thus, to compute the limit of the ratio, we proceed to the next-higher terms, both of which are non-zero:

$$\tilde{w}(\tilde{z}_T) = \frac{\hat{w}'(1)}{\tilde{z}''(1)} = \frac{2}{c}\hat{w}'(1). \tag{8.131}$$

Particularly of interest to us is the relative velocity between the fluid and the edge itself, given by (2.41) evaluated at $\tilde{z}_T$:

$$\tilde{w}(\tilde{z}_T) - \tilde{W}_b(\tilde{z}_T) = \frac{2}{c}\hat{w}'(1) - \tilde{U}_r + i\tilde{V}_r + i\Omega c/2. \tag{8.132}$$

We can readily compute the derivative of the circle-plane velocity field, $\hat{w}$, and evaluate it at the trailing edge. Doing so, and accounting for the constrained relationship between the point vortex strengths and the body motion dictated by the Kutta condition (8.119), we obtain a result that should not be surprising: The relative velocity between the fluid and body at the trailing edge is

$$\tilde{w}(\tilde{z}_T) - \tilde{W}_b(\tilde{z}_T) = \tilde{U}_\infty - \tilde{U}_r + \mathrm{Re}\left(\frac{1}{2\pi i}\sum_{J=1}^{N_v}\frac{\Gamma_J}{c/2 - \tilde{z}_J}\right). \tag{8.133}$$

This is simply the difference between the mean tangential fluid velocity at the trailing edge, given earlier in (8.43), and the tangential velocity of the plate, $\tilde{U}_r$.

As with other results, the summation over the point vortices can be replaced by the integral over a free vortex sheet emanating from the trailing edge. However, since we are evaluating the resulting integral at a point on the sheet itself, we must use the Plemelj formulae (A.125) to account for the limit as this point is approached from either side. This results in

$$\tilde{w}(\tilde{z}_T) - \tilde{W}_b(\tilde{z}_T) = \tilde{U}_\infty - \tilde{U}_r \mp \frac{1}{2}\gamma_S(0)\mathrm{Re}\,\tilde{Z}'(0) + \mathrm{Re}\left(\frac{1}{2\pi i}\fint_S \frac{\gamma_S(s)}{c/2 - \tilde{Z}(s)}\,\mathrm{d}s\right), \tag{8.134}$$

where the principal value of the integral over the free vortex sheet is used, and $\gamma_S(0)$ denotes the wake vortex sheet evaluated at the trailing edge. Note that the factor multiplying $\gamma_S(0)$ contains $\mathrm{Re}\,\tilde{Z}'(0)$. This represents the dot product between the tangent of the plate (which is along the $\tilde{x}$ axis) and the tangent of the wake vortex sheet at the trailing edge. This leads to an important question: At what angle does the free vortex sheet leave the trailing edge? This angle is determined by its transport, which is effected by the mean velocity. But this mean velocity, given by the remaining parts of (8.134), is entirely in the $\tilde{x}$ direction, so ***the free vortex sheet must leave the trailing edge parallel to the plate***, and the factor $\mathrm{Re}\,\tilde{Z}'(0)$ is unity. In fact, we already proved this result more generally by appealing to mass conservation and velocity continuity in the discussion of the Kutta condition in Sect. 5.1.2. Thus, the velocity in the vicinity of the edge is

$$\tilde{w}(\tilde{z}_T) - \tilde{W}_b(\tilde{z}_T) = \tilde{U}_\infty - \tilde{U}_r \mp \frac{1}{2}\gamma_S(0) + \mathrm{Re}\left(\frac{1}{2\pi i}\fint_S \frac{\gamma_S(s)}{c/2 - \tilde{Z}(s)}\,\mathrm{d}s\right), \tag{8.135}$$

and in particular, the transport velocity of the sheet is given by the average of this on either side of the sheet, as derived in Eq. (8.43):

$$\mu(\tilde{z}) - \tilde{U}_r = \tilde{U}_\infty - \tilde{U}_r + \mathrm{Re}\left(\frac{1}{2\pi i}\fint_S \frac{\gamma_S(s)}{c/2 - \tilde{Z}(s)}\,\mathrm{d}s\right). \tag{8.136}$$

Pressure in the Vicinity of the Trailing Edge As we proved in Result 3.6, the pressure is continuous across a free vortex sheet, and, since we have now proved that the velocity is continuous as the edge is approached from along the plate and from the sheet, the pressure must be, as well. That is, the pressure difference across the plate is zero at the trailing edge when a free vortex sheet emerges from the edge; in the parlance of aerodynamics, we say that ***there is no loading at this edge***.

The pressure difference is also continuous when the edge is assumed to generate discrete point vortices rather than a free vortex sheet. However, it is no longer zero. To understand why this is the case, let us recall the expression for pressure jump across the plate in (8.108). The vortex sheet strength is zero at the trailing edge in the absence of a free sheet rooted at this edge, so we are left with

$$[p]_{-}^{+} = \rho \frac{d\Gamma_{\mathcal{P}}}{dt}, \tag{8.137}$$

where $\Gamma_{\mathcal{P}}$ is the local circulation of the plate at this edge. In other words, there *is* loading at the trailing edge in this discrete vortex representation of the wake. How can this pressure difference extend continuously into the fluid and remain non-zero? By Kelvin's circulation theorem, the plate circulation's rate of change is equal and opposite to the rate of change of circulation shed into the fluid. But, in the context of a discrete vortex representation, this change of circulation must be borne by a single nascent vortex whose strength is changing in time. But such a variable-strength vortex must be accompanied by a discontinuity in pressure across the branch cut, as we discussed in Note 3.3.1, in this case extending from the edge to the vortex. This pressure discontinuity is exactly (8.137), as discussed in that note.

Pressure-Based Form of the Kutta Condition In our general discussion of the Kutta condition in Chap. 5, we discussed a flux form of this condition based on continuity of pressure in the region surrounding the edge. This led to a relationship (5.1) describing the flux of circulation into the wake. Let us examine that relationship here at the trailing edge in the case of a free vortex sheet rooted at the edge. We have just obtained the relative velocity between the fluid and the plate's edge in (8.135); the mean part of this is the velocity at which vorticity in the trailing-edge vortex sheet, $\mathcal{S}$, is convected away from the edge. Adapting the expression (5.1) to complex form, we have

$$\frac{d\Gamma_{\mathcal{P}}}{dt} = -\gamma_{\mathcal{S}}(0)\left[\tilde{U}_{\infty} - \tilde{U}_{r} + \mathrm{Re}\left(\frac{1}{2\pi i}\int_{\mathcal{S}} \frac{\gamma_{\mathcal{S}}(s)}{c/2 - \tilde{Z}(s)}\, ds\right)\right]. \tag{8.138}$$

As we discussed in Sect. 5.1, this relationship can also be interpreted as a form of Kelvin's circulation theorem: it accounts for the change of bound circulation about the body by the equal and opposite change (via flux at the edge) of circulation in the fluid.

8.4.2 At Both Edges

Now, let us seek to simultaneously remove the singularity at both ends of the bound vortex sheet. In other words, we seek to set σ to zero at both edges of the plate. As exhibited by the form of σ in terms of Fourier–Chebyshev coefficients (8.115), this is achieved by placing the following set of conditions on these coefficients: $2A_0 + A_1 - B_0 = 0$ and $2A_0 - A_1 + B_0 = 0$. Clearly, these conditions are satisfied if and only if

$$A_0 = 0, \qquad B_0 = A_1. \tag{8.139}$$

These equations dictate relationships between the various flow contributors,

$$\sum_{J=1}^{N_v} \frac{\Gamma_J}{2\pi} \mathrm{Re}\left(\frac{(\tilde{z}_J + c/2)^{1/2}}{(\tilde{z}_J - c/2)^{1/2}} - \frac{(\tilde{z}_J - c/2)^{1/2}}{(\tilde{z}_J + c/2)^{1/2}}\right) = (\tilde{V}_\infty - \tilde{V}_r)c, \tag{8.140}$$

$$\sum_{J=1}^{N_v} \frac{\Gamma_J}{2\pi} \mathrm{Re}\left(\frac{(\tilde{z}_J + c/2)^{1/2}}{(\tilde{z}_J - c/2)^{1/2}} + \frac{(\tilde{z}_J - c/2)^{1/2}}{(\tilde{z}_J + c/2)^{1/2}}\right) = \frac{\Omega c^2}{4}. \tag{8.141}$$

In general, it is impossible to satisfy both constraints with a single vortex, either at infinity or at finite distance. Clearly, we must have two nascent vortices—one associated with each edge—to satisfy these edge conditions.

It is also interesting to use the conditions to reformulate the solution of the Cauchy integral (8.38) with the constraints embedded. It can be confirmed that, by substituting for B_0 with A_1, and adding the null term $\tilde{z}A_0/(\tilde{z} - c/2)^{1/2}(\tilde{z} + c/2)^{1/2}$, one obtains the following alternative (and completely regular) form of the vortex sheet strength:

$$\gamma(\xi) = -\frac{2}{\pi}(1-\xi)^{1/2}(1+\xi)^{1/2} \fint_{-1}^{1} \frac{1}{(1-l)^{1/2}(1+l)^{1/2}} \frac{\Delta V_\perp(\tilde{z}(l))}{l - \xi}\, dl, \tag{8.142}$$

subject to the constraint that

$$\int_{-1}^{1} \frac{\Delta V_\perp(\tilde{z}(l))\, dl}{(1-l)^{1/2}(1+l)^{1/2}} = 0. \tag{8.143}$$

8.5 Classical Results

In this chapter we have specialized our analysis framework to flows about a two-dimensional flat plate. This geometry is invoked in many classical aerodynamic results, both steady and unsteady, involving flows at small angle of incidence to the plate. Let us see how our analysis leads to these classical results. As we will observe in Note 8.5.1, our general framework differs in one respect from the classical approach to these small-angle problems. The traditional approach is to enforce a trailing-edge Kutta condition on each basis flow field in the decomposition. In contrast, our

Fig. 8.9 Configuration of a two-dimensional flat plate translating steadily at small angle of attack, α_0. The starting vortex, of strength Γ_1, is far aft of the plate

analysis has been established independently of the application of edge constraints, since we reserve the possibility of applying at Kutta condition at multiple edges, or perhaps using alternative forms of constraint, such as bounds on the edge suction parameter, Result 5.6. As we will see, there is no difference in the final result—only a difference in the details in obtaining it.

8.5.1 Steady Flow at Fixed Small Angle of Attack

Let us consider the simplest application involving a flat plate: a steady uniform flow at speed U past the stationary plate at fixed angle of incidence, α_0. Equivalently, as depicted in Fig. 8.9, we can express this same problem in the inertial frame of reference by considering the plate in steady translation with body-fixed velocity components $(\tilde{U}_r, \tilde{V}_r) = (-U\cos\alpha_0, -U\sin\alpha_0)$ (or, in complex form, $\tilde{W}_r = -Ue^{-i\alpha_0}$). Let us set up the inertial coordinate system so that its $-x$ axis is parallel to the direction of motion of the plate. In other words, let the body-fixed system be rotated by angle $\alpha = -\alpha_0$.

The angle α_0 is presumed to be sufficiently small that the flow over the plate is unseparated and leaves smoothly from the trailing edge, where we apply the Kutta condition per Sect. 8.4.1. In this steady scenario, there is only one vortex in the fluid—the **starting vortex**, denoted by subscript 1. As the airfoil has moved steadily forward, it has left the vortex far behind—so that $|\tilde{z}_1| \to \infty$ (and similarly for ζ_1). Consequently, with no other flow contributor to act upon it, the vortex is nearly at rest in the inertial frame of reference, and relative to the plate, it moves with velocity $d\tilde{z}_1/dt = -\tilde{U}_r - i\tilde{V}_r = -\tilde{W}_r^* = Ue^{i\alpha_0}$, or, in vector form, $\tilde{\boldsymbol{v}}_1 := \dot{\boldsymbol{x}}_1 - \boldsymbol{V}_r = U\boldsymbol{e}_1$. Only this vortex's image—tending toward the center of the unit circle—contributes to the velocity at the trailing edge, and the parenthetical factor for Γ_1 in (8.119) is approximately unity: this condition (8.119) leads to the well-known result that

$$\Gamma_1 = -\pi\tilde{V}_r c = \pi U c \sin\alpha_0 > 0, \tag{8.144}$$

and the bound circulation about the plate is equal and opposite.

Now let us see how the force and moment in their vector form in Result 8.4 simplify in this scenario. Only the terms in the first line of these equations are operable when the vortex is quite distant from the plate. Thus, the force reverts to $f = \rho\Gamma_1 e_3 \times \tilde{v}_1 = \rho U \Gamma_1 e_2$. In our inertial coordinate system, the **drag**—that is, the component of force opposite the direction of motion—is f_x, and the **lift**—the component perpendicular to the motion—is f_y. These are, respectively,

$$f_x = 0, \qquad f_y = \rho U \Gamma_1. \tag{8.145}$$

This is a manifestation of the well-known **Kutta–Joukowski lift theorem**. In the body-fixed coordinate system, the tangential and normal components of force are easily obtained by rotation:

$$\tilde{f}_x = -\rho U \Gamma_1 \sin\alpha_0, \qquad \tilde{f}_y = \rho U \Gamma_1 \cos\alpha_0. \tag{8.146}$$

The first of these is directly attributable to leading-edge suction.

For the moment about the reference point, Eq. (8.104), the two vortex terms in the first line cancel each other out, so the only remaining term is $-(\pi/4)\rho c^2 \tilde{U}_r \tilde{V}_r$. Thus, we are left with

$$m_r = -\frac{\pi}{4}\rho c^2 U^2 \sin\alpha_0 \cos\alpha_0. \tag{8.147}$$

Since the Kutta condition requires that $\Gamma_1 = \pi U c \sin\alpha_0$, this implies that

$$m_r = -\frac{c}{4}\tilde{f}_y. \tag{8.148}$$

That is, the aerodynamic moment on the plate is equivalent to the force acting at the ***quarter-chord point***—midway between the leading edge and the reference point at the centroid. This is known as the **aerodynamic center** of the plate.

Note 8.5.1: Alternative Enforcement of the Trailing-Edge Kutta Condition

In Sect. 5.3 we contrasted our approach for enforcing the Kutta condition and the constraint of zero total circulation with the more traditional approach. We noted that this latter approach dictates that the bound circulation in each basis field be chosen to enforce regularity at one edge (the trailing edge) in that field. In the present context of the flat plate, this bound circulation is embodied in the B_0 coefficient in our Cauchy and Fourier–Chebyshev solutions (or, equivalently, in the set of image vortices inside the unit circle in the circle plane). Thus, if we decompose the coefficient B_0 (and hence, the bound circulation $\Gamma_{\mathcal{P}}$) into flow contributors, as we have done with the other Fourier–Chebyshev coefficients, then Eq. (8.116) imposed on each set of basis coefficients will ensure that regularity is enforced at the trailing edge ***in each basis field***. For example, for the vorticity-induced coefficients,

$$B_{v,0} = 2A_{v,0} + A_{v,1}. \tag{8.149}$$

Substituting from Eq. (8.90), it can be easily verified that this results in

$$B_{v,0} = \frac{2}{\pi c}\sum_{J=1}^{N_v} \Gamma_J \mathrm{Re}\left[\left(\frac{2\tilde{z}_J/c+1}{2\tilde{z}_J/c-1}\right)^{1/2} - 1\right]. \tag{8.150}$$

Equation (8.48) gives the proportionality between this coefficient and the bound circulation about the plate in this vorticity-induced basis field: Eq. (8.150), without the factor $2/\pi c$, is the bound circulation associated with fluid vorticity. In other words, for each vortex of strength Γ_J at $\tilde{z}_J$ (at ζ_J in the circle plane), the no-penetration condition requires that we place the usual image of strength $-\Gamma_J$ at $1/\zeta_J^*$, and regularity at the trailing edge dictates that we also place an image vortex of strength

$$\Gamma_J \mathrm{Re}\left[\left(\frac{2\tilde{z}_J/c+1}{2\tilde{z}_J/c-1}\right)^{1/2}\right] \tag{8.151}$$

at the origin of the circle plane.

For the body translation-induced basis field, for which we found $A_{bt,0} = 1$ and $A_{bt,1} = 0$, Eq. (8.116) requires that $B_{bt,0} = 2$ to ensure regularity at the trailing edge in that basis field. That is, the bound circulation in this basis field, due to translation at unit velocity in the plate-normal direction—or uniform flow of unit speed in the opposite direction—is πc. In the rotation-induced field, our previous results for $A_{br,0}$ and $A_{br,1}$ show that we require $B_{br,0} = c/2$ for smooth flow at the trailing edge, so that the bound circulation apportioned to the field induced by unit body angular velocity is $\pi c^2/4$.

The constraint of zero total circulation is still described by Eq. (8.69) in this traditional approach, but with B_0 composed from all of its basis coefficients. It is easy to verify that this still results in Eq. (8.119). Let us write that circulation constraint again, but now in terms of the total bound circulation associated with body motion, typically called the **quasi-steady circulation**:

$$\Gamma_b := \pi(\tilde{V}_r - \tilde{V}_\infty)c + \frac{\pi}{4}\Omega c^2, \tag{8.152}$$

whence the constraint (from which the strength of a new vortex element is chosen) is

$$\sum_{J=1}^{N_v} \Gamma_J \mathrm{Re}\left(\frac{(\tilde{z}_J + c/2)^{1/2}}{(\tilde{z}_J - c/2)^{1/2}}\right) = -\Gamma_b. \tag{8.153}$$

For a plate translating steadily at speed U and small angle α_0, after the starting vortex has moved far from the plate, the two image vortices associated with the vorticity-induced basis field coincide at the center of the circle and exactly cancel one another. The body translation-induced basis field has its own bound circulation of πc, so that the overall bound circulation is $\Gamma_b = \pi \tilde{V}_r c = -\pi U c \sin \alpha_0$, as expected.

8.5.2 General Background on Classical Unsteady Results

In this section, we have highlighted the reduction of our general analysis framework to classical thin airfoil theory for steady flow past an infinitely thin flat plate. The larger context of this classical theory, due largely to Munk [55] and Glauert [26], is more generally applicable to airfoils with finite thickness and camber. The foundation of the theory relies on a representation of such an airfoil as a bound vortex sheet concentrated along its mean camber line rather than on the airfoil surface. This representation achieves the zeroth-order solution of the flow field when asymptotically expanded in small ratio of thickness to chord length. The effect of thickness then enters at the next order. Our analysis framework can clearly be specialized to obtain this theory, provided we relax our assumption of flatness and return to the general formulas of our previous chapters, simplifying them by the same asymptotic expansion in thickness-to-chord ratio. We leave this as an exercise for the reader.

However, since this book is devoted primarily to *unsteady* inviscid flows, it is helpful to use our analysis framework to obtain some of the classical results in this context. There are four such results on which we will focus our attention. First, we will consider the result due to Theodorsen [68] for purely oscillatory motion, and second, to the flow analyzed by Wagner [71] for an impulsive change of motion. We will find that these results are intimately related via a Fourier transform. Then, we will consider results for two cases in which the flat plate is subjected to spatially-varying environmental disturbances, or ***gusts***, manifested in the normal component of velocity: in the first, the gust is sinusoidally distributed, while in the second, the gust is 'sharp-edged', corresponding approximately to the edge of an upward jet. These cases, too, are related to each other by Fourier transform, and their solutions are attributed to Küssner [42] and, later, to Sears [64].

All of these results are obtained from an analysis of an infinitely-thin flat plate, assumed to undergo small-amplitude excursions from (or be subjected to small-amplitude disturbances to) a mean motion along the $\tilde{x}$ axis (i.e., tangent to the plate). The analysis relies on the following simplified conceptualization of the flow:

- ***Wake vorticity convects with the 'bulk' fluid, rather than at the local velocity.*** In other words, since we are assuming that the plate's motion is of small amplitude, the flow field that it induces on the wake can be assumed negligible at leading order.

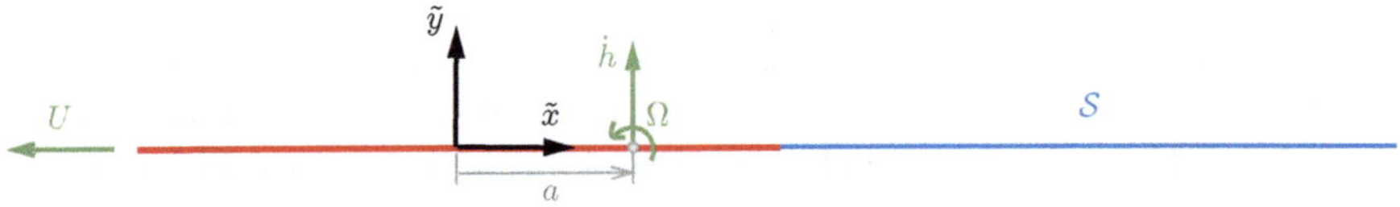

Fig. 8.10 Configuration of the flat wake model of a two-dimensional flat plate subjected to small-amplitude disturbances

- ***The wake lies along the $\tilde{x}$ axis.*** This is implied by the fact that the mean motion is along this axis, and based on the previous assumption, the wake can only convect away from the plate along this axis.

It should be noted that, in the case of gusts, the ambient flow is not strictly irrotational. However, the associated vorticity of the disturbance flow is considered so weak as to have negligible effect on our potential flow analysis.

To be consistent with the coordinate systems we established in Fig. 8.1, we will assume that the wake vortex sheet $\mathcal{S}$ extends along the positive $\tilde{x}$ axis from the trailing edge at $\tilde{z} = c/2$ ($\xi = 1$), as depicted in Fig. 8.10. We will denote its strength distribution by $\gamma_{\mathcal{S}}(\tilde{x})$. Let us suppose that the plate's motion is described about a pivot axis fixed at a distance a from the centroid along the plate. In Note 2.5.2 we discussed how to transform the motion about such an axis to a description in body-fixed components. We will use the same notation here for the pivot's velocity components, U and $\dot{h}$, but with the minor change that U denotes the component of the pivot's velocity in the negative (rather than positive) x direction. The motion is dominantly described by U, and any deviations—specifically, the vertical velocity, $\dot{h}$, and the rotationally-induced velocity along the plate, on the order of Ωc—will be presumed small compared to U; angular deviations from the mean are also presumed small, $|\alpha| \ll 1$, so that $\sin\alpha \approx \alpha$ and $\cos\alpha \approx 1$. (This angle is still defined as positive for counterclockwise rotations from the x axis, so it is opposite to what we would usually call the angle of attack.) Thus, the plate's body-fixed components of velocity are approximately

$$\tilde{U}_{\mathrm{r}} = -U, \quad \tilde{V}_{\mathrm{r}} = U\alpha + \dot{h} - a\Omega. \tag{8.154}$$

Assuming further that U is constant, the components of the rate of change of the velocity are approximately

$$\dot{\tilde{U}}_{\mathrm{r}} = 0, \quad \dot{\tilde{V}}_{\mathrm{r}} = \ddot{h} - a\dot{\Omega} + U\Omega. \tag{8.155}$$

The Kutta condition, in its trailing-edge form (8.121), becomes somewhat simpler in this flat-wake approximation. One can show that this condition becomes

$$\int_{c/2}^{\infty} \frac{(\tilde{x} + c/2)^{1/2}}{(\tilde{x} - c/2)^{1/2}} \gamma_{\mathcal{S}}(\tilde{x})\, \mathrm{d}\tilde{x} = -\Gamma_{\mathrm{b}}, \tag{8.156}$$

where Γ_{b} is the quasi-steady circulation, defined in Note 8.5.1, and here taking the specific form

$$\Gamma_b = \pi c \left[\dot{h} + U\alpha + \left(\frac{c}{4} - a\right)\Omega \right]. \tag{8.157}$$

Remember that the quasi-steady circulation is merely a convenient grouping of terms, but is not an actual circulation observed in the flow. Rather, it represents the bound circulation that would exist about the airfoil if the body suddenly stopped producing vorticity and the existing wake vorticity was allowed to progress infinitely far from the plate. It is also interesting to note that it represents πc times the normal velocity of the three-quarter chord point of the plate.

As we would expect, only deviations from the plate's mean tangential motion are invoked in this Kutta constraint; only deviations are responsible for creating new vorticity. Because these deviations are assumed small, then the wake strength itself, which is determined by this constraint, must also be small compared with U.

In this flat-wake model, each constituent vortex element in the wake vortex sheet convects relative to the plate with uniform velocity $\mathrm{d}\tilde{z}/\mathrm{d}t = U$. The force and moment on the plate in Eqs. (8.100)–(8.102) thus simplify to the following (in the small-amplitude approximation):

$$\tilde{f}_x = 0, \tag{8.158}$$

$$\tilde{f}_y = -\frac{\pi}{4}\rho c^2 \dot{\tilde{V}}_r + \rho U \int_{c/2}^{\infty} \frac{\tilde{x}\gamma_S(\tilde{x})}{(\tilde{x}^2 - c^2/4)^{1/2}}\,\mathrm{d}\tilde{x}, \tag{8.159}$$

$$m_r = -\frac{\pi}{128}\rho c^4 \dot{\Omega} + \frac{\pi}{4}\rho c^2 U \tilde{V}_r + \frac{1}{8}\rho c^2 U \int_{c/2}^{\infty} \frac{\gamma_S(\tilde{x})}{(\tilde{x}^2 - c^2/4)^{1/2}}\,\mathrm{d}\tilde{x}. \tag{8.160}$$

If we introduce the Kutta condition (8.156) into the expression for the normal component of force (multiplying and dividing the integrand by the factor $(\tilde{x}+c/2)^{1/2}$), and substitute the specific forms of the plate kinematics, (8.154) and (8.155), we can re-write this force component as

$$\tilde{f}_y = -\frac{\pi}{4}\rho c^2 \ddot{h} + \frac{\pi}{4}\rho c^2 a \dot{\Omega} - \frac{\pi}{4}\rho c^2 U \Omega - \rho U \left(\Gamma_b + \frac{c}{2} \int_{c/2}^{\infty} \frac{\gamma_S(\tilde{x})}{(\tilde{x}^2 - c^2/4)^{1/2}}\,\mathrm{d}\tilde{x} \right). \tag{8.161}$$

Note that, to within our approximation, this normal component is equivalent to the lift, f_y, on the plate. Our manipulation of the last integral over the wake vortex sheet has isolated a term involving the quasi-steady circulation from the overall vorticity-induced contribution to the lift. This isolated part is often called the **quasi-steady lift** [70], and is the lift that would be exerted on the plate if the plate's instantaneous motion were allowed to persist indefinitely. As we mentioned at the end of Sect. 6.5.1, one could apply this distinction more generally to decompose the force in flows in which a Kutta condition has been enforced at one edge.

The moment about the pivot axis is obtained from $m_{\mathrm{p}} = m_{\mathrm{r}} - a\tilde{f}_y$, which, after substituting for the kinematics, results in the expression

$$m_{\mathrm{p}} = \frac{\pi}{4}\rho c^2 a\ddot{h} - \frac{\pi}{4}\rho c^2 \left(\frac{c^2}{32} + a^2\right)\dot{\Omega} + \pi\rho c a \left(\frac{c}{4} - a\right) U\Omega$$
$$+ \rho c \left(\frac{c}{4} + a\right) U \left[\pi\left(\dot{h} + U\alpha\right) + \frac{1}{2}\int_{c/2}^{\infty} \frac{\gamma_S(\tilde{x})}{(\tilde{x}^2 - c^2/4)^{1/2}}\,\mathrm{d}\tilde{x}\right], \quad (8.162)$$

or, in terms of quasi-steady circulation,

$$m_{\mathrm{p}} = \frac{\pi}{4}\rho c^2 a\ddot{h} - \frac{\pi}{4}\rho c^2 \left(\frac{c^2}{32} + a^2\right)\dot{\Omega} - \frac{\pi}{4}\rho c^2 \left(\frac{c}{4} - a\right) U\Omega$$
$$+ \rho \left(\frac{c}{4} + a\right) U \left(\Gamma_{\mathrm{b}} + \frac{c}{2}\int_{c/2}^{\infty} \frac{\gamma_S(\tilde{x})}{(\tilde{x}^2 - c^2/4)^{1/2}}\,\mathrm{d}\tilde{x}\right). \quad (8.163)$$

As with the force, a **quasi-steady moment** has been extracted from the vorticity-induced contribution. It is interesting to observe that, if the moment is taken about the quarter-chord point (corresponding to $a = -c/4$), then the final set of terms vanishes. This generalizes the observation that we previously made in the steady case, where we identified this point as the aerodynamic center of the plate. In particular, the wake's contribution to the lift acts through this point.

We recall that Kelvin's circulation theorem requires that the material elements of the sheet have invariant strength, and since these elements convect from the plate with relative speed U, the strength distribution necessarily has the form $\gamma_S(\tilde{x} - Ut)$, a frozen pattern propagating away from the edge at speed U (i.e., a traveling wave).

It is important to emphasize that, by assuming small perturbations from the dominant motion of the plate, U, *all* of our results are linearly dependent on the perturbed components of motion, $\dot{h}$, α and Ω. This is a more thorough linearity than in the decomposition of the flow fields we have emphasized throughout this book. Here, we have also obtained linear dependence of the force and moment on the motion perturbations; in contrast, in general cases in which the motion is not limited to small amplitudes, the force and moment depend non-linearly on these components of motion. Thus, this analysis represents a ***linearization*** of the problem about the base motion.

Now, let us consider two types of motion, oscillatory and impulsive, and derive the resulting force and moment on the plate.

8.5.3 Oscillatory Motion: Theodorsen

Let us first consider a scenario in which the plate's motion is purely oscillatory and has been so for a long time—long enough that any transient effects from the initiation of motion have died out. This is the situation originally considered by Theodorsen [68]. This oscillatory motion may be contained in the vertical or angular

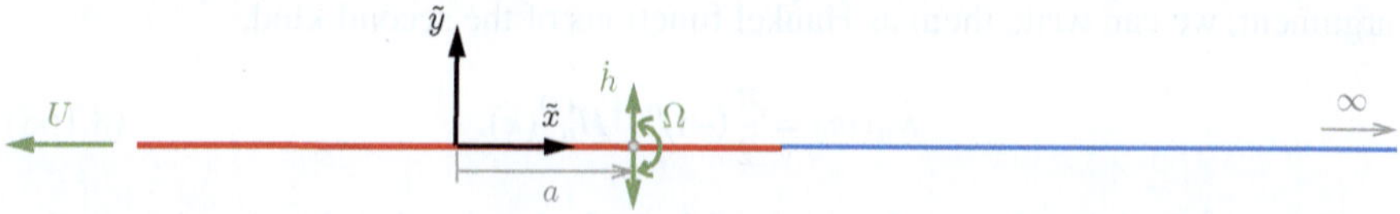

Fig. 8.11 Configuration of a two-dimensional flat plate undergoing small-amplitude oscillatory motion

excursions, $\dot{h}$ or Ω, or possibly a combination, as illustrated in Fig. 8.11. Since all of our results are linearly dependent on these components of motion, we can consider them separately. We will denote the angular frequency of these oscillations by ω (not to be confused with vorticity); the results we obtain below will naturally be dependent on the so-called **reduced frequency**, κ, a dimensionless ratio of the two important time scales of the oscillatory motion:

$$\kappa := \frac{\omega c}{2U}. \tag{8.164}$$

The oscillatory motion necessarily creates an oscillatory strength distribution in the wake. Based on our observation that this strength must also propagate as a traveling wave of speed U, it must have the form

$$\gamma_S(\tilde{x} - Ut) = g_0 \mathrm{e}^{\mathrm{i}\omega(t-\tilde{x}/U)}. \tag{8.165}$$

It is to be understood that only the real part of the expression on the right-hand side is physically meaningful. The constant g_0 is a complex amplitude, describing both the magnitude and phase of the traveling wave; we seek its value in terms of the motion, via the Kutta condition (8.156).

Oscillatory Plunging In this case, the plate's vertical velocity is oscillatory, but its angle remains fixed at zero. Thus, we can write the kinematic variables as

$$\dot{h} = V_0 \mathrm{e}^{\mathrm{i}\omega t}, \qquad \alpha = 0,\ \Omega = 0, \tag{8.166}$$

where V_0 is a complex amplitude, describing both the magnitude and phase of the plunging motion. Remember that we are assuming that $|V_0| \ll U$.

The Kutta condition (8.156) in this plunging scenario has only the term proportional to $\dot{h}$. When we introduce the exponential forms for $\dot{h}$ and γ_S, and re-scale the spatial coordinate in the familiar manner, $\tilde{x} = \xi c/2$, this condition becomes

$$g_0 \mathrm{e}^{\mathrm{i}\omega t} \frac{c}{2} \left(\int_1^\infty \frac{\mathrm{e}^{-\mathrm{i}\kappa\xi}}{(\xi^2 - 1)^{1/2}}\, \mathrm{d}\xi + \int_1^\infty \frac{\xi \mathrm{e}^{-\mathrm{i}\kappa\xi}}{(\xi^2 - 1)^{1/2}}\, \mathrm{d}\xi \right) = -\pi c \dot{h}. \tag{8.167}$$

The oscillatory factor $\mathrm{e}^{\mathrm{i}\omega t}$ cancels from both sides. Furthermore, it can be verified—by making the transformation $\xi = \cosh u$—that the two integrals inside parentheses are, respectively, $K_0(\mathrm{i}\kappa)$ and $K_1(\mathrm{i}\kappa)$, where K_0 and K_1 are the zeroth- and first-order modified Bessel functions of the second kind. To avoid the awkwardness of a complex argument, we can write them as Hankel functions of the second kind,

$$K_n(\mathrm{i}\kappa) = \frac{\pi}{2}(-\mathrm{i})^{n+1} H_n^{(2)}(\kappa). \tag{8.168}$$

Thus, we find that the amplitude g_0 of the wake is related to the plunging motion as follows:

$$g_0 \mathrm{e}^{\mathrm{i}\omega t} = \frac{4\dot{h}}{H_1^{(2)}(\kappa) + \mathrm{i}H_0^{(2)}(\kappa)}. \tag{8.169}$$

This important equation quantifies the manner in which the plunging motion of the plate is encoded into the wake. From this relationship, we can obtain expressions for the lift and moment—from Eqs. (8.161) and (8.163)—entirely in terms of the plunging motion. Both equations involve the same integral over the wake, one that we have already encountered in the expression for the Kutta condition and found it to be proportional to $H_0^{(2)}(\kappa)$. It is straightforward to show that the lift and the moment about the pivot become, respectively,

$$f_y = -\frac{\pi}{4}\rho c^2 \ddot{h} - \pi \rho c U \dot{h} C(\kappa) \tag{8.170}$$

and

$$m_{\mathrm{p}} = \frac{\pi}{4}\rho c^2 a \ddot{h} + \pi \rho c \left(\frac{c}{4} + a\right) U \dot{h} C(\kappa), \tag{8.171}$$

where

$$C(\kappa) := \frac{H_1^{(2)}(\kappa)}{H_1^{(2)}(\kappa) + \mathrm{i}H_0^{(2)}(\kappa)} \tag{8.172}$$

is often called the **Theodorsen function**. It is complex valued, and thus affects both the magnitude and phase of the time-varying quantities it multiplies. For negative values of the wavenumber, the Theodorsen function is equal to its complex conjugate, $C(-\kappa) = C^*(\kappa)$. Furthermore, it is straightforward to show that $C(\kappa) \to 1/2$ as $\kappa \to \infty$. A vector plot of the Theodorsen function, parameterized by reduced frequency, is depicted in Fig. 8.12, and separate plots of the real and imaginary parts versus reduced frequency are shown in Fig. 8.13.

Oscillatory Pitching Now, let us consider a scenario in which the plate only rotates sinusoidally about its pivot axis, with zero mean angle of attack to its steady forward travel. That is,

$$h = 0, \qquad \alpha = A_0 \mathrm{e}^{\mathrm{i}\omega t},\ \Omega = \mathrm{i}\omega A_0 \mathrm{e}^{\mathrm{i}\omega t}, \tag{8.173}$$

where $|A_0| \ll 1$ and $|\Omega| c = \omega |A_0| c \ll U$. Note that the latter restriction is equivalent to requiring that $|A_0|\kappa \ll 1$. Following the same steps as in the plunging scenario, the amplitude of the wake vortex strength is found to be

$$g_0 \mathrm{e}^{\mathrm{i}\omega t} = \frac{4\left(U\alpha + \Omega(c/4 - a)\right)}{H_1^{(2)}(\kappa) + \mathrm{i}H_0^{(2)}(\kappa)}, \tag{8.174}$$

whence the lift and moment are

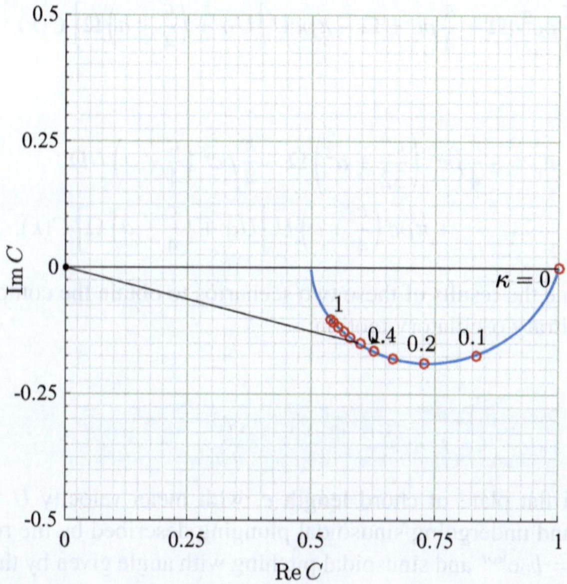

Fig. 8.12 Vector plot of the Theodorsen function (8.172), parameterized by reduced frequency κ. The length and direction of the arrow indicate the magnitude and phase of the function at the corresponding value of κ

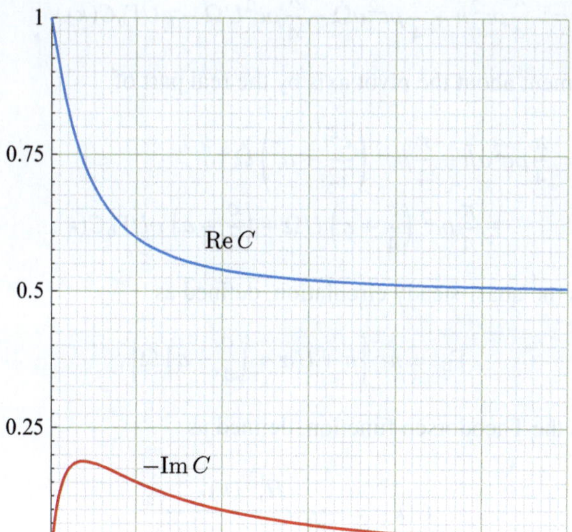

Fig. 8.13 The real and the (negative of the) imaginary parts of the Theodorsen function (8.217) versus reduced frequency κ

$$f_y = \frac{\pi}{4}\rho c^2 a\dot{\Omega} - \frac{\pi}{4}\rho c^2 U\Omega - \pi\rho c U\left[U\alpha + \left(\frac{c}{4} - a\right)\Omega\right] C(\kappa) \tag{8.175}$$

and

$$m_\text{p} = -\frac{\pi}{4}\rho c^2\left(\frac{c^2}{32} + a^2\right)\dot{\Omega} - \frac{\pi}{4}\rho c^2\left(\frac{c}{4} - a\right)U\Omega + \pi\rho c\left(\frac{c}{4} + a\right)U\left[U\alpha + \left(\frac{c}{4} - a\right)\Omega\right] C(\kappa). \tag{8.176}$$

Let us combine the results of these two scenarios to obtain the complete force and moment in arbitrary oscillatory motion:

Result 8.6: Low-Amplitude Oscillatory Motion of a Flat Plate

Consider a flat plate of chord length c, with mean velocity U tangent to the plate, and undergoing sinusoidal plunging described by the real part of velocity $\dot{h} = h_0 \mathrm{e}^{\mathrm{i}\omega t}$ and sinusoidal pitching with angle given by the real part of $\alpha = A_0 \mathrm{e}^{\mathrm{i}\omega t}$, about a pivot axis located on the plate a distance a aft of the centroid. Assume that $|\alpha| \ll 1$, $|\dot{h}| \ll U$ and $|\alpha|\omega c \ll U$. Then the lift on the plate is given by the real part of

$$f_y = -\frac{\pi}{4}\rho c^2 \ddot{h} + \frac{\pi}{4}\rho c^2 a\dot{\Omega} - \frac{\pi}{4}\rho c^2 U\Omega - \rho U \Gamma_\text{b} C(\kappa), \tag{8.177}$$

and the moment about the pivot axis by the real part of

$$m_\text{p} = \frac{\pi}{4}\rho c^2 a\ddot{h} - \frac{\pi}{4}\rho c^2\left(\frac{c^2}{32} + a^2\right)\dot{\Omega} - \frac{\pi}{4}\rho c^2\left(\frac{c}{4} - a\right)U\Omega + \left(\frac{c}{4} + a\right)\rho U \Gamma_\text{b} C(\kappa), \tag{8.178}$$

where Γ_b is the quasi-steady circulation, defined as

$$\Gamma_\text{b} := \pi c\left[\dot{h} + U\alpha + \left(\frac{c}{4} - a\right)\Omega\right], \tag{8.179}$$

and $C(\kappa)$ is the Theodorsen function, defined as

$$C(\kappa) := \frac{H_1^{(2)}(\kappa)}{H_1^{(2)}(\kappa) + \mathrm{i}H_0^{(2)}(\kappa)} \tag{8.180}$$

and $\kappa := \omega c/(2U)$ is the reduced frequency.

It is important to remember that, because the problem we have posed in this section is completely linear, we are able to superpose basic solutions to form richer ones. In particular, we can use the solution obtained here at a single unspecified frequency in order to compose the solution for motion defined over a spectrum of such frequencies. Let us encompass the two different types of oscillatory motion considered here by simply using the quasi-steady circulation to demonstrate the general procedure. For a single frequency, we could write this circulation as $\Gamma_b = G_b \delta\kappa e^{i\omega t}$, where $G_b \delta\kappa$ is the complex amplitude. It is useful to write the argument of the exponential in terms of the reduced frequency, $e^{i\omega t} \equiv e^{i2\kappa t^+}$, where

$$t^+ := tU/c \tag{8.181}$$

is sometimes called the **convected time**, a dimensionless measure of elapsed time in chord lengths of travel. Superposition of a spectrum of frequencies then leads to a Fourier integral for the quasi-steady circulation,

$$\Gamma_b(t^+) = \int_{-\infty}^{\infty} G_b(\kappa) e^{i2\kappa t^+} \, d\kappa; \tag{8.182}$$

the amplitude function $G_b(\kappa)$ is simply the inverse Fourier transform of the quasi-steady circulation,

$$G_b(\kappa) = \frac{1}{2\pi} \int_{-\infty}^{\infty} \Gamma_b(t^+) e^{-i2\kappa t^+} \, dt^+. \tag{8.183}$$

Since Γ_b must be a real-valued quantity, then its Fourier transform must have the property $G_b(-\kappa) = G_b^*(\kappa)$.

The lift response can then be immediately written down, since it is composed from the superposition of responses at each frequency, viz.

$$f_y = -\frac{\pi}{4}\rho c^2 \ddot{h} + \frac{\pi}{4}\rho c^2 a \dot{\Omega} - \frac{\pi}{4}\rho c^2 U \Omega - 2\rho U \mathrm{Re}\left(\int_0^{\infty} G_b(\kappa) C(\kappa) e^{i2\kappa t^+} \, d\kappa \right). \tag{8.184}$$

Note that the integral only extends over positive frequencies, since the negative frequencies contribute identically to the real part (hence the factor of 2). The moment response is similarly composed.

8.5.4 Impulsive Change of Motion: Wagner

In the previous case, the plate was presumed to have been in invariant motion for a very long time. It is also useful to consider a situation in which the plate's motion deviates impulsively from its baseline tangent motion to some new constant motion—in other words, a step change. This was the situation first analyzed by Wagner [71]. At first, we will not be specific about the type of motion deviation, or even about its magnitude. What we mainly seek is the response exhibited by the wake when the plate's motion suddenly undergoes a unit change; we can always re-scale this step response as needed, since the solution is linearly dependent on the deviation in motion. This response is determined by solving Eq. (8.156), as in the oscillatory

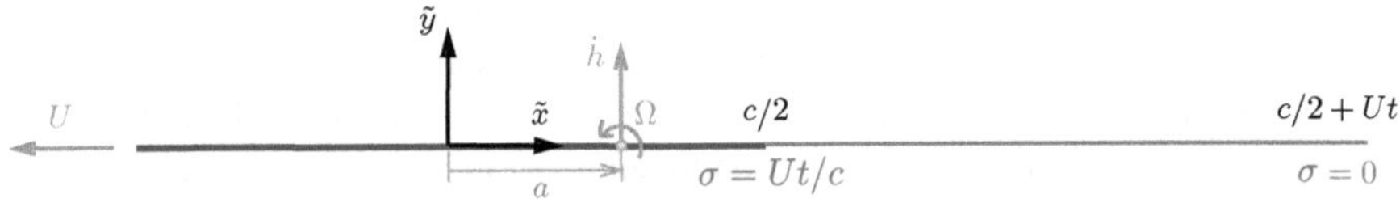

Fig. 8.14 Configuration of a two-dimensional flat plate with an impulsive change to its motion

case, but now with the right-hand side replaced by a jump at $t = 0$ from 0 to πc—a somewhat arbitrary amplitude, but chosen to correspond to a step change in the quasi-steady circulation due to unit translational motion in the $-\tilde{y}$ direction:

$$\int_{c/2}^{\infty} \frac{(\tilde{x}+c/2)^{1/2}\gamma_S(\tilde{x}-Ut)}{(\tilde{x}-c/2)^{1/2}}\,\mathrm{d}\tilde{x} = \begin{cases} 0, & t < 0, \\ \pi c, & t \geq 0. \end{cases} \tag{8.185}$$

Let us remember that the wake vortex sheet records the history of the plate's motion. However, this history only began at $t = 0$, so at time $t > 0$, the sheet extends not from $c/2$ to ∞, but only from $c/2$ to $c/2 + Ut$, as illustrated in Fig. 8.14. In recognition of this, we will define a new coordinate, σ, that measures the upstream-directed distance from the far end of the sheet toward the trailing edge (scaled by the chord length):

$$\sigma := 1/2 + Ut/c - \tilde{x}/c; \tag{8.186}$$

we will also make use of the convected time t^+, defined already in Eq. (8.181). Transforming the integral in (8.185) into these new variables, we have

$$\frac{1}{\pi}\int_0^{t^+} \frac{(1+t^+-\sigma)^{1/2}}{(t^+-\sigma)^{1/2}}\mu(\sigma)\,\mathrm{d}\sigma = \begin{cases} 0, & t^+ < 0, \\ 1, & t^+ \geq 0, \end{cases} \tag{8.187}$$

where μ denotes the vortex sheet strength, but now regarded as a function of the new scaled distance σ:

$$\mu(\sigma) := \gamma_S(\tilde{x} - Ut). \tag{8.188}$$

Equation (8.187) is to be solved for μ, such that it is satisfied at all values of scaled time, $t^+ \geq 0$.

We will discuss μ further below. But first, let us observe how the wake's contribution to lift and moment is manifested in this impulsive change of motion. Since its contribution to moment is simply a multiple of its contribution to lift, we will focus our attention on the integral term in Eq. (8.161); in our new variables, this takes the specific form

$$-\frac{1}{2}\rho c U \int_0^{t^+} \frac{\mu(\sigma)}{(1+t^+-\sigma)^{1/2}(t^+-\sigma)^{1/2}}\,\mathrm{d}\sigma. \tag{8.189}$$

Remember that we have defined μ based on a step change in quasi-steady circulation from 0 to $-\pi c$. Thus, for a general step change to quasi-steady circulation Γ_b, we would expect the wake's contribution to force to be the same, but re-scaled with the factor $-\Gamma_b/(\pi c)$. We arrive at the following result:

Result 8.7: Force and Moment due to a Sudden Change of Motion

The lift response to a sudden change in quasi-steady circulation from 0 to Γ_b—brought about by a sudden change in the plate's motion, $\Gamma_b := \pi c \left[\dot{h} + U\alpha + (c/4 - a)\,\Omega\right]$—is given as a function of chord lengths of travel, $t^+ := tU/c$, by

$$f_y = -\frac{\pi}{4}\rho c^2 \ddot{h} + \frac{\pi}{4}\rho c^2 a\dot{\Omega} - \frac{\pi}{4}\rho c^2 U\Omega - \rho U \Gamma_b \Phi(t^+), \tag{8.190}$$

and the corresponding moment about the pivot axis is

$$m_p = \frac{\pi}{4}\rho c^2 a\ddot{h} - \frac{\pi}{4}\rho c^2 \left(\frac{c^2}{32} + a^2\right)\dot{\Omega} - \frac{\pi}{4}\rho c^2 \left(\frac{c}{4} - a\right) U\Omega + \left(\frac{c}{4} + a\right)\rho U \Gamma_b \Phi(t^+). \tag{8.191}$$

In these expressions, we have used the function Φ, defined as

$$\Phi(t^+) := 1 - \frac{1}{2\pi}\int_0^{t^+} \frac{\mu(\sigma)}{(1 + t^+ - \sigma)^{1/2}(t^+ - \sigma)^{1/2}}\,d\sigma, \tag{8.192}$$

where μ, the wake vortex sheet strength, is the solution of the integral equation

$$\frac{1}{\pi}\int_0^{t^+} \frac{(1 + t^+ - \sigma)^{1/2}}{(t^+ - \sigma)^{1/2}}\mu(\sigma)\,d\sigma = 1 \tag{8.193}$$

for all $t^+ \geq 0$. The function Φ is commonly referred to as the **Wagner function.**[3]

[3] von Kármán and Sears [70] actually used the symbol Φ to denote the **lift deficiency function,** representing the fractional difference from the steady-state lift, $-\rho U \Gamma_b$. In our present notation, this would be given by $1 - \Phi$.

Solution for Φ by Expansion of μ Equation (8.193) was solved approximately by Wagner [71] by writing the wake vortex sheet strength, for all $\sigma < 1$—the section adjacent to the far end—as

$$\mu(\sigma) = \frac{1}{\sigma^{1/2}}\sum_{m=0}^{\infty} c_m^- \sigma^m. \tag{8.194}$$

Note that this expansion includes an inverse square root singularity at the far end: this is an expected behavior in the wake resulting from impulsive motion. The section farther from the end, $\sigma > 1$, is smooth and finite, and its expansion is

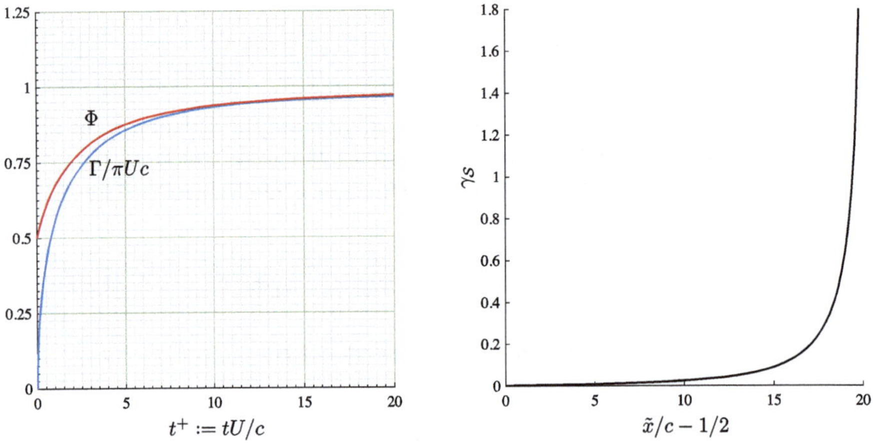

Fig. 8.15 Results for impulsive change of motion. (Left) The time histories of the Wagner function, $\Phi(t^+)$ (8.192) (red solid line), and the scaled wake circulation, $\Gamma(t^+)/\pi U c$ (blue solid line). (Right) The wake vortex sheet strength distribution, γ_S, at $t^+ = 20$, plotted versus scaled distance from the trailing edge

$$\mu(\sigma) = \frac{1}{\sigma^{1/2}} \sum_{m=0}^{\infty} c_m^+ \sigma^{-m}. \tag{8.195}$$

The coefficients c_m^- can be obtained by substituting into (8.193) and solving it explicitly for any $t^+ < 1$—that is, during the early development of the wake. This solution process is assisted by Taylor expanding the factor $(1 + t^+ - \sigma)^{1/2}$ over the variable $(t^+ - \sigma)^{1/2}$ about $t^+ - \sigma = 0$ (the trailing edge). The coefficients c_m^+ are then found by minimizing the error of Eq. (8.193) during the later stages of development, $t^+ > 1$. For this purpose, the integration is split into two intervals, since the interval from $\sigma = 0$ to 1 involves the coefficients c_m^-, now known. Wagner found an adequate two-term expansion for μ in $\sigma > 1$. However, his approximation becomes increasingly poor for $t^+ \gtrsim 6$. A more accurate result through $t^+ \approx 20$ can be found with a six-term expansion, obtained by a least-squares minimization. These coefficients are

$$\begin{aligned} c_0^+ &= -0.0496, \quad c_1^+ = 1.2301, \quad c_2^+ = 0.6987, \\ c_3^+ &= -4.3763, \quad c_4^+ = 5.1183, \quad c_5^+ = -1.9837. \end{aligned} \tag{8.196}$$

The Wagner function is computed at all t^+ by substituting the two expansions for μ, Taylor expanding the factor $(1 + t^+ - \sigma)^{1/2}$ in powers of $\sigma^{1/2}/(1 + t^+)^{1/2}$, and evaluating the resulting integrals at each power.

Solution for Φ via Theodorsen Function At the end of our discussion of the oscillatory plate motion, we observed that the solution for more general time histories of plate motion can be composed from a Fourier superposition across all frequencies. Indeed, the step change is one such time history, so we can obtain the lift response to a step change—and hence, the Wagner function—by starting from the general Fourier form (8.184). We need only one result from the theory of generalized functions, namely, that the Fourier representation of the unit step function, $H(t)$ (i.e., the Heaviside function) is

$$H(t) = \frac{1}{2} + \frac{1}{2\pi} \int_{-\infty}^{\infty} \frac{\mathrm{e}^{\mathrm{i}\omega t}}{\mathrm{i}\omega}\,\mathrm{d}\omega, \tag{8.197}$$

or, in terms of our dimensionless variables,

$$H(t^+) = \frac{1}{2} + \frac{1}{2\pi} \int_{-\infty}^{\infty} \frac{\mathrm{e}^{\mathrm{i}2\kappa t^+}}{\mathrm{i}\kappa}\,\mathrm{d}\kappa. \tag{8.198}$$

Thus, for a unit step change in quasi-steady circulation, the lift response can be composed from two parts: a part corresponding to a constant bound circulation, $\Gamma_b = 1/2$, which is simply $-\rho U/2$; and another part composed from the individual Fourier coefficients, $G_b = 1/(2\pi\mathrm{i}\kappa)$, which is given by the integral in (8.184). The Wagner function must be the sum of these parts, with the factor $-\rho U$ removed:

$$\Phi(t^+) = \frac{1}{2} + \frac{1}{\pi}\mathrm{Re}\left(\int_0^{\infty} \frac{C(\kappa)\mathrm{e}^{\mathrm{i}2\kappa t^+}}{\mathrm{i}\kappa}\,\mathrm{d}\kappa\right), \tag{8.199}$$

where C is the Theodorsen function, defined in Eq. (8.172). Recall that $C(\kappa)$ approaches $1/2$ at large frequencies, so it is helpful to isolate this behavior from the rest of the function by defining $C'(\kappa) := C(\kappa) - 1/2$. The contribution from the $1/2$ can be evaluated analytically; indeed, with reference to (8.198), it is easy to show that it is equal to $H(t^+)/2 - 1/4$. Since the Wagner function must be identically zero for $t^+ < 0$, it must be the case that

$$\frac{1}{4} + \frac{1}{\pi}\mathrm{Re}\left(\int_0^{\infty} \frac{C'(\kappa)\mathrm{e}^{-\mathrm{i}2\kappa t^+}}{\mathrm{i}\kappa}\,\mathrm{d}\kappa\right) = 0. \tag{8.200}$$

Thus, we arrive at an expression most favorable for numerical evaluation:

Note 8.5.2: The Wagner Function

At any time $t^+ > 0$, the Wagner function $\Phi(t^+)$ can be evaluated from the expression

$$\Phi(t^+) = \frac{1}{2} + \frac{2}{\pi} \int_0^{\infty} \frac{\mathrm{Re}\,(C'(\kappa))\sin 2\kappa t^+}{\kappa}\,\mathrm{d}\kappa, \tag{8.201}$$

where

$$C'(\kappa) := C(\kappa) - \frac{1}{2}, \tag{8.202}$$

and C is the Theodorsen function (8.172).

Expression (8.201) cannot be evaluated analytically and thus requires numerical integration. The results are shown in Fig. 8.15, which depicts the Wagner function along with the wake vortex sheet strength distribution, μ (or rather, γ_S) and the corresponding time history of the circulation in the wake.

There are a few important points to make regarding the solution of this problem:

- Note that $\Phi(0) = 1/2$. That is, ***exactly half of the steady-state lift is created immediately after the motion is initiated***. This is a consequence of the vorticity fluxed into the wake at that first instant, which, as we observed, has an inverse square root singularity in order to counter the finite step in plate motion.
- It takes a long time for the flow to achieve steady state. Even after 20 chords of travel, the lift is still only 97% of its steady-state value.
- A good approximation for the Wagner function, $\Phi(t^+)$, is given by

$$\Phi(t^+) \approx 1 - 0.165\mathrm{e}^{-0.091t^+} - 0.335\mathrm{e}^{-0.6t^+}. \tag{8.203}$$

This deviates less than 1% from the actual solution through $t^+ = 20$.

The Wagner function is much more significant than we have shown so far. As the solution to a step change in the motion, it represents the so-called **indicial response function** of the flow. From this response function, we can produce the response for any variation in the plate's motion—remembering, of course, that we are still restricted to small-amplitude deviations from the basic tangent motion. Otherwise, our basic premise of a flat wake originating smoothly from the trailing edge is inappropriate. This procedure works because the lift is linearly dependent on the deviations in motion, and thus, the responses to many such deviations can be superposed. Let us consider an infinitesimal perturbation to the quasi-steady circulation, $\delta\Gamma_b = \dot{\Gamma}_b(\tau^+)\delta\tau^+$, introduced at time τ^+ over an infinitesimal time interval from τ^+ to $\tau^+ + \delta\tau^+$, where $\dot{\Gamma}_b$ represents the rate of change of a smoothly-varying quasi-steady circulation history. At any time t^+ after the perturbation has been introduced, the wake's contribution to the lift due to this perturbation is obviously $-\rho U\dot{\Gamma}_b(\tau^+)\delta\tau^+\Phi(t^+ - \tau^+)$. For a continuous sequence of such perturbations, the lift is

$$f_y = -\frac{\pi}{4}\rho c^2\ddot{h} + \frac{\pi}{4}\rho c^2 a\dot{\Omega} - \frac{\pi}{4}\rho c^2 U\Omega - \rho U\left(\Gamma_b(0)\Phi(t^+) + \int_0^{t^+}\dot{\Gamma}_b(\tau^+)\Phi(t^+ - \tau^+)\,\mathrm{d}\tau^+\right), \tag{8.204}$$

The integral within the parentheses is called **Duhamel's integral**, and represents a convolution—with the Wagner function as the kernel—over the entire history of the perturbations to the quasi-steady circulation. The relationship between perturbation and lift response is ***causal*** of course: the lift response cannot precede a disturbance to the motion. Hence, the upper limit of t^+ on the integral.

In the two cases we have just considered, the disturbances were introduced through the plate's rigid-body motion. It is also of interest to consider cases in which the disturbances are *environmental*: that is, they are associated with some departure from quiescence in the fluid through which the plate is traveling. Such disturbances are

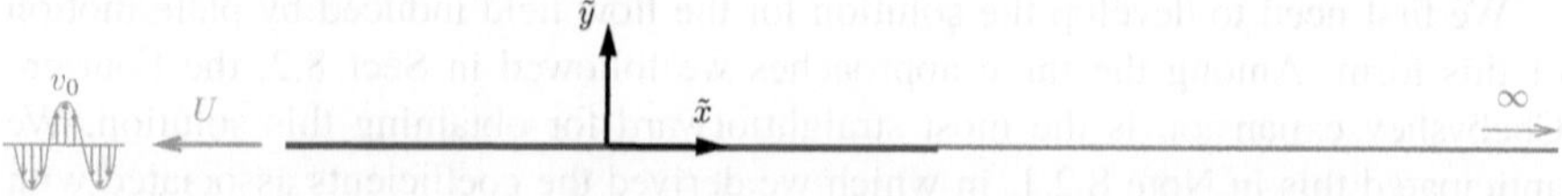

Fig. 8.16 Configuration of a two-dimensional flat plate subjected to a sinusoidal gust

conventionally known as gusts. Here, we will consider two classes, both involving a distribution of small vertical velocity in the fluid, homogeneous in the plate-normal direction but varying in the tangent direction. The gust is presumed to be 'frozen': that is, it does not vary in time relative to the mean state of the fluid. Thus, we could impose the gust via a disturbance in the free stream of the form

$$\boldsymbol{v}_{\infty} = v_0 f(\tilde{x})\tilde{\boldsymbol{e}}_2, \tag{8.205}$$

where $|v_0| \ll U$ and the function f is of order unity, and where U still represents the tangential speed at which the plate translates through the fluid. However, in order to avoid the awkwardness of straddling the windunnel and inertial reference frames, we could equivalently express the gust as a disturbance in the plate motion, $\boldsymbol{V}_{\text{b}}$. We need only note that, from the reference frame of the plate, the gust is in the form of a wave of fixed form traveling at speed U in the $\tilde{x}$ direction:

$$\boldsymbol{V}_{\text{b}} = -U\tilde{\boldsymbol{e}}_1 - v_0 f(\tilde{x} - Ut)\tilde{\boldsymbol{e}}_2. \tag{8.206}$$

Though this represents a wavy disturbance to the plate, it is presumed to be of sufficiently small amplitude that the plate can still be considered flat. We will consider two forms of the function f: in the first, it will be sinusoidal, and in the second, it will a sharp discontinuity encountered suddenly. For both, we will need to revisit some of our basic results, since we have mostly neglected non-rigid motions of the plate thus far.

8.5.5 Sinusoidal Gust Response

Let us suppose that the gust takes the form of a pure sinusoid, as illustrated in Fig. 8.16. As such, we can represent the plate motion as the real part of

$$-U\tilde{\boldsymbol{e}}_1 - v_0 \mathrm{e}^{\mathrm{i}\omega(t-\tilde{x}/U)}\tilde{\boldsymbol{e}}_2, \tag{8.207}$$

where v_0 is a constant complex amplitude. (We have used a vector, rather than complex, notation to clarify the direction of the flow, in order to avoid confusion with the complex notation of the waveform.) In terms of the reduced frequency and our dimensionless measure of spatial position, $\xi = 2\tilde{x}/c$, the disturbance can be written as

$$-U\tilde{\boldsymbol{e}}_1 - v_0 \mathrm{e}^{\mathrm{i}\omega t}\mathrm{e}^{-\mathrm{i}\kappa\xi}\tilde{\boldsymbol{e}}_2. \tag{8.208}$$

We first need to develop the solution for the flow field induced by plate motion of this form. Among the three approaches we followed in Sect. 8.2, the Fourier–Chebyshev expansion is the most straightforward for obtaining this solution. We anticipated this in Note 8.2.1, in which we derived the coefficients associated with sinusoidal plate deformation. For wavenumber κ, the coefficients of the normal velocity are the real parts of

$$A_{\kappa,n} = -v_0 e^{i\omega t} \sigma_n i^{-n} J_n(\kappa), \tag{8.209}$$

for all $n \geq 0$, where $\sigma_0 = 1$ and $\sigma_n = 2$ for all other n; the coefficients of the bound vortex sheet are the real parts of

$$B_{\kappa,n} = -v_0 e^{i\omega t} \frac{4n i^{-n-1}}{\kappa} J_n(\kappa), \tag{8.210}$$

for $n > 0$; note that $B_{\kappa,0} = 0$, since, in our approach, this coefficient is set to zero for all but the vorticity-induced basis flow field.

The Kutta condition at the trailing edge (8.116) leads to the following equation for the wake vortex sheet strength:

$$\int_{c/2}^{\infty} \frac{(\tilde{x} + c/2)\gamma_S(\tilde{x})}{(\tilde{x}^2 - c^2/4)^{1/2}} d\tilde{x} = \pi c v_0 e^{i\omega t} \left(J_0(\kappa) - iJ_1(\kappa)\right). \tag{8.211}$$

As for the case of oscillatory rigid motion, the wake vortex sheet strength takes the form of a traveling wave, $\gamma_S(\tilde{x} - Ut) = g_0 e^{i(\omega t - \kappa\xi)}$; when substituted into the integrals, we obtain the same left-hand side as we did in (8.167). Thus, it is easy to show that

$$g_0 = v_0 \frac{4i\left(J_1(\kappa) + iJ_0(\kappa)\right)}{H_1^{(2)}(\kappa) + iH_0^{(2)}(\kappa)}. \tag{8.212}$$

To determine the force and moment generated by this non-rigid disturbance to the plate motion, we need to revisit one of our earlier (general) forms for these quantities. For the present purposes, these are most easily obtained from the impulse expressions in terms of Fourier–Chebyshev coefficients in (8.99). Note that we need only address the contributions from body motion; our existing expressions for the contributions from wake vorticity still hold. Thus, we obtain

$$\tilde{f}_y = \rho \frac{\pi c^2}{8} \frac{dB_{\kappa,1}}{dt} + \rho U \int_{c/2}^{\infty} \frac{\tilde{x}\gamma_S(\tilde{x})}{(\tilde{x}^2 - c^2/4)^{1/2}} d\tilde{x}, \tag{8.213}$$

and

$$m_r = \rho \frac{\pi c^3}{64} \frac{dB_{\kappa,2}}{dt} - \rho \frac{\pi c^2}{8} U B_{\kappa,1} + \frac{1}{8} \rho c^2 U \int_{c/2}^{\infty} \frac{\gamma_S(\tilde{x})}{(\tilde{x}^2 - c^2/4)^{1/2}} d\tilde{x}. \tag{8.214}$$

respectively. The wavy deformation of the plate thus assumes the role of an added mass term. We can evaluate all of these expressions using our results thus far, and we arrive at

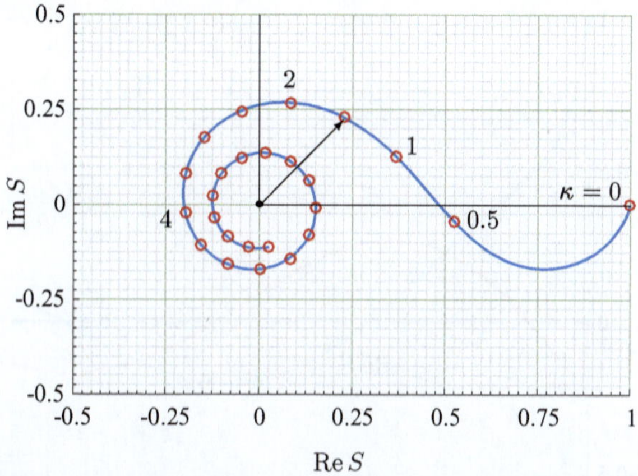

Fig. 8.17 Vector plot of the Sears function (8.217), parameterized by reduced frequency κ. The length and direction of the arrow indicate the magnitude and phase of the function at the corresponding value of κ

$$f_y = \rho\pi c U v_0 e^{i\omega t} S(\kappa) \tag{8.215}$$

for the lift, and

$$m_r = -\rho\frac{\pi c^2}{4} U v_0 e^{i\omega t} S(\kappa) \tag{8.216}$$

for the moment through the centroid, where S is known as the **Sears function** [64], defined as

$$S(\kappa) := \frac{2i}{\pi\kappa\left(H_1^{(2)}(\kappa) + iH_0^{(2)}(\kappa)\right)}. \tag{8.217}$$

Note from the expression of moment that, similar to what we have seen before, the lift due to a sinusoidal gust acts through the quarter-chord point.

The Sears function provides the transfer function between the oscillatory character of the gust and the consequent fluctuations in lift. Figure 8.17 depicts the function in a vector plot, with reduced frequency as a parameter along the curve. As can be observed from the plot, at $\kappa = 0$, the Sears function is equal to unity. This corresponds to a steady vertical component of motion, and thus, a steady angle of incidence relative to the fluid. The magnitude of the function becomes progressively weaker with increasing frequency, and the phase varies continuously. The real and imaginary parts are plotted versus reduced frequency separately in Fig. 8.18.

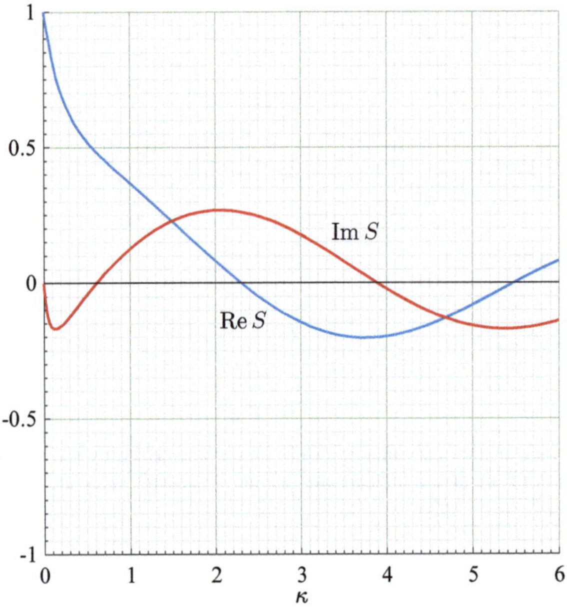

Fig. 8.18 Real and imaginary parts of the Sears function (8.217) versus reduced frequency κ

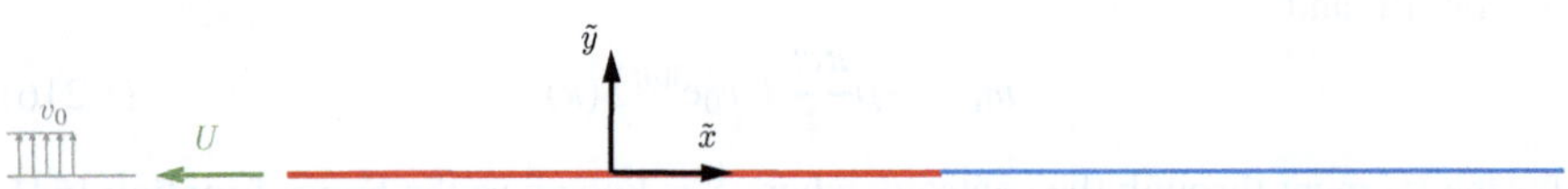

Fig. 8.19 Configuration of the flat wake model of a two-dimensional flat plate subjected to a sharp-edged gust

8.5.6 Sharp-Edge Gust

The response of the plate and its wake to a sinusoidal gust of some frequency provides us with the basic element with which to obtain, through superposition, the response to arbitrary vertical disturbances. As a simple example of this, let us consider a sharp-edged gust, depicted in Fig. 8.19, in which, at $t = 0$, the leading edge of the plate encounters a sudden change in vertical velocity from zero to a small uniform value, v_0. In other words, the gust function takes the form of a Heaviside function, $f(\tilde{x}) = H(-\tilde{x} - c/2)$ (i.e., vanishing at all points downstream of the leading edge, $\tilde{x} = -c/2$). The approach we follow is analogous to our earlier derivation of the Wagner function from the Theodorsen function, and it is straightforward to obtain the following expressions for lift and moment:

$$f_y = \rho\pi c U v_0 \Psi(t^+) \tag{8.218}$$

and

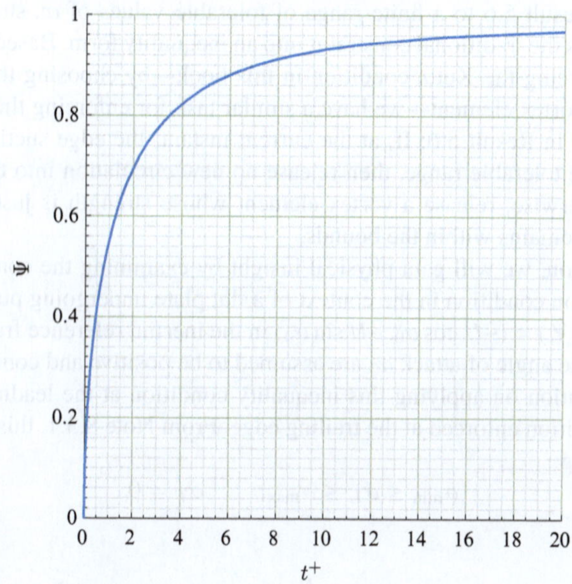

Fig. 8.20 The Küssner function (8.220)

$$m_{\rm r} = -\rho \frac{\pi c^2}{4} U v_0 \Psi(t^+), \tag{8.219}$$

where Ψ is called the **Küssner function** [42], defined by the integral

$$\Psi(t^+) := \frac{2}{\pi} \int_0^\infty \frac{\mathrm{Re}(S(\kappa)\mathrm{e}^{-\mathrm{i}\kappa}) \sin 2\kappa t^+}{\kappa} \, \mathrm{d}\kappa. \tag{8.220}$$

Like the Wagner function, the Küssner function cannot be evaluated analytically, but requires numerical evaluation. The plot of the function is shown in Fig. 8.20. It bears some resemblance to the Wagner function, with the obvious difference that it starts at zero rather than $1/2$. This difference is easily explained: If we consider the incident gust as an equivalent motion of the plate, this motion effectively creates a sharp bend in the plate—that is, a discontinuity in angle of incidence—that moves progressively from the leading to the trailing edge. Thus, the trailing edge remains initially undisturbed relative to the fluid, and consequently, its wake lacks the square-root singularity at its end, as in the wake of the Wagner flow.

8.6 Generalized Edge Conditions and Their Interpretation

In Sect. 5.4 we showed that the Kutta condition at an edge is equivalent to forcing the edge's suction parameter, σ, to be identically zero; this was verified for the case of the flat plate in Sect. 8.4. Based on this interpretation, we generalized the Kutta condition in Result 5.6 to a finite range of tolerable values of σ, straddling zero. This transforms the traditional constraint into an inequality form. Based on our usual means of enforcing the Kutta condition in this book—by choosing the strength of a newly-shed vortex element—we have a similar task for enforcing this generalized form, as stated in Result 5.6: If, at the current instant, the edge suction parameter falls within the tolerable range, then release no new circulation into the fluid from that edge; otherwise, release a vortex element whose strength is just sufficient to ensure that σ remains within the bounds.

In this section, we will gain physical insight by examining the consequences of this edge suction condition in the context of a flat plate undergoing pure translation at velocity $(\tilde{U}_r, \tilde{V}_r) = (-U\cos\alpha_0, -U\sin\alpha_0)$ in the inertial reference frame. Both the speed U and the angle of attack α_0 are assumed to be positive and constant. We will focus our attention on applying this inequality condition at the leading edge, with the Kutta condition enforced at the trailing edge. From Note 8.3.1, this is equivalent to the following:

$$\sigma_{\min} \leq \sigma_L \leq \sigma_{\max}, \qquad \sigma_T = 0, \tag{8.221}$$

where

$$\sigma_L := \sigma(-1) = \frac{1}{2}U\sin\alpha_0 + \sum_{J=1}^{N_v} \frac{\Gamma_J}{2\pi c}\mathrm{Re}\left(\frac{(\tilde{z}_J - c/2)^{1/2}}{(\tilde{z}_J + c/2)^{1/2}}\right) \tag{8.222}$$

and

$$\sigma_T := \sigma(1) = -\frac{1}{2}U\sin\alpha_0 + \sum_{J=1}^{N_v} \frac{\Gamma_J}{2\pi c}\mathrm{Re}\left(\frac{(\tilde{z}_J + c/2)^{1/2}}{(\tilde{z}_J - c/2)^{1/2}}\right). \tag{8.223}$$

For the sake of developing intuition on the role of the critical suction parameters, let us meet the conditions (8.221) with two point vortices of variable strength, one released from each edge. This is simpler than the alternative representations of shed vorticity—a vortex sheet or a discrete cloud of constant-strength point vortices released from each edge—and thus provides a more analytically tractable framework for gaining insight. Let us further suppose that the trailing-edge vortex is already far behind the plate, so that the ratio of its distances from the leading and trailing edges is nearly unity. Admittedly, this is a rather crude simplification, but it does not affect the conclusions we will draw from the model. Thus, the trailing-edge Kutta condition can be written as:

$$\Gamma_T = \pi c U\sin\alpha_0 - \Gamma_L \mathrm{Re}\left(\frac{(\tilde{z}_L + c/2)^{1/2}}{(\tilde{z}_L - c/2)^{1/2}}\right). \tag{8.224}$$

In this expression, we see the classical strength of the starting vortex due to plate motion at fixed angle of attack, but modified by the presence of a leading-edge vortex. Substituting this into the leading-edge condition on the suction parameter in order to eliminate Γ_T, we obtain

$$\sigma_{\min} \leq U\sin\alpha_0 - \frac{\Gamma_L}{2\pi}\mathrm{Re}\left(\frac{1}{(\tilde{z}_L + c/2)^{1/2}(\tilde{z}_L - c/2)^{1/2}}\right) \leq \sigma_{\max}. \tag{8.225}$$

In this form, it is easy to discern the physical meaning of the bounds $\sigma_{\min}$ and $\sigma_{\max}$. We already knew that σ has units of velocity; here, we see that we can interpret the bounds on σ as bounds on the vertical velocity of the plate, $U \sin \alpha_0$. Let us define bounds on the angle of attack, $\alpha_{\min} < 0$ and $\alpha_{\max} > 0$, such that $\sigma_{\min} \equiv U \sin \alpha_{\min}$ and $\sigma_{\max} \equiv U \sin \alpha_{\max}$. Then, we can write the criterion as

$$\sin \alpha_{\min} \leq \sin \alpha_0 - \frac{\Gamma_L}{2\pi U} \mathrm{Re}\left(\frac{1}{(\tilde{z}_L + c/2)^{1/2} (\tilde{z}_L - c/2)^{1/2}} \right) \leq \sin \alpha_{\max}. \quad (8.226)$$

The leading-edge suction condition has now been equivalently expressed in terms of angle of attack: If the angle of attack, α_0, lies between the bounds $\alpha_{\min}$ and $\alpha_{\max}$, then no leading edge vorticity need be released. However, if the angle of attack is, for example, larger than $\alpha_{\max}$, then a leading-edge vortex must develop, with strength given by

$$\Gamma_L = 2\pi U (\sin \alpha_0 - \sin \alpha_{\max}) \left[\mathrm{Re}\left(\frac{1}{(\tilde{z}_L + c/2)^{1/2} (\tilde{z}_L - c/2)^{1/2}} \right) \right]^{-1}. \quad (8.227)$$

A similar expression can be formed when the angle of attack is negative and lies below $\alpha_{\min}$.

Now that we have expressed the edge suction condition in a more recognizable form, let us contrast it with the application of the Kutta condition at the leading edge, which, we will recall, is equivalent to setting both the lower and upper bounds to zero. In other words, under the strict enforcement of the Kutta condition, the flat plate cannot adopt a non-zero angle of attack without some release of vorticity from the leading edge. The suction condition, in contrast, permits a certain range of acceptable angles without such release. If the angle is outside this range, then the strength of the vortex emerging from the leading edge is directly proportional to the amount by which the plate's normal velocity exceeds the threshold, and is thus weaker than the vortex generated in response to the Kutta condition.

Note that the strength given by (8.227) depends on the position of the vortex relative to the edges of the plate, and that this dependence seems to suggest that the vortex's strength will continue to grow as it moves further away. It should be borne in mind that this model we have proposed for the flow past a translating plate, with a leading-edge vortex represented by only one point vortex, is only reasonable in the first few chord lengths of travel, well before the travel distance of the vortex would lead to an unphysically large strength.

8.7 A Deforming Plate

Though the focus of this chapter is on a flat plate, it is interesting to consider the more general context in which we must enforce the no-penetration condition on a deforming plate. We already considered a limited version of this problem in the response to a low-amplitude sinusoidal gust in Sect. 8.5.5, which we found, by change of reference frame, to be equivalent to a deforming plate. In that case we were able to linearize the problem in order to make the problem analytically tractable. However, if we place no restriction on the amplitude of the plate's deformation, we no longer have an analytical solution, and thus must seek a numerical approximation. This numerical solution encounters difficulties unless some care is taken to formulate the mathematical problem. The following discussion is based on work by Shukla and Eldredge [66] and uses tools from Muskhelishvili's book [56]. The problem has also been studied by Alben [2], who obtained the deformation not by prescription but by coupling the fluid dynamics with the elastic mechanics of the plate.

Our starting point for this problem is, as it was for the flat plate, Eq. (4.36). However, in contrast to the flat plate, the tangent of this deformed plate is not uniform but varies along its length, c. We parameterize the plate by the signed arc length from its midpoint, $s \in [-c/2, c/2]$, so that a point on the plate can be described in inertial coordinates as

$$z(s) = Z_r + \tilde{z}(s)e^{i\alpha}. \tag{8.228}$$

The tangent is then

$$\tau(s) := \frac{dz}{ds} = e^{i\alpha}\frac{d\tilde{z}}{ds} = e^{i(\phi(s)+\alpha)}. \tag{8.229}$$

We will take the body-fixed coordinate system, $\tilde{z}$, to be tangent and normal to the plate at the reference point, Z_r. The body's deviation from flat is described by the local angle, $\phi(s)$, which, by construction, is identically zero at Z_r. By writing the deformation in this manner, through changes in this local angle, we ensure that the plate's length remains invariant. Note that the local curvature, $\kappa(s)$, is equal to $d\phi/ds$. We will assume that the deformation is smooth along the length of the plate, so that $\tilde{z}(s)$ is infinitely differentiable, though this restriction can be relaxed.

Up to this point, we have omitted the time dependence of these quantities. But the deformation is described by the time variation of the plate's coordinates in its own reference frame, $\tilde{z}(s,t)$. This time variation gives rise to a **deformation velocity**, written in the usual conjugate form in plate-fixed components as

$$\tilde{w}_b(s,t) := \partial\tilde{z}^*(s,t)/\partial t = -i\int_{-c/2}^{s} \dot{\phi}(\sigma,t)e^{-i\phi(\sigma,t)}\,d\sigma; \tag{8.230}$$

this deformation velocity adds to the plate's intrinsic rigid motion, so that the local velocity on the plate is given (also in plate-fixed components) by

$$\tilde{W}_b(s,t) = \tilde{w}_b(s,t) + \tilde{W}_r(t) - i\Omega(t)\tilde{z}^*(s,t). \tag{8.231}$$

Using this notation for the deforming plate, Eq. (4.36) can be written in a form parameterized by the arc length coordinate as

$$\mathrm{Re}\left[\frac{\mathrm{e}^{\mathrm{i}\phi(s)}}{2\pi}\fint_{-c/2}^{c/2}\frac{\gamma(\sigma)}{\tilde{z}(\sigma)-\tilde{z}(s)}\,\mathrm{d}\sigma\right]=\Delta V_{\perp}(\tilde{z}(s)),\quad s\in[-c/2,c/2],\tag{8.232}$$

where the right hand side represents, as for the flat plate, the local normal component of the difference in velocity between the uniform flow and induced by fluid vorticity and the velocity of the plate:

$$\Delta V_{\perp}(\tilde{z}(s)):=-\mathrm{Im}\left[\mathrm{e}^{\mathrm{i}\phi(s)}\left(\tilde{W}_{\infty}+\sum_{\mathrm{J}=1}^{N_{\mathrm{v}}}\frac{\Gamma_{\mathrm{J}}}{\tilde{z}(s)-\tilde{z}_{\mathrm{J}}}-\tilde{W}_{\mathrm{b}}(s)\right)\right].\tag{8.233}$$

It is useful to compare Eq. (8.232) with its corresponding form for the flat plate, Eq. (8.35). Both involve the principal value of the integral on the plate. But in the case of the deformed plate, the singular behavior of the integrand is slightly more obscure, since the evaluation point on the plate is no longer approached along a straight path, but on an arbitrarily curved path. However, we can perform a simple trick to recover a more explicit form of the singularity. In the vicinity of the evaluation point, $\tilde{z}(\sigma)\approx\tilde{z}(s)+\tilde{z}'(s)(\sigma-s)=\tilde{z}(s)+\mathrm{e}^{\mathrm{i}\phi(s)}(\sigma-s)$. The difference between the arc length parameter along the plate, σ, and its value at the evaluation point, s, is preferable to the difference in their complex coordinates. Thus, by trivially adding and subtracting a kernel with this difference in arc length, we can define a new form of the integral equation,

$$\frac{1}{2\pi}\fint_{-c/2}^{c/2}\frac{\gamma(\sigma)}{\sigma-s}\,\mathrm{d}\sigma+\frac{1}{2\pi}\int_{-c/2}^{c/2}M(s,\sigma)\gamma(\sigma)\,\mathrm{d}\sigma=\Delta V_{\perp}(\tilde{z}(s)),\quad s\in[-c/2,c/2],\tag{8.234}$$

where we have defined a second kernel, M, that is completely regular for all $s,\sigma\in[-c/2,c/2]$, including when $s=\sigma$:

$$M(s,\sigma):=\mathrm{Re}\left(\frac{\mathrm{e}^{\mathrm{i}\phi(s)}}{\tilde{z}(\sigma)-\tilde{z}(s)}\right)-\frac{1}{\sigma-s}.\tag{8.235}$$

Indeed, it is easy to verify that in the vicinity of the evaluation point, $\sigma\approx s$, this kernel has the form

$$M(s,\sigma)=-\frac{1}{12}\kappa^{2}(s)(\sigma-s)+O((\sigma-s)^{2}).\tag{8.236}$$

What we have done is map the curved contour $\mathcal{P}$ of the deformed plate onto a straight contour $s\in[-c/2,c/2]$. In the special case of a flat plate, this mapping is obviously just the identity: the regular kernel M vanishes identically and Eq. (8.234) reduces to (8.35), which we can solve by the techniques described earlier in this chapter. For a deforming plate, however, we must contend with the influence of the regular kernel, M.

Unfortunately, there is no general analytical solution for this problem, and furthermore, Eq. (8.234) constitutes a singular Fredholm equation of the first kind. In the absence of an analytical solution, such equations are well known to be challenging to solve numerically, often requiring special regularization techniques. However, we can use a trick to transform this equation into a singular Fredholm equation of the second kind, more conducive to numerical solution. The main idea, described in §114 of Muskhelishvili's book [56], is to treat the first term in (8.234) as the 'dominant' part of the integral operator and make use of the known inversion of this operator, as we did in Eq. (8.38). Here, instead of an explicit solution for γ, we get a new form of integral equation for this sheet strength. For mathematical clarity, we also express this new equation in terms of the scaled coordinate $\xi := 2s/c \in [-1, 1]$:

$$\gamma(\xi) + \frac{1}{\pi}\int_{-1}^{1} D(\xi, l)\gamma(l)\,\mathrm{d}l = F(\xi) + \frac{C}{\left(1-\xi^2\right)^{1/2}}, \tag{8.237}$$

where we have defined the right-hand side function as

$$F(\xi) := -\frac{2}{\pi}\fint_{-1}^{1} \left(\frac{1-l^2}{1-\xi^2}\right)^{1/2} \frac{\Delta V_{\perp}(\tilde{z}(l))}{l-\xi}\,\mathrm{d}l, \tag{8.238}$$

and a new kernel,

$$D(\xi, l) := -\frac{1}{\pi}\fint_{-1}^{1} \left(\frac{1-\tau^2}{1-\xi^2}\right)^{1/2} \frac{\hat{M}(\tau, l)}{\tau-\xi}\,\mathrm{d}\tau. \tag{8.239}$$

The kernel $\hat{M}(\xi, l)$ is equal to $cM(c\xi/2, cl/2)/2$, where M is the smooth kernel defined in (8.235). As in the case of the flat plate, the solution $\gamma(\xi)$ is not unique without applying edge conditions, but here, the arbitrary constant, C, enters the integral equation itself rather than the solution.

Equation (8.237) for the vortex sheet strength must be solved numerically, and we have expressed it in a standard form for this purpose [19]. We could simply apply a quadrature approximation of the integrals and evaluate the resulting equation at the quadrature points. However, we should first recognize that we have not yet applied any edge conditions to this equation, and until we do, we expect the solution for $\gamma(\xi)$ to have inverse square-root singularities at the edges $\xi = \pm 1$. A quadrature should account for this behavior. It is sensible, then, to expose this singular behavior of the solution so that our numerical approximation can concentrate on the remaining smooth structure. For the latter, we will rely on a Fourer–Chebyshev expansion, as we did for the flat plate. Thus, we first write the vortex sheet strength in the form (8.66), repeated here for reference:

$$\gamma(\tilde{z}(\xi)) = \frac{1}{(1-\xi^2)^{1/2}} \sum_{n=0}^{\infty} B_n T_n(\xi). \tag{8.66}$$

We will also expand the normal velocity distribution, $\Delta V_\perp(\tilde{z}(l))$, in the same form (8.72) as before. The kernel, $\hat{M}(\xi, l)$, is smooth in both of its independent variables along the plate, so we will expand it in a two-dimensional series,

$$\hat{M}(\xi, l) = \sum_{m,n=0}^{\infty} M_{mn} T_m(\xi) T_n(l). \tag{8.240}$$

When we substitute these expansions for γ, $\Delta V_\perp$, and $\hat{M}$ into the integral equation (8.237), the resulting integrals over the range $[-1, 1]$ can be evaluated analytically; indeed, we have already encountered all of these integrals in our discussion of the flat plate. For example, after substituting the expansion for $\Delta V_\perp$, the right-hand side function becomes

$$F(\xi) = \frac{2}{(1-\xi^2)^{1/2}} \sum_{n=0}^{\infty} A_n \fint_{-1}^{1} \frac{(1-l^2)^{1/2} T_n(l)}{l - \xi} \, \mathrm{d}l. \tag{8.241}$$

We can evaluate the resulting integrals in this expansion with identity (A.228), so that we can finally write

$$F(\xi) = \frac{1}{(1-\xi^2)^{1/2}} \sum_{n=1}^{\infty} F_n T_n(\xi), \tag{8.242}$$

where

$$F_1 = A_2 - 2A_0, \qquad F_n = A_{n+1} - A_{n-1}, \quad n > 1. \tag{8.243}$$

Similarly, the kernel D can be evaluated after substituting the expansion of $\hat{M}$; it can ultimately be written as

$$D(\xi, l) = \frac{1}{2(1-\xi^2)^{1/2}} \sum_{m=1}^{\infty} \sum_{n=0}^{\infty} D_{mn} T_m(\xi) T_n(l), \tag{8.244}$$

where, for all $n \geq 0$,

$$D_{1,n} = 2M_{0,n} - M_{2,n}, \qquad D_{mn} = M_{m-1,n} - M_{m+1,n}, \qquad m > 1. \tag{8.245}$$

The remaining integral, of $D(\xi, l)\gamma(l)$, simply relies on orthogonality of the Chebyshev polynomials, and the reader can verify that this integral becomes

$$\frac{1}{\pi} \int_{-1}^{1} D(\xi, l)\gamma(l) \, \mathrm{d}l = \frac{1}{(1-\xi^2)^{1/2}} \sum_{m=1}^{\infty} \left(\frac{1}{2} D_{m,0} B_0 + \frac{1}{4} \sum_{n=1}^{\infty} D_{mn} B_n \right) T_m(\xi). \tag{8.246}$$

Thus, assembling these expansions and eliminating the common factor of $(1 - \xi^2)^{-1/2}$, the integral equation (8.237) can be written as

$$B_0 - C + \sum_{m=1}^{\infty} \left(B_m + \frac{1}{2} D_{m,0} B_0 + \frac{1}{4} \sum_{n=1}^{\infty} D_{mn} B_n - F_m \right) T_m(\xi) = 0. \tag{8.247}$$

After applying the orthogonality of Chebyshev polynomials and truncating the series after N terms, we obtain a linear system of equations for the coefficients B_n of the vortex sheet strength:

$$B_0 = C, \quad \sum_{n=1}^{N} \left(\delta_{mn} + \frac{1}{4} D_{mn} \right) B_n = F_m - \frac{1}{2} D_{m,0} C, \quad m = 1, \ldots, N. \tag{8.248}$$

As for the flat plate, the leading coefficient in the sheet's expansion, B_0, is associated with the total circulation about the plate, $B_0 = 2\Gamma_{\mathcal{P}}/\pi c$. We see now that the arbitrary constant, C, is associated with this circulation. By Kelvin's circulation theorem, we can set this constant so that the total circulation is zero, as we did in (8.51) for the flat plate:

$$C = B_0 = -\frac{2}{\pi c} \sum_{\mathrm{J}=1}^{N_\mathrm{v}} \Gamma_\mathrm{J}. \tag{8.249}$$

The system of equations (8.248) relies on the coefficients A_n of the normal velocity distribution $\Delta V_\perp$ (to obtain F_n in (8.243)) and the coefficients M_{mn} of the smooth kernel $\hat{M}$ (to provide D_{mn} in (8.245)). In fact, if this were a flat plate, then the coefficients M_{mn}–and thus, D_{mn}—would be identically zero, and Eq. (8.248) would revert to the relations (8.75) that we derived earlier in this chapter.

One of the great advantages of the expanding the distributions in Chebyshev polynomials is that, when expressed instead in the independent variable $\theta = \arccos \xi$, the distributions are simply cosine series, and their coefficients can be obtained by **discrete cosine transform (DCT)**, a standard tool in a fast Fourier transform library. We simply evaluate the distributions ($\Delta V_\perp$, and $\hat{M}$ with respect to both variables) at the N **Chebyshev collocation points** $\xi_j = \cos \theta_j$, where $\theta_j = (j-1)\pi/(N-1)$ are uniformly spaced between 0 and π, and apply the DCT to the sampled data. The linear system of equations can be solved for the N coefficients B_n with standard methods, such as Gauss elimination. These coefficients can then be used to evaluate the vortex sheet strength, γ_j, at the sample locations ξ_j by inverse DCT.

The other computational steps of the problem all follow from this solution for B_n (or γ_j). For these purposes, it is useful to note that all of the coefficients on the right-hand side of (8.248) can be decomposed into flow contributors, so that the coefficients B_n can be decomposed, as well. (The contributors now include among them the plate's deformation velocity.) The Kutta condition, or other condition on the suction parameter, can be enforced by following the techniques in Sect. 5.2 or 5.4. The suction parameters at the trailing edge, $\xi = 1$, and leading edge, $\xi = -1$, are

$$\sigma_T := \sigma(-1) = -\frac{1}{2\sqrt{2}} \sum_{n=0}^{N} B_n, \quad \sigma_L := \sigma(1) = -\frac{1}{2\sqrt{2}} \sum_{n=0}^{N} (-1)^n B_n. \tag{8.250}$$

Unlike the flat plate, we have no natural truncation of these series of the kind that led to (8.115). This is because the deforming plate contributes in a more complicated manner to its own edge behavior. However, by the decomposition of the right-hand side, we can easily isolate the contribution made by a new vortex element to these coefficients B_n, and thence apply the formula (5.15) or (5.17).

The fluid velocity field induced by the bound vortex sheet, such as required for advancing the point vortex positions, can be computed by applying, for example, Clenshaw–Curtis quadrature of the integral in Eq. (4.15) at the Chebyshev collocations points. The impulse formulas (6.40) and (6.43) can be discretized with the same form of quadrature for purposed of calculating force and moment on the deforming plate.

Chapter 9
Examples of Two-Dimensional Flow Modeling

Contents

9.1 Time Marching 342
9.2 Co-rotating Vortex Patches 344
9.3 Interaction of Vortex Patches with a Bluff Body 348
9.4 A Translating Flat Plate at Large Angle of Attack 352
9.4.1 Free Vortex Sheet Model 352
9.4.2 Variable-Strength Vortex Model 361
9.5 Flow Past a NACA Airfoil 364

In Chap. 8 we devoted our attention to the mathematics of flow about a two-dimensional flat plate of infinitesimal thickness. Then, we utilized these tools to develop some of the well-known results from unsteady aerodynamics. These results were based on a quite restrictive assumption: that the flow was generated by a small-amplitude disturbance to the flow past the plate traveling at small angle of attack. Under this assumption, we were able to obtain analytical expressions for the flow field and the associated force and moment.

But suppose we do not have the luxury of assuming small-amplitude motions or small angles of attack. In such a situation, we will generally not be able to obtain analytical expressions, but will instead have to rely to some extent on numerical (i.e., computational) solution of the problem. How would such a numerical model be constructed? All such models must have the same basic ingredients, each of which was noted in our illustrative example in Chap. 1:

1. A choice of representation of vorticity in the fluid. If the representation is a distributed spatial structure—a sheet in two dimensions—this will require further a means of **discretizing** this representation into a finite number of constituent vortex elements. The form of regularization of the vortex elements, if any, must be chosen, as well.
2. A model for advancing the vortex elements forward in time, based on the instantaneous fluid velocity field. This fluid velocity field must be calculable from the

J. D. Eldredge, *Mathematical Modeling of Unsteady Inviscid Flows*, Interdisciplinary Applied Mathematics 50, https://doi.org/10.1007/978-3-030-18319-6_9

current collection of vortex elements in the fluid, and, if a body is present, account for the body's motion and enforce the no-penetration condition on its surface.
3. If desired, a means of enforcing conditions at an edge (or edges) of a body, for example, by introducing new vortex elements into the fluid.
4. An expression for calculating the force and moment on the body. Generally, we will rely on impulse-based calculations of these quantities in our examples.

All of these ingredients were also present in the classical analytical models of Sect. 8.5, though some appeared in trivial form. For example, elements in the free vortex sheet propagated at a constant speed away from the trailing edge, requiring no special treatment for the transport. In this chapter, we will describe, through representative examples, the practical construction of a model that relies, to some extent, on numerical treatment of these ingredients.

9.1 Time Marching

Though the examples we present below will differ in some ways, they will all share one thing in common: they require the time advancement of a **state vector**, $\mathsf{u}(t)$, that completely describes the instantaneous state of the fluid–body system. As we have discovered in the preceding chapters, such a system is sufficiently described by the vortex elements' current positions and strengths and the configuration of a body or bodies. The time evolution of this state vector is given by a set of ordinary differential equations, which generically can be written as

$$\frac{\mathrm{d}\mathsf{u}}{\mathrm{d}t} = \mathsf{F}(\mathsf{u}, t), \qquad \mathsf{u}(0) = \mathsf{u}_0, \tag{9.1}$$

where u_0 is the initial value of the state vector. The right-hand side function F describes the rate of change of the state vector. It comprises the components of the vortex element and rigid-body velocities; it also contains any model for the variation of vortex element strengths, though this might be trivially set to zero if the elements obey Helmholtz' third theorem. This state velocity function, F, depends on the system's current state, and because of prescribed body motion, might also depend explicitly on time.

Some problems will require us to make certain modifications to the state vector as time proceeds. Such modifications arise, for example, when an edge condition calls for the release of a new vortex element from the edge: the state vector's length increases by the number of degrees of freedom associated with the new element. Modifications also typically arise when we represent vorticity in the form of a sheet. The discrete elements comprising the sheet occasionally become a poor approximation of the underlying continuous structure (e.g., by becoming too far apart from a neighbor), and new elements must be locally inserted or the set of elements must be redistributed along the sheet's curve; these changes also typically alter the length of the state vector. Each of these state modifications is most naturally carried out between discrete time steps during the advancement of Eq. (9.1). To accommodate them, it is more appropriate to consider the system's integration from time level t^n to a later time level, $t^{n+1} := t^n + \Delta t$:

$$\frac{d\mathbf{u}}{dt} = \mathbf{F}(\mathbf{u}, t), \qquad \mathbf{u}(t^n) = \mathbf{u}^n. \tag{9.2}$$

The result of this single integration step, denoted by $\mathbf{u}^*$, is then modified as needed to obtain $\mathbf{u}^{n+1}$, the approximation of the state vector at the end of the step, $\mathbf{u}(t^{n+1})$. We will generically denote this modification step by $\mathbf{u}^{n+1} = \mathbf{M}(\mathbf{u}^*)$. For problems (or for steps within problems) in which no such modification is needed, $\mathbf{M}$ is simply the identity operator.

To summarize,

Result 9.1: State Update

Each time step $(\mathbf{u}^n, t^n) \mapsto (\mathbf{u}^{n+1}, t^{n+1})$ consists of the following two-part sequence:

State advancement, $(\mathbf{u}^n, t^n) \mapsto (\mathbf{u}^*, t^{n+1})$

$$\mathbf{u}^* = \mathbf{u}^n + \int_{t^n}^{t^{n+1}} \mathbf{F}(\mathbf{u}(\tau), \tau)\, d\tau, \tag{9.3}$$

accounting for vortex element convection, and possibly, strength variation; and

State modification, $(\mathbf{u}^*, t^{n+1}) \mapsto (\mathbf{u}^{n+1}, t^{n+1})$

$$\mathbf{u}^{n+1} = \mathbf{M}(\mathbf{u}^*), \tag{9.4}$$

providing introduction of new vortex elements (e.g., to enforce edge conditions) or redistribution of existing elements to improve resolution.

For the state advancement part, we generally need a method of **numerical integration**: a means of discretizing the time integral in Eq. (9.3). One of the useful features of inviscid vortex models and rigid-body mechanics is that they collectively form a **Lagrangian system**—that is, their governing equations can be derived by considering the minimum variation of a Lagrangian over the time interval of interest [65, 37, 36]. This perspective suggests that we advance the system state with a **variational integrator** [48], which has the enticing property of exactly conserving the fluid–body system's kinetic energy when such conservation is expected in the continuous system (that is, in the absence of external forces and moments). However, in this book, we will sacrifice this exact conservation property for the advantages of ease and flexibility, and instead use an **explicit Runge–Kutta method** to advance the state vector. Methods in the Runge–Kutta family advance the system from the beginning of a step to the end by approximating the integral with a sum over s stages,

$$\mathbf{u}^* = \mathbf{u}^n + \Delta t \sum_{i=1}^{s} b_i \mathbf{F}(\mathbf{u}_i, t_i), \tag{9.5}$$

where $(\mathbf{u}_i, t_i)$ denote the state vector and time level at the ith stage,

$$\mathbf{u}_i = \mathbf{u}^n + \Delta t \sum_{j=1}^{i-1} a_{ij} \mathbf{F}(\mathbf{u}_j, t_j), \quad t_i = t^n + \Delta t \sum_{j=1}^{i-1} a_{ij}. \tag{9.6}$$

Note that, to ensure that it is explicit (i.e., that the right-hand side only relies on known values of the state), only coefficients a_{ij} for which $j < i$ are used in the assembly at each stage. The most popular methods in this family are the first-order forward Euler method ($s = 1$ and $b_1 = 1$) and the classical fourth-order algorithm ($s = 4$, $a_{21} = 1/2$, $a_{32} = 1/2$, $a_{43} = 1$, and all other $a_{ij} = 0$, and $b_1 = 1/6$, $b_2 = b_3 = 1/3$, $b_4 = 1/6$).

9.2 Co-rotating Vortex Patches

In this first example, we will consider a flow attributable to vortex dynamics alone, absent of any rigid bodies. The initial condition consists of two circular **vortex patches**—regions of spatially uniform vorticity—of equal radius r_0 and strength (circulation) Γ_0, whose centers are separated by distance D_0. We seek to predict their subsequent evolution. Figure 9.1 illustrates the problem setup.

We define the **vorticity centroid**, $\boldsymbol{x}_{pk}$, of each patch, $k = 1, 2$, as

$$\boldsymbol{x}_{pk} = \frac{1}{\Gamma_0} \int_{\mathcal{V}_{pk}} \boldsymbol{x} \omega_3 \, \mathrm{d}V, \tag{9.7}$$

where $\mathcal{V}_{pk}$ denotes the region occupied by patch k. It is easy to check that the initial vorticity centroid of each patch coincides with the center of the circle. The centroid provides a useful metric for quantifying the patch evolution. Note that the centroid is closely related with linear impulse. Also, recall that we proved in Note 3.1.3 that the first moment of vorticity over the whole flow—and thus, the centroid of the full vorticity field—is invariant.

Suppose, without loss of generality, that the patches are initially centered at $\boldsymbol{x}_{p1}(0) = \frac{1}{2} D_0 \boldsymbol{e}_2$ and $\boldsymbol{x}_{p2}(0) = -\frac{1}{2} D_0 \boldsymbol{e}_2$, respectively, as shown in Fig. 9.1. The only dimensionless parameter that distinguishes the patch dynamics is the ratio of initial patch radius to initial centroid separation, r_0/D_0. We should expect smaller patches as measured by this parameter to orbit each other in counter-clockwise fashion on circular trajectories of diameter D_0, as would occur exactly for two point vortices of identical strength. However, what happens for larger patches?

To calculate the patches' evolution, we will discretize each with a collection of regularized point vortices (i.e., blobs). The blobs are arranged in a series of N_r concentric rings, distance $\Delta r = r_0/(N_r - 1/2)$ apart, with the innermost ring

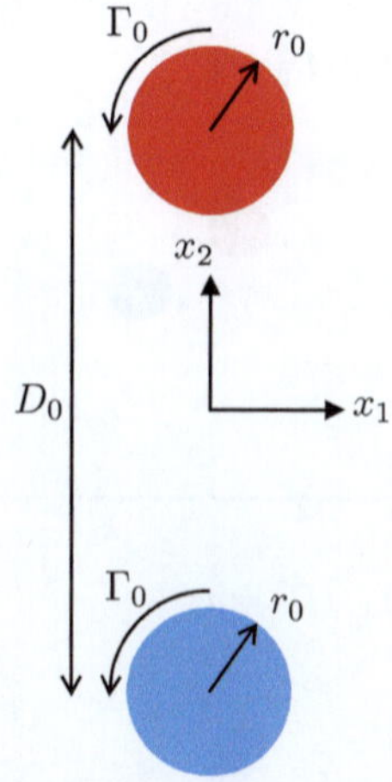

Fig. 9.1 Schematic of co-rotating vortex patches

consisting of a single blob at the center of the patch ($\boldsymbol{x}_{p1}$ or $\boldsymbol{x}_{p2}$) and $8(j-1)$ blobs distributed uniformly on the jth ring, $j = 2, \ldots, N_r$. The reader can verify that this arrangement partitions the patch into $N_{pv} = 1 + 4N_r(N_r - 1)$ identical blobs, each surrounded by a region of area $\pi\Delta r^2/4$. Each blob is assigned a strength $\Delta\Gamma = \Gamma_0/N_{pv}$.

The state vector u in this problem consists of $4N_{pv}$ entries: the two Cartesian components of each of the $N_\text{v} = 2N_{pv}$ vortex blobs' positions. The right-hand side vector, F, is composed of each blob's Kirchhoff velocity components, as described in Result 7.1. We could also include the strengths of the vortex blobs in the state vector, but as dictated by Helmholtz' third theorem these remain constant, so their corresponding entries in F would be trivially zero. The set of vortex elements of this problem constitute a ***vortex cloud***, as we defined it in Sect. 7.1.3. As such, the rate of change of the position of any member of this cloud is the Kirchhoff velocity expressed in Eq. (7.28), though obviously without the contribution from a body's bound vortex sheet. Furthermore, since we are using vortex blobs in lieu of point vortices in order to regularize the interactions in the cloud, we modify the velocity kernel accordingly:

$$\frac{\mathrm{d}\boldsymbol{x}_\mathsf{J}}{\mathrm{d}t} = \boldsymbol{v}_{\mathrm{v},-\mathsf{J}} = \sum_{\substack{\mathsf{K}=1\\ \mathsf{K}\neq\mathsf{J}}}^{N_\mathrm{v}} \boldsymbol{K}_\varepsilon(\boldsymbol{x}_\mathsf{J} - \boldsymbol{x}_\mathsf{K}) \times \Gamma_\mathsf{K}\boldsymbol{e}_3. \tag{9.8}$$

The regularized velocity kernel, $\boldsymbol{K}_\varepsilon$, can be obtained from a number of different choices of regularization function; here, we use the commonly-used algebraic form (7.65).

The computational cost of forming the right-hand side F is $O(N_\mathrm{v}^2)$, since each element's rate of change depends on contributions from all other elements. In this book, we do not use any acceleration techniques, such as the fast multipole method,

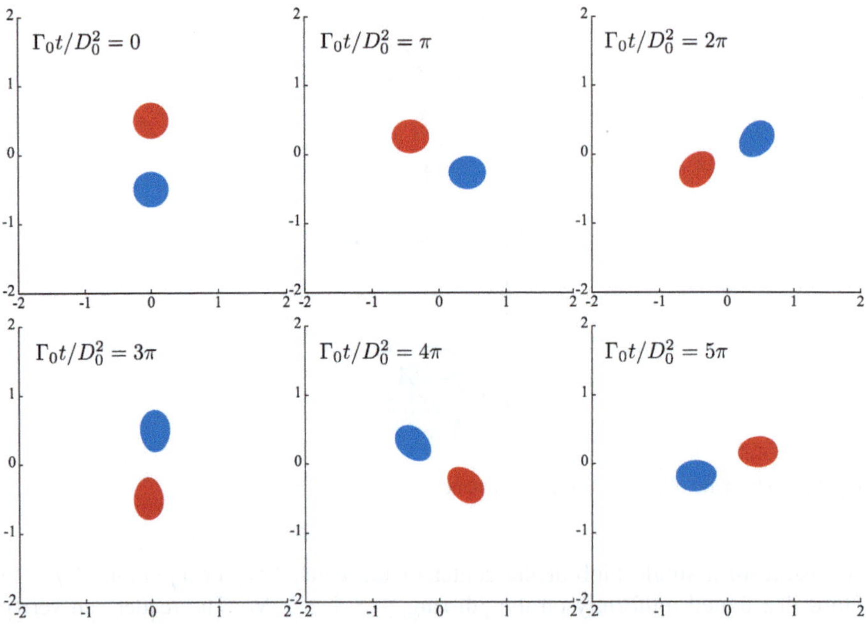

Fig. 9.2 Co-rotating vortex patches of radius-to-distance ratio $r_0/D_0 = 0.25$

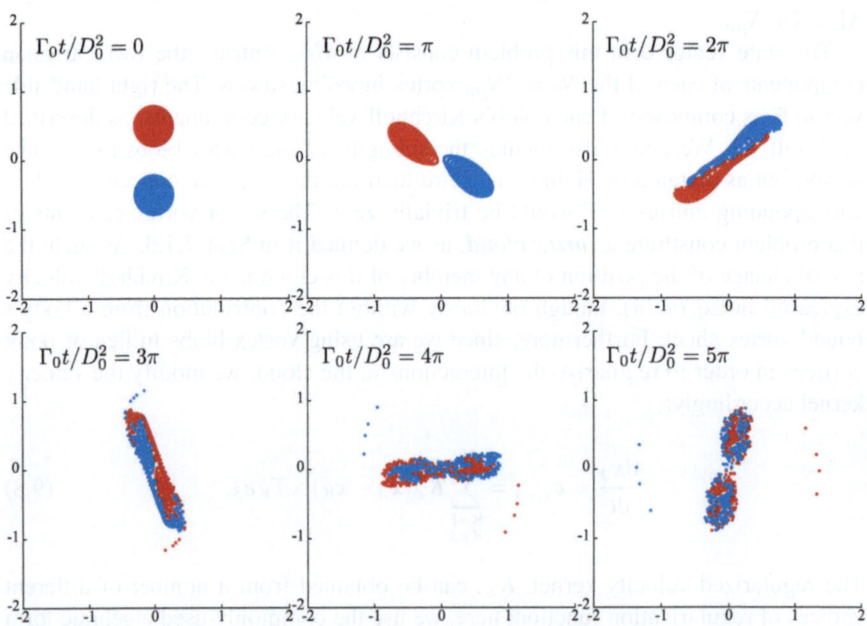

Fig. 9.3 Co-rotating vortex patches of radius-to-distance ratio $r_0/D_0 = 0.3$

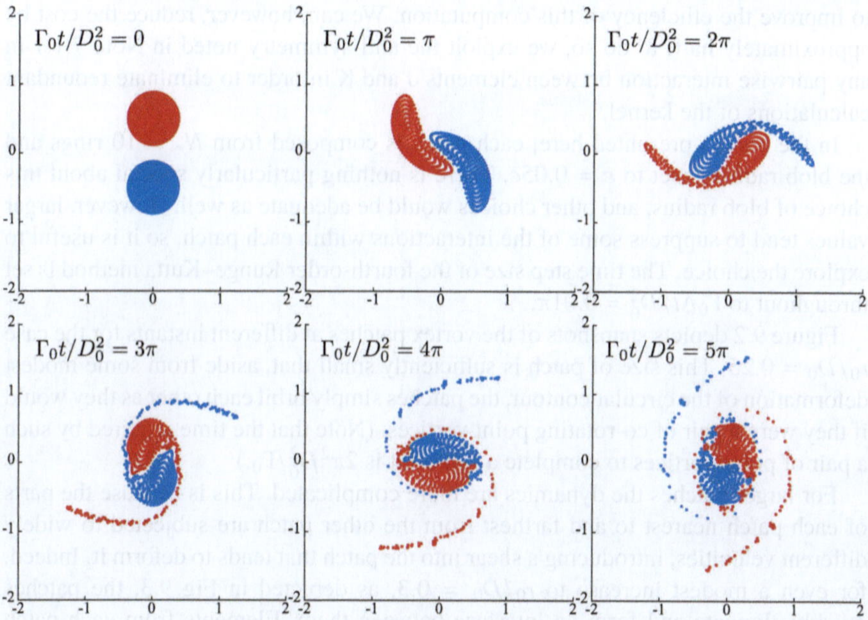

Fig. 9.4 Co-rotating vortex patches of radius-to-distance ratio $r_0/D_0 = 0.4$

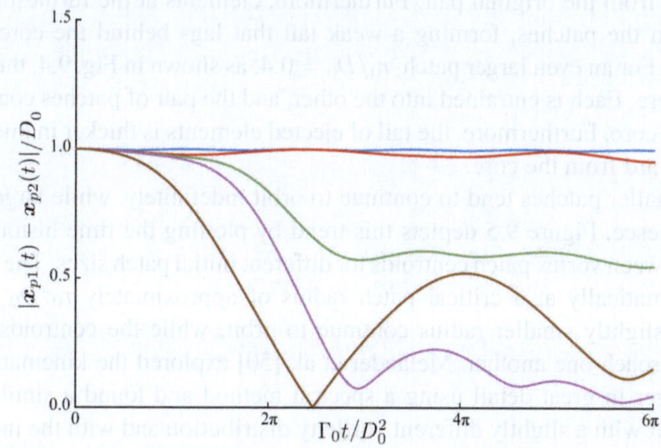

Fig. 9.5 Time history of the distance between the centroids of co-rotating vortex patches of various radii: $r_0/D_0 = 0.25$ (blue solid line), $r_0/D_0 = 0.29$ (red solid line), $r_0/D_0 = 0.295$ (green solid line), $r_0/D_0 = 0.30$ (purple solid line), $r_0/D_0 = 0.4$ (brown solid line)

to improve the efficiency of this computation. We can, however, reduce the cost by approximately half: to do so, we exploit the anti-symmetry noted in Note 7.1.1 in any pairwise interaction between elements J and K in order to eliminate redundant calculations of the kernel.

In the results presented here, each patch is composed from $N_r = 10$ rings and the blob radius is set to $\varepsilon = 0.05c$. There is nothing particularly special about this choice of blob radius, and other choices would be adequate as well. However, larger values tend to suppress some of the interactions within each patch, so it is useful to explore the choice. The time step size of the fourth-order Runge–Kutta method is set throughout to $\Gamma_0 \Delta t / D_0^2 = 0.01\pi$.

Figure 9.2 depicts snapshots of the vortex patches at different instants for the case $r_0/D_0 = 0.25$. This size of patch is sufficiently small that, aside from some modest deformation of the circular contour, the patches simply orbit each other as they would if they were a pair of co-rotating point vortices. (Note that the time required by such a pair of point vortices to complete a full orbit is $2\pi^2 D_0^2/\Gamma_0$.)

For larger patches the dynamics are more complicated. This is because the parts of each patch nearest to and farthest from the other patch are subjected to widely different velocities, introducing a shear into the patch that tends to deform it. Indeed, for even a modest increase to $r_0/D_0 = 0.3$, as depicted in Fig. 9.3, the patches quickly elongate and form an interface between them. Elements from each patch are exchanged along this interface, and two new patches are formed from a mixture of elements from the original pair. Furthermore, elements at the furthest extent are ejected from the patches, forming a weak tail that lags behind the cores along a circular arc. For an even larger patch, $r_0/D_0 = 0.4$, as shown in Fig. 9.4, the shearing is more severe. Each is entrained into the other, and the pair of patches coalesce into an elliptical core. Furthermore, the tail of ejected elements is thicker in this case and spirals outward from the core.

Thus, smaller patches tend to continue to orbit indefinitely, while larger patches tend to coalesce. Figure 9.5 depicts this trend by plotting the time histories of the distance between vortex patch centroids for different initial patch sizes. The behaviors change dramatically at a critical patch radius of approximately $r_0/D_0 = 0.295$. Patches of slightly smaller radius continue to orbit, while the centroids of larger patches approach one another. Melander et al. [50] explored the kinematics of this vortex merger in great detail using a spectral method and found a similar critical radius, albeit with a slightly different vorticity distribution and with the inclusion of a small amount of hyperviscosity to augment the Euler equations.

9.3 Interaction of Vortex Patches with a Bluff Body

In the previous example we only considered the dynamics of patches of vorticity. Now, in this example, let us introduce a stationary rigid body. The patches will be initially configured in the same manner as before, with initial radius r_0 and inter-centroid distance D_0, but in this case the lower patch will be assigned circulation

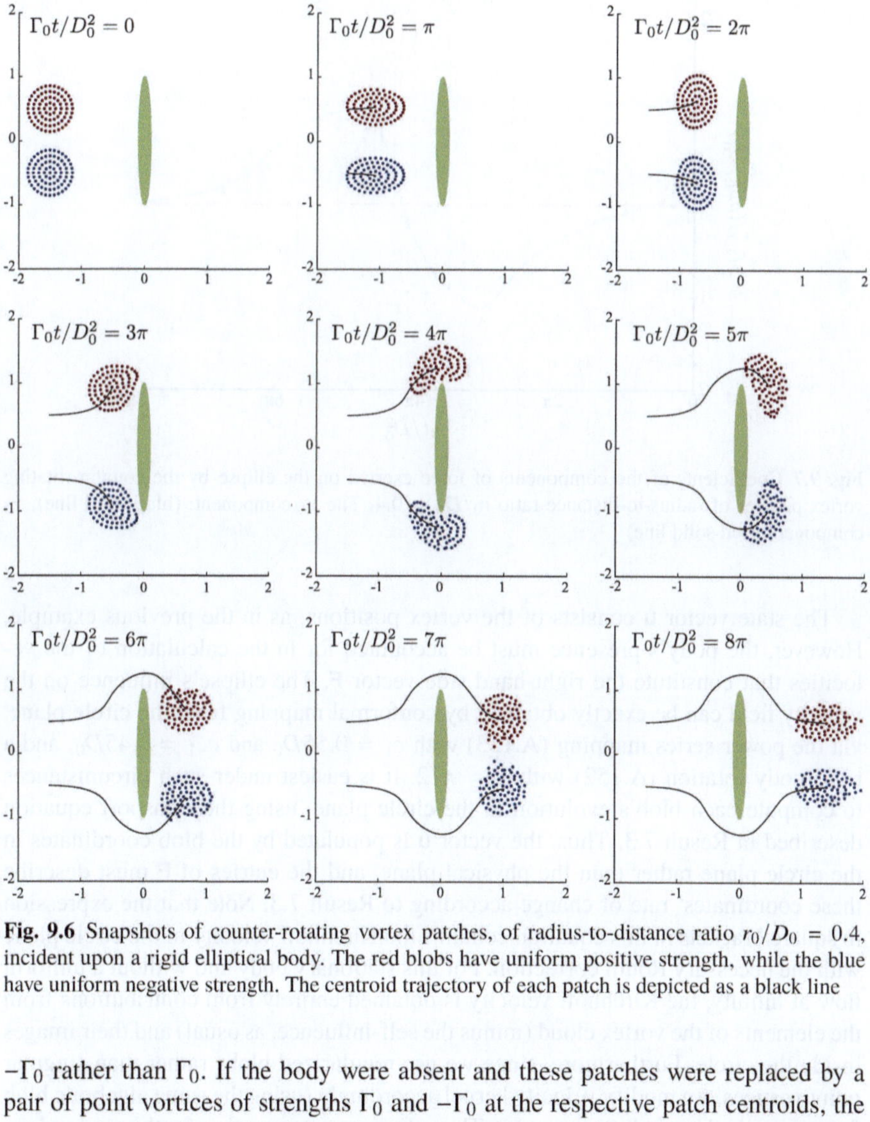

Fig. 9.6 Snapshots of counter-rotating vortex patches, of radius-to-distance ratio $r_0/D_0 = 0.4$, incident upon a rigid elliptical body. The red blobs have uniform positive strength, while the blue have uniform negative strength. The centroid trajectory of each patch is depicted as a black line

$-\Gamma_0$ rather than Γ_0. If the body were absent and these patches were replaced by a pair of point vortices of strengths Γ_0 and $-\Gamma_0$ at the respective patch centroids, the pair would propel itself in the x_1 direction at velocity $\Gamma_0/(2\pi D_0)$. A pair of patches propels itself in similar fashion, but with some degree of deformation induced by their interaction. A rigid body placed in the path would be expected to deflect the pair from this path and compel it to pass around the body.

In this example, we place the center of an elliptical body of semi-major axis length D_0 and semi-minor axis length $0.1D_0$ on the line of symmetry of the vortex pair, with its major axis perpendicular to this symmetry line. The vortex pair is initially located $1.5D_0$ units ahead of the ellipse. The patches are discretized in this case by $N_r = 5$ rings, and each constituent blob has radius $0.05c$.

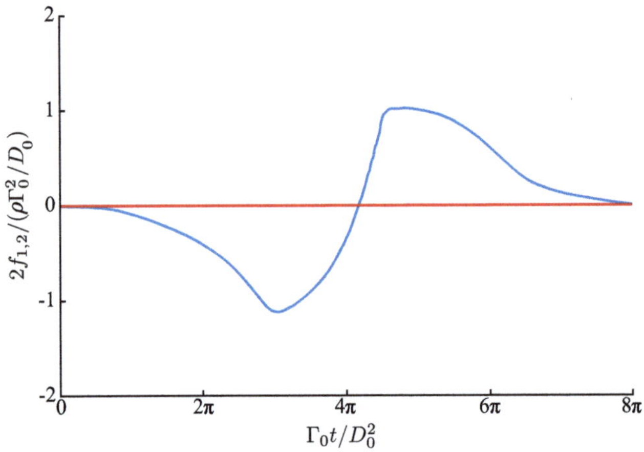

Fig. 9.7 Coefficients of the components of force exerted on the ellipse by the counter-rotating vortex patches of radius-to-distance ratio $r_0/D_0 = 0.4$. The x_1 component: (blue solid line), x_2 component: (red solid line)

The state vector $\mathbf{u}$ consists of the vortex positions, as in the previous example. However, the body's presence must be accounted for in the calculation of the velocities that constitute the right-hand side vector $\mathbf{F}$. The ellipse's influence on the velocity field can be exactly obtained by conformal mapping from the circle plane, via the power series mapping (A.153) with $c_1 = 0.55D_0$ and $c_{-1} = 0.45D_0$, and a rigid-body rotation (A.152) with $\alpha = \pi/2$. It is easiest under such circumstances to compute each blob's evolution in the circle plane, using the transport equation described in Result 7.3. Thus, the vector $\mathbf{u}$ is populated by the blob coordinates in the circle plane rather than the physical plane, and the entries of $\mathbf{F}$ must describe these coordinates' rate of change according to Result 7.3. Note that the expression in square brackets in this equation contains the Kirchhoff velocity in the circle plane with the necessary Routh correction. For this stationary body and without a uniform flow at infinity, the Kirchhoff velocity is obtained entirely from contributions from the elements of the vortex cloud (minus the self-influence, as usual) and their images inside the circle. Furthermore, since we use regularized blobs rather than singular point vortices, we modify velocity kernel accordingly (using the same algebraic blob function as in the previous example). Thus, the transport equation for the circle-plane coordinates of blob J is

$$\frac{\mathrm{d}\zeta_{\mathsf{J}}}{\mathrm{d}t} = \frac{1}{|\bar{z}'(\zeta_{\mathsf{J}})|^2}\left(-\sum_{\mathsf{K}=1}^{N_v}\frac{\Gamma_{\mathsf{K}}(\zeta_{\mathsf{J}}-\zeta_{\mathsf{K}})}{2\pi\mathrm{i}(|\zeta_{\mathsf{J}}-\zeta_{\mathsf{K}}|^2+\varepsilon^2)}+\sum_{\mathsf{K}=1}^{N_v}\frac{\Gamma_{\mathsf{K}}}{2\pi\mathrm{i}(\zeta_{\mathsf{J}}^*-1/\zeta_{\mathsf{K}})}+\frac{\Gamma_{\mathsf{J}}}{4\pi\mathrm{i}}\frac{z''(\zeta_{\mathsf{J}})^*}{z'(\zeta_{\mathsf{J}})^*}\right). \tag{9.9}$$

Note that we have not omitted K = J from the first summation, since it makes null contribution via the de-singularized kernel. Also observe that, in the second summation, we have *not* regularized the images' influence on the vortex elements; this ensures that the no-penetration condition on the body surface is enforced exactly.

The real and imaginary parts of this expression, for all $J = 1, \ldots, N_v$, form the entries of $\mathbf{F}$.

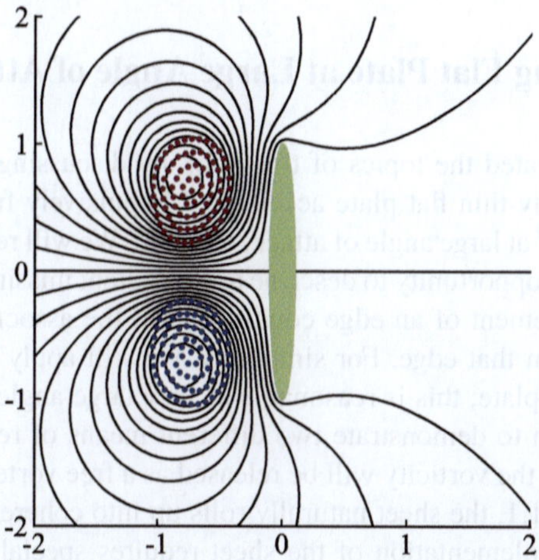

Fig. 9.8 Streamlines from counter-rotating vortex patches of radius-to-distance ratio $r_0/D_0 = 0.4$ at $\Gamma_0 t/d_0^2 = 2\pi$

Figure 9.6 depicts the subsequent dynamics for patches of radius-to-centroid distance $r_0/D_0 = 0.4$. The symmetry of the initial configuration is preserved in these dynamics. Each of the patches is deformed by its interaction with the other patch and by the obstructive influence of the body. This deformation is particularly apparent as the patches pass around the edges of the ellipse. However, the patches remain coherent and return to a similar (deforming) shape after their passage over the body; the trajectory of each patch's vorticity centroid is nearly symmetric from front to back.

We can also calculate the force exerted on the ellipse during its interaction with the vortex pair. This force is easily obtained from the rate of change of linear impulse, using the results described in Sect. 6.5.1; the derivative is approximated by a first-order backward difference. Only the vorticity-induced impulse, given by the second term in Eq. (6.91), contributes to this force. The resulting histories of these force components are shown in Fig. 9.7. Because of the symmetry maintained by the motion, only the x_1 component of force is non-zero; the moment is also identically zero. This longitudinal force component is initially negative, due to the suction exerted by the vortex patches as they approach the front of the ellipse; the force changes sign as the pair traverses to the other side of the ellipse.

It might be surprising, based on the superficial similarity between this flow and that of a jet, that the incident vortex pair does not exert a repulsive force as it approaches the ellipse. However, it must be borne in mind that the vortices do not bear momentum in the same manner as a jet, nor does the stationary ellipse redirect this momentum in the same fashion. The streamlines at one instant when the pair of vortex patches has nearly reached the ellipse are shown in Fig. 9.8. The dense streamlines near the ellipse are indicative of an accelerating flow, with a correspondingly low (i.e., suction) pressure field in their vicinity.

9.4 A Translating Flat Plate at Large Angle of Attack

In Chap. 1 we motivated the topics of this book by discussing the ingredients for modeling an infinitely-thin flat plate accelerated impulsively from rest to a steady translational speed U at large angle of attack, $\alpha = 60^\circ$. We will return to this example now, and use it as an opportunity to describe an ingredient missing from the previous example: the enforcement of an edge condition, and the associated release of new vortex elements from that edge. For simplicity, we will apply the Kutta condition at both edges of the plate; this is reasonable at very large angles of attack. We will also use the problem to demonstrate two different means of representing the fluid vorticity. In the first, the vorticity will be released as a free vortex sheet. As we have already seen in Fig. 1.1, the sheet naturally rolls up into coherent vortex structures, and the practical implementation of the sheet requires special treatment—a point insertion procedure—to ensure that the process is adequately resolved with discrete points. In the second case, the fluid vorticity will be represented as variable-strength point vortices.

9.4.1 Free Vortex Sheet Model

In this model of the translating plate, we will represent the vorticity that enters the fluid from each edge of the plate as a free vortex sheet. As we know well, a two-dimensional vortex sheet is a continuous curve, but in this problem we lack the

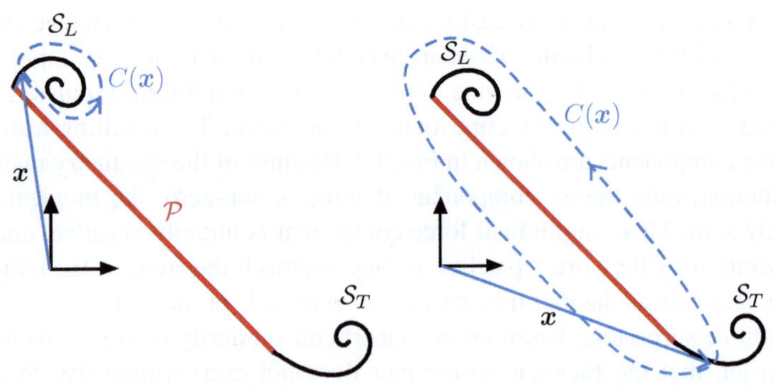

Fig. 9.9 Free vortex sheet model of flow about a flat plate. Two examples of contour $C(\boldsymbol{x})$ are shown, used to define the local circulation at any point $\boldsymbol{x}$ in the plate–sheet system. On the trailing-edge sheet $\mathcal{S}_T$, an alternative local circulation can be defined by a contour that also intersects the sheet at $\boldsymbol{x}$ but which encloses the portion of the system not enclosed by $C(\boldsymbol{x})$

ability to account analytically for the contributions and transport of every point along the curve, so we must rely on some numerical approximation. This requires that we represent the sheet discretely, with a finite number of **control points** distributed along the curve, and use these points to locally interpolate the sheet. We will address this interpolation below. How are these points distributed? As we will also discuss below when we describe the enforcement of the Kutta condition, a new control point will be released from each edge in each discrete time step. The distribution then arises naturally through the subsequent transport of these control points. As we alluded to in Sect. 9.1, we might find that this distribution provides an unacceptable approximation after some time, in which case we will need to modify the distribution. We will describe a procedure for this, as well. We note that many of the details for these procedures are adapted from work by Jones [34].

In Sect. 7.1.3 we showed that the transport of a free vortex sheet is described by the Birkhoff–Rott equation, given in Result 7.5. This equation accounts for the influences on any point of the sheet from the rest of the sheet; the remaining contributions from body motion, uniform flow, or other fluid vorticity are simply added to this. The equation also takes advantage of the local circulation's role as a material coordinate along the sheet, as we outlined in Note 3.6.1, by using it directly as an integration variable and a parameterization of the sheet's position, $\boldsymbol{X}(\Gamma)$.

Let us be definite about this local circulation in our plate–sheet system. Figure 9.9 depicts the plate and the two associated free vortex sheets: $\mathcal{S}_L$ at the leading edge and $\mathcal{S}_T$ at the trailing edge. These definitions, as well as the contour $C(\boldsymbol{x})$ depicted in the illustration, follow closely from those given in Fig. 8.2 for the plate amidst a discrete set of point vortices. Now, however, we can easily associate the set of vortex elements (which form a connected sheet) with the edge from which they were released. Furthermore, we can define coordinates and functions along the sheets—such as local circulation, position, and arc length—that remain sensible even after the sheet is discretized, since the connectivity between control points is stored.

The depicted contour in Fig. 9.9 cuts through a single point of this system and enables us to define the local circulation, $\Gamma(\boldsymbol{x})$, as a continuous quantity, starting at the free end of $\mathcal{S}_L$, where its value is zero; to the leading edge, where it is equal to Γ_L, the circulation of $\mathcal{S}_L$; then along the plate to the trailing edge, where its value includes the circulation of the plate, $\Gamma_L + \Gamma_{\mathcal{P}}$; and finally, to the free end of $\mathcal{S}_T$, where its value must return to zero to ensure that the global circulation is zero. This constraint on global circulation, Eq. (4.67), requires that the circulation of $\mathcal{S}_T$ is $\Gamma_T = -\Gamma_L - \Gamma_{\mathcal{P}}$. In our numerical treatment below, it will be helpful to perform the integration on any free vortex sheet by starting at the free end and progressing toward the releasing edge. For the trailing-edge vortex sheet, this requires an alternative definition of local circulation that starts at the free end of that sheet (where it is equal to 0) and ends at the trailing edge (where it is Γ_T). The reader can verify that the sum of this alternative circulation with the local circulation defined in Fig. 9.9 is equal to the global circulation, and thus, these definitions of circulation must be equal and opposite to each other. So, in summary, for either free vortex sheet, the local circulation used for integration is equal to 0 at the free end of the sheet and Γ_S at the releasing edge, where $\Gamma_S = \Gamma_L$ for the leading-edge sheet and $\Gamma_S = \Gamma_T$ for the trailing-edge sheet.

Before we discuss the discretization, we should acknowledge that, based on the fact that the local circulation is equal to zero at the free end of both free vortex sheets, this quantity clearly does not vary monotonically along the plate–sheet system. In fact, in problems involving more complicated plate motion than we consider here (e.g., oscillatory pitching or heaving), the circulation need not even remain monotonic along the individual sheets. The reader will recall that we insisted on such monotonicity in Note 3.6.1 when we identified local circulation as a means of parameterizing the sheet; otherwise, the same value of circulation may correspond to multiple locations on the sheet. However, we also observed in Note 3.6.1 that it is easy to circumvent this restriction by simply interpreting any sheet as a collection of subsheets, formed by dividing the original vortex sheet at points where its local strength, $\gamma = \partial\Gamma/\partial s$, vanishes. Each integral over the sheet is correspondingly decomposed. Ultimately, however, the *discretely-represented* sheet is actually parameterized by the integer index of each control point, so we never actually rely on a functional dependence of position (or any other sheet quantity) on circulation. The division of the sheet described in this scenario requires no special treatment, and the results that follow apply equally to cases with or without monotonic circulation.

Discrete Representation of Free Vortex Sheet As noted earlier, our discrete representation of any free vortex sheet consists of a set of control points. The number of such points will change from one time step to the next, and the two free sheets need not have the same number. For descriptive purposes, let us consider one of the sheets and suppose that it currently has N_v control points. Based on our observations above, these points can be described by their positions and invariant circulations: $(\boldsymbol{X}_J, \Gamma_J)$, $J = 1, \ldots, N_v$. Point 1 lies at the free end of the sheet, so that $\Gamma_1 = 0$, and N_v labels the point nearest the releasing edge, so that $\Gamma_{N_v} = \Gamma_S$; Fig. 9.10 illustrates an example of this indexing system.

The velocity field induced at some point $\boldsymbol{x}$ by this discrete approximation of the sheet comes from a numerical quadrature of the continuous form of the sheet-induced velocity field. For this, we use a trapezoidal rule over Γ,

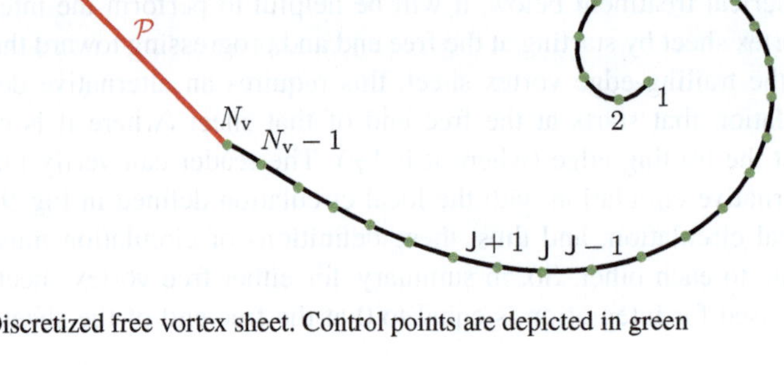

Fig. 9.10 Discretized free vortex sheet. Control points are depicted in green

$$\int_0^{\Gamma_S} \boldsymbol{K}\left(\boldsymbol{x}-\boldsymbol{X}(\hat{\Gamma})\right) \mathrm{d}\hat{\Gamma}\times \boldsymbol{e}_3 \approx \frac{1}{2}\sum_{\mathrm{J}=1}^{N_v-1} \left[\boldsymbol{K}\left(\boldsymbol{x}-\boldsymbol{X}_{\mathrm{J}+1}\right)+\boldsymbol{K}\left(\boldsymbol{x}-\boldsymbol{X}_{\mathrm{J}}\right)\right]\left(\Gamma_{\mathrm{J}+1}-\Gamma_{\mathrm{J}}\right)\times \boldsymbol{e}_3 \tag{9.10}$$

Rearranging slightly, we can write this expression in a form in which the effect of the discretized vortex sheet is replaced by a set of point vortices:

$$\sum_{\mathrm{J}=1}^{N_v} \boldsymbol{K}\left(\boldsymbol{x}-\boldsymbol{X}_{\mathrm{J}}\right)\times \Delta\Gamma_{\mathrm{J}}\boldsymbol{e}_3, \tag{9.11}$$

where the point vortex strengths are defined as

$$\Delta\Gamma_{\mathrm{J}} = \frac{1}{2}\begin{cases} \Gamma_2-\Gamma_1, & \mathrm{J}=1 \\ \Gamma_{\mathrm{J}+1}-\Gamma_{\mathrm{J}-1}, & \mathrm{J}\in[2,N_v-1] \\ \Gamma_{N_v}-\Gamma_{N_v-1}, & \mathrm{J}=N_v. \end{cases} \tag{9.12}$$

It is easy to verify that the sum of these point vortex strengths is identically the sheet strength:

$$\sum_{\mathrm{J}=1}^{N_v} \Delta\Gamma_{\mathrm{J}} = \Gamma_{N_v+1}-\Gamma_1 = \Gamma_S \tag{9.13}$$

Expression (9.11) provides the velocity field induced by either of the discretized free vortex sheets. Collectively, the free vortex sheets make their direct contribution to the velocity field through their combined sets of control points, which, from a computational point of view, is equivalent to a vortex cloud. It is important to note that, if the target point $\boldsymbol{x}$ is a control point on a sheet, we simply omit that point from the summation. This omission approximates the principal value of the integral, as called for in the Birkhoff–Rott equation. For the interactions with other vortex elements (i.e., control points) in the fluid, we also replace the singular velocity kernel, $\boldsymbol{K}$, by its regularized version, $\boldsymbol{K}_\varepsilon$, e.g., as given by Eq. (7.65). The choice of the value of ε will be discussed below.

To complete the velocity field, we need the other contributions: the indirect part of the vorticity-induced velocity field, arising from the no-penetration condition on the plate; and the body motion-induced velocity field. The construction of the overall velocity field about a flat plate was described at length in Sect. 8.2 using three different approaches: mapping from the circle plane, solution of the Cauchy integral problem, and Fourier–Chebyshev expansion. The vorticity-induced contributions in these approaches are explicitly written for discrete sets of vortex elements, and, although they can be adapted to account instead for continuous free vortex sheets, we have no need of such an adaptation now that we have discretized the sheet. Thus, any of the expressions developed through these approaches will suffice for computing the remaining contributions. Result 8.3, for example, provides one such expression for the velocity field, in complex notation and body-fixed components, using the Fourier–Chebyshev solution approach. The direct contribution from the fluid vortex elements is clearly identified in the expression. The remaining contributions are described by the expansion in extended Chebyshev polynomials. To avoid the complications of numerically evaluating the infinite sum in the vorticity-induced part of this expansion, it is prudent to use the compact form (8.55) induced by each discrete vortex element.

Enforcement of the Kutta Condition We will enforce the Kutta condition at both edges in its constraint form, expressed in Result 5.5, following the procedure outlined in Eq. (5.15). The most convenient expression for the edge suction parameters is given by Eq. (8.114). Let us suppose that we have N_v control points at the end of the state advancement step, and we wish to insert a new control point, indexed as $N_v + 1$, and the associated segment between points $N_v + 1$ and N_v. Rather than place this new point at the edge itself, we place it instead along a line that extends between the edge and control point N_v, one third of the distance to this previously-released point. The suction parameter due to the existing fluid vorticity and body motion, $\breve{\sigma}$, is already known. The unit contribution of the new segment to the suction parameter, $\sigma^v_{N_v+1}$, is given by evaluating (8.114) with each control point assigned a point vortex strength $1/2$, as dictated by the trapezoidal rule. Since we are simultaneously inserting segments into each of the two free vortex sheets, and each of these new segments makes a contribution to each edge, ± 1, we express the unit contribution of one such segment to both edges here:

$$\sigma^v_{N_v+1}(\pm 1) = \frac{1}{4\pi c}\mathrm{Re}\left(\frac{(\tilde{z}_{N_v} \pm c/2)^{1/2}}{(\tilde{z}_{N_v} \mp c/2)^{1/2}} + \frac{(\tilde{z}_{N_v+1} \pm c/2)^{1/2}}{(\tilde{z}_{N_v+1} \mp c/2)^{1/2}}\right). \tag{9.14}$$

The two constraint equations, one for each edge, are solved simultaneously for the new circulation to be released into each sheet. To be consistent with the trapezoidal approximation of the sheet, the new circulation $\delta\Gamma_{N_v+1}$ predicted by Eq. (5.15) is partitioned equally into control points $N_v + 1$ and N_v:

$$\Delta\Gamma_{N_v} = \Delta\Gamma_{N_v} + \frac{1}{2}\delta\Gamma_{N_v+1}, \qquad \Delta\Gamma_{N_v+1} = \frac{1}{2}\delta\Gamma_{N_v+1}. \tag{9.15}$$

This procedure approximates the initial advection of the new segment, and furthermore, ensures that the newly-introduced control point on this segment—with strength $1/2$—makes a finite contribution to the suction parameter at its nearest edge. Note that it does not explicitly force the new segment to lie tangent to the edge; instead, we rely on the transport of the control points to naturally effect such expected behavior.

To initialize the motion from rest, we seed each free vortex sheet with two control points along a straight line tangent to the plate, located, respectively, a distance $U\Delta t/2$ and $3U\Delta t/2$ from the edge. The circulation of these seed segments is established by the same procedure outlined above for all newly-introduced segments.

Redistribution of Control Points As mentioned earlier in the chapter, the natural straining process in a flow may cause adjacent control points to move apart from one another. To prevent degradation in the resolution of each vortex sheet, we occasionally redistribute the control points along the sheet, with uniform spacing $2U\Delta t$. This technique linearly interpolates the positions and local circulations from the old set of control points to the new set. Though the results are not very sensitive to the frequency with which this technique is applied, we utilize it in every time step.

The choice of spacing, $2U\Delta t$, is not arbitrary. Each new control point introduced at an edge travels away from the edge at the local velocity, given by the difference between the average velocity on either side of the sheet, minus the velocity of the edge itself. This local transport speed, derived for a flat plate in (8.136), scales with the translational velocity of the plate, U, and the distance traveled by the control point in a given time step is proportional to $U\Delta t$. The choice of the factor of 2 is not unique, but represents a compromise between good resolution elsewhere on the sheet and overall computational efficiency.

In place of this global redistribution procedure, some researchers have used a local treatment, based on point insertion, to target regions of the sheet that need resolution adjustment [41, 34, 66]. This procedure sets a critical separation distance between any two adjacent control points, beyond which a new control point is inserted between them. The critical distance, like the redistribution spacing, is generally chosen to be proportional to $U\Delta t$. Some researchers have also inserted points if the angle between adjacent segments exceeds a threshold, in order to ensure that emergent features with high curvature are well resolved.

Calculation of Force and Moment To compute the force and moment on the plate, we first calculate the linear and angular impulse at the end of each time step, using Eq. (8.99) to obtain these impulses from the Fourier–Chebyshev coefficients. The force and moment then follow from Eq. (6.92); we use a first-order backward difference to approximate the time derivative. These will be reported in their typical dimensionless form, as lift coefficient $C_L = 2f_y/\rho U^2 c$ and drag coefficient $C_D = 2f_y/\rho U^2 c$. The moment coefficient, not shown here, is defined as $C_M = 2m_r/\rho U^2 c^2$.

Results The results of this numerical model for the translating flat plate were already shown in Fig. 1.1. In Fig. 9.11 we depict more snapshots of the vortex sheets. The discretized vortex sheets clearly display the natural roll-up and inward spiraling inherent to the shed vorticity behind a translating bluff body. In particular, between $t^+ = 2$ and $t^+ = 3$, the sheet rooted at the trailing edge undergoes an instability from which a new coherent vortex emerges. The lift and drag coefficients are also shown in Fig. 1.1. These are initially infinite due to the inertial reaction of the fluid—the added mass—to the plate's impulsive start, but relax to finite values that vary smoothly as the plate sheds vorticity.

The results depicted in Fig. 1.1 were obtained with a time step size of $U\Delta t/c = 0.01$ and a blob radius of $\varepsilon = 0.1c$. What are the effects of these choices on the results? We already have discussed that the choice of time step size establishes the spatial resolution of the vortex sheet. Let us explore this role further by lowering the time step size to $U\Delta t/c = 0.0025$ while keeping the same value of ε. The effect of this lower time step size on the vortex sheet at $t^+ = 1$ is depicted in the left panel in Fig. 9.12. By visual comparison with the corresponding snapshot in Fig. 9.11, there is little apparent difference: the features in the vortex sheet were already well resolved with the larger time step size. But now let us explore this choice of time step size when we use a much smaller blob radius, $\varepsilon = 0.001c$. The vortex sheet resulting from this choice of parameters is shown at the same instant, $t^+ = 1$, in the right panel in Fig. 9.12. In addition to the basic spiraling structures, we now see the emergence

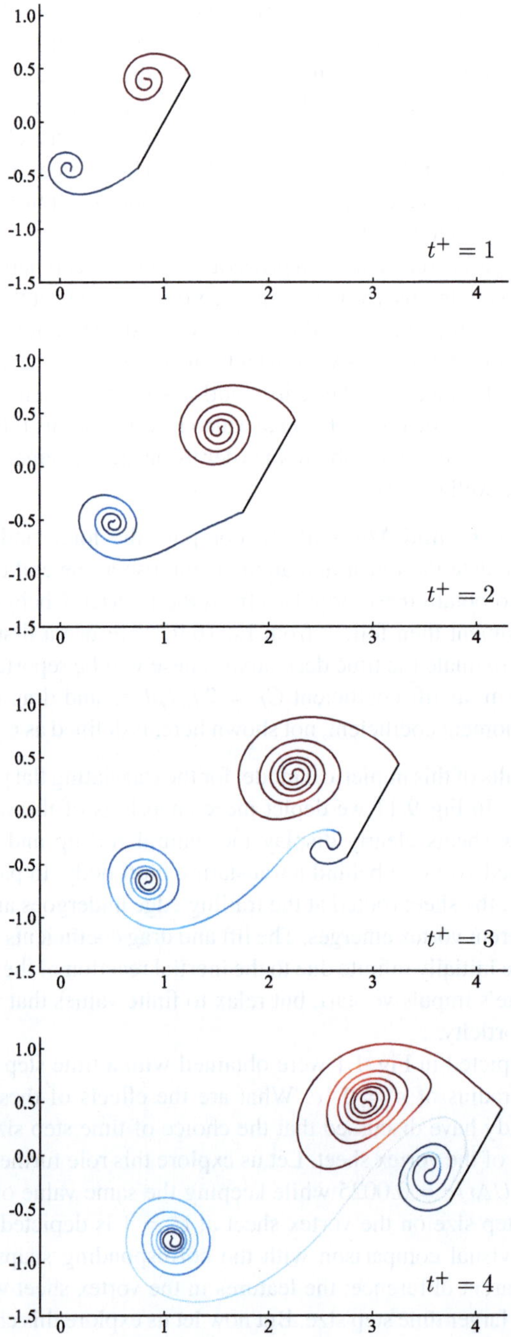

Fig. 9.11 Several snapshots of a flat plate of chord length c translating at 60° angle of attack. Red in the vortex sheet indicates positive vorticity, blue represents negative vorticity, and darker hues represent stronger values

of much smaller features: the roll-up from a Kelvin–Helmholtz instability along the outermost spirals, and a denser core of vorticity at the center of the spirals. In other words, ***the blob radius controls the minimum length scale of the flow features that emerge in the fluid vorticity, and the time step size sets the numerical resolution of these features***. It is important to observe that these small-scale flow features do not exert a significant influence on the force on the plate: the lift and drag predicted with both choices of ε, shown in the lower panel in Fig. 9.12, are nearly indistinguishable.

This revelation on the effect of blob radius raises a logical question: how small a blob radius is small enough? Let us use dimensional analysis to describe the expected range of length scales in the flow, $(L_{\min}, L_{\max})$. We will assume that we have chosen Δt such that all flow features are well resolved. Thus, the extremes of this range can only possibly depend on the geometric and flow parameters—c, U, α, ρ—and the blob radius, ε. By intuition and simple inspection of dimensions, the largest flow length scale is controlled by the largest geometric length scale, $L_{\max} \sim c$. What about the smallest flow length scale, $L_{\min}$? This length scale should be independent of the overall geometry and dependent only on local flow conditions. But the only characteristic length scale that can be formed from the remaining parameters is ε. The reader familiar with turbulence will recall that viscosity and the rate of dissipation set the smallest length scale, called the Kolmogorov length scale, in a real flow. In an inviscid context, viscosity and dissipation are absent. This leads us to a general conclusion that should be strongly emphasized:

Result 9.2: The Minimum Physical Length Scale in an Inviscid Flow

There is none.

In the absence of physical parameters that set the floor of the range of length scales, it is the role of the blob radius, ε, to artificially set this floor. This parameter therefore acts loosely as a surrogate for viscosity (but without associated dissipation). One could choose this floor however one wishes; the choice depends only on the objective. If the goal is to predict the principal flow behavior, then the blob radius can be set to a moderate fraction of the geometric length scale, e.g., $\varepsilon = 0.1c$. If smaller features are desired, then the blob radius should be reduced accordingly. As we described above, the force is not significantly affected by the choice of ε.

Before finishing our discussion of this model, we should make some further observations of the predicted force. The results depicted thus far—in Figs. 1.1 and 9.12—seem to show the force with a smooth decay from its initial infinite value to its eventual finite-valued behavior. But this initial behavior contains some artifacts of the numerical discretization, as we will now show. In Fig. 9.13, we depict the early force history predicted with several choices of time step size, with the blob radius maintained at $\varepsilon = 0.001c$ for each. We have also expanded the range on the vertical axis of these plots compared to the previous plots. On this new plot, the 'infinite' initial value of each component observed in, e.g., Fig. 9.12, is revealed to merely

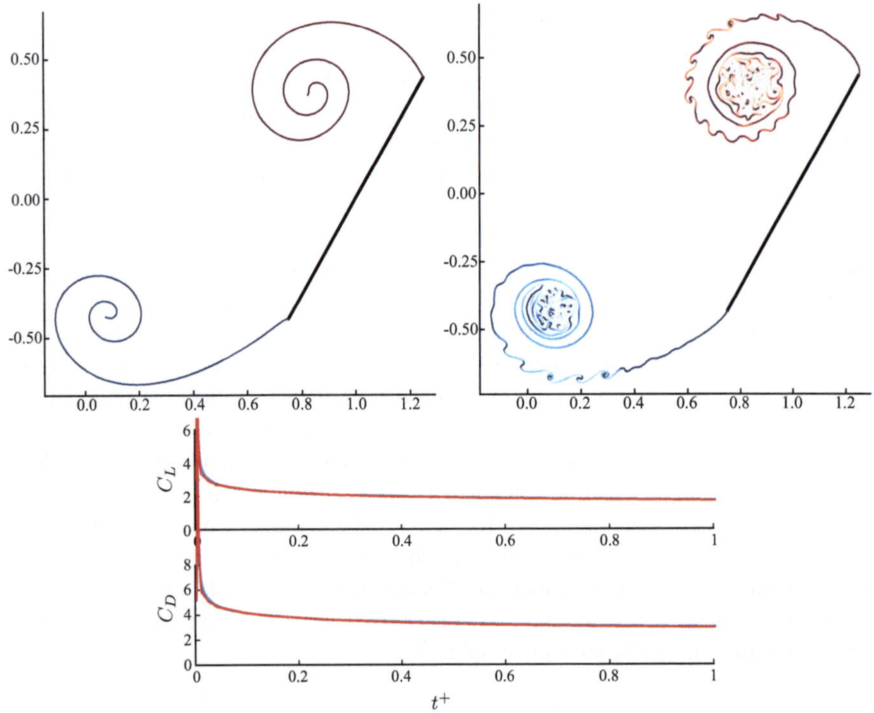

Fig. 9.12 A flat plate of chord length c translating at 60° angle of attack. The upper left panel depicts the vortex sheet at $t^+ = 1$ with the vortex sheet regularized with blob radius $\varepsilon = 0.1c$; the upper right panel shows the same instant but with $\varepsilon = 0.001c$. Red hues indicate positive vorticity and blue hues represent negative vorticity. The lower panel shows the lift and drag coefficients resulting from each choice, $\varepsilon = 0.1c$ (blue solid line) and $\varepsilon = 0.001c$ (red solid line)

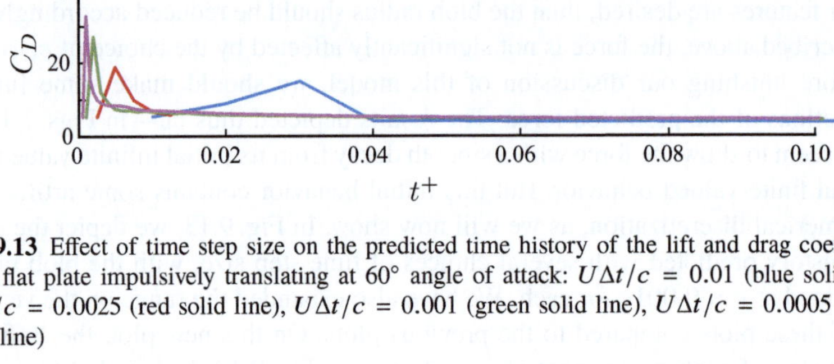

Fig. 9.13 Effect of time step size on the predicted time history of the lift and drag coefficients for a flat plate impulsively translating at 60° angle of attack: $U\Delta t/c = 0.01$ (blue solid line), $U\Delta t/c = 0.0025$ (red solid line), $U\Delta t/c = 0.001$ (green solid line), $U\Delta t/c = 0.0005$ (purple solid line)

be a large finite value. Furthermore, since we are using backward differencing to approximate the derivative of impulse, the first prediction of force is only available at the end of the first time step. As Fig. 9.13 shows, the largest predicted magnitude does not occur until the end of the *second* time step: the initial step's prediction is based on the guess we have made in seeding each vortex sheet, and only in the second step does the flow correct this sheet's guessed behavior.

Since we observe this same qualitative behavior from one step to the next for every choice of time step size, we can conclude that these results are suggestive of a sequence that converges as $\Delta t \to 0$. But converge to what? In other words, what is the expected behavior of a free vortex sheet as it first emerges from an edge at which the Kutta condition has been imposed? In the absence of an imposed length scale (i.e., with $\varepsilon = 0$), this sheet must adopt a self-similar structure: it maintains a uniform configuration that merely grows in size. Pullin [58] showed this structure to be a spiral rooted at the edge, after the early work of Kaden [35]. Graham [27] conducted a thorough analysis of the flow and the associated forces during the initial phase of an airfoil's motion, assuming vorticity to be shed only from the trailing edge; this was later extended by Pullin and Wang [59] to a plate with vorticity shed from both edges. These analyses have shown that, based on this self-similar spiral development, the initial force should scale like $t^{+-1/3}$, with the constant of proportionality dependent upon the angle of attack. Of course, this prediction is only reasonable as long as the vortex sheet development is independent of other length scales (such as the chord length of the plate). The small but non-zero blob radius also imposes itself, but only to slightly impede the development of the spiral. Thus, the early behavior is expected to be singular at $t = 0$, but smooth and finite thereafter.

Let us close with two observations on this early-time behavior:

- This prediction contrasts significantly with the classical prediction, first derived by Wagner and discussed in Sect. 8.5.4, of initially finite lift due to an impulsive change in motion. But that earlier work relied fundamentally on the notion of a vortex sheet that remains flat, even from the earliest instants. One can view the difference in these early behaviors as the force we would need to apply to the fluid to suppress the natural roll-up of the vortex sheet.
- The results in Fig. 9.13 show that, by making a simple guess of the expected initial behavior of the sheet, our prediction does not become reasonable until four or five time steps after the problem has started. Some researchers (e.g., Jones [34]) have made a greater effort to seed the flow with an analytical prediction of the early-time behavior.

9.4.2 Variable-Strength Vortex Model

One of the drawbacks of the vortex sheet model that we constructed for this example is that it becomes computationally more expensive as the sheet rolls up into coherent vortices. The inward spiraling in these structures continues indefinitely, requiring a progressively larger number of control points to resolve. But as concentrated regions

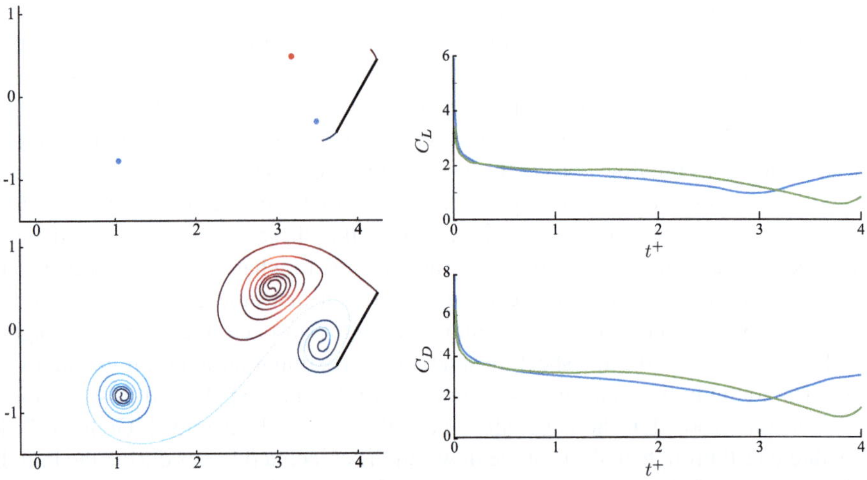

Fig. 9.14 Impulsively-translating flat plate at 60° angle of attack. The top left panel depicts the configuration of fluid vorticity—a combination of vortex sheets and point vortices—at $t^+ = 4$, predicted by a hybrid vortex model. The red point vortex (and red segments) have positive circulation; the blue have negative circulation. The bottom left panel is the result of the full vortex sheet model at $t^+ = 4$, shown for comparison. The right panels show the coefficients of lift, C_L, and drag, C_D, versus time predicted by the full vortex sheet model (blue solid line) and the hybrid model (green solid line)

of vorticity, these structures' dynamical significance—their influence on the flow and on the rate of change of impulse—can be represented nearly as accurately and much more economically with point vortices. Of course, since each structure progressively amasses circulation as it draws the sheet inward, its substitute point vortex must have time-varying strength. Wang and Eldredge [72] developed a model for flow past a flat plate at large angle of attack, using one point vortex for each edge of the plate. Several previous researchers, e.g., Cortelezzi and Leonard [13], Michelin and Llewelyn Smith [51], and Tchieu and Leonard [67] have also used variable-strength vortices for modeling the release of vorticity, generally from a single edge. In these models, the vortices' strengths are determined by the same constraint form of the Kutta condition, Result 5.5, used for the vortex sheet model. Instead of expressing this condition as the addition of a new discrete addition of circulation at the end of each time step (the state modification substep described at the beginning of the chapter), it is convenient to express it in a time-derivative form, $\dot{\sigma} = 0$, as discussed in Sect. 5.2. The resulting ordinary differential equation couples the rates of change of strength, $\dot{\Gamma}_J$, and position, $\dot{z}_J$, for the elements. This equation combines naturally with the element transport equation in the state advancement substep. As we discussed in detail in Sect. 7.1.4, this transport is modified to account for the elements' time-varying strength, e.g., in the impulse matching form (7.40): the transport equation also invokes both $\dot{z}_J$ and $\dot{\Gamma}_J$.

Wang and Eldredge [72] showed that one point vortex can adequately capture the formation of the initial vortex structure at each edge of the plate. However, as

Fig. 9.11 shows, later structures form from vorticity already in the fluid, through instability and the subsequent roll-up in the vortex sheet. It is impossible to represent this process with isolated point vortices: in other words, such a model lacks a means for determining when one vortex has gathered sufficient circulation and is deemed to be 'shed' from the edge, to allow another variable-strength vortex to take its place. Darakananda and Eldredge [17] addressed this deficiency by releasing vorticity into a vortex sheet rooted at the edge, but then siphoning the circulation from the sheet's free end into a variable-strength point vortex, called the **active vortex**. This approach preserves the essential (and irreplaceable) dynamics of the sheet, but restricts the growth in population of control points. The transport of the active vortex is governed by an extended form of the impulse matching principle, Eq. (7.45), which ensures that the force receives no ill effects from transferring circulation from one location to another.

Darakananda and Eldredge [17] showed that, rather than directly specify the rate of transfer of circulation from the sheet into its active vortex, it is preferable instead to choose a tolerance, ϵ_F, for the accumulated impulse error in each time step. Such error is inevitable from the elimination of finite-length segments of the sheet. In each time step, segments are eliminated, and their circulation transferred to the point vortex, until this error tolerance is reached. The tolerance is specified proportional to the time-step size, $\epsilon_F = B_F \Delta t$; the proportionality constant, B_F, represents a tolerance for **force discrepancy** in the hybrid model. If no segments can be transferred in a step without exceeding this threshold, then this stalemate serves as an indication that the active vortex has become dynamically distinct from the sheet: its strength is frozen and a new active vortex is originated from the end of the sheet. In order to control the creation of active vortices, the model also makes use of another tunable parameter, $T_{\min}$, that specifies the minimum amount of time that must pass before a new active vortex can be created.

As Darakananda and Eldredge [17] explained, any type of vortex model can be obtained through the choice of the two parameters, B_F and $T_{\min}$. If we set $B_F = 0$ and $T_{\min}$ to a large value, then we never transfer circulation out of the sheet, and our vortex sheet model remains intact. If we keep $B_F = 0$ but set $T_{\min} = 0$, then we transform the end of the sheet into an active vortex in every time step, leading to a discrete vortex cloud model. If, instead, we maintain a large $T_{\min}$ but set B_F to a modest non-zero value, then we permit only a small number of active point vortices to form. This latter choice leads to the most computationally efficient versions of the model. An example of this last combination—specifically, with $B_F = 0.1\rho U^2 c$ and $UT_{\min}/c = 0.2$—is depicted in Fig. 9.14. At the instant shown, $t^+ = 4$, one variable-strength point vortex is gathering circulation from each sheet; a third point vortex far behind the plate had previously held such a role, but its strength has since been frozen. The three point vortices align well with the rolled-up structures formed in the full vortex sheet model. The force coefficients agree quite well for the first few convective time units, but deviate slightly at later times.

9.5 Flow Past a NACA Airfoil

In the previous section we developed models for the flow past a flat plate at large angle of attack. These models made direct use of the tools and results that we developed in detail in Chap. 8. Furthermore, each edge of the plate has zero interior angle, and as we discussed at length in Chap. 5, there is little ambiguity in the application of the Kutta condition at such an edge. Let us now consider a geometry in which the edge has non-zero interior angle, at the trailing edge of an aerodynamically-relevant shape: an airfoil. For this, we will draw upon a member of the NACA (National Advisory Committee for Aeronautics) four-digit family of airfoil sections [1]. The four-digit designation describes the length parameters of the airfoil's shape as a percentage of chord length: the first digit represents the maximum camber; the second digit describes the location of this maximum camber from the leading edge; and the last two digits indicate the maximum thickness. For this example, we will focus on a NACA 0012: an airfoil with no camber and 12% thickness. This airfoil is therefore symmetric from top to bottom; its trailing edge has an interior angle of 16°.

We will apply a more interesting motion to this airfoil than simple translation. Instead, we will assign it oscillatory pitching and plunging while it translates steadily forward. Both of these motions will be carried out about zero mean, as they were in the analysis of Theodorsen described in Sect. 8.5.3. As we defined it in Eq. (8.164), the reduced frequency, κ, represents the ratio of the convective to oscillatory time scales. However, in contrast to that earlier analysis, we will not assume here that the amplitudes of these motions are small, and thus, shall not apply linear approximations to predict the flow behavior. Rather, we will compute the dynamics of a free vortex sheet released from the trailing edge, whose strength is determined by the Kutta condition. We will discretize this sheet in the same manner as in the example of the flat plate, and track the motions of the finite set of control points.

To account for the body's presence in the velocity of the control points, let us consider two possibilities:

Panel Method In this approach, we would use the integral equation for the bound vortex sheet strength, Eq. (4.27), and discretize the airfoil contour C_b with straight finite-length segments, called **panels**. Generally, one assumes a linear distribution of sheet strength between the end points of the panel; for N_p panels, there are $N_p + 1$ points with unknown strengths. The integral equation is enforced at the center of each panel, giving N_p equations. The set of equations is completed with the constraint of zero total circulation, Eq. (4.68). (Note that the complex sheet strength, g, can always be expressed alternatively as the real-valued sheet strength, γ, using the relationship expressed in Note 3.6.3.) The resulting velocity field follows from Result 4.2, with the contour integral discretized accordingly.

Conformal Transform There is no known conformal mapping for a general NACA airfoil shape[1] However, once the body contour is discretized into panels, as described

[1] There are, however, conformal transformations that generate airfoil shapes, e.g., the Joukowski transformation, and its generalization, the Karman–Trefftz transformation. These are described in Sect. A.2.4 in the Appendix.

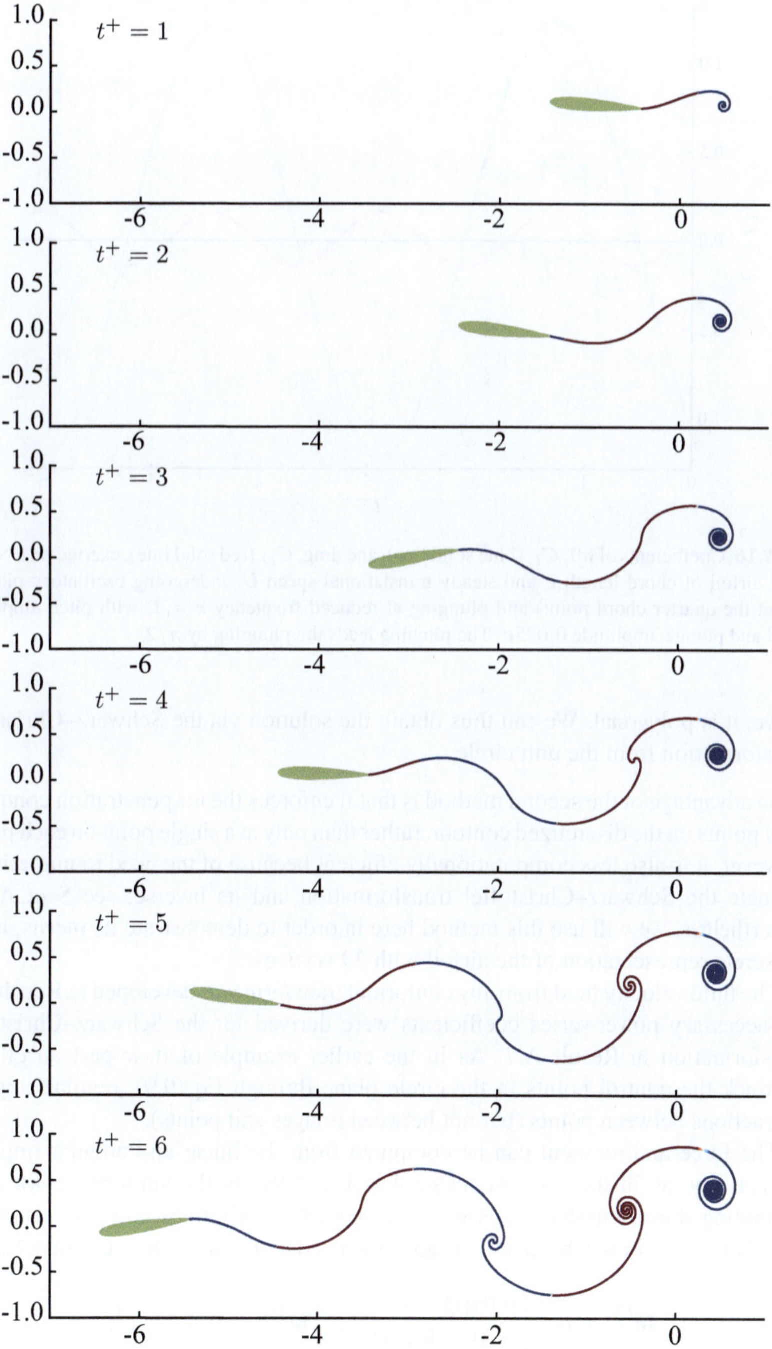

Fig. 9.15 Several snapshots of a NACA 0012 airfoil of chord length c and steady translational speed U undergoing oscillatory pitching (about the quarter chord point) and plunging at reduced frequency $\kappa = 1$, with pitch amplitude $\pi/18$ and plunge amplitude $0.025c$. The pitching leads the plunging by $\pi/2$. The vortex sheet segments are colored according to their local strength, with blue indicating negative strength and red positive

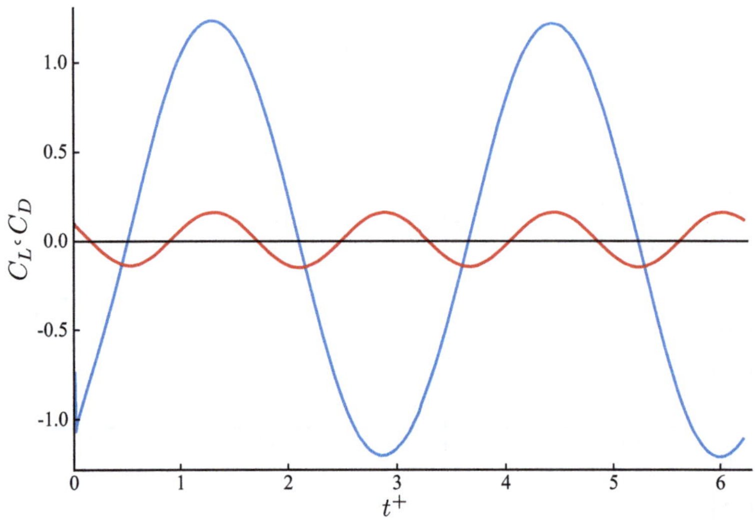

Fig. 9.16 Coefficients of lift, C_L (blue solid line), and drag, C_D (red solid line), exerted on a NACA 0012 airfoil of chord length c and steady translational speed U undergoing oscillatory pitching (about the quarter chord point) and plunging at reduced frequency $\kappa = 1$, with pitch amplitude $\pi/18$ and plunge amplitude $0.025c$. The pitching leads the plunging by $\pi/2$

above, it is polygonal. We can thus obtain the solution via the Schwarz–Christoffel transformation from the unit circle.

The advantage of the second method is that it enforces the no-penetration condition at all points on the discretized contour, rather than only at a single point on each panel. However, it is also less computationally efficient because of the need to numerically evaluate the Schwarz–Christoffel transformation and its inverse; see Sect. A.2.5. Nevertheless, we will use this method here in order to demonstrate its merits, using a discrete representation of the airfoil with 39 vertices.

The fluid velocity field from any conformal transform was developed in Result 4.5; the necessary power-series coefficients were derived for the Schwarz–Christoffel transformation in Result A.7. As in the earlier example of flow past an ellipse, we track the control points in the circle plane through Eq. (9.9), regularizing the interactions between points (but not between images and points).

The force and moment can be computed from the linear and angular impulse, respectively, as in the previous examples; Eq. (6.91) is the appropriate form for computing these impulses. It is useful to note the body-fixed components of the added mass tensor for the airfoil, as approximated by the discrete set of panels, are

$$\tilde{\mathbf{M}}^{FV} = \rho c^2 \begin{bmatrix} 0.0103 & 0 \\ 0 & 0.7760 \end{bmatrix}, \quad \tilde{\mathbf{M}}^{M\Omega} = 0.0261\rho c^4,$$
$$\tilde{\mathbf{M}}^{MV} = (\tilde{\mathbf{M}}^{F\Omega})^{\mathsf{T}} = \rho c^3 [0 \;\; 0.0414]. \tag{9.16}$$

It is interesting to compare these values with those of the flat plate in Eq. (8.32). The diagonal components are similar, except for the non-zero value of added mass associated with translation along the chord, the contribution from the airfoil's thickness. Furthermore, the airfoil lacks the fore-aft symmetry of the flat plate, and it thus admits some small coupling between angular motion and transverse force (and transverse linear motion and moment).

The Kutta condition at the trailing edge is enforced by the same procedure described in the previous example: at the end of the time step, we create a new control point one third of the distance to the last control point, then use Eq. (5.15) to determine the circulation to assign to the new segment. We developed an expression for the edge suction parameter (i.e., signed intensity) at any vertex in the Schwarz–Christoffel transform in Result 4.4. This result makes use of the tangent component of velocity in the circle plane, which we can obtain from the bracketed expression in Result 4.5.

Figure 9.15 depicts the development of the wake vortex sheet for one example of oscillatory pitching and plunging. In this example, the reduced frequency, κ, is 1, the pitching amplitude is 10°, and the plunging amplitude (the maximum transverse excursion relative to the airfoil's chord length) is 0.025: in other words, the motion is dominated by pitching. The force generated by this motion is depicted in Fig. 9.16. Both components are nearly sinusoidal, with the drag varying at twice the frequency of the lift because of its identical response to upward and downward pitches of the airfoil's nose. Unsurprisingly, in light of the airfoil's symmetry, the lift has zero mean. It is somewhat more surprising that the drag has nearly zero mean, as well. This particular reduced frequency lies in a transitional regime: at lower frequencies, the mean drag is positive, as one would expect of the drag on a body that is fixed at one angle of attack; at higher frequencies, however, the mean drag is negative, indicating that the oscillatory pitching has generated a net propulsive force on the airfoil.

Chapter 10
Rigid Motion of an Ellipsoidal Body

Contents

10.1 Ellipsoidal Coordinates and Harmonics 369
10.2 Translation of a General Ellipsoidal Body 372
10.3 Rotational Motion of a General Ellipsoidal Body 379
10.4 Added Mass 380

In contrast to the two-dimensional context, there are relatively few exact solutions for inviscid flows about three-dimensional bodies. However, there is one quite general class of problems for which analytical solution is possible: the irrotational flow about an ellipsoidal body. This might seem quite a limited set of results. But it is useful to note that the extremes of this family include infinitely thin ellipses and slender rods, both of which have meaningful roles. The flat elliptical shapes, in particular, provide the rare opportunity for analytical access to a sharp edge of a three-dimensional body.

For this geometric family, we have the benefit of **ellipsoidal coordinates**, an orthogonal curvilinear system in which the body's shape is described by a coordinate isosurface. Solution can be achieved by composing from a combination of **ellipsoidal harmonics**: basic homogeneous solutions of Laplace's equation when expressed in the ellipsoidal coordinate system. More details on ellipsoidal coordinates can be found in [73], and on the analysis that follows, in [43, 53]. We will only summarize the results and the procedures used to obtain them here.

10.1 Ellipsoidal Coordinates and Harmonics

Ellipsoidal coordinates are the roots θ of the cubic equation

J. D. Eldredge, *Mathematical Modeling of Unsteady Inviscid Flows*, Interdisciplinary Applied Mathematics 50, https://doi.org/10.1007/978-3-030-18319-6_10

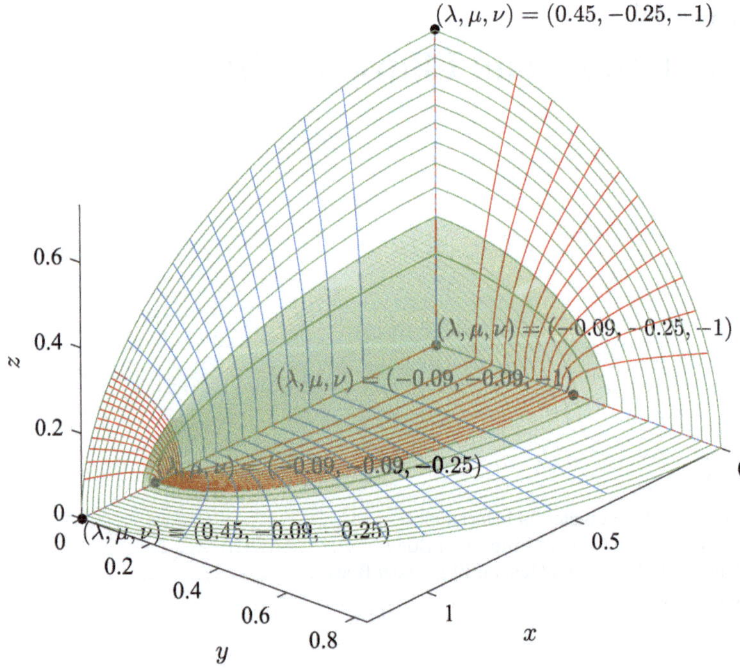

Fig. 10.1 Single octant of the ellipsoidal coordinate system based on an ellipsoid of semi-axes $a = 1$, $b = 0.5$ and $c = 0.3$ (depicted as a green surface, corresponding to $\lambda = 0$). The perimeters of surfaces of constant λ are depicted in green, surfaces of constant μ in red, and surfaces of constant ν in blue. All other octants are simply mirror images of this illustration

$$\frac{x^2}{a^2 + \theta} + \frac{y^2}{b^2 + \theta} + \frac{z^2}{c^2 + \theta} = 1, \tag{10.1}$$

where x, y, and z are the known Cartesian coordinates, and a, b, and c are fixed constants that represent the semi-axis lengths of the ellipsoidal body. Let us label the three roots θ as λ, μ and ν; these are the ellipsoidal coordinates. The isosurfaces of these coordinates are called confocal quadrics; they are the analog of the coordinate planes in a Cartesian system. It is sufficiently general to insist that $a > b > c$; cases in which any or all are dimensions are equal—corresponding to bodies of revolution—are also permissible, but require some care. Then, one can show that the ranges of the ellipsoidal coordinates are as follows:

$$\lambda > -c^2 > \mu > -b^2 > \nu > -a^2. \tag{10.2}$$

An example of the coordinate surfaces is depicted in for a single octant of the Cartesian coordinate system in Fig. 10.1; the ellipsoidal body with semi-axes a, b, and c is described by the shaded isosurface $\lambda = 0$. The surfaces of constant λ are ellipsoidal, while the surfaces of constant μ and ν are hyperboloidal: single-sheet hyperboloids for μ, and two-sheet hyperboloids for ν. The Cartesian coordinates can be expressed in terms of the ellipsoidal coordinates via the following relations:

$$x^2 = \frac{(a^2+\lambda)(a^2+\mu)(a^2+\nu)}{(a^2-b^2)(a^2-c^2)},$$
$$y^2 = \frac{(b^2+\lambda)(b^2+\mu)(b^2+\nu)}{(b^2-a^2)(b^2-c^2)}, \tag{10.3}$$
$$z^2 = \frac{(c^2+\lambda)(c^2+\mu)(c^2+\nu)}{(c^2-a^2)(c^2-b^2)}.$$

The sign chosen for each of the square roots determines which octant is described by a given set of ellipsoidal coordinates. From these, it is straightforward to show that, in the standard Cartesian basis,

$$\frac{\partial \boldsymbol{x}}{\partial \lambda} = \frac{1}{2}\left(\frac{x}{a^2+\lambda}, \frac{y}{b^2+\lambda}, \frac{z}{c^2+\lambda}\right), \tag{10.4}$$

and the local unit vector along the λ axis is simply $\boldsymbol{e}_\lambda := (\partial \boldsymbol{x}/\partial\lambda)/|\partial \boldsymbol{x}/\partial\lambda|$. Similar expressions can be found for $\partial \boldsymbol{x}/\partial\mu$ and $\partial \boldsymbol{x}/\partial\nu$, to form unit vectors in the μ and ν directions, respectively. As suggested by Fig. 10.1, the ellipsoidal coordinate system is curvilinear, so these vectors depend on position in space; however, the system is also orthogonal, so, for example, $\boldsymbol{e}_\lambda \cdot \boldsymbol{e}_\mu = 0$, etc. Furthermore, since there is a one-to-one correspondence between the ellipsoidal and Cartesian coordinates in a given octant, one can form the inverses as follows:

$$\frac{\partial\lambda}{\partial x} = \frac{\partial x/\partial\lambda}{|\partial \boldsymbol{x}/\partial\lambda|^2}, \quad \frac{\partial\lambda}{\partial y} = \frac{\partial y/\partial\lambda}{|\partial \boldsymbol{x}/\partial\lambda|^2}, \ldots \tag{10.5}$$

It can be shown that Laplace's equation for the scalar potential, φ, expressed in ellipsoidal coordinates, is

$$(\mu-\nu)k_\lambda\frac{\partial}{\partial\lambda}\left(k_\lambda\frac{\partial\varphi}{\partial\lambda}\right) + (\nu-\lambda)k_\mu\frac{\partial}{\partial\mu}\left(k_\mu\frac{\partial\varphi}{\partial\mu}\right) + (\lambda-\mu)k_\nu\frac{\partial}{\partial\nu}\left(k_\nu\frac{\partial\varphi}{\partial\nu}\right) = 0, \tag{10.6}$$

where, for example,

$$k_\lambda^2 := (a^2+\lambda)(b^2+\lambda)(c^2+\lambda), \tag{10.7}$$

and analogously for k_μ and k_ν. As Milne-Thomson [53] shows, ellipsoidal harmonics can be formed by seeking solutions to this equation of the form $\varphi(\lambda,\mu,\nu) = \chi(\lambda)\Phi(\lambda,\mu,\nu)$, where Φ is itself a basic solution of Laplace's equation. Since solutions to Laplace's equation in any coordinate system will do, we select Φ from the set of Cartesian solutions: 1, x, y, …, xyz, in order to form the harmonics. The λ dependence of Φ can always be separated from the dependencies on μ and ν: $\Phi(\lambda,\mu,\nu) = \Phi_\lambda(\lambda)f(\mu,\nu)$. Then, one can show that, when χ has the form

$$\chi(\lambda) = abc\int_\lambda^\infty \frac{\mathrm{d}\tilde{\lambda}}{\Phi_\lambda^2(\tilde{\lambda})k_\lambda(\tilde{\lambda})}, \tag{10.8}$$

the product $\chi(\lambda)\Phi(\lambda, \mu, \nu)$ is also a solution of Laplace's equation; the factor abc is provided for later convenience.

For example, starting from $\Phi = x$, we can use the relations (10.3) to write this basic solution in terms of the ellipsoidal coordinates, and thereby find that

$$\Phi_\lambda(\lambda) = (a^2 + \lambda)^{1/2}, \tag{10.9}$$

with the other factors forming x absorbed into the function $f(\mu, \nu)$. The forms of Φ_λ for $\Phi = y$ and $\Phi = z$ are straightforward to find. If, instead, we start with $\Phi = yz$, then

$$\Phi_\lambda(\lambda) = (b^2 + \lambda)^{1/2}(c^2 + \lambda)^{1/2}, \tag{10.10}$$

and the remaining factors are again absorbed into $f(\mu, \nu)$.

The no-penetration condition on the surface of the body can be written in terms of the ellipsoidal coordinates by noting that the outward normal on the body is also the local unit basis along the λ axis, perpendicular to the surface $\lambda = 0$. That is, $\boldsymbol{n} = \boldsymbol{e}_\lambda$. Thus, the no-penetration condition can be written as

$$\boldsymbol{e}_\lambda \cdot \nabla\varphi = \boldsymbol{e}_\lambda \cdot \boldsymbol{V}_\mathrm{b} \tag{10.11}$$

on the surface $\lambda = 0$. Discarding the normalization factor in $\boldsymbol{e}_\lambda$ for the sake of simplicity, and assuming that the ellipsoidal body remains rigid—so that the body's velocity, $\boldsymbol{V}_\mathrm{b}$, is described by (2.20), with the centroid at the origin—we can write this as

$$\frac{\partial\varphi(\lambda, \mu, \nu)}{\partial\lambda} = \boldsymbol{V}_\mathrm{r} \cdot \frac{\partial\boldsymbol{x}}{\partial\lambda} + \boldsymbol{\Omega} \cdot \left(\boldsymbol{x} \times \frac{\partial\boldsymbol{x}}{\partial\lambda}\right), \quad \text{on the surface } \lambda = 0. \tag{10.12}$$

Solutions for the basis fields of this general rigid-body problem are found by our usual technique of setting one velocity component to unity and all others to zero. As we will show below in specific cases, the solution for each of these basis flows is described by only a single ellipsoidal harmonic in the set represented by Eq. (10.8).

10.2 Translation of a General Ellipsoidal Body

Let's consider a problem in which an ellipsoidal body translates in the positive z direction at unit speed. Then the condition (10.12) takes the simple form

$$\frac{\partial\varphi(\lambda, \mu, \nu)}{\partial\lambda} = \frac{\partial z}{\partial\lambda}, \quad \text{on } \lambda = 0. \tag{10.13}$$

Based on the known form of $\partial z/\partial\lambda$ in (10.4), let us propose a solution based on the ellipsoidal harmonic formed from $\Phi = z$; that is, let us try a solution of the form $\varphi = Az\chi(\lambda)$, where $\chi(\lambda)$ is given by (10.8), with $\Phi_\lambda = (c^2 + \lambda)^{1/2}$, and A is an as-yet undetermined coefficient. The function, χ, takes the form

$$\chi(\lambda) = abc \int_{\lambda}^{\infty} \frac{\mathrm{d}\tilde{\lambda}}{(a^2+\tilde{\lambda})^{1/2}(b^2+\tilde{\lambda})^{1/2}(c^2+\tilde{\lambda})^{3/2}}. \tag{10.14}$$

When we substitute this proposed solution into condition (10.13), and note that $\partial\chi/\partial\lambda = -1/c^2$ at $\lambda = 0$, it is easy to show that the no-penetration condition is satisfied by choosing

$$A = \frac{1}{\chi(0)-2}, \tag{10.15}$$

so that our solution for the scalar potential is

$$\varphi(\lambda,\mu,\nu) = \frac{z\chi(\lambda)}{\chi(0)-2}. \tag{10.16}$$

The Cartesian velocity components of this potential field are then simply obtained by the chain rule:

$$u = Az\frac{\partial\chi}{\partial\lambda}\frac{\partial\lambda}{\partial x}, \quad v = Az\frac{\partial\chi}{\partial\lambda}\frac{\partial\lambda}{\partial y}, \quad w = A\left(z\frac{\partial\chi}{\partial\lambda}\frac{\partial\lambda}{\partial z} + \chi(\lambda)\right). \tag{10.17}$$

Note that we have expressed these components as a mix of Cartesian and ellipsoidal coordinates in order to ensure brevity. One should interpret each of the Cartesian coordinates in the expression as itself a function of the ellipsoidal coordinates, via (10.3).

The integral in (10.14) can be evaluated analytically in terms of elliptic integrals; the result is

$$\chi(\lambda) = \frac{2abc}{b^2-c^2}\left[\left(\frac{b^2+\lambda}{(a^2+\lambda)(c^2+\lambda)}\right)^{1/2} - \frac{1}{(a^2-c^2)^{1/2}}E\left(\arcsin\left(\frac{a^2-c^2}{a^2+\lambda}\right)^{1/2}\middle|\,\frac{a^2-b^2}{a^2-c^2}\right)\right], \tag{10.18}$$

where $E(\phi|k^2) := \int_0^{\phi}(1-k^2\sin^2\theta)^{1/2}\,\mathrm{d}\theta$ is the incomplete elliptic integral of the second kind. An example of the streamlines generated by this motion is depicted in Fig. 10.2. Note that the denominator of the expressions for φ and the velocity components is given by

$$\chi(0)-2 = \frac{2c^2}{b^2-c^2} - \frac{2abc}{(b^2-c^2)(a^2-c^2)^{1/2}}E\left(\arcsin\left(\frac{(a^2-c^2)^{1/2}}{a}\right)\middle|\,\frac{a^2-b^2}{a^2-c^2}\right). \tag{10.19}$$

It is straightforward to construct the solutions for unit translational velocity in the x and y directions; the only differences are that the boundary condition (10.13) is changed in the obvious way, and we start with the corresponding ellipsoidal harmonic generated from $\Phi = x$ and $\Phi = y$, respectively. What we are actually constructing are the basis flow fields, as defined in Sect. 4.6, associated with translation in the

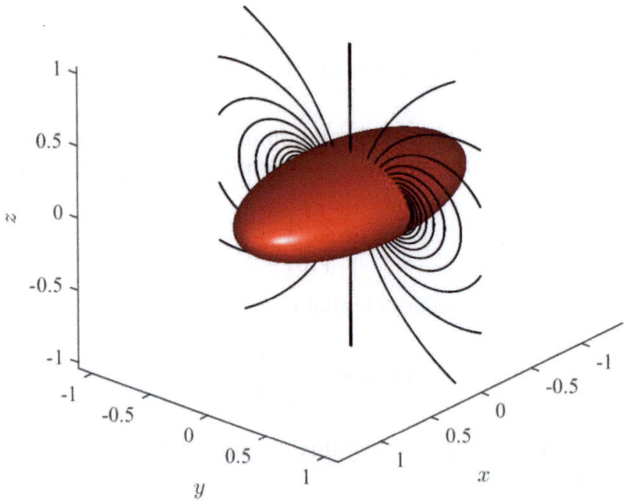

Fig. 10.2 Slice of the streamlines of the flow created by an ellipsoid with dimensions $a = 1$ and $b = 0.5$ and $c = 0.3$ translating in the z direction

three body-fixed coordinate directions, $\tilde{x}_1$, $\tilde{x}_2$ and $\tilde{x}_3$. The basis scalar potential for translation in direction $\tilde{x}_j$ is thus given by

$$\varphi_{\mathrm{bt}}^{(j)}(\lambda, \mu, \nu) = \frac{\tilde{x}_j \chi_j(\lambda)}{\chi_j(0) - 2}, \tag{10.20}$$

where no summation is implied by the repeated index j. The functions $\chi_j(\lambda)$ are given below for reference:

$$\begin{aligned}
\chi_1(\lambda) &= \frac{2abc}{(a^2 - c^2)^{3/2} k^2} \left[F\left(\phi \middle| k^2\right) - E\left(\phi \middle| k^2\right) \right], \\
\chi_2(\lambda) &= -\frac{2abc}{b^2 - c^2} \left[\left(\frac{c^2 + \lambda}{(a^2 + \lambda)(b^2 + \lambda)} \right)^{1/2} \right. \\
&\qquad \left. - \frac{1}{(a^2 - c^2)^{1/2} k^2} \left[E\left(\phi \middle| k^2\right) + (k^2 - 1) F\left(\phi \middle| k^2\right) \right] \right], \\
\chi_3(\lambda) &= \frac{2abc}{b^2 - c^2} \left[\left(\frac{b^2 + \lambda}{(a^2 + \lambda)(c^2 + \lambda)} \right)^{1/2} - \frac{1}{(a^2 - c^2)^{1/2}} E\left(\phi \middle| k^2\right) \right],
\end{aligned} \tag{10.21}$$

where $F(\phi|k^2) := \int_0^\phi (1 - k^2 \sin^2 \theta)^{-1/2} \, \mathrm{d}\theta$ is the incomplete elliptic integral of the first kind, and, for brevity, we have defined

$$\phi := \arcsin\left(\frac{a^2 - c^2}{a^2 + \lambda} \right)^{1/2}, \qquad k^2 := (a^2 - b^2)/(a^2 - c^2). \tag{10.22}$$

The ellipsoidal solution can be reduced to a number of special cases. Let's inspect four of them.

Oblate Spheroid A body that is axisymmetric about the z axis, so that $b = a > c$, is said to be an **oblate spheroid**. As shown in Fig. 10.3, the isosurfaces for λ and μ become axisymmetric, while the hyperboloidal surfaces of constant ν degenerate into meridional planes as the range of this coordinate shrinks to zero. The meridional view of the remaining coordinates is shown in the lower panel of Fig. 10.3. Thus, only two ellipsoidal coordinates (λ, μ) are needed, obtained from the roots of the equations

$$\frac{r^2}{a^2+\lambda} + \frac{z^2}{c^2+\lambda} = 1, \qquad \frac{r^2}{a^2+\mu} + \frac{z^2}{c^2+\mu} = 1, \tag{10.23}$$

where $r^2 = x^2 + y^2$.

The scalar potentials under these circumstances can be found in straightforward fashion by setting $a = b$ in Eq. (10.21). The resulting expressions for $\chi_1(\lambda) = \chi_2(\lambda)$ and $\chi_3(\lambda)$ are

$$\chi_1(\lambda) = \chi_2(\lambda) = -\frac{a^2 c}{(a^2-c^2)^{3/2}} \left[\frac{(a^2-c^2)^{1/2}(c^2+\lambda)^{1/2}}{a^2+\lambda} - \arcsin\left(\frac{a^2-c^2}{a^2+\lambda}\right)^{1/2} \right],$$

$$\chi_3(\lambda) = \frac{2a^2 c}{(a^2-c^2)^{3/2}} \left[\left(\frac{a^2-c^2}{c^2+\lambda}\right)^{1/2} - \arcsin\left(\frac{a^2-c^2}{a^2+\lambda}\right)^{1/2} \right]. \tag{10.24}$$

The streamlines generated by translation in the z direction are depicted for one choice of c/a in Fig. 10.4.

Prolate Spheroid In this case, let us consider a body that is axisymmetric about its longest axis, the x axis, so that $b = c < a$. This is known as a **prolate spheroid**. Let us see how the solution reduces in such a case. When taking the limit of $b \to c$, we should note that $k^2 \to 1$. There is some subtlety in taking this limit for $\chi_2(\lambda)$ and $\chi_3(\lambda)$, since the numerator and denominator both vanish. This requires that we expand the numerator in a Taylor series about $k^2 = 1$ (i.e., we use L'Hôpital's rule). It is useful to note that, near this limit,

$$E\left(\phi \middle| k^2\right) \approx E(\phi|1) + \frac{1}{2}(k^2-1)(E(\phi|1) - F(\phi|1)), \tag{10.25}$$

and that, at this limiting value of k^2,

$$E(\phi|1) = \sin\phi, \qquad F(\phi|1) = \log(\tan\phi + \sec\phi). \tag{10.26}$$

Using these results, we can show that

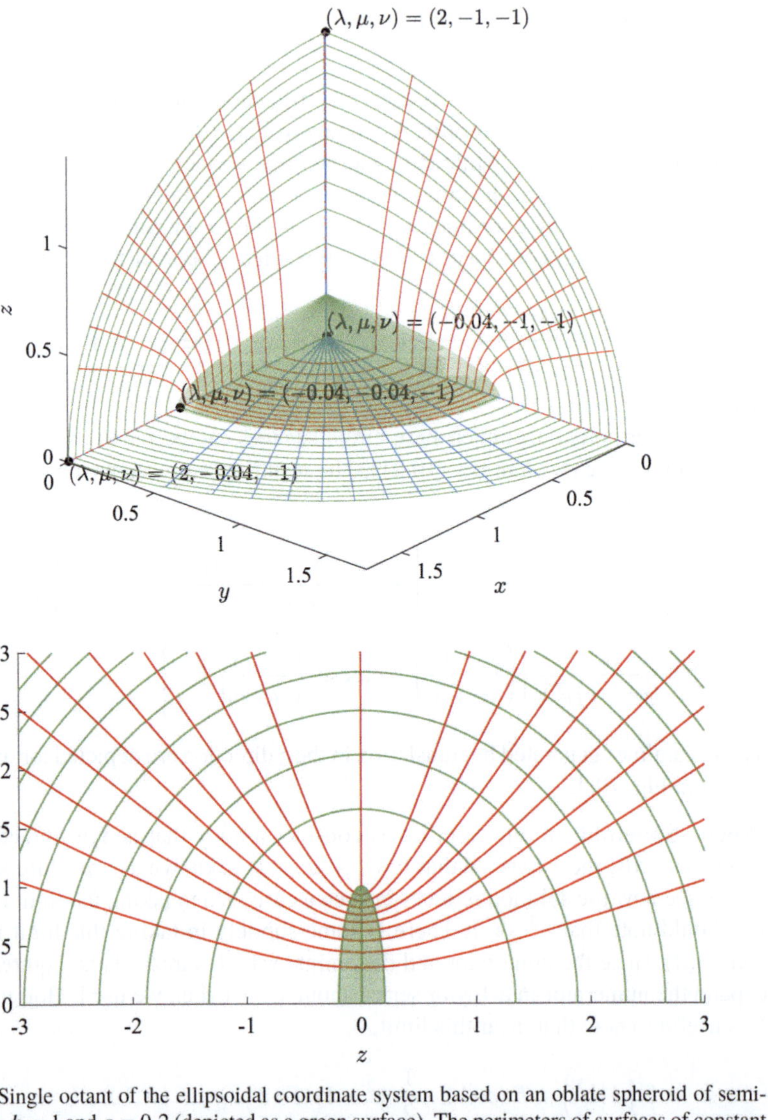

Fig. 10.3 Single octant of the ellipsoidal coordinate system based on an oblate spheroid of semi-axes $a = 1$, $b = 1$ and $c = 0.2$ (depicted as a green surface). The perimeters of surfaces of constant λ are depicted in green, surfaces of constant μ in red, and surfaces of constant ν in blue. All other octants are simply mirror images of this illustration. A meridional plane of this ellipsoid is depicted in the lower panel

$$\chi_1(\lambda) = \frac{2ac^2}{(a^2-c^2)^{3/2}} \left[\log\left(\left(\frac{a^2-c^2}{c^2+\lambda}\right)^{1/2} + \left(\frac{a^2+\lambda}{c^2+\lambda}\right)^{1/2} \right) - \left(\frac{a^2-c^2}{a^2+\lambda}\right)^{1/2} \right], \quad (10.27)$$

$$\chi_2(\lambda) = \chi_3(\lambda) =$$
$$-\frac{ac^2}{(a^2-c^2)^{3/2}} \left[\log\left(\left(\frac{a^2-c^2}{c^2+\lambda}\right)^{1/2} + \left(\frac{a^2+\lambda}{c^2+\lambda}\right)^{1/2} \right) - \frac{(a^2-c^2)^{1/2}(a^2+\lambda)^{1/2}}{c^2+\lambda} \right].$$

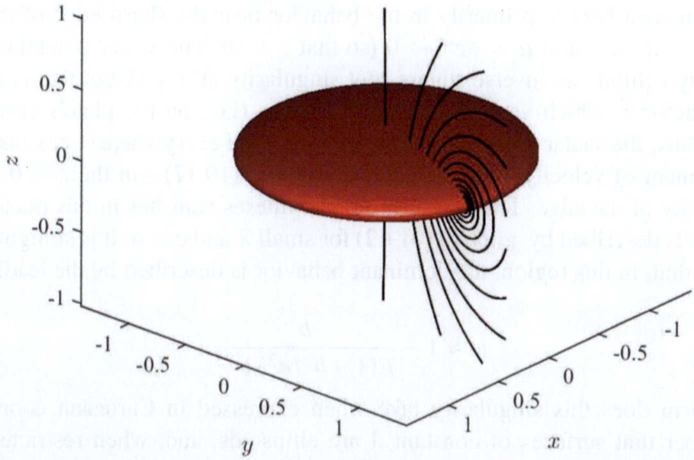

Fig. 10.4 Slice of the streamlines created by an oblate spheroid with dimensions $a = b = 1$ and $c = 0.2$ translating in the z direction

Spherical Body The limiting case of $a = b = c$ is of particular interest, since the ellipsoid simplifies in this set of dimensions to a sphere. In this case, Eq. (10.1) degenerates to a single-degree polynomial for θ, with only a single root: $\lambda = x^2 + y^2 + z^2 - a^2 = r^2 - a^2$. The function $\chi_3(\lambda)$ (as well as $\chi_1(\lambda)$ and $\chi_2(\lambda)$) can be shown to reduce from its axisymmetric form (10.24) to

$$\chi_1(\lambda) = \chi_2(\lambda) = \chi_3(\lambda) = \frac{2}{3}\frac{a^3}{(a^2+\lambda)^{3/2}} \tag{10.28}$$

in the limit $c \to a$, so that the scalar potential field is

$$\varphi = -\frac{a^3}{2r^3} z. \tag{10.29}$$

This is the expected result, obtained, e.g., from a potential dipole centered at the origin, with its axis in the z direction.

Flat Elliptical Plate How does this flow field behave when the dimension of the ellipsoid in the direction of translation is zero (i.e., when $c = 0$)? The resulting scalar potential for this case is easily found by setting $c = 0$ in our previous expressions for an oblate spheroid, after first eliminating the common factor of c from the numerator and denominator:

$$\varphi(\lambda, \mu, \nu) = -\frac{az}{E(1 - b^2/a^2)} \left[\left(\frac{b^2 + \lambda}{(a^2 + \lambda)\lambda} \right)^{1/2} - \frac{1}{a} E\left(\arcsin\left(\frac{a^2}{a^2 + \lambda} \right)^{1/2} \middle| \, 1 - \frac{b^2}{a^2} \right) \right], \tag{10.30}$$

where $E(k^2) \equiv E(\frac{1}{2}\pi | k^2)$ is the complete elliptic integral of the second kind.

Our interest here is primarily in the behavior near the sharp edge of the plate, where $0 < \lambda \ll 1$ and $\mu \approx -c^2 \to 0$ (so that $z \approx 0$). The scalar potential (10.30) ostensibly exhibits an inverse square root singularity at $\lambda = 0$, but this is removed by the factor z, which vanishes wherever λ does (i.e., on the plate's surface and edge). Thus, the scalar potential field is well behaved everywhere. Let's inspect the z component of velocity—given by the last of Eq. (10.17)—in the $z = 0$ plane in the vicinity of the edge. The first term in parentheses vanishes in this plane, so the behavior is described by $\chi(\lambda)/(\chi(0) - 2)$ for small λ and $c = 0$. It is straightforward to show that, in this region, the dominant behavior is described by the leading term in $\chi(\lambda)$,

$$w \approx 1 - \frac{b}{E(1 - b^2/a^2)\lambda^{1/2}}. \tag{10.31}$$

What form does this singularity take when expressed in Cartesian coordinates? Remember that surfaces of constant λ are ellipsoids, and, when restricted to the $z = 0$ plane, are ellipses of semi-axes $(a^2 + \lambda)^{1/2}$ and $(b^2 + \lambda)^{1/2}$, as can be confirmed from Eq. (10.1). The λ coordinate lines are perpendicular to these ellipses. Thus, for small λ, the distance from a point $\boldsymbol{x}_b$ on the body's edge ($\lambda = 0, \mu = 0$) to an evaluation point $\boldsymbol{x}$ in the $z = 0$ plane, along one of these coordinate lines, is given by

$$|\boldsymbol{x} - \boldsymbol{x}_b| \approx \left|\frac{\partial \boldsymbol{x}}{\partial \lambda}(0, 0, \nu)\right| \lambda = \frac{1}{2}\left(\frac{x_b^2}{a^4} + \frac{y_b^2}{b^4}\right)^{1/2} \lambda. \tag{10.32}$$

Thus, we can write the z component of velocity in the plane of the plate, near its edge, as

$$w \approx 1 - \frac{C(x_b, y_b)}{|\boldsymbol{x} - \boldsymbol{x}_b|^{1/2}}, \tag{10.33}$$

where $C(x_b, y_b)$ is a function only of position along the edge, given by

$$C(x_b, y_b) := \frac{b}{2^{1/2}E(1 - b^2/a^2)}\left(\frac{x_b^2}{a^4} + \frac{y_b^2}{b^4}\right)^{1/4}. \tag{10.34}$$

The expression (10.33) clearly exhibits the inverse square root singularity that we have come to expect for the inviscid flow near a sharp edge. Let's inspect the form this singular behavior takes in two special cases. Before doing so, it is useful to note that $4aE(1 - b^2/a^2)$ is the perimeter of the elliptical plate. First, consider the case of a circular flat plate, for which $b = a$. Then the perimeter is $2\pi a$, and $x_b^2 + y_b^2 = a^2$ along the edge. In this case, since $|\boldsymbol{x} - \boldsymbol{x}_b| = r - a$, we can easily show that

$$w \approx 1 - \frac{(2a)^{1/2}}{\pi(r - a)^{1/2}}. \tag{10.35}$$

In contrast, when $b \ll 1$, the plate approaches an infinite rectangular strip in the vicinity of the $x = 0$ plane, so we expect the result to approach the known behavior of a two-dimensional flat plate near its edge. Indeed, since the perimeter approaches

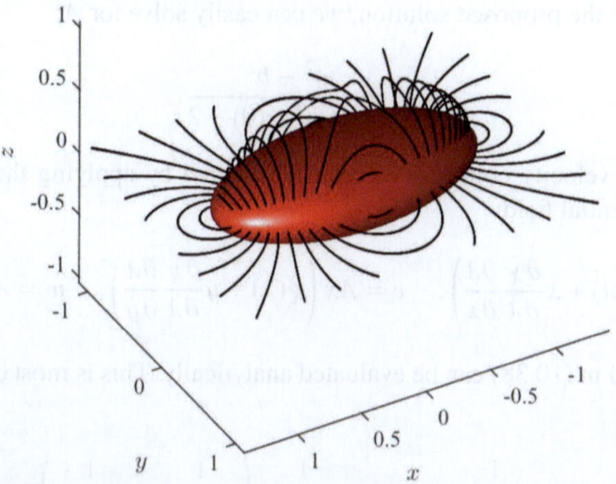

Fig. 10.5 Streamlines of the flow created by an ellipsoid with dimensions $a = 1$ and $b = 0.5$ and $c = 0.3$ rotating about the z axis

4, we can show that, near the edge at $x_b = 0$ and $y_b = b$ (and in the $z = 0$ plane), where $|\boldsymbol{x} - \boldsymbol{x}_b| = y - b$, the z component of velocity behaves as follows:

$$w \approx 1 - \frac{b^{1/2}}{2^{1/2}(y-b)^{1/2}} \approx 1 - \frac{y}{y^2 - b^2}, \tag{10.36}$$

which agrees with the result shown in (8.57), for example (after accounting for the difference in coordinate systems, of course).

10.3 Rotational Motion of a General Ellipsoidal Body

When the same ellipsoidal body is rotating about one of the Cartesian coordinate axes, the solution can be obtained by a similar procedure. Let us suppose the rotation to be with unit angular velocity about the z axis. We start with the correspondingly simplified form of the boundary condition (10.12), namely,

$$\frac{\partial \varphi(\lambda, \mu, \nu)}{\partial \lambda} = x\frac{\partial y}{\partial \lambda} - y\frac{\partial x}{\partial \lambda}, \quad \text{on } \lambda = 0. \tag{10.37}$$

The form of this condition, with the metrics $\partial x/\partial \lambda$ and $\partial y/\partial \lambda$ obtained from (10.4), suggests that we form a solution from the ellipsoidal harmonic obtained by starting with $\Phi = xy$, i.e., we propose the solution $\varphi = Axy\chi(\lambda)$, where

$$\chi(\lambda) = abc\int_{\lambda}^{\infty} \frac{d\tilde{\lambda}}{(a^2+\tilde{\lambda})^{3/2}(b^2+\tilde{\lambda})^{3/2}(c^2+\tilde{\lambda})^{1/2}}. \tag{10.38}$$

By substituting the proposed solution, we can easily solve for A:

$$A = \frac{a^2 - b^2}{(a^2 + b^2)\chi(0) - 2}. \tag{10.39}$$

The Cartesian velocity components are easily found by applying the chain rule to this scalar potential field:

$$u = Ay\left(\chi(\lambda) + x\frac{\partial\chi}{\partial\lambda}\frac{\partial\lambda}{\partial x}\right), \quad v = Ax\left(\chi(\lambda) + y\frac{\partial\chi}{\partial\lambda}\frac{\partial\lambda}{\partial y}\right), \quad w = Axy\frac{\partial\chi}{\partial\lambda}\frac{\partial\lambda}{\partial z}. \tag{10.40}$$

The integral in (10.38) can be evaluated analytically. This is most easily achieved by noting that

$$\frac{1}{(a^2 + \lambda)(b^2 + \lambda)} = -\frac{1}{a^2 - b^2}\left(\frac{1}{a^2 + \lambda} - \frac{1}{b^2 + \lambda}\right). \tag{10.41}$$

Then, we can just write $\chi(\lambda)$ in terms of the functions (10.21) obtained for translation:

$$\chi(\lambda) = -\frac{1}{a^2 - b^2}\left(\chi_1(\lambda) - \chi_2(\lambda)\right). \tag{10.42}$$

A sample of the streamlines generated by this motion is depicted in Fig. 10.5.

The flows generated by rotational motion about the x and y axes can be obtained by a similar procedure, adapted in the obvious ways, using the same partial fractions technique (10.41) to write $\chi(\lambda)$ for rotation about an axis in terms of the $\chi_j(\lambda)$ corresponding to translation about the other two axes. We can write the three basis scalar potentials, corresponding to rotation about the three axes $\tilde{x}_1$, $\tilde{x}_2$ and $\tilde{x}_3$, as

$$\begin{aligned}
\varphi_{\text{br}}^{(1)}(\lambda, \mu, \nu) &= \frac{(b^2 - c^2)\tilde{x}_2\tilde{x}_3\left(\chi_2(\lambda) - \chi_3(\lambda)\right)}{(b^2 + c^2)\left(\chi_2(0) - \chi_3(0)\right) + 2(b^2 - c^2)}, \\
\varphi_{\text{br}}^{(2)}(\lambda, \mu, \nu) &= \frac{(c^2 - a^2)\tilde{x}_3\tilde{x}_1\left(\chi_3(\lambda) - \chi_1(\lambda)\right)}{(c^2 + a^2)\left(\chi_3(0) - \chi_1(0)\right) + 2(c^2 - a^2)}, \\
\varphi_{\text{br}}^{(3)}(\lambda, \mu, \nu) &= \frac{(b^2 - c^2)\tilde{x}_1\tilde{x}_2\left(\chi_1(\lambda) - \chi_2(\lambda)\right)}{(a^2 + b^2)\left(\chi_1(0) - \chi_2(0)\right) + 2(a^2 - b^2)}.
\end{aligned} \tag{10.43}$$

The functions $\chi_j(\lambda)$ can be applied in their general form (10.21), or in their specific forms for special cases, as given in the previous section.

10.4 Added Mass

We can use one of the several formulas derived in Sects. 4.6.3 or 6.5.2 for added mass components to calculate this matrix for an ellipsoid. The form that is easiest for this task is given by Eqs. (4.160), (4.163) and (4.167), which we related to the unit impulses in (6.118) and (6.124). This form relies only on the surface values of the basis scalar potential field, and the resulting integrals take the form of general geometric identities. In the results that follow, let us note that our analysis thus far has been naturally carried out in a body-fixed coordinate system whose origin is at the centroid of the ellipsoid (so that $\boldsymbol{X}_\mathrm{r} \equiv \boldsymbol{X}_\mathrm{c}$), and whose axes are identifiable as $\tilde{x}_1 \equiv x$, $\tilde{x}_2 \equiv y$ and $\tilde{x}_3 \equiv z$.

For example, for unit translation in the z ($\equiv \tilde{x}_3$) direction, we found that the scalar potential field is given by Eq. (10.16). The value of this basis potential on the surface $\lambda = 0$ is thus

$$\varphi_\mathrm{bt}^{(3)}(0, \mu, \nu) = \frac{\tilde{x}_3 \chi_3(0)}{\chi_3(0) - 2}, \tag{10.44}$$

where χ_3 was defined in (10.21). Then, from the added mass formulas referenced above, we can calculate $\tilde{M}_{3k}^{FV}$ and $\tilde{M}_{3k}^{F\Omega}$. For example,

$$\tilde{M}_{3k}^{FV} = \left(-\rho \int_{S_\mathrm{b}} \varphi_\mathrm{bt}^{(3)} \boldsymbol{n}\, \mathrm{d}S \right) \cdot \tilde{\boldsymbol{e}}_k = -\frac{\rho \chi_3(0)}{\chi_3(0) - 2} \left(\int_{S_\mathrm{b}} \tilde{x}_3 \boldsymbol{n}\, \mathrm{d}S \right) \cdot \tilde{\boldsymbol{e}}_k. \tag{10.45}$$

However, the remaining integral is, by (A.23) in the Appendix, $\mathcal{V}_\mathrm{b} \tilde{\boldsymbol{e}}_3$, where $\mathcal{V}_\mathrm{b} \equiv \frac{4}{3}\pi abc$ is the ellipsoid's volume. Thus, $\tilde{M}_{3k}^{FV} = \rho \mathcal{V}_\mathrm{b} \delta_{3k} \chi_3(0)/(2 - \chi_3(0))$. Following similar steps, and using identity (A.25) in the Appendix, it is easy to show that the rotational-translational added mass tensor components, $\tilde{M}_{3k}^{F\Omega}$, are identically zero for this unit motion. In other words, force on an ellipsoid in the $\tilde{x}_3$ direction is only effected by translational motion in that same direction, a result that we could have expected from symmetry.

For our analysis of unit rotation about the z ($\tilde{x}_3$) axis, the basis scalar potential on the ellipsoid surface is given by

$$\varphi_\mathrm{br}^{(3)}(0, \mu, \nu) = \frac{(a^2 - b^2)\tilde{x}_1 \tilde{x}_2 \chi(0)}{(a^2 + b^2)\chi(0) - 2} = \frac{(a^2 - b^2)\tilde{x}_1 \tilde{x}_2 \left(\chi_1(0) - \chi_2(0)\right)}{(a^2 + b^2)\left(\chi_1(0) - \chi_2(0)\right) + 2(a^2 - b^2)}, \tag{10.46}$$

where the χ_j are the translational functions defined in (10.21). This can be used in (4.163) to obtain the rotational added mass component, $\tilde{M}_{3k}^{M\Omega}$, viz.,

$$\begin{aligned} \tilde{M}_{3k}^{M\Omega} &= \left(-\rho \int_{S_\mathrm{b}} (\boldsymbol{x} - \boldsymbol{X}_\mathrm{r}) \times \varphi_\mathrm{br}^{(3)} \boldsymbol{n}\, \mathrm{d}S \right) \cdot \tilde{\boldsymbol{e}}_k, \\ &= \frac{\rho (a^2 - b^2)\left(\chi_1(0) - \chi_2(0)\right)}{(a^2 + b^2)\left(\chi_1(0) - \chi_2(0)\right) + 2(a^2 - b^2)} \int_{S_\mathrm{b}} \tilde{x}_1 \tilde{x}_2 \left((\boldsymbol{x} - \boldsymbol{X}_\mathrm{r}) \times \boldsymbol{n}\right) \cdot \tilde{\boldsymbol{e}}_k\, \mathrm{d}S. \end{aligned} \tag{10.47}$$

The surface integral is purely geometric, and we can use the divergence theorem (A.18) to rewrite it as a volume integral, which can be evaluated, for example, by transforming to spherical coordinates (not shown):

$$\int_{\mathcal{S}_b} \tilde{x}_1 \tilde{x}_2 \left((\boldsymbol{x} - \boldsymbol{X}_r) \times \boldsymbol{n} \right) \cdot \tilde{\boldsymbol{e}}_k \, dS = \delta_{3k} \int_{\mathcal{V}_b} \left(\tilde{x}_1^2 - \tilde{x}_2^2 \right) dV = \frac{1}{5} \mathcal{V}_b (a^2 - b^2) \delta_{3k}. \quad (10.48)$$

Thus, the rotational added mass corresponding to rotation about the $\tilde{x}_3$ axis is

$$\tilde{M}^{M\Omega}_{3k} = \frac{1}{5} \rho \mathcal{V}_b \delta_{3k} \frac{(a^2 - b^2)^2 (\chi_1(0) - \chi_2(0))}{(a^2 + b^2)(\chi_1(0) - \chi_2(0)) + 2(a^2 - b^2)}. \quad (10.49)$$

By similar steps, it is easy to show that the translational-rotational added mass, $\tilde{M}^{MV}_{3k}$, is zero. Thus, moment about the $\tilde{x}_3$ axis is only brought about by angular motion about that axis.

We can apply the same steps to compute the added mass associated with translation in and rotation about the other coordinate axes. The results can be determined by inspection, however, and we summarize them here.

Result 10.1: The Added Mass of an Ellipsoid

Consider an ellipsoidal body of semi-axes $a \geq b \geq c$, aligned, respectively, with the $\tilde{x}_1$, $\tilde{x}_2$ and $\tilde{x}_3$ axes, and centered at the origin. The rotational added mass matrix is

$$\tilde{\mathbf{M}}^{M\Omega} = \begin{bmatrix} \tilde{M}^{M\Omega}_{11} & 0 & 0 \\ 0 & \tilde{M}^{M\Omega}_{22} & 0 \\ 0 & 0 & \tilde{M}^{M\Omega}_{33} \end{bmatrix}, \quad (10.50)$$

where the diagonal components are given by

$$\tilde{M}^{M\Omega}_{11} = -\frac{1}{5} \rho \mathcal{V}_b \frac{(b^2 - c^2)^2 (\chi_2(0) - \chi_3(0))}{(b^2 + c^2)(\chi_2(0) - \chi_3(0)) + 2(b^2 - c^2)}, \quad (10.51)$$

$$\tilde{M}^{M\Omega}_{22} = -\frac{1}{5} \rho \mathcal{V}_b \frac{(c^2 - a^2)^2 (\chi_3(0) - \chi_1(0))}{(c^2 + a^2)(\chi_3(0) - \chi_1(0)) + 2(c^2 - a^2)}, \quad (10.52)$$

$$\tilde{M}^{M\Omega}_{33} = -\frac{1}{5} \rho \mathcal{V}_b \frac{(a^2 - b^2)^2 (\chi_1(0) - \chi_2(0))}{(a^2 + b^2)(\chi_1(0) - \chi_2(0)) + 2(a^2 - b^2)}, \quad (10.53)$$

and the functions $\chi_j(\lambda)$ are defined in Eq. (10.21); the translational added mass matrix is

$$\tilde{\mathbf{M}}^{FV} = \begin{bmatrix} \tilde{M}^{FV}_{11} & 0 & 0 \\ 0 & \tilde{M}^{FV}_{22} & 0 \\ 0 & 0 & \tilde{M}^{FV}_{33} \end{bmatrix}, \quad (10.54)$$

where the diagonal components are given by

$$\tilde{M}_{11}^{FV} = \rho \mathcal{V}_b \frac{\chi_1(0)}{2 - \chi_1(0)}, \tag{10.55}$$

$$\tilde{M}_{22}^{FV} = \rho \mathcal{V}_b \frac{\chi_2(0)}{2 - \chi_2(0)}, \tag{10.56}$$

$$\tilde{M}_{33}^{FV} = \rho \mathcal{V}_b \frac{\chi_3(0)}{2 - \chi_3(0)}; \tag{10.57}$$

and the rotational-translational and translational-rotational matrices, $\tilde{\mathbf{M}}^{F\Omega}$ and $\tilde{\mathbf{M}}^{MV}$, respectively, are both zero.

These general formulas for the added mass matrices of an ellipsoid can be used to obtain the added masses of several simple three-dimensional objects. Here, we provide a few examples:

Oblate Spheroid, Axisymmetric About the $\tilde{x}_3$ Axis If we let $a = b$, then χ_j take the form given by (10.24). For such a body, if we denote the thickness ratio as $\sigma = c/a < 1$, then the added mass coefficients for translation in the $\tilde{x}_1$ and $\tilde{x}_2$ directions reduce to

$$\tilde{M}_{11}^{FV} = \tilde{M}_{22}^{FV} = \frac{4}{3}\rho\pi a^2 c \frac{\chi_1(0)}{2 - \chi_1(0)} = \frac{4}{3}\rho\pi a^3 \sigma^2 \frac{\arccos\sigma - \sigma(1-\sigma^2)^{1/2}}{(2-\sigma^2)(1-\sigma^2)^{1/2} - \sigma\arccos\sigma}, \tag{10.58}$$

and the translational added mass on the axis of symmetry becomes

$$\tilde{M}_{33}^{FV} = \frac{4}{3}\rho\pi a^2 c \frac{\chi_3(0)}{2 - \chi_3(0)} = \frac{4}{3}\rho\pi a^3 \frac{(1-\sigma^2)^{1/2} - \sigma\arccos\sigma}{\arccos\sigma - \sigma(1-\sigma^2)^{1/2}}. \tag{10.59}$$

The rotational added mass coefficient along the axis of symmetry, $\tilde{M}_{33}^{M\Omega}$, vanishes identically, and the other two components become

$$\tilde{M}_{11}^{M\Omega} = \tilde{M}_{22}^{M\Omega} = \frac{4}{15}\rho\pi a^5 (1-\sigma^2)^2 \frac{(\sigma^2+2)(1-\sigma^2)^{1/2} - 3\sigma\arccos\sigma}{\sigma(\sigma^2-7)(1-\sigma^2)^{1/2} + 3(1+\sigma^2)\arccos\sigma}. \tag{10.60}$$

These added mass coefficients are plotted in Fig. 10.6 versus the thickness ratio.

Now we consider the two limits of the oblate spheroid, as thickness ratio approaches 1 and 0, respectively:

Sphere of Radius *a* In this case, $a = b = c$ (or $\sigma = 1$, using the notation for the oblate spheroid), and the χ_j all revert to the form (10.28). Thus, $\chi_j(0)/(2 - \chi_j(0)) = 1/2$, and the translational added mass in all directions is $\tilde{M}_{11}^{FV} = \tilde{M}_{22}^{FV} = \tilde{M}_{33}^{FV} = \frac{1}{2}\rho\mathcal{V}_b$, where $\mathcal{V}_b = \frac{4}{3}\pi a^3$; that is, the translational added mass is half of the displaced fluid mass. The rotational added mass is identically zero, since an isotropic object cannot impart motion to the fluid by rotating.

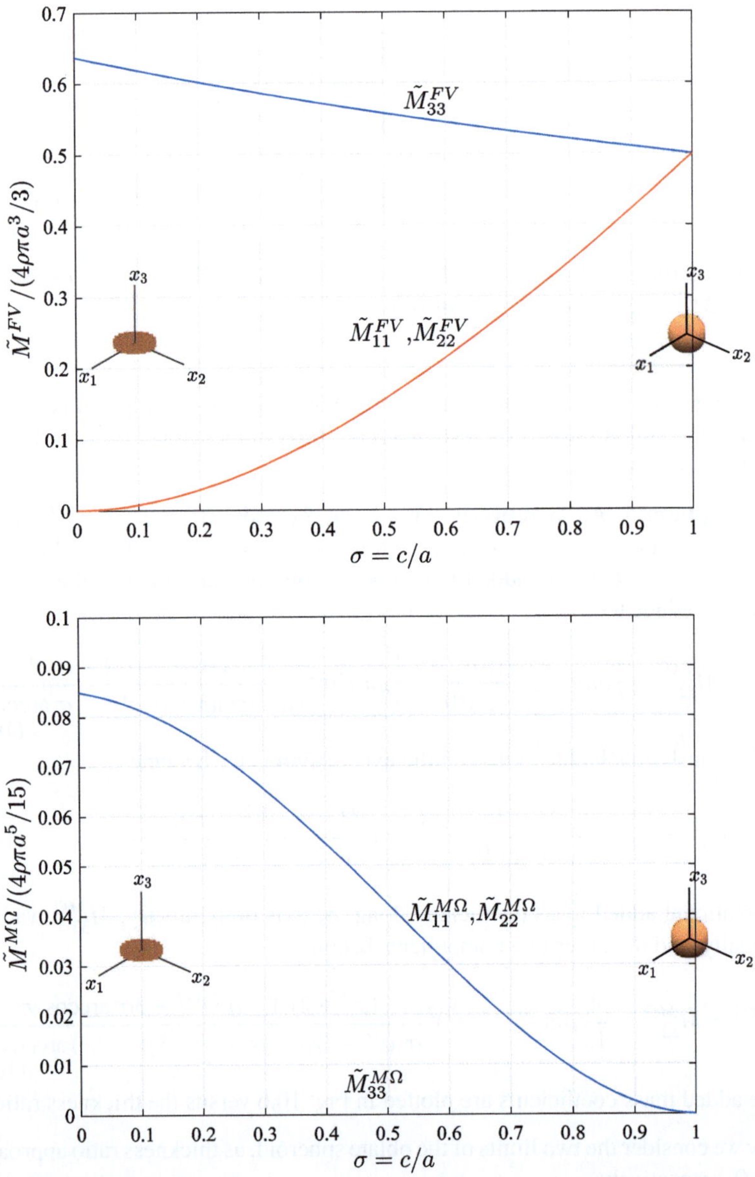

Fig. 10.6 Added mass coefficients for an oblate spheroid: translational (top) and rotational (bottom)

Circular Disk of Radius a, Axisymmetric About the $\tilde{x}_3$ Axis Now, let the thickness ratio σ approach zero. Then, it is straightforward to show, using $\arccos\sigma \approx \pi/2 - \sigma$, that

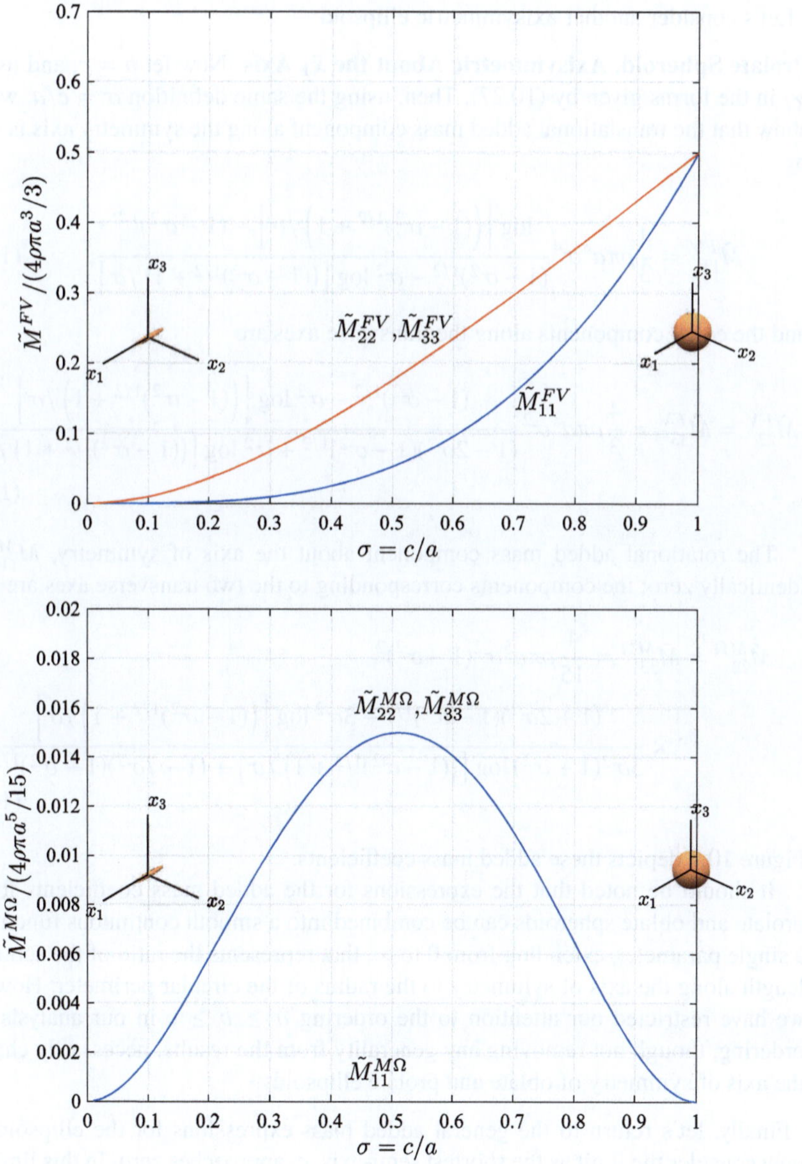

Fig. 10.7 Added mass coefficients for a prolate spheroid: translational (top) and rotational (bottom)

$$\tilde{M}^{FV}_{11} = \tilde{M}^{FV}_{22} = 0, \quad \tilde{M}^{FV}_{33} = \frac{8}{3}\rho a^3, \qquad \tilde{M}^{M\Omega}_{11} = \tilde{M}^{M\Omega}_{22} = \frac{16}{45}\rho a^5, \quad \tilde{M}^{M\Omega}_{33} = 0. \tag{10.61}$$

It is not surprising, of course, that the translational added mass coefficients vanish along the axes in the plane of the disk, since the infinitely-thin shape cannot induce motion in such directions.

Let's consider another axisymmetric ellipsoid:

Prolate Spheroid, Axisymmetric About the $\tilde{x}_1$ Axis Now let $b = c$, and use the χ_j in the forms given by (10.27). Then, using the same definition $\sigma = c/a$, we can show that the translational added mass component along the symmetry axis is given by

$$\tilde{M}_{11}^{FV} = \frac{4}{3}\rho\pi a^3\sigma^4 \frac{\log\left[\left((1-\sigma^2)^{1/2}+1\right)/\sigma\right] - (1-\sigma^2)^{1/2}}{(1-\sigma^2)^{1/2} - \sigma^2\log\left[\left((1-\sigma^2)^{1/2}+1\right)/\sigma\right]}, \tag{10.62}$$

and the equal components along the transverse axes are

$$\tilde{M}_{22}^{FV} = \tilde{M}_{33}^{FV} = \frac{4}{3}\rho\pi a^3\sigma^2 \frac{(1-\sigma^2)^{1/2} - \sigma^2\log\left[\left((1-\sigma^2)^{1/2}+1\right)/\sigma\right]}{(1-2\sigma^2)(1-\sigma^2)^{1/2} + \sigma^2\log\left[\left((1-\sigma^2)^{1/2}+1\right)/\sigma\right]}. \tag{10.63}$$

The rotational added mass component about the axis of symmetry, $\tilde{M}_{11}^{M\Omega}$, is identically zero; the components corresponding to the two transverse axes are

$$\tilde{M}_{22}^{M\Omega} = \tilde{M}_{33}^{M\Omega} = \frac{4}{15}\rho\pi a^5\sigma^2(1-\sigma^2)^2 \times \frac{(1+2\sigma^2)(1-\sigma^2)^{1/2} - 3\sigma^2\log\left[\left((1-\sigma^2)^{1/2}+1\right)/\sigma\right]}{3\sigma^2(1+\sigma^2)\log\left[\left((1-\sigma^2)^{1/2}+1\right)/\sigma\right] + (1-7\sigma^2)(1-\sigma^2)^{1/2}}. \tag{10.64}$$

Figure 10.7 depicts these added mass coefficients.

It should be noted that the expressions for the added mass coefficients for the prolate and oblate spheroids can be combined into a smooth continuous function of a single parameter, extending from 0 to ∞, that represents the ratio of the semi-axis length along the axis of symmetry to the radius of the circular perimeter. However, we have restricted our attention to the ordering $a \geq b \geq c$ in our analysis; this ordering, though not removing any generality from the results, necessarily changes the axis of symmetry of oblate and prolate ellipsoids.

Finally, let's return to the general added mass expressions for the ellipsoid, but now consider the limit as the shortest semi-axis, c, approaches zero. In this limit, we obtain the added mass of

Infinitely Thin Elliptical Disk of Semi-axes a and $b \leq a$ In this case, $\chi_1(\lambda)$ and $\chi_2(\lambda)$ are both identically zero, so

$$\tilde{M}_{11}^{FV} = \tilde{M}_{22}^{FV} = \tilde{M}_{33}^{M\Omega} = 0. \tag{10.65}$$

Let us repurpose our symbol σ to denote $\sigma := b/a \leq 1$. Thus, $k^2 = 1 - \sigma^2$, and the expression for $\chi_3(0)$ reduces to $2 - 2E(1 - \sigma^2)$. When we evaluate the general expressions in Result 10.1 in the limit $c \to 0$, we find that the remaining added mass coefficients are

$$\tilde{M}_{33}^{FV} = \frac{4}{3}\rho\pi a^3 \frac{\sigma^2}{E(k^2)}, \tag{10.66}$$

and

$$\begin{aligned}\tilde{M}_{11}^{M\Omega} &= \frac{4}{15}\rho\pi a^5 \frac{\sigma^4 k^2}{(k^2+1)E(k^2) - \sigma^2 K(k^2)},\\ \tilde{M}_{22}^{M\Omega} &= \frac{4}{15}\rho\pi a^5 \frac{\sigma^2 k^2}{(k^2-\sigma^2)E(k^2) + \sigma^2 K(k^2)}.\end{aligned} \tag{10.67}$$

These expressions approach the expected limit of a circular disk, provided it is noted that $E(k^2) \approx \pi/2 - \pi k^2/8$ and $K(k^2) \approx \pi/2 + \pi k^2/8$ for $k^2 \ll 1$.

Appendix
Mathematical Tools

Contents

A.1 Some Vector Calculus Results 389
A.1.1 Cartesian Index Notation 389
A.1.2 Derivatives of Fundamental Solutions of Laplace's Equation 392
A.1.3 The Divergence Theorem and Some Relevant Uses 393
A.1.4 Stokes' Theorem and Some Relevant Uses 405
A.1.5 Change of Variables Theorem 411
A.1.6 Field Quantities and Their Rate of Change 411
A.1.7 Time Differentiation of Spatial Integrals 412
A.2 Useful Tools from Complex Analysis 416
A.2.1 Basic Properties 417
A.2.2 The Cauchy Integral and Residue Theorem 420
A.2.3 Conformal Mapping 425
A.2.4 The Joukowski and Kármán–Trefftz Airfoils 431
A.2.5 The Schwarz–Christoffel Transformation 436
A.3 Mathematical Results for the Infinitely-Thin Plate 443
A.3.1 Notes on an Important Factor 443
A.3.2 Properties of Chebyshev Polynomials 444
A.3.3 Contour Integrals of Interest 452

A.1 Some Vector Calculus Results

This section provides a number of useful identities from vector calculus.

A.1.1 Cartesian Index Notation

Before we go any further, let's review the rules of index notation, a useful tool in proving identities. The convention we use here is a stripped-down version of the actual convention, which, for example, distinguishes covariant and contravariant

J. D. Eldredge, *Mathematical Modeling of Unsteady Inviscid Flows*, Interdisciplinary Applied Mathematics 50, https://doi.org/10.1007/978-3-030-18319-6

components. But since we only will write the indexed expressions in a Cartesian coordinate system, we have no need for such distinctions and can therefore keep it simple. First, the rules:

1. If a given letter index appears ***only once*** in each term of an equation, it is called a **free index** and the equation holds for all possible values of the index. If this index is 'relabeled' with a new letter, it must be relabeled consistently in every term.
2. If a letter index appears ***exactly twice*** in a term, then summation over all possible values of the index is implied. Such an index is called a **dummy index**. This index can be relabeled with a different letter (as long as the new letter does not conflict with another index), without affecting the other terms in the equation.
3. No letter index may appear ***more than twice*** in a given term.

Some examples help to demonstrate the application of these rules:

- The equation $\boldsymbol{a} = \boldsymbol{b} + \boldsymbol{c}$ can be represented in component form as $a_i = b_i + c_i$. This obviously implies that the equation holds for all values of the free index i (1, 2, 3 in three-dimensional space, or just 1 and 2 in the plane). It would be a violation of rule 1 to change this to $a_i = b_i + c_j$; however, it is perfectly okay to change it to $a_j = b_j + c_j$.
- The dot product of two vectors, $\boldsymbol{a} \cdot \boldsymbol{b}$, can be represented as $a_i b_i$, which implies $a_1 b_1 + a_2 b_2 + a_3 b_3$ by rule 2 (when the dummy index i extends over the range 1, 2, 3). This can be relabeled as $a_j b_j$, provided j does not already appear as a free index or another dummy index in the term.
- Note that multiplication of components is commutative, as it is for any scalar quantities, so we can reorder the components within a term however we find convenient. Thus, $a_i b_i = b_i a_i$.
- By rule 3, the expression $a_i b_i c_i$ is meaningless. For example, the vector expression $\boldsymbol{a} = (\boldsymbol{b} \cdot \boldsymbol{c})\boldsymbol{d}$ may be represented in component form as $a_i = b_j c_j d_i$. One may also relabel the dummy index j with another letter, such as k, to form $a_i = b_k c_k d_i$, but cannot relabel it with the letter i. Such relabeling can often be helpful for proving identities. For example, expression $a_i b_i c_j d_j f_k - a_m b_m c_l d_l f_k$ vanishes, which is only obvious once the dummy indices in the second term are replaced as follows: $m \longrightarrow i$ and $l \longrightarrow j$.

The components of higher-rank tensors have the same number of indices as their rank; for example, A_{ij} are the 9 (4 in two dimensions) components of a rank-two tensor, $\boldsymbol{A}$. There are two tensors of special significance. The first is the **Kronecker delta**, defined as follows:

$$\delta_{ij} := \begin{cases} 1, & i = j \\ 0, & i \neq j \end{cases} \tag{A.1}$$

Note that the Kronecker delta is symmetric with respect to its indices, $\delta_{ij} = \delta_{ji}$. It has the property of 'picking off' a single component of a vector, e.g., $\delta_{ij} v_j = v_i$. To see this clearly, write out the sum completely, when, for example, $i = 2$:

$$\delta_{2j} v_j = \delta_{21} v_1 + \delta_{22} v_2 + \delta_{23} v_3. \tag{A.2}$$

Only the middle term survives, and is equal to v_2. This makes it clear that the Kronecker delta just describes the components of the **identity tensor, 1**. This tensor is isotropic, which means its components are the same in every Cartesian coordinate system.

The cross (or vector) product between two vectors, denoted vectorially by $\boldsymbol{c} = \boldsymbol{a}\times\boldsymbol{b}$, uses another special isotropic tensor, this time of rank 3, called the **permutation symbol,** ϵ_{ijk}. This symbol has the definition

$$\epsilon_{ijk} := \begin{cases} 1, & \text{if } ijk \text{ is an even permutation of } 123 \\ -1, & \text{if } ijk \text{ is an odd permutation of } 123 \\ 0, & \text{if any of } ijk \text{ are equal} \end{cases} \tag{A.3}$$

Note that, for any three values of i, j and k, $\epsilon_{ijk} = \epsilon_{jki} = \epsilon_{kij} = -\epsilon_{ikj} = -\epsilon_{kji} = -\epsilon_{jik}$.

By this definition, the cross-product operation can be denoted in index notation as follows: The ith component of $\boldsymbol{c}$ is

$$c_i = \epsilon_{ijk}a_j b_k. \tag{A.4}$$

The easiest way to see how this gives the usual relationship between components in the cross product is to write it out in full. Note that j and k are dummy indices in this expression. For example, when $i = 1$,

$$\begin{aligned} c_1 = \epsilon_{111}a_1b_1 + \epsilon_{112}a_1b_2 + \epsilon_{113}a_1b_3 \\ + \epsilon_{121}a_2b_1 + \epsilon_{122}a_2b_2 + \epsilon_{123}a_2b_3 \\ + \epsilon_{131}a_3b_1 + \epsilon_{132}a_3b_2 + \epsilon_{133}a_3b_3. \end{aligned} \tag{A.5}$$

By the definition of the permutation symbol, only the sixth and the eighth terms in this expression survive, and the result simplifies considerably to $c_1 = a_2b_3 - a_3b_2$, as expected.

A very useful relationship between the Kronecker delta and the permutation symbol is

$$\epsilon_{ijk}\epsilon_{klm} = \delta_{il}\delta_{jm} - \delta_{im}\delta_{jl}. \tag{A.6}$$

This relationship holds in both two and three dimensions, and is generally used to derive identities involving two cross products: for example, $\boldsymbol{a}\times(\boldsymbol{b}\times\boldsymbol{c}) = (\boldsymbol{a}\cdot\boldsymbol{c})\boldsymbol{b} - (\boldsymbol{a}\cdot\boldsymbol{b})\boldsymbol{c}$. In two dimensions, it is often used with at least one of the three vectors or vector operators oriented in the out-of-plane direction. In such cases, one must take care to keep track of the range of each of the indices in (A.6).

The reader is invited to use index notation to prove the following useful identity for two differentiable vector fields, $\boldsymbol{a}$ and $\boldsymbol{b}$:

$$\nabla\times(\boldsymbol{a}\times\boldsymbol{b}) = \nabla\cdot(\boldsymbol{ba}) - \nabla\cdot(\boldsymbol{ab}) = \boldsymbol{a}(\nabla\cdot\boldsymbol{b}) - \boldsymbol{b}(\nabla\cdot\boldsymbol{a}) + \boldsymbol{b}\cdot\nabla\boldsymbol{a} - \boldsymbol{a}\cdot\nabla\boldsymbol{b}. \tag{A.7}$$

A.1.2 Derivatives of Fundamental Solutions of Laplace's Equation

We can use index notation to develop some useful results concerning the derivatives of the distance r (or its inverse) from the origin to some point at $\boldsymbol{x}$. First, note that we can write the square of this distance, in any dimension, as $r^2 = |\boldsymbol{x}|^2 = x_k x_k$, where k is a dummy index. The most basic result we need in this context is

$$\frac{\partial x_i}{\partial x_j} = \delta_{ij}. \tag{A.8}$$

In other words, the derivative of a coordinate with respect to another coordinate is either 0 if their corresponding axes are different, or 1 if their axes are the same. Then, we can easily obtain the following:

$$\frac{\partial r^2}{\partial x_i} = \frac{\partial}{\partial x_i}(x_k x_k) = 2x_k \frac{\partial x_k}{\partial x_i} = 2x_k \delta_{ki} = 2x_i. \tag{A.9}$$

But we can also use the product rule on this derivative,

$$\frac{\partial r^2}{\partial x_i} = 2r\frac{\partial r}{\partial x_i}. \tag{A.10}$$

Combining these two results, we get the very useful identity

$$\frac{\partial r}{\partial x_i} = \frac{x_i}{r}. \tag{A.11}$$

The right-hand side comprises the components of a unit vector from the origin toward the point.

All of our other results in this discussion make use of (A.11) and the chain rule. For example, in two dimensional contexts, the fundamental solutions of Laplace's equation in external regions are $\log r$ and its derivatives, so it is helpful to calculate a few of these:

$$\frac{\partial}{\partial x_i}(\log r) = \frac{1}{r}\frac{\partial r}{\partial x_i} = \frac{x_i}{r^2}, \tag{A.12}$$

followed by

$$\frac{\partial^2}{\partial x_i \partial x_j}(\log r) = \frac{\partial}{\partial x_i}\left(\frac{x_j}{r^2}\right) = \frac{\delta_{ij}}{r^2} - \frac{2x_i x_j}{r^4}, \tag{A.13}$$

and

$$\frac{\partial^3}{\partial x_i \partial x_j \partial x_k}(\log r) = \frac{\partial}{\partial x_i}\left(\frac{\delta_{jk}}{r^2} - \frac{2x_j x_k}{r^4}\right) = -\frac{2x_i\delta_{jk}}{r^4} - \frac{2x_j\delta_{ki}}{r^4} - \frac{2x_k\delta_{ij}}{r^4} + \frac{8x_i x_j x_k}{r^6}. \tag{A.14}$$

Analogously, in three dimensions, $1/r$ and its derivatives form the fundamental solutions, so we calculate the first few of these:

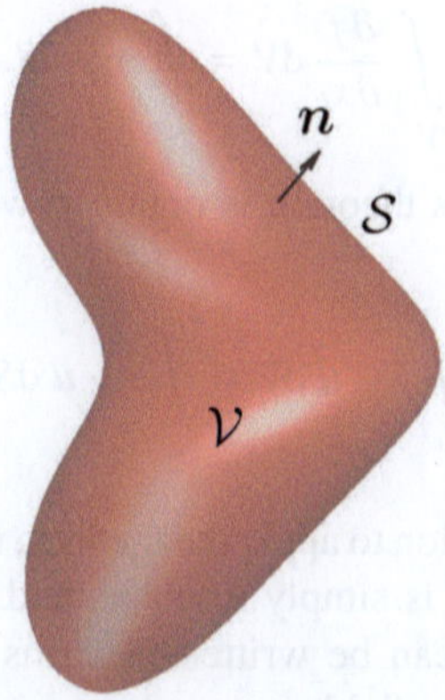

Fig. A.1 Schematic of a volume $\mathcal{V}$ and enclosing surface S

$$\frac{\partial}{\partial x_i}\left(\frac{1}{r}\right) = -\frac{1}{r^2}\frac{\partial r}{\partial x_i} = -\frac{x_i}{r^3}, \tag{A.15}$$

and then,

$$\frac{\partial^2}{\partial x_i \partial x_j}\left(\frac{1}{r}\right) = -\frac{\partial}{\partial x_i}\left(\frac{x_j}{r^3}\right) = -\frac{\delta_{ij}}{r^3} + \frac{3x_i x_j}{r^5}, \tag{A.16}$$

and again,

$$\frac{\partial^3}{\partial x_i \partial x_j \partial x_k}\left(\frac{1}{r}\right) = \frac{\partial}{\partial x_i}\left(-\frac{\delta_{jk}}{r^3} + \frac{3x_j x_k}{r^5}\right) = \frac{3x_i\delta_{jk}}{r^5} + \frac{3x_j\delta_{ki}}{r^5} + \frac{3x_k\delta_{ij}}{r^5} - \frac{15x_i x_j x_k}{r^5}. \tag{A.17}$$

A.1.3 The Divergence Theorem and Some Relevant Uses

Consider a volume $\mathcal{V}$ and an enclosing surface S with outward unit normal $\boldsymbol{n}$, as depicted in Fig. A.1. The divergence theorem allows us to relate integrals over $\mathcal{V}$, involving some differentiable field quantity defined over that volume, to fluxes through the enclosing surface. This is an extremely powerful tool.

The Generalized Divergence Theorem Though we are used to thinking of the divergence theorem applied to vector fields, in the most general form of the divergence theorem this field quantity is a tensor field of any rank. Thus, omitting the indices of this tensor for brevity and simply denoting the field by f, we can state it succinctly as[1]

[1] Strictly speaking, f must be a tensor of rank at least one and have one index that contracts with i. However, if f is a scalar field, one can define an associated vector field fc_i, where c_i is a constant vector, and show that the theorem holds for the scalar for any choice of c_i. A similar approach can be used to extend it to tensors for which there is no contraction with index i.

$$\int_{\mathcal{V}} \frac{\partial f}{\partial x_i}\, \mathrm{d}V = \int_{S} n_i f\, \mathrm{d}S. \tag{A.18}$$

The most common use of this theorem is obtained when f is a vector field, $\boldsymbol{u}$, so that it takes the familiar form

$$\int_{\mathcal{V}} \nabla \cdot \boldsymbol{u}\, \mathrm{d}V = \int_{S} \boldsymbol{n} \cdot \boldsymbol{u}\, \mathrm{d}S. \tag{A.19}$$

However, we often have occasion to apply the theorem in less familiar circumstances. For example, suppose that f is simply a scalar field. Then (A.18) shows that the gradient of this scalar field can be written in terms of a vector field $f\boldsymbol{n}$ on the surface. If f is the rank-2 tensor field $\epsilon_{kij}u_j$, where u_j are the components of a vector field, then, when the divergence theorem is applied, it is easy to show that the result is the kth component of the vector-valued identity

$$\int_{\mathcal{V}} \nabla \times \boldsymbol{u}\, \mathrm{d}V = \int_{S} \boldsymbol{n} \times \boldsymbol{u}\, \mathrm{d}S. \tag{A.20}$$

For example, this can be used to relate the vorticity over the region $\mathcal{V}$ to the tangential components of velocity on the enclosing surface. This can be extended to higher moments of the vorticity. Let's apply the divergence theorem to the rank-$n+2$ tensor $x_{i_1}x_{i_2}\cdots x_{i_n}\epsilon_{kij}u_j$. The result is the rank-$(n+1)$ identity

$$\int_{\mathcal{V}} x_{i_1}x_{i_2}\cdots x_{i_n}(\nabla \times \boldsymbol{u})_k\, \mathrm{d}V = \int_{S} x_{i_1}x_{i_2}\cdots x_{i_n}(\boldsymbol{n}\times\boldsymbol{u})_k\, \mathrm{d}S + \epsilon_{ki_1j}\int_{\mathcal{V}} x_{i_2}\cdots x_{i_n}u_j\, \mathrm{d}V$$
$$+ \cdots + \epsilon_{ki_nj}\int_{\mathcal{V}} x_{i_1}x_{i_2}\cdots x_{i_{n-1}}u_j\, \mathrm{d}V. \tag{A.21}$$

Some Useful Geometric Results of Volumes and Surfaces The divergence theorem can be used to prove some useful results on integrals over volumes or their enclosing surfaces. Most basically, the integral of the unit normal $\boldsymbol{n}$ of a surface is zero when integrated over the whole surface, as can be proved by applying (A.18) with $f \to 1$:

$$\int_{S} \boldsymbol{n}\, \mathrm{d}S = \int_{\mathcal{V}} \nabla(1)\, \mathrm{d}V = 0. \tag{A.22}$$

If, instead, f is, say, x_j, the jth component of the position vector, $\boldsymbol{x}$, in the inertial coordinate system, then the same form of the divergence theorem leads to

$$\int_{S} x_j\boldsymbol{n}\, \mathrm{d}S = \int_{\mathcal{V}} \nabla x_j\, \mathrm{d}V = \mathcal{V}\boldsymbol{e}_j. \tag{A.23}$$

Similarly, using result (A.20) with $\boldsymbol{u} \to \boldsymbol{x}$, it is simple to show that the moment of the normal over the surface is also zero:

$$\int_{S} \boldsymbol{x} \times \boldsymbol{n} \, \mathrm{d}S = -\int_{\mathcal{V}} \nabla \times \boldsymbol{x} \, \mathrm{d}V = 0, \tag{A.24}$$

since $\nabla \times \boldsymbol{x} \equiv 0$. If we apply the same identity with $\boldsymbol{u} \to x_j \boldsymbol{x}$, where x_j is the jth component of $\boldsymbol{x}$, then we get

$$\int_{S} x_j \boldsymbol{x} \times \boldsymbol{n} \, \mathrm{d}S = -\int_{\mathcal{V}} \nabla \times (x_j \boldsymbol{x}) \, \mathrm{d}V = -\boldsymbol{e}_j \times \int_{\mathcal{V}} \boldsymbol{x} \, \mathrm{d}V. \tag{A.25}$$

Note that the final result contains the definition of the centroid of $\mathcal{V}$, given in Eq. (2.2).

Now, suppose we use result (A.19) with $\boldsymbol{u} \to \boldsymbol{x}$. Then, if we note that $\nabla \cdot \boldsymbol{x} = n_{\mathrm{d}}$, where n_{d} is the number of spatial dimensions, we obtain

$$\mathcal{V} = \frac{1}{n_{\mathrm{d}}} \int_{S} \boldsymbol{x} \cdot \boldsymbol{n} \, \mathrm{d}S, \tag{A.26}$$

where $\mathcal{V}$ is used here to denote the volume (or area in two dimensions) of the region of the same symbol. This formula is useful for computing the volume of a body in terms of only its surface structure. This can be generalized: The nth moment tensor of $\mathcal{V}$ can be calculated by replacing f in (A.18) with the rank-$n+1$ tensor $x_i x_{i_1} x_{i_2} \cdots x_{i_n}$. It is easy to show that

$$\frac{\partial}{\partial x_i}(x_i x_{i_1} x_{i_2} \cdots x_{i_n}) = (n_{\mathrm{d}} + n) x_{i_1} x_{i_2} \cdots x_{i_n}, \tag{A.27}$$

from which we can obtain

$$\int_{\mathcal{V}} x_{i_1} x_{i_2} \cdots x_{i_n} \, \mathrm{d}V = \frac{1}{n_{\mathrm{d}} + n} \int_{S} x_{i_1} x_{i_2} \cdots x_{i_n} \boldsymbol{n} \cdot \boldsymbol{x} \, \mathrm{d}S. \tag{A.28}$$

This generalized formula provides the centroid of $\mathcal{V}$ when $n = 1$.

A slightly different surface integral for the centroid can be obtained when f in (A.18) represents the scalar field $\boldsymbol{x} \cdot \boldsymbol{x}$. Then, it is easy to show that

$$\int_{\mathcal{V}} \boldsymbol{x} \, \mathrm{d}V = \frac{1}{2} \int_{S} (\boldsymbol{x} \cdot \boldsymbol{x}) \boldsymbol{n} \, \mathrm{d}S, \tag{A.29}$$

since $\nabla(\boldsymbol{x} \cdot \boldsymbol{x}) = 2\boldsymbol{x}$. Though the integrand of the surface integral in (A.29) is different from (A.28) with $n = 1$, the result of the integral is the same.

The identity (A.28) can also be used to write the moment of inertia tensor of a body in a surface integral form (provided the body's density is uniform). We invite the reader to show that this leads to the following expression for the ijth component, based on the volume integral definition in (2.4):

$$I_{O,ij} = -\boldsymbol{e}_i \cdot \frac{\rho_\mathrm{b}}{n_\mathrm{d}+2} \int_{\mathcal{S}_\mathrm{b}} \boldsymbol{x} \times \left[\boldsymbol{x} \times \left(\boldsymbol{n} \times (\boldsymbol{e}_j \times \boldsymbol{x})\right)\right] \,\mathrm{d}S. \tag{A.30}$$

The Divergence Theorem and Green's Function A powerful use of the divergence theorem is achieved by applying it to differential operations of f on $\mathcal{V}$ in combination with the so-called **Green's function** (or simply, fundamental solution) of the operator. This enables us to develop a corollary (or, in a sense, a solution) of a differential equation in region $\mathcal{V}$ in terms of integrals over $\mathcal{V}$ and its enclosing surface $\mathcal{S}$. Our interest here is focused only on the negative Laplace operator, $-\nabla^2$, and its scalar-valued Green's function G is defined as the solution of

$$-\nabla^2 G(\boldsymbol{x}) = \delta(\boldsymbol{x}) \tag{A.31}$$

in an unbounded region, where $\delta(\boldsymbol{x})$ is the Dirac delta function. The Green's function, $G(\boldsymbol{x})$, represents the influence at a 'target' (or 'evaluation') point $\boldsymbol{x}$ of a singular 'source' based at the origin. Its specific form is given by (3.67) and (3.110) in two and three dimensions, respectively. If the source is shifted to any other point $\boldsymbol{y}$, the influence is simply described by $G(\boldsymbol{x}-\boldsymbol{y})$. It is straightforward to show that the Green's function is invariant to a switch of the source and target points, $G(\boldsymbol{x}-\boldsymbol{y}) = G(\boldsymbol{y}-\boldsymbol{x})$, since it only depends on the distance between them. The basic property of the delta function that we will make use of is

$$\int_{\mathcal{V}} f(\boldsymbol{y})\delta(\boldsymbol{x}-\boldsymbol{y})\,\mathrm{d}V(\boldsymbol{y}) = \begin{cases} f(\boldsymbol{x}), & \boldsymbol{x} \in \mathcal{V}, \\ 0, & \boldsymbol{x} \notin \mathcal{V} \cup \mathcal{S}, \end{cases} \tag{A.32}$$

where f is a function that is continuous and has compact support. Note that we use the notation $\mathrm{d}V(\boldsymbol{y})$ to signify that $\boldsymbol{y}$ is the integration variable. We will use this throughout this work, for spatial integrals of any sort, where the integration variable might be otherwise ambiguous.

First, let's use this Green's function to derive an integral corollary of the scalar-valued Poisson equation

$$\nabla^2 \varphi = \Theta \tag{A.33}$$

in $\mathcal{V}$, for a scalar forcing field Θ. Using index notation, we start with the basic identity

$$\frac{\partial}{\partial x_j}\left(G(\boldsymbol{x}-\boldsymbol{y})\frac{\partial \varphi}{\partial x_j}(\boldsymbol{x}) - \varphi(\boldsymbol{x})\frac{\partial G}{\partial x_j}(\boldsymbol{x}-\boldsymbol{y})\right) = \\ \frac{\partial^2 \varphi}{\partial x_j \partial x_j}(\boldsymbol{x}) G(\boldsymbol{x}-\boldsymbol{y}) - \varphi(\boldsymbol{x})\frac{\partial^2 G}{\partial x_j \partial x_j}(\boldsymbol{x}-\boldsymbol{y}). \tag{A.34}$$

The Laplacians on the right-hand side can be replaced by the respective forcing functions. We integrate over the region $\mathcal{V}$ and apply the divergence theorem to the integral on the left-hand side, arriving at

$$\int_S n_j \left(G(\boldsymbol{x}-\boldsymbol{y}) \frac{\partial \varphi}{\partial x_j}(\boldsymbol{x}) - \varphi(\boldsymbol{x}) \frac{\partial G}{\partial x_j}(\boldsymbol{x}-\boldsymbol{y}) \right) \mathrm{d}S(\boldsymbol{x}) =$$

$$\int_{\mathcal{V}} \Theta(\boldsymbol{x}) G(\boldsymbol{x}-\boldsymbol{y})\, \mathrm{d}V(\boldsymbol{x}) + \int_{\mathcal{V}} \varphi(\boldsymbol{x}) \delta(\boldsymbol{x}-\boldsymbol{y})\, \mathrm{d}V(\boldsymbol{x}). \tag{A.35}$$

Using the property (A.32) and switching the names of the source and target points, $\boldsymbol{x}$ and $\boldsymbol{y}$, we arrive at our desired result, which we write in vector notation (with ∇_y to denote the gradient with respect to the source point $\boldsymbol{y}$ and $\boldsymbol{n}_y$ to denote the local unit normal at $\boldsymbol{y}$):

$$\left.\begin{array}{ll} \boldsymbol{x} \in \mathcal{V}, & \varphi(\boldsymbol{x}) \\ \boldsymbol{x} \notin \mathcal{V} \cup \mathcal{S}, & 0 \end{array}\right\} = -\int_{\mathcal{V}} \Theta(\boldsymbol{y}) G(\boldsymbol{x}-\boldsymbol{y})\, \mathrm{d}V(\boldsymbol{y})$$
$$+ \int_{\mathcal{S}} \boldsymbol{n}_y \cdot \left[G(\boldsymbol{x}-\boldsymbol{y}) \nabla_y \varphi(\boldsymbol{y}) - \varphi(\boldsymbol{y}) \nabla_y G(\boldsymbol{x}-\boldsymbol{y}) \right] \mathrm{d}S(\boldsymbol{y}). \tag{A.36}$$

This identity, sometimes called **Green's theorem**, shows us that the value of φ at any target point in $\mathcal{V}$ can be obtained by adding the effects from three contributions: the forcing field Θ in $\mathcal{V}$ as well as φ and its normal derivative on the surface. The Green's function and its gradient measure the influences of these contributions at the target: monopole sources distributed throughout $\mathcal{V}$ (of local strength $\Theta\, \mathrm{d}V$) and over the surface (of strength $(\partial\varphi/\partial n)\, \mathrm{d}S$) and dipolar sources on the surface (of strength $\varphi \boldsymbol{n}\, \mathrm{d}S$). Equation (A.36) also shows us that, at any target point outside of $\mathcal{V}$, the effect of the volumetric forcing field Θ can be substituted with monopoles and dipoles distributed over the enclosing surface $\mathcal{S}$.

Equation (A.36) can also be used for a vector field $\boldsymbol{u}$ instead of a scalar field, since it clearly holds on a component-wise basis:

$$\left.\begin{array}{ll} \boldsymbol{x} \in \mathcal{V}, & u_i(\boldsymbol{x}) \\ \boldsymbol{x} \notin \mathcal{V} \cup \mathcal{S}, & 0 \end{array}\right\} = -\int_{\mathcal{V}} f_i(\boldsymbol{y}) G(\boldsymbol{x}-\boldsymbol{y})\, \mathrm{d}V(\boldsymbol{y})$$
$$+ \int_{\mathcal{S}} \boldsymbol{n}_y \cdot \left[G(\boldsymbol{x}-\boldsymbol{y}) \nabla_y u_i(\boldsymbol{y}) - u_i(\boldsymbol{y}) \nabla_y G(\boldsymbol{x}-\boldsymbol{y}) \right] \mathrm{d}S(\boldsymbol{y}), \tag{A.37}$$

where f_i is the ith component of the forcing function in the vector-valued Poisson equation for $\boldsymbol{u}$ in $\mathcal{V}$:

$$\nabla^2 \boldsymbol{u} = \boldsymbol{f}. \tag{A.38}$$

Using (A.37) as a starting point, we can obtain a particularly useful integral corollary to the vector identity

$$\nabla^2 \boldsymbol{u} = -\nabla \times (\nabla \times \boldsymbol{u}) + \nabla(\nabla \cdot \boldsymbol{u}), \tag{A.39}$$

where the terms on the right-hand side are interpreted as the forcing term $\boldsymbol{f}$. To derive this integral form, note that the volume integral of the forcing term in (A.37), substituted with this particular form of the forcing, can be integrated by parts, which we carry out here in index notation:

$$\int_{\mathcal{V}} \left(\epsilon_{ijk}\epsilon_{klm} \frac{\partial}{\partial y_j}\frac{\partial}{\partial y_l} u_m - \frac{\partial}{\partial y_i}\frac{\partial}{\partial y_j} u_j \right) G(\boldsymbol{x}-\boldsymbol{y})\,\mathrm{d}V(\boldsymbol{y})$$
$$= \epsilon_{ijk} \int_{\mathcal{V}} \frac{\partial}{\partial y_j}\left(\epsilon_{klm}\frac{\partial u_m}{\partial y_l} G(\boldsymbol{x}-\boldsymbol{y}) \right) \mathrm{d}V(\boldsymbol{y}) - \int_{\mathcal{V}} \frac{\partial}{\partial y_i}\left(\frac{\partial u_j}{\partial y_j} G(\boldsymbol{x}-\boldsymbol{y}) \right) \mathrm{d}V(\boldsymbol{y})$$
$$- \epsilon_{ijk} \int_{\mathcal{V}} \epsilon_{klm} \frac{\partial u_m}{\partial y_l}\frac{\partial G}{\partial y_j}(\boldsymbol{x}-\boldsymbol{y})\,\mathrm{d}V(\boldsymbol{y}) + \int_{\mathcal{V}} \frac{\partial u_j}{\partial y_j}\frac{\partial G}{\partial y_i}(\boldsymbol{x}-\boldsymbol{y})\,\mathrm{d}V(\boldsymbol{y}). \tag{A.40}$$

To the first two terms on the right-hand side of (A.40), we apply the divergence theorem to transform them into surface integrals. These are combined with the other surface integrals in (A.37), resulting in the following:

$$\int_{S} \left[\left(\epsilon_{ijk}\epsilon_{klm} n_{y,j} \frac{\partial u_m}{\partial y_l} - n_{y,i}\frac{\partial u_j}{\partial y_j} + n_{y,j}\frac{\partial u_i}{\partial y_j} \right) G(\boldsymbol{x}-\boldsymbol{y}) \right.$$
$$\left. - n_{y,j} u_i(\boldsymbol{y}) \frac{\partial G}{\partial y_j}(\boldsymbol{x}-\boldsymbol{y}) \right] \mathrm{d}S(\boldsymbol{y}). \tag{A.41}$$

Among the terms multiplying $G(\boldsymbol{x}-\boldsymbol{y})$, the double cross product can be expanded and combined with the other two terms. The result of this is manipulated further with a product rule in order to transfer the derivatives to G, which leads to

$$\int_{S} \left(-n_{y,j} u_j(\boldsymbol{y}) \frac{\partial G}{\partial y_i}(\boldsymbol{x}-\boldsymbol{y}) + n_{y,i} u_j(\boldsymbol{y}) \frac{\partial G}{\partial y_j}(\boldsymbol{x}-\boldsymbol{y}) \right.$$
$$\left. - n_{y,j} u_i(\boldsymbol{y}) \frac{\partial G}{\partial y_j}(\boldsymbol{x}-\boldsymbol{y}) \right) \mathrm{d}S(\boldsymbol{y}). \tag{A.42}$$

In each of these terms, as well as the remaining volume integrals on the right-hand side of (A.40), we replace the gradient of G with respect to the source point with the gradient with respect to the target: $\partial G(\boldsymbol{x}-\boldsymbol{y})/\partial y_i = -\partial G(\boldsymbol{x}-\boldsymbol{y})/\partial x_i$. And using the notation applied elsewhere in this book, we denote this gradient of the Green's function with respect to the target, ∇G, by $\boldsymbol{K}$. Thus, we arrive at our desired result, which we express in vector form:

$$\left.\begin{array}{ll} x \in \mathcal{V}, & u(x) \\ x \notin \mathcal{V} \cup \mathcal{S}, & 0 \end{array}\right\} = \int_{\mathcal{V}} \left[K(x-y) \times (\nabla_y \times u(y)) - K(x-y)(\nabla_y \cdot u(y)) \right] \mathrm{d}V(y)$$
$$- \int_{\mathcal{S}} \left[K(x-y) \times (n_y \times u(y)) - K(x-y)(n_y \cdot u(y)) \right] \mathrm{d}S(y). \tag{A.43}$$

Thus, the vector field u in $\mathcal{V}$ can be obtained from the collective contributions of its curl and divergence throughout the region, plus its tangential and normal components distributed over the enclosing surface. And at target points outside $\mathcal{V}$, the effects of the curl and divergence distributions are equivalently represented by surface distributions of the components of u.

This result (A.43) is quite general and has broad uses. When u is the velocity field of the fluid, then it states that this field can be recovered from the vorticity and rate of dilatation in the fluid, plus contributions from any bounding surfaces. This is a generalization of the Biot–Savart integral. Note that the surface integral is a composition of vortices of strength $-n \times u\, \mathrm{d}S$ and monopole sources of strength $-n \cdot u\, \mathrm{d}S$.

If the vector field is a potential flow (i.e., can be written as the gradient of a scalar potential that satisfies Laplace's equation everywhere), then the vector field anywhere in the region $\mathcal{V}$ can be recovered entirely from its values on the enclosing surface. In particular, if the vector field is spatially uniform, c, then it leads to the special result

$$\left.\begin{array}{ll} x \in \mathcal{V}, & c \\ x \notin \mathcal{V} \cup \mathcal{S}, & 0 \end{array}\right\} = - \int_{\mathcal{S}} \left[K(x-y) \times (n_y \times c) - K(x-y)(n_y \cdot c) \right] \mathrm{d}S(y). \tag{A.44}$$

Evaluation on the Surface In application of these integral theorems, it is important to know how the surface integrals behave as the target point approaches the surface. We distinguish approaches from the **positive side** (the side into which the normal is directed) and the **negative side** (the side opposite the positive) by the superscripts + and −, respectively. In the process of the target approaching a surface point $x \in \mathcal{S}$, the surface integral is naturally split into two parts: one part due to a small circular patch $\mathcal{S}_\epsilon$ of vanishingly small radius ϵ surrounding x and another part due to the remainder of the surface. This latter part has no issues with singular behavior as the patch's radius shrinks to zero and has a value that is independent of the direction from which the surface is approached; it is called the (Cauchy) principal value of the integral. In contrast, the first part has a singular, but integrable, behavior, whose effect can be shown to be

$$\lim_{\epsilon \to 0} \lim_{x \to x^\pm} \int_{\mathcal{S}_\epsilon} G(x-y)\, \mathrm{d}S(y) = 0, \qquad \lim_{\epsilon \to 0} \lim_{x \to x^\pm} \int_{\mathcal{S}_\epsilon} \nabla G(x-y)\, \mathrm{d}S(y) = \mp \frac{1}{2} n(x). \tag{A.45}$$

Thus, for some continuous function f on the surface $\mathcal{S}$,

$$\lim_{\boldsymbol{x}\to\boldsymbol{x}^{\pm}} \int\limits_{\mathcal{S}} \boldsymbol{K}(\boldsymbol{x}-\boldsymbol{y}) f(\boldsymbol{y})\, \mathrm{d}S = \mp\frac{1}{2}\boldsymbol{n}(\boldsymbol{x}) f(\boldsymbol{x}) + \fint\limits_{\mathcal{S}} \boldsymbol{K}(\boldsymbol{x}-\boldsymbol{y}) f(\boldsymbol{y})\, \mathrm{d}S, \tag{A.46}$$

where the integral with the dash through it on the right-hand side is our notation for principal value integral. It can be seen that the principal value of the integral represents the simple average of the limits obtained by approaching the surface from either side. This interpretation is sometimes useful for calculation purposes.

We can apply these general results to inspect the limits of the integral expressions developed above. For example, when the limit is applied to the surface integral in Eq. (A.36), the leading term is $\mp\boldsymbol{n}\cdot\boldsymbol{n}\varphi/2$ (where we used the fact that $\nabla_y G(\boldsymbol{x}-\boldsymbol{y}) = -\nabla G(\boldsymbol{x}-\boldsymbol{y})$). Thus, the surface form is

$$\mp\frac{1}{2}\varphi(\boldsymbol{x}) + \fint\limits_{\mathcal{S}} \boldsymbol{n}_y \cdot \left[G(\boldsymbol{x}-\boldsymbol{y})\nabla_y\varphi(\boldsymbol{y}) - \varphi(\boldsymbol{y})\nabla_y G(\boldsymbol{x}-\boldsymbol{y})\right] \mathrm{d}S(\boldsymbol{y}). \tag{A.47}$$

When this limiting form is used in the full expression (A.36), we get the following equation on the surface, independent of the direction of approach:

$$\begin{aligned}\frac{1}{2}\varphi(\boldsymbol{x}) = &-\int\limits_{\mathcal{V}} \Theta(\boldsymbol{y}) G(\boldsymbol{x}-\boldsymbol{y})\, \mathrm{d}V(\boldsymbol{y}) \\ &+ \fint\limits_{\mathcal{S}} \boldsymbol{n}_y \cdot \left[G(\boldsymbol{x}-\boldsymbol{y})\nabla_y\varphi(\boldsymbol{y}) - \varphi(\boldsymbol{y})\nabla_y G(\boldsymbol{x}-\boldsymbol{y})\right] \mathrm{d}S(\boldsymbol{y}),\end{aligned} \tag{A.48}$$

for all $\boldsymbol{x} \in \mathcal{S}$. This could be used, for example, to solve for the surface distribution of φ when the forcing field Θ in $\mathcal{V}$ and the normal derivative $\boldsymbol{n}\cdot\nabla\varphi$ on the surface are known, and the results would then be available to determine φ throughout $\mathcal{V}$ with (A.36). (Note, however, that the solution to this so-called **Neumann problem** for φ is not unique, since we can always add a constant to φ and still satisfy the governing equation and boundary condition.) Let's now combine our results from the previous two discussions into a single useful form:

Result A.1: Scalar Form of Green's Theorem

Consider a region $\mathcal{V}$ bounded by surface $\mathcal{S}$, on which the unit normal vector $\boldsymbol{n}$ is directed away from $\mathcal{V}$. Then, for a twice-differentiable scalar field φ defined throughout $\mathcal{V}$, in which $\nabla^2\varphi = \Theta$, the following identity holds

$$\left.\begin{array}{ll} x \in \mathcal{V}, & \varphi(x) \\ x \in \mathcal{S}, & \frac{1}{2}\varphi(x) \\ x \notin \mathcal{V} \cup \mathcal{S}, & 0 \end{array}\right\} = -\int_{\mathcal{V}} \Theta(y) G(x-y)\, \mathrm{d}V(y)$$

$$+ \oint_{\mathcal{S}} n_y \cdot \left[G(x-y)\nabla_y \varphi(y) - \varphi(y)\nabla_y G(x-y)\right] \mathrm{d}S(y). \tag{A.49}$$

Note that the Cauchy principal value of the surface integral is taken when x lies on $\mathcal{S}$.

When the limit (A.46) is applied to the surface integral in (A.43), the result is

$$\mp \frac{1}{2}\left[-n \times (n \times u) + n(n \cdot u)\right]_x$$
$$- \oint_{\mathcal{S}} \left[K(x-y) \times (n_y \times u(y)) - K(x-y)(n_y \cdot u(y))\right] \mathrm{d}S(y). \tag{A.50}$$

The leading term is simply $\mp u(x)/2$. Thus, we find that, regardless of the direction of approach, the value of the vector field u at a point on the surface can be related to the curl and divergence in the adjacent volume and to the values of u elsewhere on the surface:

$$\frac{1}{2}u(x) = \int_{\mathcal{V}} \left[K(x-y) \times (\nabla_y \times u(y)) - K(x-y)(\nabla_y \cdot u(y))\right] \mathrm{d}V(y)$$
$$- \oint_{\mathcal{S}} \left[K(x-y) \times (n_y \times u(y)) - K(x-y)(n_y \cdot u(y))\right] \mathrm{d}S(y). \tag{A.51}$$

In particular, for a constant vector field c, we now have the useful result that the principal value of the surface integral of this field is equal to $c/2$:

$$\frac{1}{2}c = -\oint_{\mathcal{S}} \left[K(x-y) \times (n_y \times c) - K(x-y)(n_y \cdot c)\right] \mathrm{d}S(y). \tag{A.52}$$

Let's consolidate these results for a vector field into one compact statement:

Result A.2: Vector Form of Green's Theorem

Consider a region $\mathcal{V}$ bounded by surface $\mathcal{S}$, on which the unit normal vector n is directed away from $\mathcal{V}$. Then, for a twice-differentiable vector field u defined throughout $\mathcal{V}$, the following identity holds

$$\left.\begin{array}{ll} x \in \mathcal{V}, & u(x) \\ x \in \mathcal{S}, & \frac{1}{2}u(x) \\ x \notin \mathcal{V} \cup \mathcal{S}, & 0 \end{array}\right\} = \tag{A.53}$$

$$\int_{\mathcal{V}} \left[K(x-y)\times(\nabla_y \times u(y)) - K(x-y)(\nabla_y \cdot u(y))\right] \mathrm{d}V(y)$$

$$-\oint_{\mathcal{S}} \left[K(x-y)\times(n_y \times u(y)) - K(x-y)(n_y \cdot u(y))\right] \mathrm{d}S(y).$$

Note that the Cauchy principal value of the surface integral is taken when x lies on $\mathcal{S}$.

Combining Green's Theorem in Interior and Exterior Regions In the previous discussion, we focused on obtaining a scalar or vector field in a single volumetric region from the data on the bounding surface. It should be clear that, although Fig. A.1 suggests that this region $\mathcal{V}$ constitutes the interior of the surface $\mathcal{S}$, we never actually needed to specify this. Indeed, Results A.1 and A.2 hold even if the volumetric region is exterior to the surface, as long as the normal on $\mathcal{S}$ is directed away from the region; if the normal is directed inward, the sign on the surface integral terms simply switches. The only caveat of applying Green's theorem in the exterior region is that we must apply a condition on the solution at infinity. In this work, ***we will always insist that the solution must vanish at infinity***. In other words, we do not admit uniform non-zero solutions to problems in exterior regions. This requirement is suitably general, however.

By applying Green's theorem in both the interior and the exterior regions relative to a surface $\mathcal{S}$, we can obtain new and useful forms of the theorem. Let's label these regions as shown in Fig. A.2, with the surface normal directed into $\mathcal{V}^+$. Then, we can immediately obtain the following results for scalar and vector fields in the extended region $\mathcal{V}^+ \cup \mathcal{S} \cup \mathcal{V}^-$ by simply adding the identities from the respective regions:

Result A.3: Scalar Form of Extended Green's Theorem

Consider a closed surface $\mathcal{S}$, between a region $\mathcal{V}^-$ and another region $\mathcal{V}^+$, and on which the unit normal vector n is directed toward $\mathcal{V}^+$. In each region there is a twice-differentiable scalar field, φ^- and φ^+, respectively, each of which satisfies $\nabla^2\varphi^\pm = \Theta^\pm$ in its respective region. Then, the following identity holds

$$\overline{\varphi}(\boldsymbol{x}) = -\int_{\mathcal{V}^-} \Theta^-(\boldsymbol{y})G(\boldsymbol{x}-\boldsymbol{y})\,\mathrm{d}V(\boldsymbol{y}) - \int_{\mathcal{V}^+} \Theta^+(\boldsymbol{y})G(\boldsymbol{x}-\boldsymbol{y})\,\mathrm{d}V(\boldsymbol{y}) \quad \text{(A.54)}$$

$$-\oint_{\mathcal{S}} \boldsymbol{n}_y \cdot \left[G(\boldsymbol{x}-\boldsymbol{y})\nabla_y[\varphi]_-^+(\boldsymbol{y}) - [\varphi]_-^+(\boldsymbol{y})\nabla_y G(\boldsymbol{x}-\boldsymbol{y})\right]\,\mathrm{d}S(\boldsymbol{y})$$

where $[\varphi]_-^+ := \varphi^+ - \varphi^-$ and

$$\overline{\varphi}(\boldsymbol{x}) := \begin{cases} \varphi^-(\boldsymbol{x}), & \boldsymbol{x} \in \mathcal{V}^-, \\ \frac{1}{2}\left(\varphi^-(\boldsymbol{x}) + \varphi^+(\boldsymbol{x})\right), & \boldsymbol{x} \in \mathcal{S}, \\ \varphi^+(\boldsymbol{x}), & \boldsymbol{x} \in \mathcal{V}^+. \end{cases} \quad \text{(A.55)}$$

Note that the Cauchy principal value of the surface integral is taken when $\boldsymbol{x}$ lies on $\mathcal{S}$.

Similarly, we can generate a vector form of this theorem from Result A.2:

Result A.4: Vector Form of Extended Green's Theorem

Consider a closed surface $\mathcal{S}$, between a region $\mathcal{V}^-$ and another region $\mathcal{V}^+$, and on which the unit normal vector $\boldsymbol{n}$ is directed toward $\mathcal{V}^+$. In each region there is a twice-differentiable vector field, $\boldsymbol{u}^-$ and $\boldsymbol{u}^+$, respectively. The following identity holds

$$\overline{\boldsymbol{u}}(\boldsymbol{x}) = \int_{\mathcal{V}^-} \left[\boldsymbol{K}(\boldsymbol{x}-\boldsymbol{y}) \times (\nabla_y \times \boldsymbol{u}^-(\boldsymbol{y})) - \boldsymbol{K}(\boldsymbol{x}-\boldsymbol{y})(\nabla_y \cdot \boldsymbol{u}^-(\boldsymbol{y}))\right]\,\mathrm{d}V(\boldsymbol{y})$$

$$+\int_{\mathcal{V}^+} \left[\boldsymbol{K}(\boldsymbol{x}-\boldsymbol{y}) \times (\nabla_y \times \boldsymbol{u}^+(\boldsymbol{y})) - \boldsymbol{K}(\boldsymbol{x}-\boldsymbol{y})(\nabla_y \cdot \boldsymbol{u}^+(\boldsymbol{y}))\right]\,\mathrm{d}V(\boldsymbol{y})$$

$$+\oint_{\mathcal{S}} \left[\boldsymbol{K}(\boldsymbol{x}-\boldsymbol{y}) \times (\boldsymbol{n}_y \times [\boldsymbol{u}]_-^+(\boldsymbol{y})) - \boldsymbol{K}(\boldsymbol{x}-\boldsymbol{y})(\boldsymbol{n}_y \cdot [\boldsymbol{u}]_-^+(\boldsymbol{y}))\right]\,\mathrm{d}S(\boldsymbol{y})$$

(A.56)

where

$$\overline{\boldsymbol{u}}(\boldsymbol{x}) := \begin{cases} \boldsymbol{u}^-(\boldsymbol{x}), & \boldsymbol{x} \in \mathcal{V}^-, \\ \frac{1}{2}\left(\boldsymbol{u}^-(\boldsymbol{x}) + \boldsymbol{u}^+(\boldsymbol{x})\right), & \boldsymbol{x} \in \mathcal{S}, \\ \boldsymbol{u}^+(\boldsymbol{x}), & \boldsymbol{x} \in \mathcal{V}^+. \end{cases} \quad \text{(A.57)}$$

Note that the Cauchy principal value of the surface integral is taken when $\boldsymbol{x}$ lies on $\mathcal{S}$.

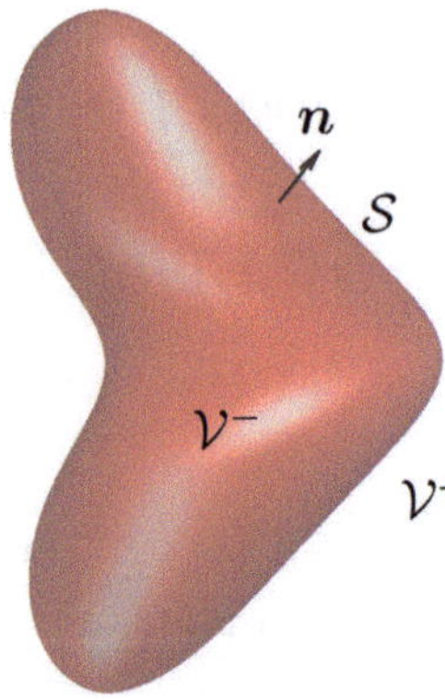

Fig. A.2 Schematic of an interior volume $\mathcal{V}^-$, exterior volume $\mathcal{V}^+$ and bounding surface $\mathcal{S}$

These forms of Green's theorem offer some flexibility in solving problems. Since we are often focused on solving a Poisson problem (A.33) or (A.39) in only one of the two regions bordering $\mathcal{S}$, we have some freedom in choosing the solution in the other region, and in particular, what boundary conditions, if any, it satisfies on $\mathcal{S}$. For example, such a condition might be applied on the relationship between the solutions in each domain. In a Poisson problem, we can only apply at most one condition on the solution on a given boundary. However, since there are two solutions adjacent to $\mathcal{S}$, then we can apply up to two conditions on their relationship on $\mathcal{S}$, provided there are no other conditions enforced on either of them on $\mathcal{S}$.

Let us use this flexibility to develop forms of Green's theorem for a surface $\mathcal{P}$ that isn't closed. We can always consider such a surface to be a portion of a larger surface $\mathcal{S}$ that is closed and that separates two regions $\mathcal{V}^+$ and $\mathcal{V}^-$, as in Fig. A.2. On the remainder of this larger surface, $\mathcal{S} \setminus \mathcal{P}$, we are free to set two conditions on the relationship between the solutions in each region.

Helmholtz Decomposition Suppose we have a known vector field $\boldsymbol{u}$ in the usual region $\mathcal{V}$ bounded by surface $\mathcal{S}$. The Helmholtz (or Helmholtz–Hodge) decomposition theorem states that this vector field can be decomposed into two parts: the gradient of a scalar potential field, φ, and the curl of a vector potential field, $\boldsymbol{\Psi}$, viz.

$$\boldsymbol{u} = \nabla\varphi + \nabla \times \boldsymbol{\Psi}. \tag{A.58}$$

The proof of this theorem follows in straightforward fashion from Eq. (A.43). That equation, expressing $\boldsymbol{u}(\boldsymbol{x})$ at any point $\boldsymbol{x}$ in $\mathcal{V}$, can readily be rewritten as

$$u(x) = \nabla\left[-\int_V G(x-y)(\nabla_y \cdot u(y))\,\mathrm{d}V(y) + \int_S G(x-y)(n_y \cdot u(y))\,\mathrm{d}S(y)\right] +$$
$$\nabla \times \left[\int_V G(x-y)(\nabla_y \times u(y))\,\mathrm{d}V(y) - \int_S G(x-y)(n_y \times u(y))\,\mathrm{d}S(y)\right]. \tag{A.59}$$

This form makes it obvious that φ can be defined as the first term in brackets and $\boldsymbol{\Psi}$ the second term in brackets.

A.1.4 Stokes' Theorem and Some Relevant Uses

Just as we relate an integral over a volume to one over an enclosing surface, there are many opportunities in this work to interchange an integral over a surface S with an integral over an enclosing contour C. Consider the schematic of such a situation in the left panel of Fig. A.3. It is important to note that the surface here is not closed, but rather, has an edge defined by C. A two-dimensional analog of the surface and contour is depicted in the right panel of Fig. A.3, in which the surface is a curve and the enclosing contour consists of two end points with respective tangents $\pm e_3$, as shown.

Generalized Stokes' Theorem Stokes' theorem relates integrals over S and C. Consider a continuous vector field u on the surface S. Then Stokes' theorem is

$$\int_S (\nabla \times u) \cdot n\,\mathrm{d}S = \int_C u \cdot \mathrm{d}l, \tag{A.60}$$

where, as shown in Fig. A.3, n is the unit normal of S and $\mathrm{d}l := \tau\,\mathrm{d}s$ is the unit tangent τ multiplied by the differential arc length $\mathrm{d}s$ along C. Equation (A.60) relates

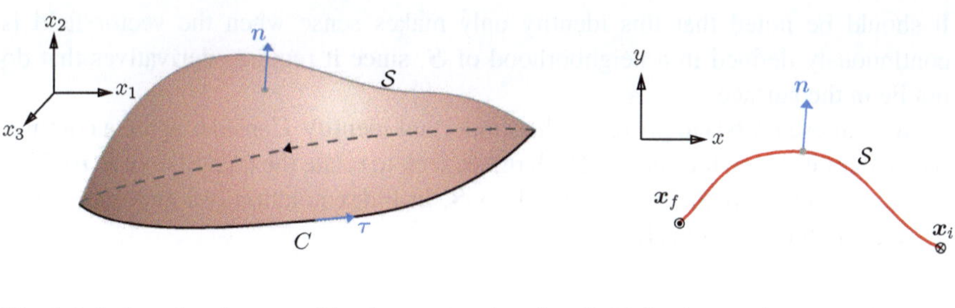

Fig. A.3 Left: a closed contour C and an enclosed surface S. Right: the two-dimensional version of S is a curve, for which the enclosing contour comprises endpoints x_f and x_i, where x_f is associated with a tangent $\tau = e_3$ and x_i with tangent $\tau = -e_3$

the flux of the curl of vector field $\boldsymbol{u}$ through $\mathcal{S}$ to the circulation of $\boldsymbol{u}$ about C; this is the classical form of the theorem. When applied to the two-dimensional analog geometry in the right panel of Fig. A.3, this reduces (per unit length into the page) to the fundamental theorem of calculus,

$$\int_{s_-}^{s^+} \frac{\partial u}{\partial s}\, \mathrm{d}s = u(s^+) - u(s^-), \tag{A.61}$$

where u is the component of $\boldsymbol{u}$ in the $\boldsymbol{e}_3$ direction.

To facilitate our discussion of generalizing theorem (A.60), let us write it in index notation:

$$\int_{\mathcal{S}} \epsilon_{ijk} n_i \frac{\partial u_k}{\partial x_j}\, \mathrm{d}S = \int_{C} u_i\, \mathrm{d}l_i. \tag{A.62}$$

Various other results can be obtained if we substitute $\boldsymbol{u}$ with other fields. For example, if we let $\boldsymbol{u} = \boldsymbol{c}\varphi$, where $\boldsymbol{c}$ is a constant vector and φ is a continuous scalar field, then it is easy to show that Stokes' theorem becomes

$$\int_{\mathcal{S}} \boldsymbol{n} \times \nabla\varphi\, \mathrm{d}S = \int_{C} \varphi\, \mathrm{d}\boldsymbol{l}, \tag{A.63}$$

where we have used the fact that each of the components of $\boldsymbol{c}$ is arbitrarily chosen. This allows us to relate the gradient of φ in the surface to the distribution of φ along the contour.

Similar to how the divergence theorem (A.18) was generalized by interpreting f more abstractly, Stokes' theorem can be generalized by allowing u_k to denote the kth component of some dimension of a rank-n tensor. For example, suppose we replace u_k with the rank-2 tensor $\epsilon_{klm} a_m$, where a_m is the mth component of a continuous vector field $\boldsymbol{a}$ on $\mathcal{S}$. From (A.62), we obtain the vector-valued integral identity

$$\int_{\mathcal{S}} [(\nabla \cdot \boldsymbol{a})\boldsymbol{n} - \nabla\boldsymbol{a} \cdot \boldsymbol{n}]\, \mathrm{d}S = \int_{C} \boldsymbol{a} \times \mathrm{d}\boldsymbol{l}. \tag{A.64}$$

It should be noted that this identity only makes sense when the vector field is continuously defined in a neighborhood of $\mathcal{S}$, since it requires derivatives that do not lie in the surface.

We can use (A.64) to prove another important identity. Consider again a continuous scalar field φ on the surface $\mathcal{S}$. Here, we seek to relate the quantities $\boldsymbol{x} \times (\boldsymbol{n} \times \nabla\varphi)$ and $\varphi\boldsymbol{n}$, where $\boldsymbol{n}$ is the usual normal on $\mathcal{S}$. In index notation, we can manipulate these using the product rule,

$$\begin{aligned}\epsilon_{ijk}x_j\epsilon_{klm}n_l\frac{\partial\varphi}{\partial x_m} &= x_jn_i\frac{\partial\varphi}{\partial x_j} - x_jn_j\frac{\partial\varphi}{\partial x_i}\\ &= n_i\frac{\partial(x_j\varphi)}{\partial x_j} - n_j\frac{\partial(x_j\varphi)}{\partial x_i} - (n_\mathrm{d}-1)n_i\varphi.\end{aligned} \tag{A.65}$$

Now, let's integrate both sides over the surface S. The first two terms on the right-hand side are of the form of the surface integrand in (A.64), with $\boldsymbol{a} \to \boldsymbol{x}\varphi$. Thus, we arrive at the identity

$$\int_S \boldsymbol{x}\times(\boldsymbol{n}\times\nabla\varphi)\,\mathrm{d}S = -(n_\mathrm{d}-1)\int_S \varphi\boldsymbol{n}\,\mathrm{d}S + \int_C \boldsymbol{x}\times\varphi\,\mathrm{d}\boldsymbol{l}. \tag{A.66}$$

Note that, in a two dimensional context ($n_\mathrm{d} = 2$) in which the geometry on the right side of Fig. A.3 is operative, the integral over C reduces to the difference $[\boldsymbol{x}\varphi]_i^f \times \boldsymbol{e}_3$ between the two endpoints of the curve S.

A related identity can be obtained, with only marginally more effort:

$$\int_S \boldsymbol{x}\times[\boldsymbol{x}\times(\boldsymbol{n}\times\nabla\varphi)]\,\mathrm{d}S = -n_\mathrm{d}\int_S \boldsymbol{x}\times\varphi\boldsymbol{n}\,\mathrm{d}S + \int_C \boldsymbol{x}\times(\boldsymbol{x}\times\varphi\,\mathrm{d}\boldsymbol{l}). \tag{A.67}$$

Here, the integral over C reduces to $[\boldsymbol{x}\times(\boldsymbol{x}\times\varphi\boldsymbol{e}_3)]_i^f$ in the two-dimensional context.

Generalizing Further, to Discontinuous Fields Stokes' theorem is useful when we are interested in an integral over a surface on which the field to be integrated is discontinuous along some contours in the surface. Such a surface can always be composed from a finite collection of surface patches on which the field is continuous. The adjacent contours of neighboring patches necessarily run in opposite directions, so, where the field is continuous across these contours, the integrals along them cancel. The net effect is that only integrals over contours of field discontinuity remain. Accounting for multiple such contours of discontinuity, $C_{I,j}$, Stokes' theorem (A.62) over the composite surface is

$$\int_S \epsilon_{ijk}n_i\frac{\partial u_k}{\partial x_j}\,\mathrm{d}S = \int_C u_i\,\mathrm{d}l_i + \sum_j\int_{C_{I,j}}[u_i]_-^+\,\mathrm{d}l_i, \tag{A.68}$$

where $[u_i]_-^+$ denotes the jump in the ith component of $\boldsymbol{u}$ from the $-$ to the $+$ side of the contour of discontinuity, where the $+$ side is defined to the left with respect to the integration direction $\mathrm{d}\boldsymbol{l}$ (that is, the side into which $\boldsymbol{n}\times\mathrm{d}\boldsymbol{l}$ is directed). If the surface is closed, then the first integral on the right-hand side vanishes, and only the integrals along the $C_{I,j}$ remain.

Some Useful Geometric Results of Surfaces and Contours The various forms of Stokes' theorem can be used to obtain some valuable geometric identities. For example, if we let $\varphi \to 1$ in (A.66), then it is straightforward to show that

$$\int_S \boldsymbol{n}\,\mathrm{d}S = \frac{1}{n_\mathrm{d}-1}\int_C \boldsymbol{x}\times\mathrm{d}\boldsymbol{l}. \tag{A.69}$$

This identity looks very similar to (A.22), obtained from the divergence theorem. However, it is important to note that the surface here is not closed, but rather, has an edge described by the closed contour C. Clearly, we can obtain (A.22) from (A.69) by composing a closed surface from a patchwork of bounded ones and noting that the contributions from each pair of adjacent contour segments cancel. In the two-dimensional analog geometry in which $\mathcal{S}$ is a finite-length curve, the contour integral simply becomes the difference in values at the endpoints: $(\boldsymbol{x}^+ - \boldsymbol{x}^-)\times \boldsymbol{e}_3$.

Similarly, using (A.67) with $\varphi \to 1$,

$$\int_S \boldsymbol{x}\times\boldsymbol{n}\,\mathrm{d}S = \frac{1}{n_\mathrm{d}}\int_C \boldsymbol{x}\times(\boldsymbol{x}\times\mathrm{d}\boldsymbol{l}), \tag{A.70}$$

which reduces to $-\boldsymbol{e}_3(|\boldsymbol{x}^+|^2 - |\boldsymbol{x}^-|^2)/2$ in a two-dimensional setting.

A Useful Result on Tangent Components of Vector Fields The next result is important for establishing certain conditions at the interface between two regions.

Result A.5: Continuity of Tangent Component of Vector Fields

Suppose there are two regions $\mathcal{V}^+$ and $\mathcal{V}^-$ as in Fig. A.2, divided by surface $\mathcal{S}$ with normal $\boldsymbol{n}$ directed into $\mathcal{V}^+$, and that there is a continuously-differentiable vector field in each, denoted respectively by $\boldsymbol{u}^+$ and $\boldsymbol{u}^-$. Let their tangent components be equal on the common surface, $\mathcal{S}$, so that

$$\boldsymbol{n}\times\boldsymbol{u}^+ = \boldsymbol{n}\times\boldsymbol{u}^- \tag{A.71}$$

at any $\boldsymbol{x}\in\mathcal{S}$. Then the normal components of the curl of these vector fields must match at all points on $\mathcal{S}$:

$$(\nabla\times\boldsymbol{u}^+)\cdot\boldsymbol{n} = (\nabla\times\boldsymbol{u}^-)\cdot\boldsymbol{n}. \tag{A.72}$$

Proof Suppose instead that (A.72) is violated, so that $(\nabla\times\boldsymbol{u}^+)\cdot\boldsymbol{n} = (\nabla\times\boldsymbol{u}^-)\cdot\boldsymbol{n} + f$, where f is a scalar field on $\mathcal{S}$. Now, since Eq. (A.71) holds throughout $\mathcal{S}$, it clearly must hold in any arbitrary subset (i.e., patch) $\mathcal{S}_s$ of the surface $\mathcal{S}$, with bounding contour C_s, so that

$$\int_{C_s}(\boldsymbol{n}\times\boldsymbol{u}^+)\cdot\mathrm{d}\boldsymbol{l} = \int_{C_s}(\boldsymbol{n}\times\boldsymbol{u}^-)\cdot\mathrm{d}\boldsymbol{l}. \tag{A.73}$$

By Stokes' theorem (A.60), we can write these as integrals over S_s:

$$\int_{S_s} (\nabla \times \boldsymbol{u}^+) \cdot \boldsymbol{n} \, \mathrm{d}S = \int_{S_s} (\nabla \times \boldsymbol{u}^-) \cdot \boldsymbol{n} \, \mathrm{d}S. \tag{A.74}$$

But since we have supposed that these integrands differ by f, then this equality implies that

$$\int_{S_s} f \, \mathrm{d}S = 0. \tag{A.75}$$

Since S_s is arbitrary, the only way for this integral to vanish for all possible choices is if f is everywhere zero. Thus, our theorem is proved. □

One might justifiably ask if the inverse of this is also true: If the normal components of the curls match, does this imply that the tangent components must match, as well? But a simple thought experiment shows that this cannot be true, since we can add any constant vector field to either $\boldsymbol{u}^+$ or $\boldsymbol{u}^-$, with no effect on its curl, but which clearly causes the tangent components to disagree on S. In fact, one can show that if (A.72) is true, then $\boldsymbol{n} \times (\boldsymbol{u}^+ - \boldsymbol{u}^-) = \boldsymbol{a}$ on S, where $\boldsymbol{a}$ is an arbitrary divergence-free vector field that lies entirely in S; a particular example of $\boldsymbol{a}$ is $\boldsymbol{n} \times \boldsymbol{c}$, where $\boldsymbol{c}$ is a constant vector. We leave this proof to the reader.

Use of Stokes' Theorem to Modify Green's Theorem Earlier, we used Green's theorem to develop expressions for scalar and vector fields in terms of surface and volume distributions. We can use Stokes' theorem to modify these and obtain alternative forms of Green's theorem. Let us first prove a useful result:

Result A.6: Equivalence of Two Forms of Surface Singularity Distributions

Let us consider a surface S bounded by contour C, such as that shown in Fig. A.3, in which there exists a continuous scalar field μ. Then, at any point $\boldsymbol{x}$ not on the surface or contour,

$$-\nabla \times \int_S G(\boldsymbol{x} - \boldsymbol{y}) \boldsymbol{n}_y \times \nabla_y \mu \, \mathrm{d}S(\boldsymbol{y}) = \tag{A.76}$$

$$- \int_C \mu(\boldsymbol{y}) \nabla G(\boldsymbol{x} - \boldsymbol{y}) \times \mathrm{d}\boldsymbol{l}(\boldsymbol{y}) + \nabla \int_S \mu(\boldsymbol{y}) \boldsymbol{n}_y \cdot \nabla G(\boldsymbol{x} - \boldsymbol{y}) \, \mathrm{d}S(\boldsymbol{y}),$$

where G is the Green's function of the negative Laplacian, and $\boldsymbol{n}_y$ is the unit normal at point $\boldsymbol{y} \in S$.

Proof Let us first write the ith component of the left-hand side in index notation:

$$-\epsilon_{ijk}\frac{\partial}{\partial x_j}\int_S G(\boldsymbol{x}-\boldsymbol{y})\epsilon_{klm}n_{y,l}\frac{\partial \mu}{\partial y_m}\,\mathrm{d}S(\boldsymbol{y}). \tag{A.77}$$

The derivative with respect to the target point $\boldsymbol{x}$ can be brought inside the integral and applied directly to G, which is the only part of the integrand dependent on this point. Using the product rule and re-arranging factors a bit, we can write the integral as

$$-\int_S \epsilon_{klm}n_{y,l}\frac{\partial}{\partial y_m}\left(\epsilon_{ijk}\mu(\boldsymbol{y})\frac{\partial G}{\partial x_j}(\boldsymbol{x}-\boldsymbol{y})\right)\mathrm{d}S(\boldsymbol{y})$$
$$+\,\epsilon_{ijk}\epsilon_{klm}\int_S n_{y,l}\mu(\boldsymbol{y})\frac{\partial}{\partial x_j}\frac{\partial G}{\partial y_m}(\boldsymbol{x}-\boldsymbol{y})\,\mathrm{d}S(\boldsymbol{y}). \tag{A.78}$$

The first integral is in the form of the surface integral in Stokes' theorem (A.68), with the generic u_k replaced by the tensor-valued function inside parentheses. In the second integral, after we switch the derivative of G from the source to the target point, we use the identity (A.6), which gives rise to two terms. One of these involves $\nabla^2 G$, which vanishes at all points not on S. The reader can verify that these steps lead to the following:

$$-\int_C \epsilon_{ijk}\mu(\boldsymbol{y})\frac{\partial G}{\partial x_j}(\boldsymbol{x}-\boldsymbol{y})\,\mathrm{d}l_k(\boldsymbol{y})+\frac{\partial}{\partial x_i}\int_S \mu(\boldsymbol{y})n_{y,j}\frac{\partial G}{\partial x_j}(\boldsymbol{x}-\boldsymbol{y})\,\mathrm{d}S(\boldsymbol{y}). \tag{A.79}$$

This is just the ith component of the right-hand side of (A.76), written in index notation. □

We can generalize this result to allow for discontinuities in the field μ on S. As we described earlier in this section, we can always subdivide an extensive surface S into a set of smaller surface patches, and obviously, such patches can be made of arbitrary size and shape. Equation (A.76) applies to any surface patch and its own bounding contour; the net effect from S is simply the sum over the patches. We can always choose the patches so that μ is continuous within each patch, and any discontinuities in μ lie along bounding contours. Wherever μ is continuous across the boundary between adjacent patches, the contributions from their line integrals, which run in opposite directions, will exactly cancel. The effect of discontinuities in μ is embodied in any line integrals that remain. Let us denote a contour segment along which μ is discontinuous by C_I; the jump in μ across this contour is denoted by $[\mu]_-^+$, where the positive side is to our left as we follow the contour. Then, the result (A.76) for the entire surface S can be slightly amended to

$$-\nabla \times \int_S G(\boldsymbol{x}-\boldsymbol{y})\boldsymbol{n}_y \times \nabla_y \mu \,\mathrm{d}S(\boldsymbol{y}) = \tag{A.80}$$

$$-\int_{C_I} [\mu]_-^+(\boldsymbol{y}) \nabla G(\boldsymbol{x}-\boldsymbol{y}) \times \mathrm{d}\boldsymbol{l}(\boldsymbol{y}) + \nabla \int_S \mu(\boldsymbol{y})\boldsymbol{n}_y \cdot \nabla G(\boldsymbol{x}-\boldsymbol{y})\,\mathrm{d}S(\boldsymbol{y}).$$

Of course, there can be more than one such contour of discontinuity, C_I; we can always replace the line integral by a sum over all such contours.

A.1.5 Change of Variables Theorem

A really useful tool in integral calculus is the so-called **change of variables theorem**. Suppose we are interested in computing a volume integral of some field quantity, f, over some region $\mathcal{V}$ that has been 'deformed' from a region $\mathcal{V}_0$ by a differentiable map, $\boldsymbol{\phi}$. That is, every point $\boldsymbol{\xi} \in \mathcal{V}_0$ in the original region is mapped uniquely to some other point $\boldsymbol{x} = \boldsymbol{\phi}(\boldsymbol{\xi}) \in \mathcal{V}$, and furthermore, this map can be differentiated, so that, in particular, there is a Jacobian, J, which measures how much infinitesimal volumes change, $J := \det(\partial \boldsymbol{\phi}/\partial \boldsymbol{\xi})$. Then the change of variables theorem provides a way to transform the integral over $\mathcal{V}$ to one over $\mathcal{V}_0$:

$$\int_{\mathcal{V}} f(\boldsymbol{x})\,\mathrm{d}V(\boldsymbol{x}) = \int_{\mathcal{V}_0} f[\boldsymbol{\phi}(\boldsymbol{\xi})]\, J(\boldsymbol{\xi})\mathrm{d}V_0(\boldsymbol{\xi}). \tag{A.81}$$

This formula is useful for simplifying integrals over complex regions (for example, when a complicated shape can be mapped from a unit sphere).

Often in fluid dynamics, the flow map is smoothly parameterized by time. This means that the location to which a point labeled by $\boldsymbol{\xi}$ is mapped depends on how much time has elapsed. We will use the notation $\boldsymbol{\phi}_t$ for this time-parameterized flow map, so that the point's position at time t is described by $\boldsymbol{x}(\boldsymbol{\xi}, t) := \boldsymbol{\phi}_t(\boldsymbol{\xi})$. By definition, $\boldsymbol{\xi} = \boldsymbol{\phi}_0(\boldsymbol{\xi})$. Thus, the region $\mathcal{V}_0$ is smoothly mapped to a time-dependent region $\mathcal{V}(t)$, where $\mathcal{V}(0) = \mathcal{V}_0$. The rate at which the mapped point labeled by $\boldsymbol{\xi}$ moves is denoted by $\boldsymbol{u}(\boldsymbol{\xi}, t) := \partial \boldsymbol{x}(\boldsymbol{\xi}, t)/\partial t$. In fact, the flow map is defined by the integral curves of this vector field. Note that we will *not* always take this vector field to be the velocity of the fluid, as sometimes, it will be useful to construct flow maps that achieve other objectives. We will only assume that the flow map is differentiable with respect to time.

A.1.6 Field Quantities and Their Rate of Change

Throughout the book, a field quantity f will have its dependence on space and time denoted by $f(\boldsymbol{x}, t)$, where $\boldsymbol{x}$ represents a certain position in inertial space and t a certain instant in time. The meaning of $\partial f/\partial t$ is unambiguous in this notation: it represents the rate of change of f when observed at a fixed location in inertial space.

What happens when the spatial position at which we evaluate f is moving? Let's denote this moving point by $\boldsymbol{x}(t)$ and its velocity by $\mathrm{d}\boldsymbol{x}/\mathrm{d}t$. Then, the rate of change of the value of f we observe at this moving evaluation point is described by the chain rule:

$$\frac{\mathrm{d}f}{\mathrm{d}t}(\boldsymbol{x}(t),t) = \frac{\partial f}{\partial t}(\boldsymbol{x}(t),t) + \frac{\mathrm{d}\boldsymbol{x}}{\mathrm{d}t} \cdot \nabla f(\boldsymbol{x}(t),t). \tag{A.82}$$

The first term on the right-hand side represents the rate of change of f due to the field's unsteadiness at the fixed location at which it is instantaneously evaluated. The second term embodies the change in f that we observe by moving around in space and sampling f at different points. For this term—the convective rate of change—to make a contribution, the field must have a non-zero spatial gradient in the direction in which our evaluation point moves.

Let's assume that this evaluation point's motion is described by a time-parameterized flow map, $\boldsymbol{\phi}_t$, defined over a larger region $\mathcal{V}_0$. As we discussed in the previous section, this flow map acts on points $\boldsymbol{\xi} \in \mathcal{V}_0$ and transforms them to new locations $\boldsymbol{x}(\boldsymbol{\xi},t) = \boldsymbol{\phi}_t(\boldsymbol{\xi}) \in \mathcal{V}(t)$. The quantity f, in addition to its usual dependence on position in the instantaneous region $\mathcal{V}(t)$, can be alternatively regarded as a function of points $\boldsymbol{\xi}$ in the initial region $\mathcal{V}_0$; we will denote this latter dependence by $\tilde{f}(\boldsymbol{\xi},t)$. At any given instant $t \geq 0$, these fields are equivalent at corresponding points: $f(\boldsymbol{\phi}_t(\boldsymbol{\xi}),t) \equiv \tilde{f}(\boldsymbol{\xi},t)$. This equivalence allows us to reconcile their rates of change:

$$\frac{\mathrm{d}f}{\mathrm{d}t}(\boldsymbol{\phi}_t(\boldsymbol{\xi}),t) = \frac{\partial f}{\partial t}(\boldsymbol{\phi}_t(\boldsymbol{\xi}),t) + \boldsymbol{u}(\boldsymbol{\xi},t) \cdot \nabla f(\boldsymbol{\phi}_t(\boldsymbol{\xi}),t) \equiv \frac{\partial \tilde{f}}{\partial t}(\boldsymbol{\xi},t). \tag{A.83}$$

When the flow map is based on the fluid velocity field, $\boldsymbol{v}$, then $\boldsymbol{\xi}$ is the label of a **material point**, and the rate of change of f observed when following such a point is the **material derivative**. This is conventionally denoted by

$$\frac{Df}{Dt} := \frac{\partial f}{\partial t} + \boldsymbol{v} \cdot \nabla f. \tag{A.84}$$

A.1.7 Time Differentiation of Spatial Integrals

We are often interested in this work in computing the time derivative of an integral over a spatial region that is evolving in time. If the integral is one-dimensional, then the Leibniz rule allows us to perform this differentiation:

$$\frac{\mathrm{d}}{\mathrm{d}t}\int_{a(t)}^{b(t)} f(x,t)\,\mathrm{d}x = \int_{a(t)}^{b(t)} \frac{\partial f}{\partial t}(x,t)\,\mathrm{d}x + f(a(t),t)\frac{\mathrm{d}a}{\mathrm{d}t} - f(b(t),t)\frac{\mathrm{d}b}{\mathrm{d}t}. \tag{A.85}$$

How do we extend this operation when the region in question is multi-dimensional: a volume, a surface or a curve? Here, we will provide a summary of identities for abstractions of the Leibniz rule.

Volume Integrals The change of variables theorem (A.81) is particularly helpful when the map is smoothly parameterized by time, because it allows us to very easily and unambiguously compute time derivatives of integrals. The Jacobian of this map is clearly time dependent, and it can be shown that its time derivative is proportional to the divergence of the velocity field:

$$\frac{\mathrm{d}J}{\mathrm{d}t} = J(\boldsymbol{\xi},t)\nabla_{\boldsymbol{\xi}} \cdot \boldsymbol{u}(\boldsymbol{\xi},t). \tag{A.86}$$

Now, we can write the change of variables theorem as

$$\int_{\mathcal{V}(t)} f(\boldsymbol{x},t)\,\mathrm{d}V(\boldsymbol{x}) = \int_{\mathcal{V}_0} f[\boldsymbol{\phi}_t(\boldsymbol{\xi}),t]\,J(\boldsymbol{\xi},t)\mathrm{d}V_0(\boldsymbol{\xi}). \tag{A.87}$$

Then, if one seeks the time derivative of this integral, it is easily calculated from the derivative of the integral over $\mathcal{V}_0$ (where we can readily differentiate the integrand with the help of the chain rule) and transforming the resulting integral back to $\mathcal{V}(t)$. The reader can verify that the result is

$$\frac{\mathrm{d}}{\mathrm{d}t}\int_{\mathcal{V}(t)} f\,\mathrm{d}V = \int_{\mathcal{V}(t)} \left[\frac{\partial f}{\partial t} + \nabla\cdot(\boldsymbol{u}f)\right]\mathrm{d}V, \tag{A.88}$$

where the velocity, $\boldsymbol{u}$, is that of the flow that deforms $\mathcal{V}_0$ to $\mathcal{V}(t)$: the velocity of the fluid when $\mathcal{V}(t)$ is a material volume, for example. If we go a step further and apply the divergence theorem (A.19) to this result, for a bounding surface $\mathcal{S}(t)$ with outward-directed normal $\boldsymbol{n}$, we arrive at

$$\frac{\mathrm{d}}{\mathrm{d}t}\int_{\mathcal{V}(t)} f\,\mathrm{d}V = \int_{\mathcal{V}(t)} \frac{\partial f}{\partial t}\,\mathrm{d}V + \int_{\mathcal{S}(t)} \boldsymbol{n}\cdot\boldsymbol{u}f\,\mathrm{d}S. \tag{A.89}$$

This equation accounts for the rate of change of 'total f' in $\mathcal{V}(t)$ via the accumulation of local changes of f and the net amount of f 'folded into' $\mathcal{V}(t)$ by movement of its bounding surface. It is very useful for proving the Reynolds transport theorem. Furthermore, the forms derived here can be adapted very naturally to any n-dimensional space. In fact, the one-dimensional version of (A.89) is nothing more than Leibniz's rule (A.85), where the bounding surface is simply the pair of endpoints.

Surface Integrals The machinery of differential geometry provides a powerful and elegant set of tools to compute time derivatives of integrals. In fact, in the language of that discipline, the change of variables theorem (A.81) is a **pull-back** of the volume form to $\mathcal{V}_0$ [24]. The subsequent differentiation yields a Lie derivative of the volume form with respect to the vector field $\boldsymbol{u}$ (or, more precisely, the vector field $\boldsymbol{u}$ plus a unit vector along the time axis, to account for the time variation of the integrand). This is actually a more general result that can be applied to all time-dependent integrals. In particular, we can use it to compute the rate of change of the **flux** of some time-varying vector field $\boldsymbol{a}(\boldsymbol{x},t)$ through a time-varying surface $\mathcal{S}(t)$

with local normal $\boldsymbol{n}$:

$$\frac{\mathrm{d}}{\mathrm{d}t}\int\limits_{S(t)} \boldsymbol{a}\cdot\boldsymbol{n}\,\mathrm{d}S = \int\limits_{S(t)} \left[\frac{\partial \boldsymbol{a}}{\partial t} + (\nabla\cdot\boldsymbol{a})\boldsymbol{u} - \nabla\times(\boldsymbol{u}\times\boldsymbol{a})\right]\cdot\boldsymbol{n}\,\mathrm{d}S. \tag{A.90}$$

Here, the velocity, $\boldsymbol{u}$, is that of points in the moving surface. How do we make intuitive sense of this identity? Before we address that, let's reformulate the last term. Let $C(t)$ be the contour that encloses the surface $S(t)$, in the sense of Fig. A.3. If we then apply Stokes' theorem (A.60) to the last term, we get

$$\frac{\mathrm{d}}{\mathrm{d}t}\int\limits_{S(t)} \boldsymbol{a}\cdot\boldsymbol{n}\,\mathrm{d}S = \int\limits_{S(t)} \left[\frac{\partial \boldsymbol{a}}{\partial t} + (\nabla\cdot\boldsymbol{a})\boldsymbol{u}\right]\cdot\boldsymbol{n}\,\mathrm{d}S + \int\limits_{C(t)} \boldsymbol{a}\cdot(\boldsymbol{u}\times\mathrm{d}\boldsymbol{l}). \tag{A.91}$$

This form of the equation indicates that there are three ways in which the flux of $\boldsymbol{a}$ through the surface might change. The first is because of inherent unsteadiness in the field $\boldsymbol{a}$. The second and third, however, are present whether the $\boldsymbol{a}$ field is unsteady or not. The second term includes motion of the surface normal to itself, so that the changes in the flux arise as the surface moves toward regions in which $\boldsymbol{a}$ is different; the divergence of this field measures the relevant difference. The final term accounts for changes in the flux that occur as the surface expands outward, perpendicular to the enclosing contour, since $\boldsymbol{u}\times\mathrm{d}\boldsymbol{l}$ is the local rate of increase of the differential contour element.

A particular form of this equation can be obtained by setting $\boldsymbol{a} = \boldsymbol{c}\varphi$, where $\boldsymbol{c}$ is an arbitrary constant vector and φ is a scalar field. Then, it can be shown that the resulting equation for φ is

$$\frac{\mathrm{d}}{\mathrm{d}t}\int\limits_{S(t)} \varphi\boldsymbol{n}\,\mathrm{d}S = \int\limits_{S(t)} \left[\left(\frac{\partial \varphi}{\partial t} + \nabla\cdot(\boldsymbol{u}\varphi)\right)\boldsymbol{n} - \varphi\nabla\boldsymbol{u}\cdot\boldsymbol{n}\right]\mathrm{d}S, \tag{A.92}$$

where $\nabla\boldsymbol{u}\cdot\boldsymbol{n}$ denotes a vector field with ith component $(\partial u_j/\partial x_i)n_j$, where the summation convention has been used. The term in which this appears accounts for the deformation and rotation of the surface.[2] Similar to (A.91), the right-hand side in (A.92) can be rewritten with the help of the vector-valued form of Stokes' theorem (A.64), so that another form of the rate of change is given by

$$\frac{\mathrm{d}}{\mathrm{d}t}\int\limits_{S(t)} \varphi\boldsymbol{n}\,\mathrm{d}S = \int\limits_{S(t)} \left(\frac{\partial \varphi}{\partial t}\boldsymbol{n} + \nabla\varphi\,\boldsymbol{u}\cdot\boldsymbol{n}\right)\mathrm{d}S + \int\limits_{C(t)} \varphi\boldsymbol{u}\times\mathrm{d}\boldsymbol{l}. \tag{A.93}$$

In a planar context, the surface S is a curve (see the right panel of Fig. A.3), and the enclosing contour C intersects the plane at the set of endpoints of the curve $\{\boldsymbol{x}_i, \boldsymbol{x}_f\}$, so that the contour integral (per unit depth into the page) reduces to the difference $[\varphi\boldsymbol{u}]_i^f \times \boldsymbol{e}_3$ between the endpoints.

[2] In the particular case that the surface $S(t)$ is that of a rigid body, then the velocity $\boldsymbol{u}$ is divergence free, and $\nabla\boldsymbol{u}\cdot\boldsymbol{n} = -\boldsymbol{\Omega}\times\boldsymbol{n}$, where $\boldsymbol{\Omega}$ is the body's angular velocity.

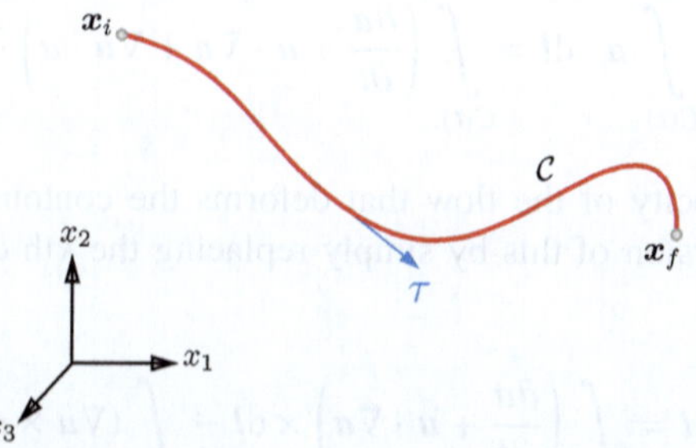

Fig. A.4 Illustration of a curve C in space, with end points $\boldsymbol{x}_i$ and $\boldsymbol{x}_f$ and unit tangent $\boldsymbol{\tau}$

As we discussed in conjunction with Stokes' theorem in Sect. A.1.4 when we constructed the generalized form (A.68), a surface in which a field is discontinuous can be composed from a collection of patches, each with a continuous field. In this composition, the contour integrals in (A.93) cancel along adjacent contours, except where discontinuities in the φ field intersect the surface; examples of such discontinuities, such as a branch cut in two dimensions, are illustrated in Fig. 6.2. Along such an intersection (a point $\boldsymbol{x}_i$ in two dimensions or a finite-length curve C_i in three dimensions), the φ in the contour integral is replaced by its jump $[\varphi]_-^+$ across the discontinuity, where the + and − sides of the intersection are labeled in Fig. 6.2.

Another useful rate of change can be obtained if we let $\boldsymbol{a} = \boldsymbol{c} \times \boldsymbol{x}\varphi$. Then, this results in

$$\frac{\mathrm{d}}{\mathrm{d}t}\int_{S(t)} \boldsymbol{x} \times \varphi \boldsymbol{n}\, \mathrm{d}S = \int_{S(t)} \left[\boldsymbol{x} \times \left(\frac{\partial \varphi}{\partial t} + \nabla \cdot (\boldsymbol{u}\varphi) \right) \boldsymbol{n} - \boldsymbol{x} \times (\varphi \nabla \boldsymbol{u} \cdot \boldsymbol{n}) + \boldsymbol{u} \times \varphi \boldsymbol{n} \right] \mathrm{d}S. \tag{A.94}$$

This, too, can be written in a different form with the help of a version of Stokes' theorem similar to (A.64):

$$\frac{\mathrm{d}}{\mathrm{d}t}\int_{S(t)} \boldsymbol{x} \times \varphi \boldsymbol{n}\, \mathrm{d}S = \int_{S(t)} \boldsymbol{x} \times \left(\frac{\partial \varphi}{\partial t}\boldsymbol{n} + \nabla \varphi \boldsymbol{u} \cdot \boldsymbol{n} \right) \mathrm{d}S + \int_{C(t)} \boldsymbol{x} \times (\boldsymbol{u}\varphi \times \mathrm{d}\boldsymbol{l}). \tag{A.95}$$

Similar to the previous case, the contour integral reduces to the difference $[\boldsymbol{x} \times (\varphi \boldsymbol{u} \times \boldsymbol{e}_3)]_-^+$ between the endpoints. And as before, in either dimension, the integral over a closed surface can be composed from a coverage of adjacent open surfaces, so that, as for (A.93), the contour integral is restricted to the intersections of discontinuities in φ, on which it is replaced by $[\varphi]_-^+$.

Line Integrals over Curves Thus far, we have obtained expressions for the time derivatives of volume and surface integrals. For completeness, let us now express the rate of change of a line integral, namely, the integral of a time-varying vector field $\boldsymbol{a}$ projected along a time-varying curve $C(t)$, not necessarily closed, as illustrated in Fig. A.4. The tools of differential geometry also enable us to obtain the following:

$$\frac{\mathrm{d}}{\mathrm{d}t}\int\limits_{C(t)} \boldsymbol{a}\cdot \mathrm{d}\boldsymbol{l} = \int\limits_{C(t)} \left(\frac{\partial \boldsymbol{a}}{\partial t} + \boldsymbol{u}\cdot\nabla\boldsymbol{a} + \nabla\boldsymbol{u}\cdot\boldsymbol{a}\right)\cdot \mathrm{d}\boldsymbol{l}, \tag{A.96}$$

where $\boldsymbol{u}$ is the velocity of the flow that deforms the contour. We can also write a vector product version of this by simply replacing the kth component of $\boldsymbol{a}$ with $\epsilon_{ijk}a_j$:

$$\frac{\mathrm{d}}{\mathrm{d}t}\int\limits_{C(t)} \boldsymbol{a}\times \mathrm{d}\boldsymbol{l} = \int\limits_{C(t)} \left(\frac{\partial \boldsymbol{a}}{\partial t} + \boldsymbol{u}\cdot\nabla\boldsymbol{a}\right)\times \mathrm{d}\boldsymbol{l} - \int\limits_{C(t)} (\nabla\boldsymbol{u}\times\boldsymbol{a})\cdot \mathrm{d}\boldsymbol{l}. \tag{A.97}$$

This last result can be used to prove an identity of high value in this work. Suppose that $\boldsymbol{a} \rightarrow \boldsymbol{x}$. Then the integrand of the first integral on the right-hand side of (A.97) reduces to $\boldsymbol{u}\times$. The second integral can be integrated by parts, noting that $(\nabla\boldsymbol{u}\times\boldsymbol{x})\cdot\mathrm{d}\boldsymbol{l} = \mathrm{d}\boldsymbol{l}\cdot\nabla(\boldsymbol{u}\times\boldsymbol{x}) - \boldsymbol{u}\times\mathrm{d}\boldsymbol{l}$. Then, the resulting identity is

$$\frac{\mathrm{d}}{\mathrm{d}t}\int\limits_{C(t)} \boldsymbol{x}\times \mathrm{d}\boldsymbol{l} = 2\int\limits_{C(t)} \boldsymbol{u}\times\mathrm{d}\boldsymbol{l} + [\boldsymbol{x}\times\boldsymbol{u}]_i^f, \tag{A.98}$$

where the last term represents the difference of $\boldsymbol{u}\times\boldsymbol{x}$ between the end (f) and beginning (i) of the curve, as labeled in Fig. A.4. In some cases, we wish to apply this result in a planar context, in which the curve C is an infinite straight line in the out-of-plane direction, and the curve's velocity, $\boldsymbol{u}$, is restricted to the plane. In such a case, this identity simplifies to the motion of the single point $\boldsymbol{x}_C(t)$ at which the curve intersects the plane:

$$\frac{\mathrm{d}}{\mathrm{d}t}(\boldsymbol{x}_C\times\boldsymbol{e}_3) = \boldsymbol{u}_C\times\boldsymbol{e}_3, \tag{A.99}$$

where $\boldsymbol{u}_C := \dot{\boldsymbol{x}}_C$.

A related identity, also valuable, can be obtained by a similar approach:

$$\frac{\mathrm{d}}{\mathrm{d}t}\int\limits_{C(t)} \boldsymbol{x}\times(\boldsymbol{x}\times \mathrm{d}\boldsymbol{l}) = 3\int\limits_{C(t)} \boldsymbol{x}\times(\boldsymbol{u}\times\mathrm{d}\boldsymbol{l}) + [\boldsymbol{x}\times(\boldsymbol{x}\times\boldsymbol{u})]_i^f, \tag{A.100}$$

whose two-dimensional version is simply

$$\frac{\mathrm{d}}{\mathrm{d}t}[\boldsymbol{x}_C\times(\boldsymbol{x}_C\times\boldsymbol{e}_3)] = 2\boldsymbol{x}_C\times(\boldsymbol{u}_C\times\boldsymbol{e}_3). \tag{A.101}$$

A.2 Useful Tools from Complex Analysis

Throughout the book, we denote the imaginary unit by i and complex conjugate with $()^*$. The real and imaginary parts of a complex number are obtained by operating on it with Re() and Im(), respectively. The complex coordinate will often be denoted by $z = x + \mathrm{i}y$.

A.2.1 Basic Properties

Holomorphic Functions A complex-valued function $F(z)$ is called **holomorphic** (or sometimes called **analytic**) in an open region U if it is uniquely differentiable with respect to z at every point in U. This means, in particular, that the real and imaginary components satisfy the Cauchy–Riemann equations. Denoting these components as φ and ψ, respectively,

$$\frac{\partial \varphi}{\partial x} = \frac{\partial \psi}{\partial y}, \qquad \frac{\partial \varphi}{\partial y} = -\frac{\partial \psi}{\partial x}. \tag{A.102}$$

It is easy to show from these equations that the Laplacian of each component must necessarily be zero.

Vector-Like Operations in Complex Analysis We can naturally identify a planar vector $\boldsymbol{a}$ with a complex quantity a, whose real and imaginary parts are simply the x and y components of $\boldsymbol{a}$:

$$\boldsymbol{a} \Longleftrightarrow a. \tag{A.103}$$

Then, it is useful to express common vector operations in complex form. For example, an inner (i.e., dot) product between two planar vectors $\boldsymbol{a}$ and $\boldsymbol{b}$ can be written in terms of their complex equivalents as

$$\boldsymbol{a} \cdot \boldsymbol{b} = \mathrm{Re}(a^* b) = \mathrm{Re}(a b^*). \tag{A.104}$$

The sole component of the cross product of these two vectors can be written as

$$(\boldsymbol{a} \times \boldsymbol{b}) \cdot \boldsymbol{e}_3 = \mathrm{Im}(a^* b) = -\mathrm{Im}(a b^*). \tag{A.105}$$

If we take the cross product of any planar vector with the out-of-plane unit vector $\boldsymbol{e}_3$, this is equivalent to a clockwise rotation of the vector by $\pi/2$, or equivalently, multiplication of the complex form of the vector by $-\mathrm{i}$:

$$\boldsymbol{a} \times \boldsymbol{e}_3 \Longleftrightarrow -\mathrm{i}a \tag{A.106}$$

On a contour C in the plane, with arc length parameter s, the unit tangent (in complex form) is given by $\tau = \mathrm{d}z/\mathrm{d}s$. We can define the unit normal, n as shown in Fig. A.5, directed to the right as the contour is traversed in the direction of increasing s. It is easy to see that this is obtained from the tangent by clockwise rotation by $\pi/2$:

$$\boldsymbol{n} = (n_x, n_y) \Longleftrightarrow n := n_x + \mathrm{i}n_y = -\mathrm{i}\tau \equiv -\mathrm{i}\frac{\mathrm{d}z}{\mathrm{d}s} = \frac{\mathrm{d}y}{\mathrm{d}s} - \mathrm{i}\frac{\mathrm{d}x}{\mathrm{d}s}. \tag{A.107}$$

Alternatively, we could define the unit normal as directed to the *left* as s increases; this is often the convention in our analysis of thin bodies, for example. This choice simply results in a change of sign in (A.107): $n = \mathrm{i}\tau \equiv \mathrm{i}\mathrm{d}z/\mathrm{d}s$.

Conjugate Notation Frequently, we encounter complex-valued functions that depend on a complex variable. Let us denote one such function and its dependence by $F(z)$. When we need to take the conjugate of this function, usually we apply this conjugate after the function has been evaluated at some value of z. However, occasionally we will need to take the conjugate *first*, and then evaluate it at its (unconjugated) argument. To make this distinction clear in our notation, we use the following throughout this book: Let $F(z)^*$ denote the application of the conjugate after the function has been evaluated at z, and $F^*(z)$ the application before evaluation at z. Note that $F(z)^* \equiv F^*(z^*)$; that is, the conjugated function is evaluated at the argument's conjugate.

Let us examine the real and imaginary parts of the function and its argument in both cases: $F(z) = \varphi(x, y) + \mathrm{i}\psi(x, y)$. Then, $F(z)^* \equiv \varphi(x, y) - \mathrm{i}\psi(x, y)$, whereas $F^*(z) \equiv \varphi(x, -y) - \mathrm{i}\psi(x, -y)$. These might look inconsistent with our definitions of the notation. However, it is instructive to consider the simple example, $F(z) = z$, so $\varphi(x, y) = x$ and $\psi(x, y) = y$. Then $F(z)^* = z^* = x - \mathrm{i}y = \varphi(x, y) - \mathrm{i}\psi(x, y)$, while $F^*(z) = z = x + \mathrm{i}y = \varphi(x, -y) - \mathrm{i}\psi(x, -y)$.

Derivatives Consider a function, $f(x, y)$, which may be complex-valued (but not necessarily holomorphic). Any such function can be alternatively regarded as a function of the independent variables $z = x+\mathrm{i}y$ and $z^* = x-\mathrm{i}y$. Then the derivatives of this function with respect to x and y can be written instead in terms of derivatives with respect to these alternative arguments, using the chain rule:

$$\frac{\partial f}{\partial x} = \frac{\partial f}{\partial z}\frac{\mathrm{d}z}{\mathrm{d}x} + \frac{\partial f}{\partial z^*}\frac{\mathrm{d}z^*}{\mathrm{d}x} = \frac{\partial f}{\partial z} + \frac{\partial f}{\partial z^*}, \qquad \frac{\partial f}{\partial y} = \mathrm{i}\left(\frac{\partial f}{\partial z} - \frac{\partial f}{\partial z^*}\right). \tag{A.108}$$

These equations can be solved for $\partial f/\partial z$ and $\partial f/\partial z^*$ in order to express these derivatives in terms of those with respect to x and y:

$$\frac{\partial f}{\partial z} = \frac{1}{2}\left(\frac{\partial f}{\partial x} - \mathrm{i}\frac{\partial f}{\partial y}\right), \qquad \frac{\partial f}{\partial z^*} = \frac{1}{2}\left(\frac{\partial f}{\partial x} + \mathrm{i}\frac{\partial f}{\partial y}\right). \tag{A.109}$$

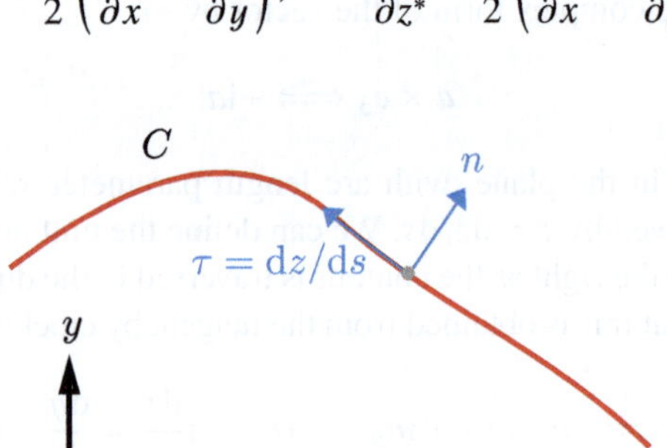

Fig. A.5 Contour C and the conventions for the unit normal n and tangent τ

What about second derivatives? Starting from (A.108), let's take derivatives with respect to x and y. We get

$$\frac{\partial^2 f}{\partial x^2} = \frac{\partial^2 f}{\partial z^2} + 2\frac{\partial^2 f}{\partial z \partial z^*} + \frac{\partial^2 f}{\partial z^{*2}} \tag{A.110}$$

$$\frac{\partial^2 f}{\partial x \partial y} = \mathrm{i}\left(\frac{\partial^2 f}{\partial z^2} - \frac{\partial^2 f}{\partial z^{*2}}\right) \tag{A.111}$$

$$\frac{\partial^2 f}{\partial y^2} = -\frac{\partial^2 f}{\partial z^2} + 2\frac{\partial^2 f}{\partial z \partial z^*} - \frac{\partial^2 f}{\partial z^{*2}} \tag{A.112}$$

Now, let's suppose that $F(z) = \varphi(x, y) + \mathrm{i}\psi(x, y)$ is a holomorphic function, and φ and ψ are its real and imaginary parts, respectively. Then, the Cauchy-Riemann equations (A.102) can be expressed succinctly as

$$\frac{\partial F}{\partial z^*} = 0. \tag{A.113}$$

From this and from (A.109), we can also easily show that

$$\frac{\partial \varphi}{\partial z} = \frac{1}{2}\frac{\partial F}{\partial z}, \quad \frac{\partial \psi}{\partial z} = -\frac{\mathrm{i}}{2}\frac{\partial F}{\partial z}. \tag{A.114}$$

If we add Eqs. (A.110) and (A.112) when applied to either the real or imaginary component of F, we obtain that component's Laplacian, which is necessarily zero for this type of function. Therefore, from the right-hand side of this sum of equations, we can conclude that,

$$\frac{\partial^2 f}{\partial z \partial z^*} = 0, \tag{A.115}$$

for f either φ or ψ. Some useful results can be obtained from this. For example, if a and b are complex constants, then

$$\left(a\frac{\partial}{\partial z} + b\frac{\partial}{\partial z^*}\right)^n f = a^n \frac{\partial^n f}{\partial z^n} + b^n \frac{\partial^n f}{\partial z^{*n}}, \tag{A.116}$$

again, for f either φ or ψ. If $\boldsymbol{a}$ is a vector-valued constant, then the **directional derivative** is defined as the operator $\boldsymbol{a} \cdot \nabla$. Using the results we've obtained so far, it is easy to show that this operator can be written alternatively as

$$\boldsymbol{a} \cdot \nabla \Longleftrightarrow \mathrm{Re}\left[a\left(\frac{\partial}{\partial x} - \mathrm{i}\frac{\partial}{\partial y}\right)\right] = \left(a\frac{\partial}{\partial z} + a^*\frac{\partial}{\partial z^*}\right), \tag{A.117}$$

where a is the complex equivalent of $\boldsymbol{a}$. From (A.115), the nth directional derivative of f (identified again as either φ or ψ) is easily shown to be

$$(\boldsymbol{a} \cdot \nabla)^n f = a^n \frac{\partial^n f}{\partial z^n} + a^{*n} \frac{\partial^n f}{\partial z^{*n}}. \tag{A.118}$$

Furthermore, using (A.113), we can show that

$$(\boldsymbol{a} \cdot \nabla)^n \varphi = \mathrm{Re}\left(a^n \frac{\mathrm{d}^n F}{\mathrm{d}z^n}\right), \qquad (\boldsymbol{a} \cdot \nabla)^n \psi = \mathrm{Im}\left(a^n \frac{\mathrm{d}^n F}{\mathrm{d}z^n}\right). \tag{A.119}$$

A special case of this result is obtained when $\boldsymbol{a}$ is either of the unit coordinate vectors. Then, it is straightforward to show that

$$\frac{\partial^n \varphi}{\partial x^m \partial y^{n-m}} = \mathrm{Re}\left(\mathrm{i}^{n-m}\frac{\partial^n}{\partial z^n}\right)\varphi = \mathrm{Re}\left(\mathrm{i}^{n-m}\frac{\mathrm{d}^n F}{\mathrm{d}z^n}\right), \tag{A.120}$$

$$\frac{\partial^n \psi}{\partial x^m \partial y^{n-m}} = \mathrm{Re}\left(\mathrm{i}^{n-m}\frac{\partial^n}{\partial z^n}\right)\psi = \mathrm{Im}\left(\mathrm{i}^{n-m}\frac{\mathrm{d}^n F}{\mathrm{d}z^n}\right). \tag{A.121}$$

Another identity, useful for transforming integrals between vector and complex notation, is obtained by considering $\boldsymbol{n} \times \nabla\mu\, \mathrm{d}s$, where μ is a real-valued function defined along a contour, and $\boldsymbol{n}$ is the unit normal vector defined as in Fig. A.5. Based on the results thus far, it can be shown that this is equivalent to the differential change of μ along the contour:

$$\boldsymbol{e}_3 \cdot (\boldsymbol{n} \times \nabla\mu\, \mathrm{d}s) \Longleftrightarrow \mathrm{d}\mu. \tag{A.122}$$

Along a contour for which the normal is defined in the opposite direction, then this relationship only changes in sign.

A.2.2 The Cauchy Integral and Residue Theorem

Cauchy Integral An important integral form that we encounter often in potential flow in the plane is the **Cauchy integral,**

$$F(z) := \frac{1}{2\pi \mathrm{i}} \int_C \frac{f(\lambda)}{\lambda - z}\, \mathrm{d}\lambda, \tag{A.123}$$

where C is a smooth curve, not necessarily closed.

Plemelj Formulae When the point z approaches curve C in (A.123), we must take special care of the singular behavior in the denominator. The trick is to break the curve into two parts: a symmetric interval surrounding z of vanishingly small length, ϵ, and the remaining portion of the curve once this interval has been removed. For this remaining portion there is no difficulty; when $\epsilon \to 0$, this is the **Cauchy principal value** of the integral, as in the vector case. For the integral over the small interval, we will show that it converges to a value that depends on the direction from which the curve is approached.

As shown in Fig. A.6, we define the + side to be to the left as we traverse the curve, and the − side to be the right. If C is a closed counterclockwise contour, for example, then the + side would be the finite interior of the contour. To evaluate the contribution of the segment of length ϵ to the overall integral over C, we can

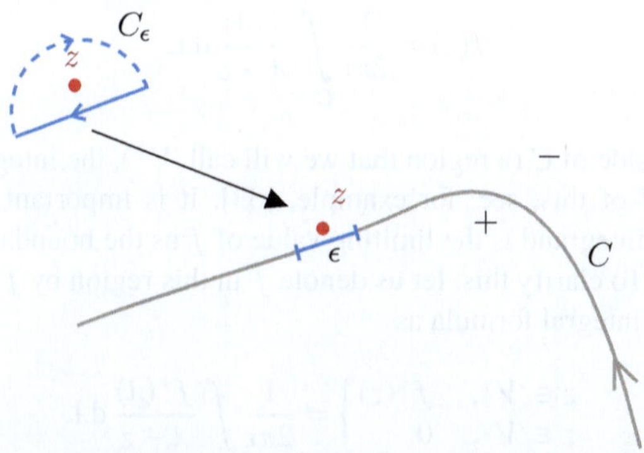

Fig. A.6 Evaluation of Cauchy integral on contour C as the evaluation point z approaches the contour. In the illustration, the point is approaching the contour from the − side. The adjacent segment of length ϵ is shown in an expanded view in the inset, as part of a closed contour C_ϵ that completes the circuit with a semicircular portion of radius $\epsilon/2$

close the segment with a semicircular portion of radius $\epsilon/2$, as illustrated in Fig. A.6 when z is on the − side of C; this forms a closed contour which we will call C_ϵ. On the semicircular portion, with z approaching the straight segment, we have that $\lambda \approx z + \epsilon e^{i\theta}/2$, where θ runs from π to 0, and $d\lambda = i\epsilon e^{i\theta}\, d\theta/2$; the function $f(\lambda)$ is approximately equal to its value at z: $f(\lambda) \approx f(z)$. The resulting integral over the semicircular contour is easily evaluated to be $-f(z)/2$. Thus, the integral over the segment of length ϵ can be written as

$$\frac{1}{2\pi i}\int_{\epsilon} \frac{f(\lambda)}{\lambda - z}\, d\lambda = \frac{1}{2\pi i}\int_{C_\epsilon} \frac{f(\lambda)}{\lambda - z}\, d\lambda + \frac{1}{2} f(z). \tag{A.124}$$

However, the integral over the closed contour C_ϵ can be evaluated with the Cauchy integral formula, described below, to be $-f(z)$: the negative sign is due to the contour's clockwise orientation. The result is that the integral over ϵ with z approaching C from the − side is $-f(z)/2$. If, instead, z approaches from the + side, then C_ϵ is completed on the other side of C with a counterclockwise semicircle; the result over ϵ is consequently the same, but of opposite sign: $f(z)/2$.

Thus, we have proved the so-called Plemelj formulae [56],

$$F^{\pm}(z) = \pm\frac{1}{2} f(z) + \frac{1}{2\pi i}⨍_{C} \frac{f(\lambda)}{\lambda - z}\, d\lambda, \quad z \in C, \tag{A.125}$$

where the integral over C is now denoted as the principal value. Note that this is the complex version of the vector result (A.46). The principal value of the integral, as in the vector case, is seen to represent the simple average of the limits, $(F^+(z)+F^-(z))/2$.

Cauchy Integral Formula Consider a complex-valued function $f(z)$ that is holomorphic and single-valued within a region $\mathcal{V}^+$ bounded by a ***closed counterclockwise contour*** C. Then, for z in the region $\mathcal{V}^+$,

$$f(z) = \frac{1}{2\pi\mathrm{i}} \int_C \frac{f(\lambda)}{\lambda - z}\, \mathrm{d}\lambda; \tag{A.126}$$

whereas for z outside of C (a region that we will call $\mathcal{V}^-$), the integral is identically zero. For a proof of this, see, for example, [33]. It is important to note that the function f in the integrand is the limiting value of f as the boundary is approached from within $\mathcal{V}^+$. To clarify this, let us denote f in this region by f^+, so that we can write the Cauchy integral formula as

$$\left.\begin{array}{ll} z \in \mathcal{V}^+, & f^+(z) \\ z \in \mathcal{V}^-, & 0 \end{array}\right\} = \frac{1}{2\pi\mathrm{i}} \int_C \frac{f^+(\lambda)}{\lambda - z}\, \mathrm{d}\lambda. \tag{A.127}$$

We can prove a similar relation for the other side of the contour, for a function f holomorphic in region $\mathcal{V}^-$, where now f is denoted by f^-:

$$\left.\begin{array}{ll} z \in \mathcal{V}^+, & 0 \\ z \in \mathcal{V}^-, & f^-(z) \end{array}\right\} = -\frac{1}{2\pi\mathrm{i}} \int_C \frac{f^-(\lambda)}{\lambda - z}\, \mathrm{d}\lambda. \tag{A.128}$$

The sign change is due to the fact that, when viewed from $\mathcal{V}^-$, the contour C is reversed in direction. It is useful to observe that these identities are the complex equivalent of the vector equation (A.43), albeit with the volumetric forcing terms set to zero.

It should be clear that we can define a function f that is holomorphic in both regions by simply adding these identities. This function will undergo a jump in passing from $\mathcal{V}^+$ to $\mathcal{V}^-$. What about when z lies on the contour C itself? For this, we can use the Plemelj formulae to obtain the result. For example, in applying the formula to Eq. (A.127), we know that the limiting value of the Cauchy integral over the closed contour C is f^+ on the $\mathcal{V}^+$ side and 0 on the $\mathcal{V}^-$ side. Thus, applying (A.125), we get

$$\frac{1}{2} f^+(z) = \frac{1}{2\pi\mathrm{i}} ⨍_C \frac{f^+(\lambda)}{\lambda - z}\, \mathrm{d}\lambda, \quad z \in C. \tag{A.129}$$

This result holds for whichever direction we approach the boundary. It is the complex analog of the vector formula (A.51), without the volumetric forcing terms. Similarly, when applied to Eq. (A.128), in which the limiting values are 0 and $f^-(z)$ from the two regions, we obtain

$$\frac{1}{2} f^-(z) = -\frac{1}{2\pi\mathrm{i}} ⨍_C \frac{f^-(\lambda)}{\lambda - z}\, \mathrm{d}\lambda, \quad z \in C. \tag{A.130}$$

The results can be summarized as follows: for any z in the complex plane,

$$f(z) = \frac{1}{2\pi\mathrm{i}} \int_C \frac{f^+(\lambda) - f^-(\lambda)}{\lambda - z}\, \mathrm{d}\lambda, \tag{A.131}$$

where the function is now defined at all z:

$$f(z) = \begin{cases} f^+(z), & z \in \mathcal{V}^+, \\ (f^+(z) + f^-(z))/2, & z \in C, \\ f^-(z), & z \in \mathcal{V}^-. \end{cases} \tag{A.132}$$

It is useful to note that we can apply these identities to a non-closed contour by closing such a contour in an arbitrary fashion, with the jump $f^+ - f^-$ set to zero on the closure.

We can extend the identity (A.126) to cases in which there is a simple pole at z_0 in $\mathcal{V}^+$, so that

$$f(z) = \frac{A}{z - z_0} + \tilde{f}(z), \tag{A.133}$$

where A is a complex constant and $\tilde{f}(z)$ is holomorphic in $\mathcal{V}^+$. Then, it can be shown that, for $z \in \mathcal{V}^+$,

$$f(z) = \frac{A}{z - z_0} + \frac{1}{2\pi \mathrm{i}} \int_C \frac{f(\lambda)}{\lambda - z}\, \mathrm{d}\lambda, \tag{A.134}$$

whereas the right-hand side is identically zero for $z \in \mathcal{V}^-$. In other words, when z lies inside C, the integral evaluates to $\tilde{f}(z)$, the portion that is holomorphic inside this contour. But when z lies outside, the result of the integral is $-A/(z - z_0)$, because this portion of $f(z)$ is holomorphic over all of $\mathcal{V}^-$. This allows us to apply Cauchy's formula to cases in which there are isolated singularities, such as point vortices or sources.

The results of the limits (A.129) and (A.130) are naturally extended to this formula,

$$\frac{1}{2}\left(-\frac{A}{z - z_0} + \tilde{f}(z) \right) = \frac{1}{2\pi \mathrm{i}} \int_C \frac{f(\lambda)}{\lambda - z}\, \mathrm{d}\lambda, \quad z \in C. \tag{A.135}$$

This gives us a more thorough analogy with the vector formula (A.51), since the simple pole corresponds to the volumetric terms evaluated at a singularity at point z_0.

Cauchy Residue Theorem The residue theorem is a powerful tool for evaluating integrals about closed contours in complex analysis. It represents a generalization of Cauchy's integral formula. We use it extensively in this work, so we summarize the basic results here. Suppose we have a function $f(z)$ that is holomorphic in a region except at a finite number of singularities. Let z_k, $k = 1, \ldots, N$ be the locations of these singularities inside the closed (counterclockwise) contour C; the contour itself is free of singularities. The main result of the residue theorem is that the integral of f on the contour is equal to the sum of the **residues** of f at the N singular points z_k:

$$\int_C f(z)\, \mathrm{d}z = 2\pi \mathrm{i} \sum_{k=1}^{N} \mathrm{Res}(f, z_k). \tag{A.136}$$

It is important to remember that only singularities inside of the contour are relevant to the integral; those outside the contour make no contribution to the integral. There are a number of ways to find the residue of the function at a singular point z_k. The most straightforward is to obtain the **Laurent series expansion** about z_k:

$$f(z) = \sum_{m=-\infty}^{\infty} a_m (z - z_k)^m. \tag{A.137}$$

The residue of f at z_k is simply the coefficient a_{-1} in this expansion—that is, the coefficient on the term $1/(z - z_k)$. This provides one means of calculating the residue. There are a few other approaches, based on the type of singularity. For example, if the singularity is a simple pole, then the residue can be calculated from

$$\mathrm{Res}(f, z_k) = \lim_{z \to z_k} (z - z_k) f(z). \tag{A.138}$$

In other words, the value of the integral is determined by the rest of the function f, after the singular factor is removed. If the singularity is a higher-order pole, then the residue involves derivatives of the non-singular part of the function. In particular, for a pole of order l,

$$\mathrm{Res}(f, z_k) = \frac{1}{(l-1)!} \lim_{z \to z_k} \frac{\mathrm{d}^{l-1}}{\mathrm{d}z^{l-1}} \left[(z - z_k)^l f(z) \right]. \tag{A.139}$$

Note that this equation includes the result for the simple pole, when $l = 1$.

To better understand the residue theorem, it is helpful to see it used in examples. The examples chosen here are relevant to the work in this book. First, let us consider $f(z) = z^{-n}$, where n is a positive integer, integrated on a closed contour C that encloses the origin. In this case, there is a pole of order n at the origin, which is inside the contour. In this case, the function is its own Laurent series, and the residue is only non-zero if $n = 1$, in which case it is equal to 1. Thus, the integral is

$$\int_C \frac{\mathrm{d}z}{z^n} = \begin{cases} 2\pi \mathrm{i}, & n = 1, \\ 0, & n \neq 1. \end{cases} \tag{A.140}$$

Now consider $f(z) = z^{-n}(z - z_0)^{-1}$, again on a contour C that wraps around the origin. The result depends on whether z_0 is inside or outside of C. In both cases, there is a pole of order n at the origin; if z_0 is also inside of the contour, then there is also a simple pole at that location. First, let's consider the case when z_0 is outside C, so that there is only a single residue (at the origin). Note that

$$\frac{\mathrm{d}^{n-1}}{\mathrm{d}z^{n-1}} \left(\frac{1}{z - z_0} \right) = \frac{(-1)^{n-1}(n-1)!}{(z - z_0)^n}. \tag{A.141}$$

Thus, following (A.139), the result of the integral is

$$\int_C \frac{dz}{z^n(z-z_0)} = 2\pi i \,\text{Res}(f,0) = \lim_{z\to 0} \frac{(-1)^{n-1}}{(z-z_0)^n} = -\frac{2\pi i}{z_0^n}. \tag{A.142}$$

for any positive integer n. If, however, z_0 is inside of the contour, then there is a second residue, at z_0, due to the simple pole there. That residue is easily calculated to be $1/z_0^n$. Thus, in this case, the integral is identically zero:

$$\int_C \frac{dz}{z^n(z-z_0)} = 2\pi i \,[\text{Res}(f,0) + \text{Res}(f,z_0)] = -\frac{2\pi i}{z_0^n} + \frac{2\pi i}{z_0^n} = 0. \tag{A.143}$$

If z_0 is on the contour, then we must indent the contour with a semicircle around z_0, from which we get a result that is half of the residue. The resulting integral is the average of the results when z_0 approaches the contour from either side: $-\pi i/z_0^n$.

Occasionally, it is necessary to compute a contour integral in which the 'interior' of the contour is the region external to C, extending to infinity; a positively-oriented contour of this region would be clockwise rather than counterclockwise. The residue theorem remains valid in such a case, but here we must consider the possibility of a residue at infinity. This is obtained by reflecting z to the interior of C and calculating instead the residue at the origin:

$$\text{Res}(f,\infty) = -\text{Res}\left(\frac{1}{z^2} f(1/z), 0\right). \tag{A.144}$$

For example, if $f(z) = 1/z$, then $f(1/z)/z^2 = 1/z$, for which the residue at the origin is 1, so the residue of f at infinity is -1. This should not be a surprise, since, by considering a clockwise contour that bounds the entire outer region, we should simply get the negative of the result obtained on a counterclockwise contour bounding the interior (in which the origin is a simple pole).

A.2.3 Conformal Mapping

Holomorphic functions can be used to great advantage for solving problems in two dimensions, because we can often find one that allows us to map to a complicated region in the physical plane from a relatively simple shape in another plane. Provided the function and its derivative are non-zero everywhere in the mapped region, then it is said to be a **conformal map**. A crucial property of conformal maps is that they locally preserve angles. For example, grid lines that are orthogonal in one plane are mapped to orthogonal grid lines in the other. This is illustrated in Fig. A.7, in which the region exterior to a circle of unit radius has been mapped into a region outside of a more complicated shape. The radial grid lines, orthogonal to the unit circle, are mapped to curved lines that remain orthogonal to the contour of the object (and to the mapped version of every circle of larger radius).

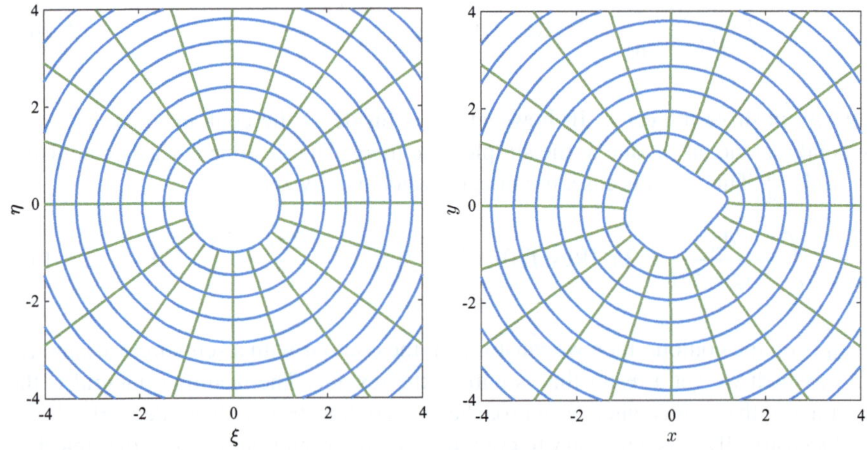

Fig. A.7 Example of conformal mapping from ζ plane (left) to the z plane (right). Orthogonality of grid lines is preserved under conformal mappings, as illustrated here

The usual objective of conformal mapping is to shift the burden of solving the underlying problem (the Laplace equation in the physical plane) to a geometry in which the solution is straightforward. Thus, it is common to seek maps from a straight line or a circle of unit radius. A few other properties of conformal maps should be mentioned. Let us denote z as the complex coordinate in the physical plane and ζ as the coordinate in the other plane of some simpler geometry. The conformal map we denote by $z(\zeta)$ and its derivative (the **Jacobian**) by $z'(\zeta)$.

- Because the map and its derivative are non-zero in the region to which they are applied, the map is invertible. We denote this by $\zeta(z)$. Furthermore, the Jacobian of the inverse is simply the inverse of the Jacobian: $\zeta'(z(\zeta)) = 1/z'(\zeta)$. (We will encounter situations in which we relax this so that the Jacobian or its inverse vanish at a finite number of points on the boundary. However, the map remains invertible in these situations.)
- If a function $\hat{F}(\zeta)$ satisfies Laplace's equation in a region of the ζ plane, then the composition $F(z) = \hat{F}(\zeta(z))$ also satisfies Laplace's equation in the corresponding region of the z plane. Thus, we are guaranteed to obtain the solution we desire in the physical plane if we first obtain it in the ζ plane.
- The Riemann mapping theorem states that any simply connected subset of the complex plane (i.e., a finite region with no holes) can be conformally mapped from the unit disk. A corollary of this theorem is that the exterior of the enclosing boundary of this region can be mapped from the region exterior to the unit circle. Thus, by studying the problem outside a circle, we can obtain the solution outside *any* closed object!

Geometric Properties It is particularly easy to obtain formulas for the geometric properties of the physical shape of the body from the conformal map. Let us suppose—as we only do in this work—that we are mapping the unit circle C_c in the ζ plane to a shape described by contour C_b in the z plane. The unit circle is naturally parameterized by angle $\theta \in [0, 2\pi)$, and its shape is described by $\zeta = \exp(i\theta)$. Through the map, this parameterization by θ naturally carries over to the shape C_b, which is described by the curve

$$Z(\theta) = z(\exp(i\theta)). \tag{A.145}$$

If we differentiate the curve $Z(\theta)$ with respect to θ and normalize the result, we obtain the unit (counterclockwise) tangent, τ, on C_b; the unit normal, n, is simply a clockwise rotation by $\pi/2$:

$$\tau = Z'(\theta)/|Z'(\theta)|, \quad n = -iZ'(\theta)/|Z'(\theta)|. \tag{A.146}$$

Note that the magnitude in the denominator, $|Z'(\theta)|$, represents the differential arc length along C_b corresponding a differential change in angle along C_c; that is, $|Z'(\theta)| = ds/d\theta$. Now, we can relate this to the Jacobian of the conformal map by differentiating the right side of (A.145) with respect to θ with the help of the chain rule. The result is

$$Z'(\theta) = i\zeta z'(\zeta), \tag{A.147}$$

where $\zeta = \exp(i\theta) \in C_c$. Note that $i\zeta$ is simply the counterclockwise tangent on the unit circle. Thus, Eq. (A.147) reveals an important role of the Jacobian of the map: it rotates vectors from the ζ plane to the z plane. The length has been altered, however, so we still must normalize the result to obtain the unit tangent and normal on C_b, noting that $|\zeta| = 1$ on the unit circle:

$$\tau = i\zeta\frac{z'(\zeta)}{|z'(\zeta)|}, \quad n = \zeta\frac{z'(\zeta)}{|z'(\zeta)|}, \quad \zeta = \exp(i\theta) \in C_c. \tag{A.148}$$

Note that the Jacobian of the map has performed the same role in rotating both the normal and tangent on the circle to their counterparts on C_b. This illustrates one of the defining properties of the conformal map: ***the angle between vectors (in this case, between the tangent and normal on the body surface) is preserved***.

The area, $\mathcal{V}_b$, the centroid, Z_c, and the second area moment, $\mathcal{J}$, of a closed shape in the plane can be obtained by adapting, respectively, the vector identities (A.26), (A.29) and (A.28) (for $n = 2$) to complex form. These allow us to transform the area, centroid and second area moment of a two-dimensional shape into integrals over its surface contour. The complex versions of these can be shown to be, respectively,

$$\mathcal{V}_b = \frac{1}{2}\mathrm{Im}\int_{C_b} z^*\,dz, \quad Z_c = -\frac{i}{2\mathcal{V}_b}\int_{C_b} |z|^2\,dz, \quad \mathcal{J} = \frac{1}{4}\mathrm{Im}\int_{C_b} |z|^2 z^*\,dz. \tag{A.149}$$

The expression for the centroid is only sensible if $\mathcal{V}_b > 0$. These can be transformed into integrals over the unit circle, noting that $\zeta^* = 1/\zeta$ on this circle:

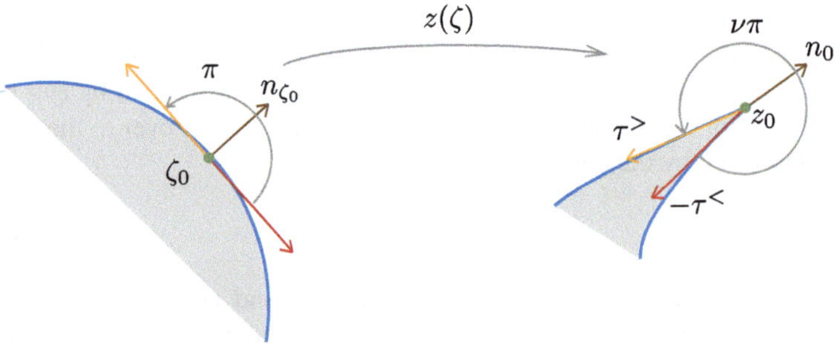

Fig. A.8 Map from a point ζ_0 on the boundary contour at which the Jacobian $z'(\zeta)$ vanishes

$$\mathcal{V}_b = \frac{1}{2}\mathrm{Im}\int_{C_c} z^*(1/\zeta)z'(\zeta)\,\mathrm{d}\zeta, \quad Z_c = -\frac{\mathrm{i}}{2\mathcal{V}_b}\int_{C_c} z^*(1/\zeta)z(\zeta)z'(\zeta)\,\mathrm{d}\zeta. \quad \text{(A.150)}$$

and

$$\mathcal{J} = \frac{1}{4}\mathrm{Im}\int_{C_c} [z^*(1/\zeta)]^2\, z(\zeta)z'(\zeta)\,\mathrm{d}\zeta. \quad \text{(A.151)}$$

The Power Series Transformation Let us consider a broad class of conformal transformations described by

$$z(\zeta) = Z_r + \mathrm{e}^{\mathrm{i}\alpha}\tilde{z}(\zeta), \quad \text{(A.152)}$$

where the body-fixed coordinates $\tilde{z}(\zeta)$ are described by the power series map of the circle plane:

$$\tilde{z}(\zeta) = c_1\zeta + c_0 + \sum_{k=1}^{\infty}\frac{c_{-k}}{\zeta^k}. \quad \text{(A.153)}$$

The unit circle is transformed to the shape of the body by (A.153). This shape is then rigidly transformed by Eq. (A.152) with a rotation by angle α and translation to Z_r. The first coefficient, c_1, defines the re-scaling of distance between the circle plane and physical plane, since $\tilde{z}(\zeta) \sim c_1\zeta$ as $|\zeta| \to \infty$. The second coefficient, c_0, represents a uniform translation of the shape, and can either be set to zero or chosen to enforce a constraint on the map. We generally must restrict the coefficients so that the Jacobian,

$$z'(\zeta) = -\mathrm{e}^{\mathrm{i}\alpha}\sum_{k=-1}^{\infty}\frac{kc_{-k}}{\zeta^{k+1}}, \quad \text{(A.154)}$$

does not vanish at any point outside of the unit circle. However, we do allow for a finite number of zeros of the Jacobian to lie on the circle itself: these become edges in the physical plane, as will be discussed later in this section. Using the residue theorem, it is straightforward to show that the area, centroid and second moment of area of the body generated by a transformation in this class are given by

$$\mathcal{V}_b = -\pi \sum_{k=-1}^{\infty} k|c_{-k}|^2, \quad Z_c = Z_r - e^{i\alpha} \frac{\pi}{\mathcal{V}_b} \sum_{k=-1}^{\infty} k c_{-k} d_k, \tag{A.155}$$

and

$$\mathcal{J} = -\frac{\pi}{2} \sum_{k,l=-1}^{\infty} l c^*_{-k} c_{-l} d_{-k+l} - |Z_c - Z_r|^2 \mathcal{V}_b + |Z_c|^2 \mathcal{V}_b, \tag{A.156}$$

where the coefficients d_{-k} are defined from the Cauchy product of the map with its conjugate in evaluating $|z(\zeta)|^2$:

$$d_{-k} = \begin{cases} \sum_{l=-1}^{\infty} c^*_{-l} c_{-l-k}, & k \geq 0, \\ \sum_{l=-1}^{\infty} c_{-l} c^*_{-l+k} = d^*_k, & k \leq -1. \end{cases} \tag{A.157}$$

Clearly, the coefficients must be chosen so that $\mathcal{V}_b \geq 0$. Also, note that Z_r does not represent the body centroid unless the summation in the second of the equations (A.155) vanishes. However, this can be ensured by choosing the coefficient c_0 appropriately. Therefore, in our use of this map, we will always assume that the coefficients have been selected in this fashion, so that $Z_c = Z_r$.

Any closed body shape can be constructed from a suitable choice of the coefficients, a general discussion of which we will omit here. But as an example, if we set $Z_r = 0$, $c_1 = (a + b)/2$, $c_{-1} = (a - b)/2$, and all other coefficients to zero, then we generate an ellipse with semi-major axis length a and semi-minor axis length b centered at the origin. The only non-zero d_{-k} are $d_0 = c_1^2 + c_{-1}^2 = (a^2 + b^2)/2$ and $d_{-2} = d_2 = c_1 c_{-1} = (a^2 - b^2)/4$. The area, centroid, and second moment, are easily found from (A.155) and (A.156) to be $\mathcal{V}_b = \pi ab$, $Z_c = 0$ and $\mathcal{J} = \pi ab(a^2 + b^2)/4$, as expected. If we let $b \to 0$ and $a \to c/2$, then we obtain an infinitely-thin flat plate of length c:

$$z(\zeta) = \frac{c}{4}\left(\zeta + \frac{1}{\zeta}\right). \tag{A.158}$$

These examples with only the c_1 and c_{-1} coefficients nonzero comprise the **Joukowski transformation**.

Corners As stated earlier, we allow for the Jacobian $z'(\zeta)$ or its inverse $\zeta'(z)$ to vanish at a finite number of isolated points on the boundary of the mapped region. What is the consequence of such a point? In the vicinity of any such point ζ_0, the Jacobian has the form

$$z'(\zeta) \approx (\zeta - \zeta_0)^{\nu-1} f(\zeta), \tag{A.159}$$

where $0 \leq \nu \leq 2$ and $f(\zeta)$ is smooth and non-zero throughout the neighborhood of ζ_0. For $\nu > 1$, the Jacobian z' vanishes, whereas for $\nu < 1$, its inverse vanishes.

From a simple integration, the map and the inverse map locally take the form

$$z - z_0 \approx \frac{1}{\nu} f(\zeta_0)(\zeta - \zeta_0)^\nu, \qquad \zeta - \zeta_0 \approx \left(\frac{z - z_0}{f(\zeta_0)/\nu} \right)^{1/\nu}, \tag{A.160}$$

where $z_0 = z(\zeta_0)$ is the corresponding point in the physical plane. Written another way, in the vicinity of the edge,

$$z - z_0 \approx \frac{1}{\nu} z'(\zeta)(\zeta - \zeta_0), \tag{A.161}$$

which shows that the Taylor expansion of the mapping in this vicinity is modified slightly from the smooth behavior ($\nu = 1$). By differentiating the approximate expression for $\zeta(z)$, the Jacobian of the inverse transformation is seen to be

$$\zeta'(z) \approx \frac{1}{\nu} \left(\frac{\nu}{f(\zeta_0)} \right)^{1/\nu} \frac{1}{(z - z_0)^{1-1/\nu}}. \tag{A.162}$$

An important consequence of (A.160) is that the arguments about ζ_0 and z_0 are related by

$$\arg(z - z_0) = \nu \arg(\zeta - \zeta_0) + \arg(f(z_0)/\nu), \tag{A.163}$$

from which we can see that any rotation through angle ϕ in the circle plane corresponds to rotation through $\nu\phi$ in the physical plane. In other words, the preservation of angles by conformal mapping does not hold at points at which the Jacobian or its inverse vanishes. An example of such a point ζ_0 on the unit circle is shown in Fig. A.8. Clearly, the boundary tangents that extend from either side of ζ_0 turn through angle π. When the region surrounding this point is mapped to the physical plane, the tangents are no longer co-linear, but instead, turn through angle $\nu\pi$, forming a corner on the mapped surface. The interior angle of the corner is $(2 - \nu)\pi$, which, by our earlier restriction on ν, must be between 0 and 2π. Thus, ***zeros of the Jacobian of the map and its inverse form corners on the mapped geometry***. In the most extreme cases, $\nu = 2$ and $\nu = 0$, the boundary forms a cusp on one side or the other.

In the diagram in Fig. A.8, we have included the unit normal on the circle, n_{ζ_0}, and its image, n_0. As it would be at any point on the unit circle, the unit normal at ζ_0 is simply equal to the point itself: $n_{\zeta_0} \equiv \zeta_0$. To see how the mapped normal, n_0, relates to other quantities, let's first determine the descriptions of the tangents on either side; the normal is simply the negative of the mean of these. As Eq. (A.146) indicates, we should inspect the manner in which $z'(\zeta)/|z'(\zeta)|$ behaves as ζ_0 is approached from either side. Let us denote the regions of the contour just before and just after the corner, when traversed in counterclockwise fashion, by $<$ and $>$, respectively. In either region, we can write $\zeta = \zeta_0 e^{\pm i\epsilon}$, where $\epsilon > 0$ and the + or − sign is chosen according to whether ζ is on the $>$ or $<$ side of ζ_0. Since we will allow ϵ to shrink to zero, we can write $(\zeta - \zeta_0)^> \approx i\epsilon\zeta_0$ and $(\zeta - \zeta_0)^< \approx -i\epsilon\zeta_0$. Using the form of the Jacobian (A.159) near the corner, it is straightforward to show that

$$\lim_{\epsilon \to 0} \frac{z'(\zeta)}{|z'(\zeta)|} = \frac{f(\zeta_0)}{|f(\zeta_0)|} (\pm i\zeta_0)^{\nu-1} = e^{\pm i\pi(\nu-1)/2} \zeta_0^{\nu-1} \frac{f(\zeta_0)}{|f(\zeta_0)|}, \tag{A.164}$$

and thus, by (A.148), the tangents on either side of z_0 are

$$\tau^{>}(\zeta_0) = \mathrm{e}^{\mathrm{i}\pi\nu/2}\zeta_0^{\nu}\frac{f(\zeta_0)}{|f(\zeta_0)|}, \quad \tau^{<}(\zeta_0) = -\mathrm{e}^{-\mathrm{i}\pi\nu/2}\zeta_0^{\nu}\frac{f(\zeta_0)}{|f(\zeta_0)|}, \tag{A.165}$$

and the normal is given by rotating either $\tau^{>}$ clockwise by angle $\nu\pi/2$, or $-\tau^{<}$ counterclockwise by the same angle:

$$n_0 := n(\zeta_0) = \zeta_0^{\nu}\frac{f(\zeta_0)}{|f(\zeta_0)|}. \tag{A.166}$$

Thus, the normal ζ_0 in the circle plane is rotated both by the factor $f(\zeta_0)$, which acts as it would along a smooth section of contour, as well as by the exponent ν, which 'pinches' the contour and its local vectors at the corner. Note that the values of $z - z_0$ just before (<) and just after (>) the corner are related by

$$(z - z_0)^{<} = \mathrm{e}^{-\mathrm{i}\pi\nu}(z - z_0)^{>}, \tag{A.167}$$

indicative of a turn through angle $\nu\pi$.

A.2.4 The Joukowski and Kármán–Trefftz Airfoils

The Joukowski transformation defined in (A.158) mapped the unit circle to a flat plate, a body with two sharp edges. Suppose we apply the same transformation to a circle whose center is off of the origin—to what shape does this shifted circle map? Before we answer this, let us remember that the sharp edges arise on the flat plate because the Jacobian of the mapping vanishes at two points, ±1. Here, we will restrict our attention to a circle that passes through one of these two points but encloses the other: in other words, the circle has not only shifted its center but also expanded its radius. The right panel in Fig. A.9 depicts the geometry of this circle, passing through the point 1 but enclosing the point −1. In this way, we ensure that the shape to which the circle maps has one edge and that the other zero of the Jacobian is safely inside the circle, so that the transformation remains conformal in the region exterior to the circle.

As the geometry in the right panel of Fig. A.9 illustrates, the radius of the circle, a, must necessarily be larger than unity to enclose the zero of the Jacobian at −1. Furthermore, the circle center, at $\epsilon\mathrm{e}^{\mathrm{i}\delta}$, must lie in the left-half plane. We will restrict our attention to cases in which this center lies in the upper left quadrant or on the bounding axes, $\pi/2 \leq \delta \leq \pi$; the lower left quadrant is simply a mirror image. It is easy to confirm that the labeled parameters of the circle are related by

$$a\mathrm{e}^{-\mathrm{i}\beta} = 1 - \epsilon\mathrm{e}^{\mathrm{i}\delta}, \tag{A.168}$$

or

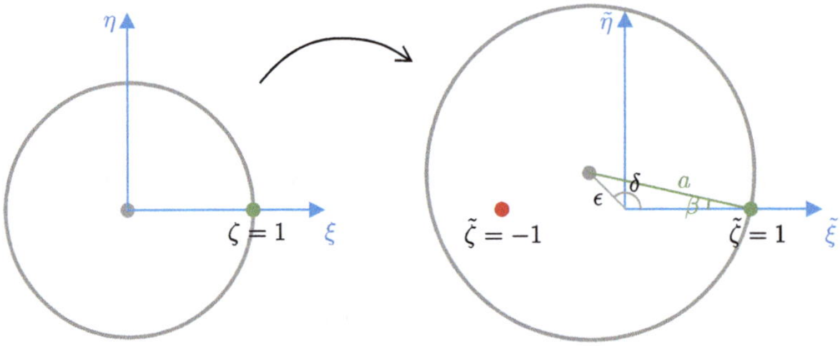

Fig. A.9 Scaling, rotation and translation of the unit circle in the ζ plane to the $\tilde{\zeta}$ plane by Eq. (A.170). The green dots represent the pre-images of the edge in each plane

$$a = \left(1 - 2\epsilon\cos\delta + \epsilon^2\right)^{1/2}, \qquad \beta = \arctan\left(\frac{\epsilon\sin\delta}{1 - \epsilon\cos\delta}\right). \tag{A.169}$$

Though the Joukowski mapping will be applied to this shifted and expanded circle, we still would like our mapping to originate from a unit circle centered at the origin, since many of our results in this book depend on this basic partitioning of the complex plane. But this desire is easy to accommodate: we simply first map the unit circle—in our usual ζ plane—to an expanded, rotated and shifted circle in an intermediate plane we will denote with coordinates $\tilde{\zeta}$, as depicted in Fig. A.9, with the transformation

$$\tilde{\zeta}(\zeta) = \epsilon e^{i\delta} + a e^{-i\beta}\zeta = \epsilon e^{i\delta} + (1 - \epsilon e^{i\delta})\zeta. \tag{A.170}$$

The unit circle is described, as always, by $\zeta = e^{i\theta}$, where $\theta \in [0, 2\pi)$. Then we map the transformed circle to a shape in the physical plane, z, with the Joukowski transformation:

$$z(\tilde{\zeta}) = \frac{c}{4}\left(\tilde{\zeta} + \frac{1}{\tilde{\zeta}}\right). \tag{A.171}$$

Note that the rotation of the unit circle by $-\beta$ in Eq. (A.170) ensures that the edge of the body at $z = c/2$ is mapped from $\zeta = 1$ via $\tilde{\zeta} = 1$. The composite mapping is

$$z(\zeta) = \frac{c}{4}\left(\epsilon e^{i\delta} + (1 - \epsilon e^{i\delta})\zeta + \frac{1}{\epsilon e^{i\delta} + (1 - \epsilon e^{i\delta})\zeta}\right). \tag{A.172}$$

So let us return to the question: To what does this scaled and shifted circle map? We can gain insight by assuming $\epsilon \ll 1$ and expanding the mapping in powers of this parameter. For this task, we isolate a factor $1/\zeta$ from the last term, and on the result, we apply the very useful Taylor expansion,

$$(1 + \mu)^p = 1 + p\mu + \frac{p(p-1)}{2}\mu^2 + \ldots + \binom{p}{n}\mu^n + \ldots, \tag{A.173}$$

for $|\mu| < 1$, where in this case, $\mu = -\epsilon e^{i\delta}(1 - 1/\zeta)$. Thus, after some additional manipulation, the transformation (A.172) can be written as

$$z(\zeta) = \frac{c}{4}\left[\zeta + \zeta^{-1} - \epsilon e^{i\delta}\zeta(1+\zeta^{-1})(1-\zeta^{-1})^2 + \epsilon^2 e^{i2\delta}\zeta^{-1}(1-\zeta^{-1})^2 + O(\epsilon^3)\right], \tag{A.174}$$

and its Jacobian as

$$z'(\zeta) = \frac{c}{4}(\zeta - 1)\left[\zeta^{-1} + \zeta^{-2} - \epsilon e^{i\delta}(\zeta^{-1} + \zeta^{-2} + 2\zeta^{-3}) - \epsilon^2 e^{i2\delta}(\zeta^{-3} - 3\zeta^{-4}) + O(\epsilon^3)\right], \tag{A.175}$$

which, even in this expanded form, clearly vanishes at $\zeta = 1$.

Let us obtain the approximate shape mapped from the unit circle by setting $\zeta = e^{i\theta}$ and separating the real and imaginary parts. With some manipulation, these are

$$x(\theta) = \frac{c}{2}\cos\theta + \frac{1}{2}\epsilon c\,(\cos(\delta - \theta) + \cos\delta)\,(1 - \cos\theta), \tag{A.176}$$

$$y(\theta) = \frac{1}{2}\epsilon c\,(\sin(\delta - \theta) + \sin\delta)\,(1 - \cos\theta). \tag{A.177}$$

This form reveals that the leading behavior of the transformation is a flat plate of chord length c. The deviation from the flat plate is an airfoil-like shape, called the **Joukowski airfoil**, primarily borne by the y component. We will illustrate the behavior with the two extreme values, $\delta = \pi$ and $\delta = \pi/2$.

Symmetric Airfoil, $\delta = \pi$ In this case, the circle in the $\tilde{\zeta}$ plane is shifted along the negative real axis: its radius is $a = 1 + \epsilon$ and the angle β is zero. From Eqs. (A.176) and (A.177), the contour of the mapped body shape is given approximately by

$$x(\theta) = \frac{c}{2}\cos\theta, \quad y(\theta) = \frac{1}{2}\epsilon c\sin\theta(1 - \cos\theta). \tag{A.178}$$

This describes an airfoil that is symmetric about the real axis and achieves its largest deviations from this axis at $\theta = 2\pi/3$ and $\theta = 4\pi/3$, where $y = 3\sqrt{3}\epsilon c/8$ and $y = -3\sqrt{3}\epsilon c/8$, respectively, and $x = -c/4$ for both. In other words, ***the thickness of this symmetric airfoil is*** $3\sqrt{3}\epsilon c/4$, ***achieved at the quarter chord***. An example of this case is shown in the upper left panel in Fig. A.10.

Arced Airfoil, $\delta = \pi/2$ For this choice, the circle is shifted along the positive imaginary axis, its radius is $a = (1 + \epsilon^2)^{1/2} \approx 1$, and $\beta \approx \epsilon$. The outline of the mapped body shape is approximately

$$x(\theta) = \frac{c}{2}\cos\theta, \quad y(\theta) = \frac{1}{2}\epsilon c\sin^2\theta. \tag{A.179}$$

The upper and lower surfaces of the airfoil are identical and lie along a curve that is arced (or **cambered**) above the x axis, reaching its ***maximum camber***, $\epsilon c/2$, ***at*** $\theta = \pi/2$, ***the midpoint of the arc***. The upper right panel of Fig. A.10 depicts an example of an arced shape generated in this manner.

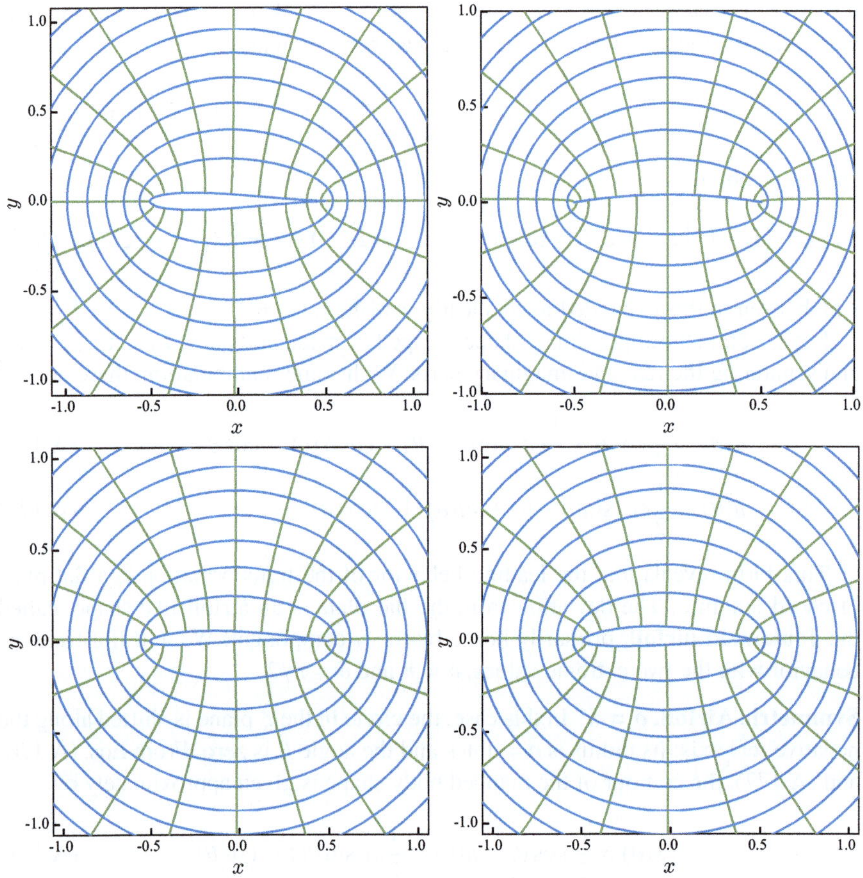

Fig. A.10 Airfoils, generated with transformation (A.171) or (A.184), with $\epsilon = 0.08$. (Upper left) Symmetric Joukowski airfoil, with $\delta = \pi$. (Upper right) Arced Joukowski airfoil, with $\delta = \pi/2$. (Lower left) Cambered thick Joukowski airfoil, with $\delta = 3\pi/4$. (Lower right) A cambered thick Kármán–Trefftz airfoil, with $\nu = 1.94$ and $\delta = 3\pi/4$

From these two extremes, we can infer that values of δ that lie between them generate an airfoil with both thickness and camber, as illustrated in the lower left panel in Fig. A.10. Both of these geometric properties are proportional to ϵc. There are a few other observations we can make.

- The power series representation of the transformation is easily obtained from the expanded form in (A.174) by grouping terms in like powers of ζ. Through order ϵ^2, the coefficients of this representation are

$$c_1 = \frac{c}{4}(1 - \epsilon e^{i\delta}), \quad c_0 = \frac{c}{4}\epsilon e^{i\delta}, \quad c_{-1} = \frac{c}{4}(1 + \epsilon e^{i\delta} + \epsilon^2 e^{i2\delta})$$
$$c_{-2} = -\frac{c}{4}(\epsilon e^{i\delta} + 2\epsilon^2 e^{i2\delta}), \quad c_{-3} = \frac{c}{4}\epsilon^2 e^{i2\delta}. \tag{A.180}$$

From these coefficients, we can calculate the geometric properties of the airfoil from Eqs. (A.155) and (A.156). In particular, it can be confirmed that the area enclosed by the airfoil contour is

$$\mathcal{V}_b = -\frac{\pi c^2}{4}\epsilon(1 + \epsilon \cos\delta)\cos\delta, \tag{A.181}$$

which is identically zero, as expected, for the arced airfoil ($\delta = \pi/2$) and sensibly positive for all choices $\pi/2 < \delta \leq \pi$ for which the airfoil has thickness.

- The angle of the trailing edge, at $\zeta_T = 1$ and $z_T = c/2$, can be found by noting the behavior of the Jacobian (A.175) in the vicinity of the edge and comparing it with Eq. (A.159). The term in brackets in (A.175) is smooth and does not vanish in the vicinity of the edge, and thus, represents the function $f(\zeta)$ in Eq. (A.159). The power ν, by inspection, is 2, and thus, the interior angle of the edge is zero. In other words, ***the trailing edge of a Joukowski airfoil is cusped***.

This last observation, that the airfoil generated by this transformation has a cusped trailing edge, is perhaps undesirable for modeling a realistic airfoil shape. Thus, it is reasonable to ask whether the transformation can be modified in some fashion to allow a trailing edge with a non-zero angle. Let us first observe that the original Joukowski transformation (A.171) can be alternatively written in a manner that emphasizes the angle of the edge:

$$\frac{z - c/2}{z + c/2} = \frac{(\tilde{\zeta} - 1)^2}{(\tilde{\zeta} + 1)^2}. \tag{A.182}$$

The power 2 on the right-hand side represents the parameter ν in the near-edge behavior (A.160). Thus, we can generalize this transformation to accommodate edges of other angles by writing

$$\frac{z - \nu c/4}{z + \nu c/4} = \frac{(\tilde{\zeta} - 1)^\nu}{(\tilde{\zeta} + 1)^\nu}. \tag{A.183}$$

This transformation, called the **Kármán–Trefftz airfoil**, maps our usual shifted, expanded circle to an airfoil shape whose trailing edge has interior angle $(2 - \nu)\pi$. Generally, ν is chosen to be only slightly less than 2. The terms $\nu c/4$ in the ratio on the left-hand side ensure that this transformation preserves the same rescaling of distance as in the Joukowski transformation, $z \approx c\tilde{\zeta}/4$ as $|\tilde{\zeta}| \to \infty$. Written in a form suitable for calculation, the transformation is

$$z(\tilde{\zeta}) = \frac{\nu c}{4}\frac{(\tilde{\zeta} + 1)^\nu + (\tilde{\zeta} - 1)^\nu}{(\tilde{\zeta} + 1)^\nu - (\tilde{\zeta} - 1)^\nu}, \tag{A.184}$$

and this is combined, as for the Joukowski transformation, with the mapping (A.170) from the unit circle to the intermediate $\tilde{\zeta}$ plane. The Jacobian (of the mapping from the intermediate $\tilde{\zeta}$ plane) is

$$z'(\tilde{\zeta}) = \nu^2 c \frac{(\tilde{\zeta}^2 - 1)^{\nu-1}}{\left[(\tilde{\zeta}+1)^\nu - (\tilde{\zeta}-1)^\nu\right]^2}; \tag{A.185}$$

the Jacobian of the composite mapping from the ζ plane is obtained by multiplying this by $\mathrm{d}\tilde{\zeta}/\mathrm{d}\zeta = 1 - \epsilon e^{i\delta}$.

The lower right panel in Fig. A.10 depicts an example of a Kármán–Trefftz airfoil, with $\nu = 1.94$ (so that the trailing-edge angle is 10.8°). By visual comparison with the Joukowski airfoil with the same choice of ϵ and δ, the Kármán–Trefftz is noticeably thicker. In fact, a Kármán–Trefftz airfoil still has thickness when $\delta = \pi/2$, in contrast to the Joukowski airfoil.

A.2.5 The Schwarz–Christoffel Transformation

In our discussion above, we noted that corners are formed on the boundary of the mapped region wherever the Jacobian of the conformal transformation vanishes. In fact, this is the foundation for a useful class of conformal maps, called the **Schwarz–Christoffel (S-C) transformation**. In its most basic form, the S-C transformation maps the upper half plane (we will call this the σ plane, to distinguish it from the circle plane) to the interior of a polygon (in the z plane). To do so, it relies on the fact we observed earlier: if the Jacobian of the transformation has the local form $\tilde{z}'(\sigma) \approx (\sigma - \sigma_j)^{\alpha_j - 1} f(\sigma)$ near a point σ_j, where f is locally holomorphic, then an angle π in the σ plane—as one would get for any smooth curve passing through σ_j—maps to an angle $\alpha_j \pi$ in the z plane. The S-C transformation exploits this behavior with a set of **pre-vertices** σ_j, $j = 1, \ldots, n$, along the real axis, arranged in order of increasing value, which map to the n vertices of the polygon, $\tilde{z}_j$, with interior angles $\alpha_j \pi$, $j = 1, \ldots, n$. The real σ axis is the pre-image of the bounding contour of the polygon, traversed in counter-clockwise fashion.

To be consistent with our focus on mappings from the unit circle to the exterior of bodies of complicated shape, we will rely here on a composite mapping that takes the exterior of the unit circle to the *exterior* of the polygon. This is given as a composition of two mappings: the first is a mapping from the exterior (in the ζ plane) to the interior (in an intermediate plane, labeled as λ) of the unit circle:

$$\lambda = h(\zeta) := -\frac{C}{|C|\zeta}; \tag{A.186}$$

the second is a mapping from the interior of the unit circle (in the λ plane) to the exterior of the polygon in the physical ($\tilde{z}$) plane [21]:

$$\tilde{z} = g(\lambda) := A + C \int^{\lambda} s^{-2} \prod_{j=1}^{n} (1 - s/\lambda_j)^{1-\alpha_j}\, \mathrm{d}s. \tag{A.187}$$

The λ_j are the pre-vertices on the unit circle in the λ plane, and A and C are complex constants that translate, scale and rotate the polygon. The factor $-C/|C|$ in the first mapping simply rotates the overall mapping, in order to preserve the orientation between the ζ and $\tilde{z}$ planes.

The composition, $\tilde{z}(\zeta) := g(h(\zeta))$—our desired mapping—is given by

$$\tilde{z}(\zeta) = A + |C| \int^{\zeta} \prod_{j=1}^{n} (1 - \zeta_j/s)^{\beta_j}\, \mathrm{d}s, \tag{A.188}$$

and its Jacobian by

$$\tilde{z}'(\zeta) = |C| \prod_{j=1}^{n} (1 - \zeta_j/\zeta)^{\beta_j}, \tag{A.189}$$

where we have defined $\beta_j = 1 - \alpha_j$ for the exterior turning angles, $\beta_j \pi$, on the polygon, and $\zeta_j := -C/(|C|\lambda_j)$ as the pre-image of the pre-vertices in the λ plane (which we might call the pre-pre-vertices, but for brevity will simply refer to as pre-vertices). As a note, the second derivative of the composite mapping is

$$\tilde{z}''(\zeta) = \tilde{z}'(\zeta)\zeta^{-2} \sum_{j=1}^{n} \frac{\beta_j \zeta_j}{1 - \zeta_j/\zeta}. \tag{A.190}$$

In order to map to polygons, we allow the interior angles to be in the range $0 \le \alpha_j \le 2$, which includes both concave ($1 < \alpha_j \le 2$) and convex ($0 \le \alpha_j < 1$) corners on the polygonal body. We only insist that

$$\sum_{j=1}^{n} \alpha_j = n - 2, \tag{A.191}$$

so that the total interior angle of the polygon is 2π.

For a general shape specified by its vertices $\tilde{z}_j$, the evaluation of the integral in (A.188), and thus, the determination of the pre-vertices ζ_j and the coefficient C, must be carried out numerically, through quadrature. Fortunately, there is a very useful Matlab-based numerical library available for this task, called *SC Toolbox*, developed by Driscoll [20]. A few examples generated with the S-C mapping are depicted in Fig. A.11. These illustrate that a 'polygon' is actually a broader class than one might imagine, since it includes infinitely thin plates (with vertices specified once for the upper side, and then repeated, in reverse order, to specify the lower side), as well as many shapes that approach continuous outlines as the number of vertices increases. A NACA 0012 airfoil, discretized with 38 segments, is depicted in the lower right panel.

Since the S-C mapping can be used to create closed polygons in the z plane, with n vertices z_j, it is useful to have expressions for the basic geometric properties of such polygons. These are obtained by evaluating the expressions (A.149), for a closed contour C_b consisting of straight segments between the vertices; they are easily found to be

$$\mathcal{V}_\mathrm{b} = \frac{1}{2}\mathrm{Im}\sum_{j=1}^{n} z^*_{j+1/2}\Delta z_{j+1/2}, \quad Z_\mathrm{c} = -\frac{\mathrm{i}}{2\mathcal{V}_\mathrm{b}}\sum_{j=1}^{n}\Delta z_{j+1/2}\left(|z_{j+1/2}|^2 + \frac{1}{12}|\Delta z_{j+1/2}|^2\right)$$
$$\mathcal{J} = \frac{1}{4}\mathrm{Im}\sum_{j=1}^{n} z^*_{j+1/2}\Delta z_{j+1/2}\left(|z_{j+1/2}|^2 + \frac{1}{12}|\Delta z_{j+1/2}|^2\right) \tag{A.192}$$

in which $z_{j+1/2} := (z_j + z_{j+1})/2$, $\Delta z_{j+1/2} := z_{j+1} - z_j$, and one simply interprets $z_{n+1} := z_1$.

For various calculations, it is useful to form the power-series representation of our S-C mapping (A.188). This is provided by the following result:

Result A.7: Power Series Representation of Schwarz–Christoffel Map

For a closed polygon with n vertices, described with pre-vertices ζ_j in the circle plane, and exterior turning angles $\beta_j\pi$, $j = 1, \ldots, n$, the S-C mapping (A.188) from the exterior of the unit circle to the exterior of the polygon can be alternatively written as a power series (A.153), with the coefficients set as follows:

$$c_1 = |C|, \quad c_0 = A, \quad c_{-k} = -|C| \sum_{\|\mathbf{I}\|_1 = k+1} (-1)^{|\mathbf{I}|} \prod_{l=1}^{k+1} \frac{1}{i_l!}\frac{M_l^{i_l}}{k l^{i_l}}, \ k \geq 1, \tag{A.193}$$

where $\mathbf{I} = (i_1, i_2, \ldots, i_\infty)$ is an infinite multi-index of integers, $|\mathbf{I}|$ denotes the sum of the multi-index, $\|\mathbf{I}\|_1$ denotes the weighted sum $\sum_{l=1}^{\infty} l i_l$, and M_l is the lth moment of turning angles,

$$M_l := \sum_{j=1}^{n} \beta_j \zeta_j^l. \tag{A.194}$$

The summation in the expression for the c_{-k} coefficient is taken over all combinations of multi-index $\mathbf{I}$ for which $\|\mathbf{I}\|_1 = k + 1$. For a given k, the multi-index can be exactly truncated after index i_{k+1}.

Proof Write the Jacobian (A.189) using the property of the logarithm:

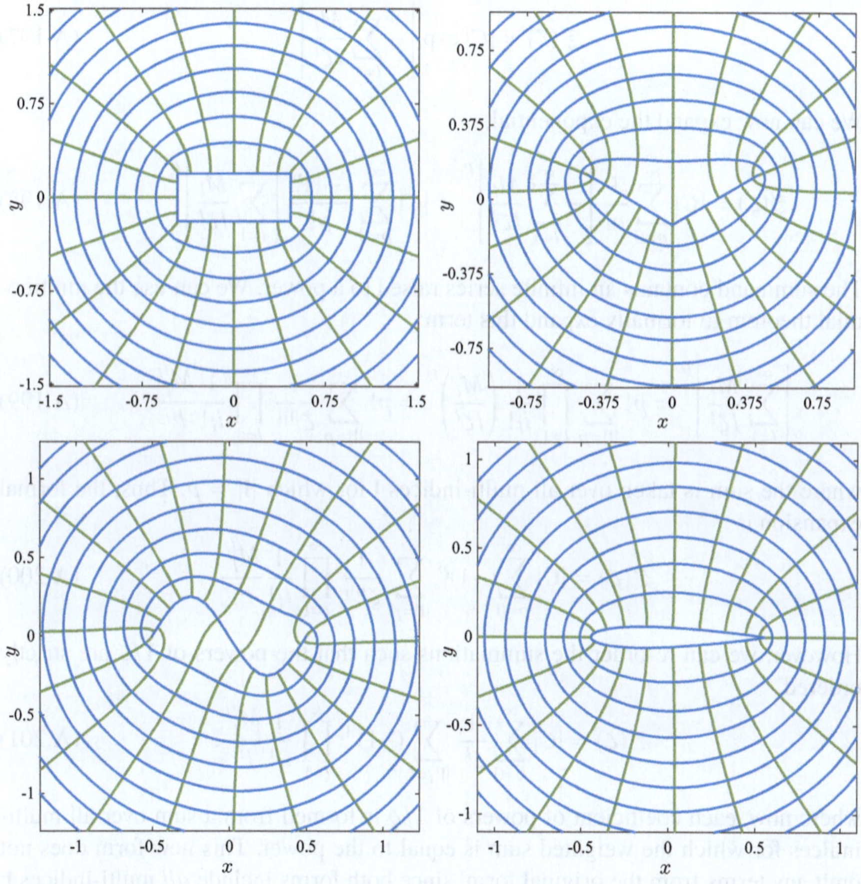

Fig. A.11 Examples of shapes achieved by a Schwarz–Christoffel mapping from the unit circle. The one on the lower left is a sine wave discretized with 20 segments. The lower right example is a NACA 0012 airfoil with 38 segments

$$\tilde{z}'(\zeta) = |C| \exp\left[\sum_{j=1}^{n} \beta_j \log(1 - \zeta_j/\zeta)\right], \tag{A.195}$$

and then use the Taylor expansion for $\log(1 - \epsilon)$ (for $|\epsilon| < 1$):

$$\tilde{z}'(\zeta) = |C| \exp\left[-\sum_{j=1}^{n} \beta_j \sum_{l=1}^{\infty} \frac{1}{l} \frac{\zeta_j^l}{\zeta^l}\right]. \tag{A.196}$$

We can switch the order of the summations inside the exponential. Then, using the definition for the moment M_l, we have

$$\tilde{z}'(\zeta) = |C| \exp\left[-\sum_{l=1}^{\infty} \frac{M_l}{l\zeta^l}\right]. \tag{A.197}$$

We can now expand the exponential,

$$\tilde{z}'(\zeta) = |C| \sum_{p=0}^{\infty} \frac{1}{p!}\left[-\sum_{l=1}^{\infty} \frac{M_l}{l\zeta^l}\right]^p = |C| \sum_{p=0}^{\infty} \frac{(-1)^p}{p!}\left[\sum_{l=1}^{\infty} \frac{M_l}{l\zeta^l}\right]^p. \tag{A.198}$$

The summand contains an infinite series raised to a power. We can use the multinomial theorem to formally expand this term:

$$\left[\sum_{l=1}^{\infty} \frac{M_l}{l\zeta^l}\right]^p = p! \sum_{|\mathbf{I}|=p} \prod_{l=1}^{\infty} \frac{1}{i_l!}\left(\frac{M_l}{l\zeta^l}\right)^{i_l} = p! \sum_{|\mathbf{I}|=p} \frac{1}{\zeta^{\|\mathbf{I}\|_1}} \prod_{l=1}^{\infty} \frac{1}{i_l!}\frac{M_l^{i_l}}{l^{i_l}}, \tag{A.199}$$

where the sum is taken over all multi-indices $\mathbf{I}$ for which $|\mathbf{I}| = p$. Thus, the formal expansion is

$$\tilde{z}'(\zeta) = |C| \sum_{p=0}^{\infty} (-1)^p \sum_{|\mathbf{I}|=p} \frac{1}{\zeta^{\|\mathbf{I}\|_1}} \prod_{l=1}^{\infty} \frac{1}{i_l!}\frac{M_l^{i_l}}{l^{i_l}}. \tag{A.200}$$

However, we can re-order the summations such that the powers of $1/\zeta$ are strictly ordered,

$$\tilde{z}'(\zeta) = |C| \sum_{k=0}^{\infty} \frac{1}{\zeta^k} \sum_{\|\mathbf{I}\|_1=k} (-1)^{|\mathbf{I}|} \prod_{l=1}^{k} \frac{1}{i_l!}\frac{M_l^{i_l}}{l^{i_l}}, \tag{A.201}$$

where now, each coefficient of powers of $1/\zeta$ is formed from a sum over all multi-indices for which the weighted sum is equal to the power. This new form does not omit any terms from the original form, since both forms include *all* multi-indices $\mathbf{I}$; we have only changed the ordering. Note that, since $(k+p)i_{k+p} > k$ for all $p > 0$ if $i_{k+p} > 0$, then the infinite multi-index can be truncated to length k for the kth term in the summation, since the finite-length multi-index will have the same value of $\|\mathbf{I}\|_1$ as its infinite counterpart. From (A.201), we can easily obtain the expansion of $\tilde{z}(\zeta)$ by integrating, and noting that $M_1 \equiv 0$ for a closed body. □

The first few coefficients in the expansion can be easily obtained (with the help of a recursive numerical script to find multi-indices that satisfy $\|\mathbf{I}\|_1 = m$ for any m):

$$c_1 = |C|, \quad c_0 = A, \quad c_{-1} = \frac{1}{2}|C|M_2, \quad c_{-2} = \frac{1}{6}|C|M_3, \quad c_{-3} = \frac{1}{24}|C|(2M_4 - M_2^2). \tag{A.202}$$

Note A.2.1: Schwarz–Christoffel Transformation for Rectangular Bodies

For rectangular bodies with four vertices, then from symmetry, the pre-vertices have the relationship $\zeta_2 = -\zeta_1^*$, $\zeta_3 = -\zeta_1$, and $\zeta_4 = \zeta_1^*$ (and all lie on the unit circle). The interior angles are all equal to $\pi/2$, so that $\beta_j = 1/2$ for all j. Thus, it is easy to show that

1. All odd-numbered moments are zero: $M_{2p+1} = 0$ for all $p \geq 0$
2. All even-numbered moments are equal to $M_{2p} = 2\cos(2p\theta_1)$, where θ_1 is the argument of ζ_1.

Also, the argument of C is equal to $\theta_1 - \pi$.

As a result of these moment properties, only terms with odd powers of $1/\zeta$ survive. This is because we are restricted to multi-indices that only have non-zero indices in every 2nd entry, so that $|\mathbf{I}|_1 = 2(i_2 + 2i_4 + 3i_6 + \cdots)$, which can therefore only be a multiple of 2. This suggests using a new multi-index, $\mathbf{J}$, consisting only of the even entries of the original: $|\mathbf{I}|_1 = 2|\mathbf{J}|_1 = 2(j_1 + 2j_2 + 3j_3 + \cdots)$. Thus,

$$\tilde{z}(\zeta) = |C|\zeta + \sum_{k=1}^{\infty} \frac{c'_{-k}}{\zeta^{2k-1}}, \tag{A.203}$$

where

$$c'_{-k} = -\frac{|C|}{2k-1} \sum_{|\mathbf{J}|_1 = k} (-1)^{|\mathbf{J}|} \prod_{l=1}^{k} \frac{\cos^{j_l}(2l\theta_1)}{l^{j_l} j_l!}. \tag{A.204}$$

Note that the multi-index $\mathbf{J}$ is of dimension k. The first few coefficients are (after manipulating them to express only in terms of $\cos(2\theta_1)$ and $\sin(2\theta_1)$):

$$\begin{aligned}
c'_{-1} &= \ |C|\cos(2\theta_1) \\
c'_{-2} &= -\frac{1}{6}|C|\sin^2(2\theta_1) \\
c'_{-3} &= -\frac{1}{10}|C|\sin^2(2\theta_1)\cos(2\theta_1) \\
c'_{-4} &= \ \frac{1}{56}|C|\sin^2(2\theta_1)(1 - 5\cos^2(2\theta_1)) \\
c'_{-5} &= \ \frac{1}{72}|C|\sin^2(2\theta_1)\cos(2\theta_1)(3 - 7\cos^2(2\theta_1)) \\
c'_{-6} &= -\frac{1}{176}|C|\sin^2(2\theta_1)(1 - 14\cos^2(2\theta_1) + 21\cos^4(2\theta_1))
\end{aligned} \tag{A.205}$$

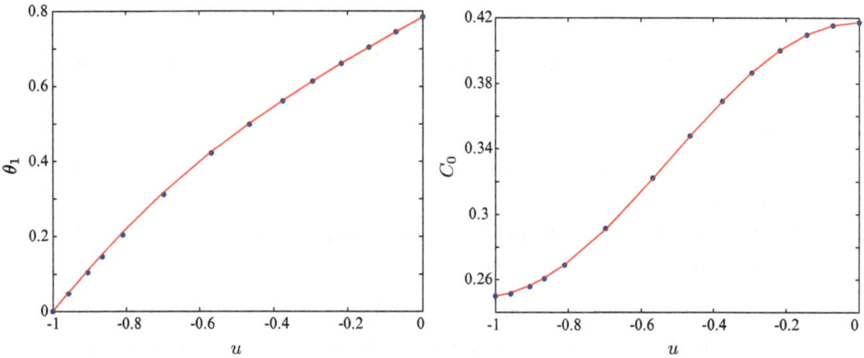

Fig. A.12 Computed values (dotted points) and numerical fits (solid line) of θ_1 (left) and C_0 (right) versus u. Only half of the u interval is shown.

If we define the rectangle by the aspect ratio h/l and the diameter D, then the angle $\gamma = \arctan(h/l) \in [0, \pi/2]$ can parameterize the rectangle for fixed diameter. The two ends of this range correspond to an infinitely-thin flat plate of length D, oriented horizontally and vertically, respectively; the pre-vertices ζ_1 and ζ_2 (and, similarly, ζ_3 and ζ_4) coalesce into a single pre-vertex in these cases. (Note that $\sin\gamma = h/D$ and $\cos\gamma = l/D$.) The angle θ_1 is independent of the diameter, while the constant $|C|$ scales linearly with D. If we define $u := (2/\pi)\arcsin(4\gamma/\pi - 1) \in [-1, 1]$, then good approximations for θ_1 and $|C|$ are found with the fits $\theta_1 = (\pi/4)\left(1 + a_1 u + a_3 u^3\right)$ and $|C| = DC_0(u)$, where $C_0(u) = C_m\left(1 - \cos(b\pi u)/(4\cos(b\pi))\right)$, and the fitting coefficients are $(a_1, a_3) \approx (229/320, 91/320)$ and $(C_m, b) = (1/3, 0.9586)$. These fits are found numerically, but ensure that both parameters have the correct behavior as $\gamma \to 0, \pi/2$: $|C| \to D/4$ and $\theta_1 \to 0, \pi/2$. These fits are plotted in Fig. A.12.

As an example, consider the case of an infinitely thin flat plate of length c. Then, $\theta_1 = 0$ (so $\zeta_1 = 1, \zeta_3 = -1$) and $|C| = c/4$. Thus, by virtue of the $\sin^2(2\theta_1)$ factor, all coefficients c'_{-k} with $k > 1$ vanish, while $c'_{-1} = |C| = c/4$. This results in an exact expression of the Schwarz–Christoffel mapping—the familiar Joukowski mapping:

$$\tilde{z}(\zeta) = \frac{c}{4}\left(\zeta + 1/\zeta\right). \tag{A.206}$$

As another example, consider a square of side length 1 (so that its diameter is $\sqrt{2}$). Then, $\theta_1 = \pi/4$, so that ζ_j are arranged symmetrically on the unit circle at odd integer multiples of $\pi/4$, and $|C| \approx 0.5902$. It is apparent that all of the odd-numbered coefficients c'_{-k} vanish, due to the $\cos(2\theta_1)$ factor. This is consistent with the fourfold symmetry of the body. We are left with the approximate mapping,

$$\tilde{z}(\zeta) \approx 0.5902\left(\zeta - \frac{1}{6\zeta^3} + \frac{1}{56\zeta^7} - \frac{1}{176\zeta^{11}}\right). \tag{A.207}$$

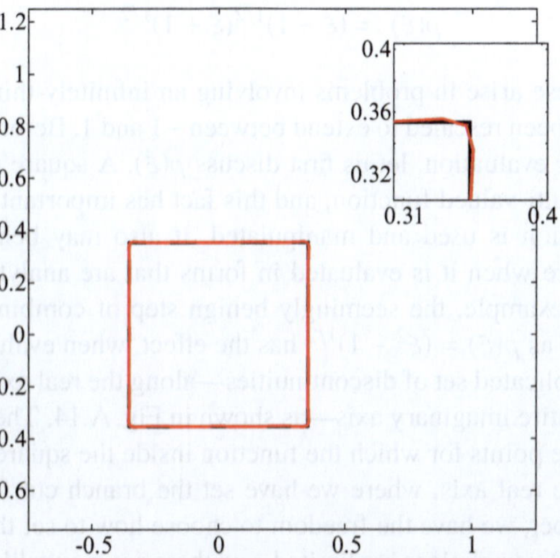

Fig. A.13 Schwarz–Christoffel (black solid line) and six-term approximate (red solid line) mappings for $\gamma = \pi/4$

In fact, because of the apparent drop in the coefficient values, this example suggests that it is safe to truncate the general series (A.203) after around six terms, as coefficients of higher terms are all less than 0.5% of the coefficient of the leading term. We can assess the quality of the mapping by inspecting the shape that the unit circle is mapped to using the six-term expansion. An example of $\gamma = \pi/4$ is shown in Fig. A.13. This plots demonstrates the high quality of the approximate mapping: the L_2 norm of the difference in the shapes (normalized by the norm of the Schwarz–Christoffel shape) is 0.25%. However, it is important to note that the approximate shape does not exactly preserve the zero of the Jacobian at the pre-vertices, $\tilde{z}'(\zeta_j) = 0$.

Infinitely Thin Multi-Body Plates In addition to polygons, the S-C transformation allows us to obtain multi-segmented infinitely thin plates, such as those shown in the upper right and lower left panels of Fig. A.13. We obtain these mappings simply by repeating the vertices on one side (except for the two ends) in reverse order to specify the other side. For example, if the ends are designated by $\tilde{z}_1$ and $\tilde{z}_m$, then the list of vertices for an infinitely thin system of plates would be $\{\tilde{z}_1, \tilde{z}_2, \ldots, \tilde{z}_{m-1}, \tilde{z}_m, \tilde{z}_{m-1}, \ldots \tilde{z}_2\}$. The total number of vertices is $n = 2(m-1)$.

A.3 Mathematical Results for the Infinitely-Thin Plate

A.3.1 Notes on an Important Factor

Several of the integrals considered in this work contain factors of the form

$$\rho(\xi) := (\xi - 1)^{1/2}(\xi + 1)^{1/2} \tag{A.208}$$

or its inverse. These arise in problems involving an infinitely-thin plate, where the plate's length has been rescaled to extend between -1 and 1. Before we discuss these integrals and their evaluation, let us first discuss $\rho(\xi)$. A square root of a complex expression is a multi-valued function, and this fact has important consequences for how such a function is used and manipulated. It also may behave differently in numerical software when it is evaluated in forms that are analytically identical to one another. For example, the seemingly benign step of combining the factors to write the function as $\rho(\xi) = (\xi^2 - 1)^{1/2}$ has the effect, when evaluated numerically, of creating a complicated set of discontinuities—along the real axis between -1 and 1 and along the entire imaginary axis—as shown in Fig. A.14. These discontinuities represent all of the points for which the function inside the square root takes values along the negative real axis, where we have set the branch cut for the square root function. Remember, we have the freedom to choose how to set this branch cut, but in this form of $\rho(\xi)$, our choices are limited, and there are inevitably a lot of annoying discontinuities for any choice.[3]

On the other hand, if we keep the factors in $\rho(\xi)$ separate, then we can control the branch cuts of each constituent square root. For example, as long as we choose a branch cut for both of the roots that lies along the negative real ξ axis, then the discontinuities of the roots partially cancel each other, and $\rho(\xi)$ is single-valued everywhere except along a discontinuity on the real ξ axis between -1 and 1, that is, along the plate. This can be seen in the left panel of Fig. A.15. With the branch cut so chosen, the function $\rho(\xi)$ takes different values as the plate is approached from the + and − sides. On the + side, $\rho(\xi) \to \rho^+(\xi) \equiv \mathrm{i}(1 - \xi^2)^{1/2}$, while on the − side, $\rho(\xi) \to \rho^-(\xi) \equiv -\mathrm{i}(1 - \xi^2)^{1/2}$. Note that $\rho^-(\xi) = -\rho^+(\xi)$. If instead, we choose the branch cut of the roots to have angle $\pi/2$, we get the less desirable set of discontinuities shown in the right panel of Fig. A.15.

A.3.2 Properties of Chebyshev Polynomials

Chebshev polynomials are a valuable tool in the analysis of a flat plate in potential flow, since they form an orthogonal basis for functions on the domain $[-1, 1]$. Here, we summarize their definition and relevant properties.

Definitions In the following, let $\xi(\theta) := \cos\theta$ (or, equivalently, $\theta(\xi) := \arccos\xi$), where $\theta \in [0, \pi]$ and, correspondingly (in reverse direction), $\xi \in [-1, 1]$. There are two kinds of Chebyshev polynomials. The Chebyshev polynomials of the **first kind**

---

[3] Note that, in a numerical evaluation, the branch cut of the operand of a multi-valued function can be rotated from $-\pi$ to some new angle τ by multiplying by $\mathrm{e}^{-\mathrm{i}(\tau+\pi)}$ and its inverse. For example, when evaluating $z^{1/2}$, one can instead evaluate $(z\mathrm{e}^{-\mathrm{i}(\tau+\pi)})^{1/2}\mathrm{e}^{\mathrm{i}(\tau+\pi)/2}$, which places the branch cut at angle τ. This works because the square root operates on the rotated z vector, for which the branch cut is at τ, and the subsequent multiplication by $\mathrm{e}^{\mathrm{i}(\tau+\pi)}$ corrects the result but leaves the branch cut unaffected.

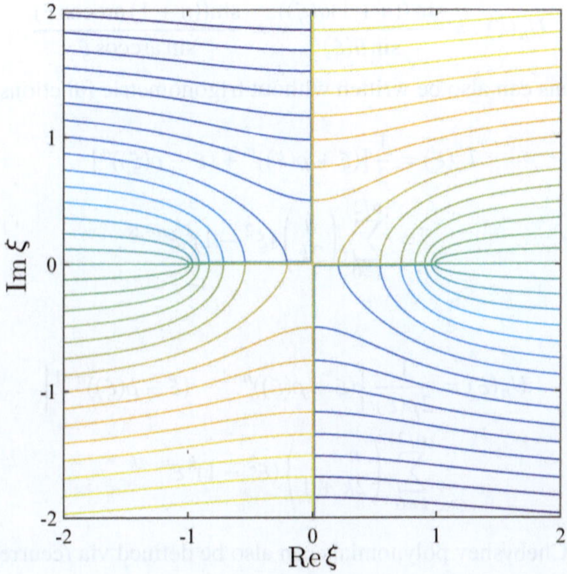

Fig. A.14 Discontinuities of $\rho(\xi) = (\xi^2 - 1)^{1/2}$ when the branch cut of the square root lies along the negative real axis

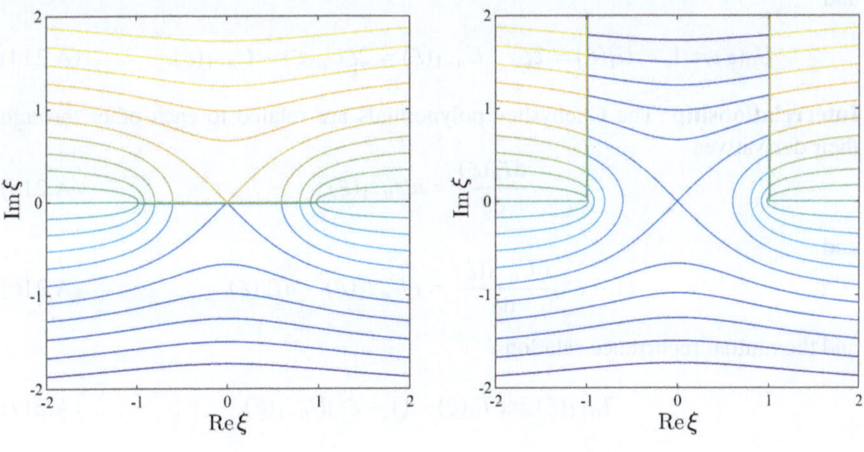

Fig. A.15 Branch cut of $\rho(\xi) = (\xi - 1)^{1/2}(\xi + 1)^{1/2}$ at π (left) and $\pi/2$ (right). In the case of π, the branch cut lies along the real ξ axis between -1 and 1

are defined (trigonometrically) as

$$T_n(\xi) := \cos(n\theta(\xi)) = \cos(n \arccos \xi), \tag{A.209}$$

where n is a non-negative integer. The polynomials of the **second kind** are defined as

$$U_n(\xi) := \frac{\sin((n+1)\theta(\xi))}{\sin\theta(\xi)} = \frac{\sin((n+1)\arccos\xi)}{\sin\arccos\xi}. \tag{A.210}$$

These definitions can also be written without trigonometric functions,

$$\begin{aligned} T_n(\xi) &= \frac{1}{2}\left[(\xi+\rho(\xi))^n + (\xi-\rho(\xi))^n\right] \\ &= \sum_{k=0}^{\lfloor n/2\rfloor} \binom{n}{2k} (\xi^2-1)^k \xi^{n-2k}, \end{aligned} \tag{A.211}$$

and

$$\begin{aligned} U_n(\xi) &= \frac{1}{2\rho(\xi)}\left[(\xi+\rho(\xi))^{n+1} - (\xi-\rho(\xi))^{n+1}\right] \\ &= \sum_{k=0}^{\lfloor n/2\rfloor} \binom{n+1}{2k+1} (\xi^2-1)^k \xi^{n-2k}. \end{aligned} \tag{A.212}$$

Both kinds of Chebyshev polynomials can also be defined via recurrence relations

$$T_0(\xi) = 1, \quad T_1(\xi) = \xi, \quad T_{n+1}(\xi) = 2\xi T_n(\xi) - T_{n-1}(\xi), \tag{A.213}$$

and

$$U_0(\xi) = 1, \quad U_1(\xi) = 2\xi, \quad U_{n+1}(\xi) = 2\xi U_n(\xi) - U_{n-1}(\xi). \tag{A.214}$$

Interrelationship The Chebyshev polynomials are related to each other through their derivatives

$$\frac{\mathrm{d}T_n(\xi)}{\mathrm{d}\xi} = nU_{n-1}(\xi) \tag{A.215}$$

and

$$(1-\xi^2)\frac{\mathrm{d}U_{n-1}(\xi)}{\mathrm{d}\xi} = \xi U_{n-1}(\xi) - nT_n(\xi), \tag{A.216}$$

and the mutual recurrence relations

$$T_{n+1}(\xi) = \xi T_n(\xi) - (1-\xi^2)U_{n-1}(\xi) \tag{A.217}$$

and

$$U_n(\xi) = \xi U_{n-1}(\xi) + T_n(\xi). \tag{A.218}$$

Note that, in using these relations for any non-negative n, it is to be understood that $U_{-1}(\xi) = 0$. Other identities that can be derived from these recurrence relations are

$$(1-\xi^2)U_{n-1}(\xi) = \frac{1}{2}\left(T_{n-1}(\xi) - T_{n+1}(\xi)\right), \quad n \geq 1, \tag{A.219}$$

and

$$T_0(\xi) = U_0(\xi), \qquad T_{n+1}(\xi) = \frac{1}{2}\left(U_{n+1}(\xi) - U_{n-1}(\xi)\right), \quad n \geq 0. \tag{A.220}$$

Orthogonality The Chebyshev polynomials of both kind are orthogonal sequences with respect to weights $(1-\xi^2)^{-1/2}$ and $(1-\xi^2)^{1/2}$, respectively:

$$\int_{-1}^{1} T_n(\xi)T_m(\xi)(1-\xi^2)^{-1/2}\,\mathrm{d}\xi = \begin{cases} 0, & n \neq m, \\ \pi, & n = m = 0, \\ \pi/2, & n = m \neq 0. \end{cases} \tag{A.221}$$

and

$$\int_{-1}^{1} U_n(\xi)U_m(\xi)(1-\xi^2)^{1/2}\,\mathrm{d}\xi = \begin{cases} 0, & n \neq m, \\ \pi/2, & n = m. \end{cases} \tag{A.222}$$

Integrals Certain indefinite integrals of the Chebyshev polynomials are useful. The following can be obtained from (A.216):

$$\int_{-1}^{\xi} T_n(l)(1-l^2)^{-1/2}\,\mathrm{d}l = \begin{cases} \pi - \arccos\xi, & n = 0, \\ -\dfrac{1}{n}(1-\xi^2)^{1/2}U_{n-1}(\xi), & n \geq 1, \end{cases} \tag{A.223}$$

and this can, in turn, be used to show that

$$\int_{-1}^{\xi} U_{n-1}(l)(1-l^2)^{1/2}\,\mathrm{d}l = \begin{cases} \dfrac{1}{2}(\pi - \arccos\xi) + \dfrac{1}{4}(1-\xi^2)^{1/2}U_1(\xi), & n = 1, \\ \dfrac{1}{2}(1-\xi^2)^{1/2}\left(\dfrac{U_n(\xi)}{n+1} - \dfrac{U_{n-2}(\xi)}{n-1}\right), & n \geq 2. \end{cases} \tag{A.224}$$

In these results, note that only T_0 and U_0 have non-zero results when $\xi = 1$; these are π and $\pi/2$, respectively.

Fourier Modes Sometimes, instead of needing the correlation between two Chebyshev modes, as the orthogonality relations provides, we need the correlation between a Chebyshev mode and a standard Fourier mode. In this case, it will be useful to know that

$$\int_{-1}^{1} T_n(\xi)\mathrm{e}^{-\mathrm{i}\kappa\xi}(1-\xi^2)^{-1/2}\,\mathrm{d}\xi = \pi\mathrm{i}^{-n}J_n(\kappa), \tag{A.225}$$

where J_n is the nth-order Bessel function of the first kind. This result can be proved by making the transformation $\xi = \cos\theta$, which transforms the integral into a standard definition of the Bessel function.

Integral Relations In this work, we make use of the standard integral relations

$$\frac{1}{\pi}⨍_{-1}^{1} \frac{T_n(l)}{(1-l^2)^{1/2}(l-\xi)}\,\mathrm{d}l = U_{n-1}(\xi), \tag{A.226}$$

and

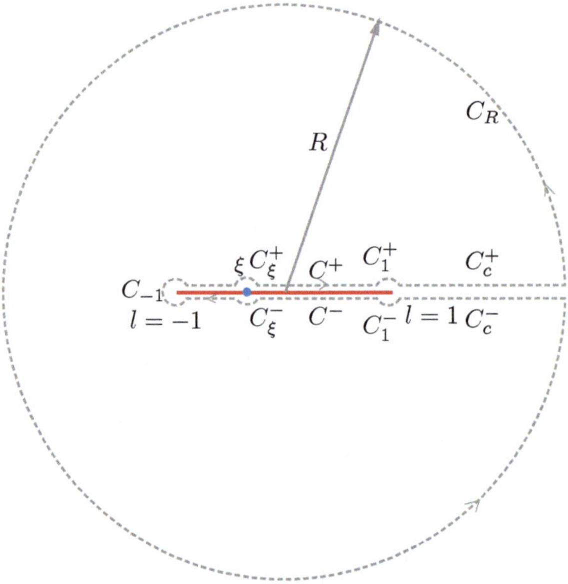

Fig. A.16 Dashed line denotes contour C used in integral in Eq. (A.250)

$$\frac{1}{\pi}⨍_{-1}^{1} \frac{(1-l^2)^{1/2} U_{n-1}(l)}{l-\xi}\, \mathrm{d}l = -T_n(\xi), \tag{A.227}$$

where $\xi \in [-1, 1]$, and the integrals are meant to be interpreted in a principal value sense. The second of these, along with the recurrence relations, can be used to show that

$$\frac{1}{\pi}⨍_{-1}^{1} \frac{(1-l^2)^{1/2} T_n(l)}{l-\xi}\, \mathrm{d}l = \begin{cases} -T_1(\xi), & n=0, \\ -\frac{1}{2}T_2(\xi), & n=1, \\ -\frac{1}{2}\left(T_{n+1}(\xi) - T_{n-1}(\xi)\right), & n>1. \end{cases} \tag{A.228}$$

Proof For later purposes, it is useful to show how these integrals are calculated. We will demonstrate it on (A.226) for the case $n = 1$, and denote the integral we seek by $I_1(\xi)$. We note that, in what follows, ξ cannot lie on the end points of the plate. Remember that we are interested here in the principal value of the integral—the integral that excludes a vanishingly small region surrounding the evaluation point. Thus, we use the contour shown in Fig. A.16, in which the point ξ is excluded. This contour consists of two segments, C^+ and C^-, that comprise the + and − side of the plate, respectively. Note that, on these segments, the factor $(1-l^2)^{1/2}$ in the denominator is related to the limiting forms of $\rho(l)$ on each segment: equal to $-\mathrm{i}\rho^+(l)$ and $\mathrm{i}\rho^-(l)$, respectively. Because these have opposite sign, and the contour segments are traversed in opposite directions, their contributions to the overall contour integral

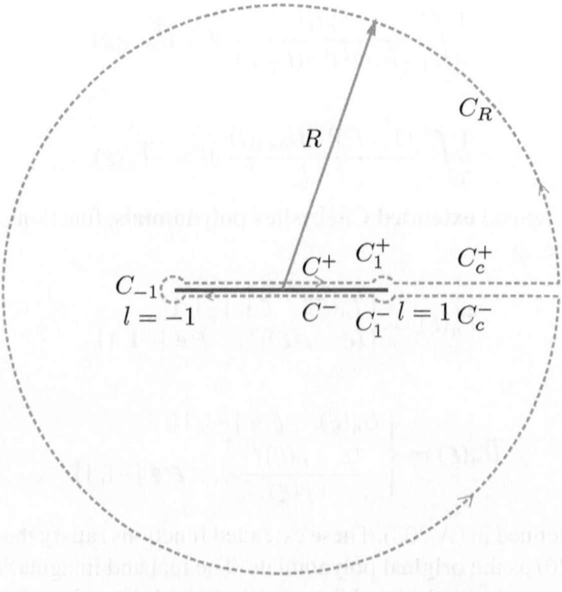

Fig. A.17 Dashed line denotes contour C used in integral in Eq. (A.245)

of $l/[\pi i\rho(l)(l-\xi)]$ are equal and additive. Now, let's inspect the remaining portions of the contour. Let the contour segment C_{-1} be a circle of vanishingly small radius ϵ, so that $l=-1+\epsilon e^{-i\beta}$, $0<\beta\le 2\pi$, and $dl=-i\epsilon e^{-i\beta}d\beta$. The factor $(l+1)^{1/2}$ is approximately equal to $\epsilon^{1/2}e^{-i\beta/2}$, while the other factor $(l-1)^{1/2}$ is approximately $i\sqrt{2}$, and $l-\xi$ approaches -1. Thus, the integral is proportional to $\epsilon^{1/2}$, so in the limit $\epsilon\to 0$, the integral over C_{-1} vanishes. This happens in a similar manner for C_1^+ and C_1^-. On the contours C_c^+ and C_c^-, the integrand is single-valued, so these integrals cancel.

The outer contour C_R is expanded to enclose the point at infinity, where we must account for the residue of the integral, evaluated with the help of Eq. (A.144). Since there are no other poles inside the contour C, our desired integral is obtained from

$$2I_1(\xi)=2\pi i\mathrm{Res}\left(-\frac{1}{\pi i}\frac{l}{\rho(l)(l-\xi)},\infty\right)=2\mathrm{Res}\left(\frac{1}{l(1-l^2)^{1/2}(1-\xi l)},0\right)=2. \tag{A.229}$$

Thus, $I_1(\xi)=1=U_0(\xi)$, as expected. □

Extended Chebyshev Polynomials What if $\xi\notin[-1,1]$? Let's generalize the integral relations (A.226) and (A.227) by allowing ξ to be a complex number. The contour in Fig. A.17 will suit this calculation. We note that ξ now lies inside the contour, so the integrals get an additional contribution from a residue at ξ; the results are $U_{n-1}(\xi)-T_n(\xi)/\rho(\xi)$ and $-T_n(\xi)+\rho(\xi)U_{n-1}(\xi)$, respectively, for the integrals in (A.226) and (A.227). The general results can be written as

$$\frac{1}{\pi}\int_{-1}^{1} \frac{T_n(l)}{(1-l^2)^{1/2}(l-\xi)}\, \mathrm{d}l = \overline{U}_{n-1}(\xi) \tag{A.230}$$

and

$$\frac{1}{\pi}\int_{-1}^{1} \frac{(1-l^2)^{1/2}U_{n-1}(l)}{l-\xi}\, \mathrm{d}l = -\overline{T}_n(\xi), \tag{A.231}$$

where we have defined **extended Chebyshev polynomials**, functions over the whole complex ξ plane, as

$$\overline{T}_n(\xi) := \begin{cases} T_n(\xi), & \xi \in [-1,1], \\ (\xi - \rho(\xi))^n, & \xi \notin [-1,1], \end{cases} \tag{A.232}$$

and

$$\overline{U}_n(\xi) := \begin{cases} U_n(\xi), & \xi \in [-1,1], \\ -\dfrac{(\xi - \rho(\xi))^{n+1}}{\rho(\xi)}, & \xi \notin [-1,1], \end{cases} \tag{A.233}$$

where $\rho(\xi)$ is defined in (A.208). These extended functions satisfy the same relations (A.213)–(A.220) as the original polynomials. The real and imaginary parts of these extended polynomials are depicted for $n = 5$ in Figs. A.18 and A.19. The imaginary part is discontinuous across interval $[-1,1]$ for both kinds. Along the interval $[-1,1]$ itself, where the extended polynomial reverts to its original form, the function takes the mean of the values immediately above and below the interval. Note that $\overline{U}_n$ is also defined for $n = -1$, where it vanishes in $[-1,1]$, but is equal to $-1/\rho(\xi)$ at all other points.

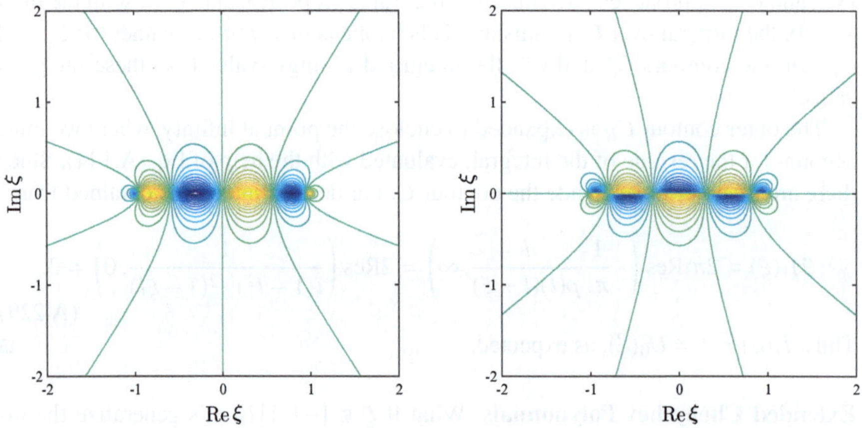

Fig. A.18 Real (left) and imaginary (right) parts of $\overline{T}_5(\xi)$

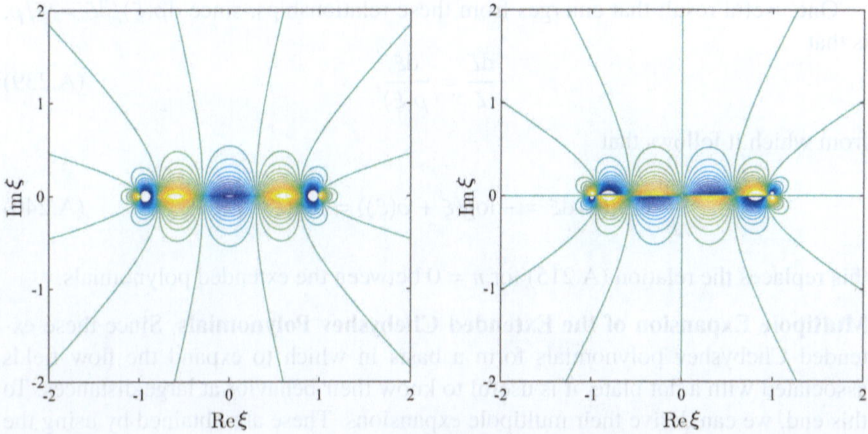

Fig. A.19 Real (left) and imaginary (right) parts of $\overline{U}_5(\xi)$

Extended Chebyshev Polynomials in the Circle Plane The Joukowski transform, from the exterior of the unit circle (in the ζ plane) to the exterior of the flat plate of length 2 (in the ξ plane), corresponds to the power series transform (A.153) with $c_1 = c_{-1} = 1/2$ and all other coefficients equal to zero:

$$\xi(\zeta) = \frac{1}{2}(\zeta + 1/\zeta). \tag{A.234}$$

The inverse of this conformal transformation is

$$\zeta(\xi) = \xi + \rho(\xi), \tag{A.235}$$

and it is easy to verify that

$$\frac{1}{\zeta(\xi)} = \xi - \rho(\xi). \tag{A.236}$$

Thus, the extended Chebyshev polynomials, when viewed as functions of ζ, are

$$\overline{T}_n(\xi(\zeta)) := \begin{cases} \frac{1}{2}(\zeta^n + \zeta^{-n}), & \zeta \in C_c, \\ \zeta^{-n}, & |\zeta| > 1, \end{cases} \tag{A.237}$$

and

$$\overline{U}_n(\xi(\zeta)) := \begin{cases} \dfrac{\zeta^{n+1} - \zeta^{-n-1}}{\zeta - \zeta^{-1}}, & \zeta \in C_c, \\ -\dfrac{\zeta^{-n-1}}{\frac{1}{2}(\zeta - \zeta^{-1})}, & |\zeta| > 1. \end{cases} \tag{A.238}$$

Thus, the extended Chebyshev polynomials of the first kind provide a particularly simple foundation in the circle plane.

One useful result that emerges from these relationships, since $\mathrm{d}\rho(\xi)/\mathrm{d}\xi = \xi/\rho$, is that

$$\frac{\mathrm{d}\zeta}{\zeta} = \frac{\mathrm{d}\xi}{\rho(\xi)}, \tag{A.239}$$

from which it follows that

$$\int \overline{U}_{-1}(\xi)\,\mathrm{d}\xi = -\log(\xi + \rho(\xi)) = \log(\xi - \rho(\xi)); \tag{A.240}$$

this replaces the relation (A.215) for $n = 0$ between the extended polynomials.

Multipole Expansion of the Extended Chebyshev Polynomials Since these extended Chebyshev polynomials form a basis in which to expand the flow fields associated with a flat plate, it is useful to know their behavior at large distances. To this end, we can derive their multipole expansions. These are obtained by using the standard form of the Laurent series expansion (about $\xi = 0$) of a complex function,

$$f(\xi) = \sum_{k=-\infty}^{\infty} a_k \xi^k, \tag{A.241}$$

where the coefficients are obtained from

$$a_k = \frac{1}{2\pi \mathrm{i}} \int_C \frac{f(l)}{l^{k+1}}\,\mathrm{d}l. \tag{A.242}$$

Here, the contour of integration is any that encloses the region $\xi \in [-1, 1]$. The easiest way to compute the integral is to transform the plate to the unit circle, using the Joukowski transformation $\xi = (\zeta + 1/\zeta)/2$. Then, it can be shown that the multipole expansions are, respectively,

$$\overline{T}_n(\xi) = \sum_{k=0}^{\infty} \frac{n}{n+k} \binom{n+2k-1}{k} \frac{1}{(2\xi)^{n+2k}} \sim \frac{1}{(2\xi)^n} \tag{A.243}$$

and

$$\overline{U}_n(\xi) = -2 \sum_{k=1}^{\infty} \binom{n+2k-1}{k-1} \frac{1}{(2\xi)^{n+2k}} \sim -\frac{1}{2^{n+1}\xi^{n+2}}. \tag{A.244}$$

where ξ is a complex number, outside the real-valued range $\xi \in [-1, 1]$. Note that $\overline{U}_n(\xi)$ decays faster with distance than $\overline{T}_n(\xi)$.

A.3.3 Contour Integrals of Interest

We can use the results of the previous section to evaluate some useful integrals on flat plates. For a plate of length c, we simply rescale the coordinate system with $\tilde{z} = c\xi/2$, and integration over the plate, $\mathcal{P}$, proceeds from $-c/2$ to $c/2$. The integration factor $\rho(\xi)$ arises in all of these integrals, in its scaled form $c\rho(2\tilde{z}/c)/2$.

First, using the orthogonality relation for Chebyshev polynomials of the first kind (A.221), then we can show that

$$\frac{1}{\pi \mathrm{i}} \int_{\mathcal{P}} \frac{\mathrm{d}\tilde{z}}{(\tilde{z}-c/2)^{1/2}\,(\tilde{z}+c/2)^{1/2}} = -1. \tag{A.245}$$

Similarly,

$$\frac{1}{\pi \mathrm{i}} \int_{\mathcal{P}} \frac{\tilde{z}\,\mathrm{d}\tilde{z}}{(\tilde{z}-c/2)^{1/2}\,(\tilde{z}+c/2)^{1/2}} = 0. \tag{A.246}$$

and

$$\frac{1}{\pi \mathrm{i}} \int_{\mathcal{P}} \frac{\tilde{z}^2\,\mathrm{d}\tilde{z}}{(\tilde{z}-c/2)^{1/2}\,(\tilde{z}+c/2)^{1/2}} = -c^2/8. \tag{A.247}$$

(And all integrals with odd-numbered powers of $\tilde{z}$ in the numerator vanish due to symmetry.)

Now we consider integrals that arise in the context of rigid-body motion of a plate. Using relation (A.231), with $n = 1$, it is straightforward to show that

$$\frac{1}{\pi \mathrm{i}} \int_{\mathcal{P}} \frac{(\tilde{\lambda}-c/2)^{1/2}(\tilde{\lambda}+c/2)^{1/2}}{\tilde{\lambda}-\tilde{z}}\,\mathrm{d}\tilde{\lambda} = -\tilde{z} + \begin{cases} 0, & \tilde{z} \in \mathcal{P}, \\ (\tilde{z}+c/2)^{1/2}(\tilde{z}-c/2)^{1/2}, & \tilde{z} \notin \mathcal{P}. \end{cases} \tag{A.248}$$

and, with $n = 2$,

$$\frac{1}{\pi \mathrm{i}} \int_{\mathcal{P}} \frac{\tilde{\lambda}(\tilde{\lambda}-c/2)^{1/2}(\tilde{\lambda}+c/2)^{1/2}}{\tilde{\lambda}-\tilde{z}}\,\mathrm{d}\tilde{\lambda} = -\tilde{z}^2 + c^2/8 + \begin{cases} 0, & \tilde{z} \in \mathcal{P}, \\ \tilde{z}(\tilde{z}+c/2)^{1/2}(\tilde{z}-c/2)^{1/2}, & \tilde{z} \notin \mathcal{P}. \end{cases} \tag{A.249}$$

Similarly, using Eq. (A.230) with $n = 0$,

$$\frac{1}{\pi \mathrm{i}} \int_{\mathcal{P}} \frac{\mathrm{d}\tilde{\lambda}}{(\tilde{\lambda}-c/2)^{1/2}(\tilde{\lambda}+c/2)^{1/2}(\tilde{\lambda}-\tilde{z})} = \begin{cases} 0, & \tilde{z} \in \mathcal{P}, \\ \dfrac{1}{(\tilde{z}+c/2)^{1/2}(\tilde{z}-c/2)^{1/2}}, & \tilde{z} \notin \mathcal{P}. \end{cases} \tag{A.250}$$

with $n = 1$,

$$\frac{1}{\pi \mathrm{i}} \int_{\mathcal{P}} \frac{\tilde{\lambda}\,\mathrm{d}\tilde{\lambda}}{(\tilde{\lambda}-c/2)^{1/2}(\tilde{\lambda}+c/2)^{1/2}(\tilde{\lambda}-\tilde{z})} = -1 + \begin{cases} 0, & \tilde{z} \in \mathcal{P} \\ \dfrac{\tilde{z}}{(\tilde{z}-c/2)^{1/2}(\tilde{z}+c/2)^{1/2}}, & \tilde{z} \notin \mathcal{P}. \end{cases} \tag{A.251}$$

and with a combination of $n = 0$ and 2,

$$\frac{1}{\pi \mathrm{i}} \int_{\mathcal{P}} \frac{\tilde{\lambda}^2\,\mathrm{d}\tilde{\lambda}}{(\tilde{\lambda}-c/2)^{1/2}(\tilde{\lambda}+c/2)^{1/2}(\tilde{\lambda}-\tilde{z})} = -\tilde{z} + \begin{cases} 0, & \tilde{z} \in \mathcal{P} \\ \dfrac{\tilde{z}^2}{(\tilde{z}-c/2)^{1/2}(\tilde{z}+c/2)^{1/2}}, & \tilde{z} \notin \mathcal{P}. \end{cases} \tag{A.252}$$

Other integrals we need in this work involve a vortex at some position in the fluid. The first we will consider is

$$I_{v,1}(\tilde{z}) := \frac{1}{\pi \mathrm{i}} \int_{\mathcal{P}} \frac{(\tilde{\lambda}-c/2)^{1/2}(\tilde{\lambda}+c/2)^{1/2}\mathrm{d}\tilde{\lambda}}{(\tilde{\lambda}-\tilde{z})(\tilde{\lambda}-\tilde{z}_v)}, \tag{A.253}$$

where $\tilde{z}_v \notin \mathcal{P}$. When $\tilde{z} \in \mathcal{P}$, we can use the contour of Fig. A.16; we still get null contributions from the contour segments at the edges of the plate and from those that extend outward along the real axis. In this case, there are two non-zero contributions: First, on contour C_R, the integrand approaches $1/\pi$ as $R \longrightarrow \infty$, so that the integral on this contour is 2. Furthermore, the contour C encloses a simple pole at $\tilde{\lambda} = \tilde{z}_v$ which contributes a residue at that point. If $\tilde{z} \notin \mathcal{P}$, then there is a further contribution from the residue at $\tilde{z}$. Thus,

$$I_{v,1}(\tilde{z}) = -1 + \frac{(\tilde{z}_v - c/2)^{1/2}(\tilde{z}_v + c/2)^{1/2}}{\tilde{z}_v - \tilde{z}} - \begin{cases} 0, & \tilde{z} \in \mathcal{P} \\ \dfrac{(\tilde{z}-c/2)^{1/2}(\tilde{z}+c/2)^{1/2}}{\tilde{z}_v - \tilde{z}}, & \tilde{z} \notin \mathcal{P}. \end{cases} \tag{A.254}$$

Let us also consider the integral

$$I_{v,2}(\tilde{z}) := \frac{1}{\pi \mathrm{i}} \int_{\mathcal{P}} \frac{\mathrm{d}\tilde{\lambda}}{(\tilde{\lambda}-c/2)^{1/2}(\tilde{\lambda}+c/2)^{1/2}(\tilde{\lambda}-\tilde{z})(\tilde{\lambda}-\tilde{z}_v)}, \tag{A.255}$$

also for $\tilde{z}_v \notin \mathcal{P}$. In this case, there are no contributions from segments of C (aside from the usual ones along the plate). Thus, we have

$$I_{v,2}(\tilde{z}) = \frac{1}{\tilde{z}_v - \tilde{z}} \left[\frac{1}{(\tilde{z}_v - c/2)^{1/2}(\tilde{z}_v + c/2)^{1/2}} - \begin{cases} 0, & \tilde{z} \in \mathcal{P} \\ \dfrac{1}{(\tilde{z}-c/2)^{1/2}(\tilde{z}+c/2)^{1/2}}, & \tilde{z} \notin \mathcal{P} \end{cases} \right]. \tag{A.256}$$

Note that all of the results compiled here involving integrands with the factor $1/(\tilde{\lambda}-\tilde{z})$ are consistent with the Plemelj formulae (A.125) in their limiting behaviors as the plate is approached from either side.

References

1. I.H. Abbott, A.E. von Doenhoff, *Theory of Wing Sections, Including a Summary of Airfoil Data* (Dover, New York, 1959)
2. S. Alben, Simulating the dynamics of flexible bodies and vortex sheets. J. Comput. Phys. **228**(7), 2587–2603 (2009)
3. S. Alben, Regularizing a vortex sheet near a separation point. J. Comput. Phys. **229**, 5280–5298 (2010)
4. J. Anderson, *Fundamentals of Aerodynamics*, 6th edn. (McGraw-Hill, New York, 2016)
5. B.C. Basu, G.J. Hancock, The unsteady motion of a two-dimensional aerofoil in incompressible inviscid flow. J. Fluid Mech. **87**(1), 159–178 (1978)
6. J.T. Beale, A. Majda, High order accurate vortex methods with explicit velocity kernels. J. Comput. Phys. **58**, 188–208 (1985)
7. G. Birkhoff, Helmholtz and Taylor instability, in *Hydrodynamic Instability*, ed. by G. Birkhoff, R. Bellman, C.C. Lin. Proceedings of Symposia in Applied Mathematics, vol. 13 (1962), pp. 63–84
8. R.L. Bisplinghoff, H. Ashley, R.L. Halfman, *Aeroelasticity* (Dover, New York, 1996)
9. S.N. Brown, P.G. Daniels, On the viscous flow about the trailing edge of a rapidly oscillating plate. J. Fluid Mech. **67**, 743–761 (1975)
10. C.E. Brown, W.H. Michael, Effect of leading edge separation on the lift of a delta wing. J. Aero. Sci. **21**, 690–694 (1954)
11. S.N. Brown, K. Stewartson, Trailing-edge stall. J. Fluid Mech. **42**(3), 561–584 (1970)
12. A.J. Chorin, Numerical study of slightly viscous flow. J. Fluid Mech. **57**(4), 785–796 (1973)
13. L. Cortelezzi, A. Leonard, Point vortex model of the unsteady separated flow past a semi-infinite plate with transverse motion. Fluid Dyn. Res. **11**, 263–295 (1993)
14. G.H. Cottet, P. Koumoutsakos, *Vortex Methods: Theory and Practice* (Cambridge University Press, Cambridge, 2000)

J. D. Eldredge, *Mathematical Modeling of Unsteady Inviscid Flows*, Interdisciplinary Applied Mathematics 50, https://doi.org/10.1007/978-3-030-18319-6

15. D.G. Crighton, The Kutta condition in unsteady flow. Annu. Rev. Fluid Mech. **17**, 411–445 (1985)
16. S.C. Crow, Stability theory for a pair of trailing vortices. AIAA J. **8**(12), 2172–2179 (1970)
17. D. Darakananda, J.D. Eldredge, A versatile taxonomy of low-dimensional vortex models for unsteady aerodynamics. J. Fluid Mech. **858**, 917–948 (2019). https://doi.org/10.1017/jfm.2018.792
18. A.C. DeVoria, M. Mohseni, The vortex-entrainment sheet in an inviscid fluid: theory and separation at a sharp edge. J. Fluid Mech. **866**, 660–688 (2019)
19. M.L. Dow, D. Elliott, The numerical solution of singular integral equations over (-1,1). SIAM J. Numer. Anal. **16**(1), 115–134 (1979)
20. T.A. Driscoll, SC Toolbox [Software package]. http://www.math.udel.edu/~driscoll/SC/index.html
21. T.A. Driscoll, L.N. Trefethen, *Schwarz–Christoffel Mapping* (Cambridge University Press, Cambridge, 2002)
22. R.H. Edwards, Leading edge separation from delta wings. J. Aero. Sci. **21**, 134–135 (1954)
23. J.D. Eldredge, A.R. Jones, Leading-edge vortices: mechanics and modeling. Annu. Rev. Fluid Mech. **51**, 75–104 (2019). https://doi.org/10.1146/annurev-fluid-010518-040334
24. T. Frankel, *The Geometry of Physics: An Introduction*, 2nd edn. (Cambridge University Press, Cambridge, 2004)
25. J.P. Giesing, Vorticity and Kutta condition for unsteady multienergy flows. ASME Trans. J. Appl. Mech. **36**(3), 608–613 (1969)
26. H. Glauert, *The Elements of Aerofoil and Airscrew Theory*, 2nd edn. (Cambridge University Press, Cambridge, 1948)
27. J.M.R. Graham, The lift on an aerofoil in starting flow. J. Fluid Mech. **133**, 413–425 (1983)
28. C. Greengard, The core spreading vortex method approximates the wrong equation. J. Comput. Phys. **61**, 345–348 (1985)
29. F.R. Hama, Progressive deformation of a curved vortex filament by its own induction. Phys. Fluids **5**(10), 1156–1162 (1962)
30. H. Helmholtz, Über Integrale der hydrodynamischen Gleichungen, welche den Wirbelbewegungen entsprechen. J. Reine Angew. Math. **55**, 25–55 (1858)
31. M.S. Howe, On unsteady surface forces, and sound produced by the normal chopping of a rectilinear vortex. J. Fluid Mech. **206**, 131–153 (1989)
32. M.S. Howe, On the force and moment on a body in an incompressible fluid, with application to rigid bodies and bubbles at high and low Reynolds numbers. Q. J. Mech. Appl. Math. **48**(3), 401–426 (1995)
33. H. Jeffreys, B. Jeffreys, *Methods of Mathematical Physics*, 3rd edn. (Cambridge University Press, Cambridge, 1972)
34. M.A. Jones, The separated flow of an inviscid fluid around a moving flat plate. J. Fluid Mech. **496**, 405–441 (2003)
35. H. Kaden, Aufwicklung einer unstabilen Unstetigkeitsfläche. Ing. Arch. **2**(2), 140–168 (1931)

36. E. Kanso, B.G. Oskouei, Stability of a coupled body-vortex system. J. Fluid Mech. **600**, 77–94 (2008). https://doi.org/10.1017/S0022112008000359
37. E. Kanso, J.E. Marsden, C.W. Rowley, J. Melli-Huber, Locomotion of articulated bodies in a perfect fluid. J. Nonlinear Sci. **15**(4), 255–289 (2005)
38. K. Karamcheti, *Principles of Ideal-Fluid Aerodynamics* (R. E. Krieger Publishing, Malabar, 1980)
39. J. Katz, A. Plotkin, *Low-Speed Aerodynamics*, 2nd edn. (Cambridge University Press, Cambridge, 2001)
40. G. Kirchhoff, *Vorlesungen Über Mathematische Physik* (Teubner, Leipzig, 1876)
41. R. Krasny, *Vortex Sheet Computations: Roll-Up, Wake, Separation*. Lectures in Applied Mathematics, vol. 28 (AMS, New York, 1991), pp. 385–402
42. H.G. Küssner, Zusammenfassender Bericht über den instationären Auftrieb von Flügeln. Luftfahrtforschung **13**(12), 410–429 (1936)
43. H. Lamb, *Hydrodynamics*, 6th edn. (Cambridge University Press, Cambridge, 1932)
44. C. Lanczos, *The Variational Principles of Mechanics*, 4th edn. (Dover Publications, New York, 1986)
45. A. Leonard, Computing three-dimensional incompressible flows with vortex elements. Annu. Rev. Fluid Mech. **17**, 523–559 (1985)
46. H.W. Liepmann, A. Roshko, *Elements of Gasdynamics* (Wiley, London, 1957)
47. M.J. Lighthill, Introduction: boundary layer theory, in *Laminar Boundary Layers*, ed. by L. Rosenhead (Clarendon Press, Oxford, 1961)
48. J.E. Marsden, M. West, Discrete mechanics and variational integrators. Acta Numer. **10**, 357–514 (2001)
49. E.C. Maskell, On the Kutta–Joukowski condition in two-dimensional unsteady flow. Tech. rep., Royal Aircraft Establishment, Farnborough (1971)
50. M.V. Melander, N.J. Zabusky, J.C. McWilliams, Symmetric vortex merger in two dimensions: causes and conditions. J. Fluid Mech. **195**, 303–340 (1988)
51. S. Michelin, S.G. Llewelyn Smith, An unsteady point vortex method for coupled fluid–solid problems. Theor. Comput. Fluid Dyn. **23**, 127–153 (2009)
52. L.M. Milne-Thomson, *Theoretical Aerodynamics*, 4th edn. (Dover, New York, 1973)
53. L.M. Milne-Thomson, *Theoretical Hydrodynamics*, 5th edn. (Dover, New York, 1996)
54. D.W. Moore, P.G. Saffman, The motion of a vortex filament with axial flow. Philos. Trans. R. Soc. Lond. A **272**(1226), 403–429 (1972)
55. M.M. Munk, General theory of thin wing sections. Tech. Rep. 142, NACA (1923)
56. N.I. Muskhelishvili, *Singular Integral Equations: Boundary Problems of Function Theory and Their Application to Mathematical Physics* (P. Noordhoff N. V., Groningen, Holland, 1953)
57. D.R. Poling, D.P. Telionis, The response of airfoils to periodic disturbances – the unsteady Kutta condition. AIAA J. **24**(2), 193–199 (1986)
58. D.I. Pullin, The large-scale structure of unsteady self-similar rolled-up vortex sheets. J. Fluid Mech. **88**(3), 401–430 (1978)

59. D.I. Pullin, Z.J. Wang, Unsteady forces on an accelerating plate and application to hovering insect flight. J. Fluid Mech. **509**, 1–21 (2004)
60. K. Ramesh, A. Gopalarathnam, K. Granlund, M.V. Ol, J.R. Edwards, Discrete-vortex method with novel shedding criterion for unsteady aerofoil flows with intermittent leading-edge vortex shedding. J. Fluid Mech. **751**, 500–548 (2014)
61. L.F. Rossi, Resurrecting core spreading vortex methods: a new scheme that is both deterministic and convergent. SIAM J. Sci. Comput. **17**(2), 370–397 (1996)
62. N. Rott, Diffraction of a weak shock with vortex generation. J. Fluid Mech. **1**(1), 111–128 (1956)
63. P.G. Saffman, *Vortex Dynamics* (Cambridge University Press, Cambridge, 1992)
64. W.R. Sears, Some aspects of non-stationary airfoil theory and its practical application. J. Aeronautical Sci. **8**(3), 104–108 (1941)
65. B.N. Shashikanth, J.E. Marsden, J.W. Burdick, S.D. Kelly, The Hamiltonian structure of a two-dimensional rigid circular cylinder interacting dynamically with N point vortices. Phys. Fluids **14**(3), 1214–1227 (2002). https://doi.org/10.1063/1.1445183
66. R.K. Shukla, J.D. Eldredge, An inviscid model for vortex shedding from a deforming body. Theor. Comput. Fluid Dyn. **21**, 343–368 (2007). https://doi.org/10.1007/s00162-007-0053-2
67. A.A. Tchieu, A. Leonard, A discrete-vortex model for the arbitrary motion of a thin airfoil with fluidic control. J. Fluids Struct. **27**(5–6), 680–693 (2011)
68. T. Theodorsen, General theory of aerodynamic instability and the mechanism of flutter. Tech. Rep. TR-496, NACA (1935)
69. W. Thomson, On vortex motion. Trans. R. Soc. Edinb. **25**, 217–260 (1869)
70. T. von Kármán, W.R. Sears, Airfoil theory for non-uniform motion. J. Aeronautical Sci. **5**(10), 379–390 (1938)
71. H. Wagner, Über die Entstehung des dynamischen Auftriebes von Tragflügeln. Z. Angew. Math. Mech. **5**(1), 17–35 (1925)
72. C. Wang, J.D. Eldredge, Low-order phenomenological modeling of leading-edge vortex formation. Theor. Comput. Fluid Dyn. **27**(5), 577–598 (2013). https://doi.org/10.1007/s00162-012-0279-5
73. E.T. Whittaker, G.N. Watson, *A Course of Modern Analysis* (Cambridge University Press, Cambridge, 1920)
74. S.E. Widnall, D. Bliss, A. Zalay, Theoretical and experimental study of the stability of a vortex pair, in *Aircraft Wake Turbulence and Its Detection*, ed. by J. Olsen, A. Goldburg, M. Rogers (Plenum, New York, 1971), pp. 305–338
75. G.S. Winckelmans, A. Leonard, Contributions to vortex particle methods for the computation of three-dimensional incompressible unsteady flows. J. Comput. Phys. **109**(2), 247–273 (1993)
76. J.C. Wu, Theory of aerodynamic force and moment in viscous flows. AIAA J. **195**(4), 432–441 (1981)
77. X. Xia, K. Mohseni, Unsteady aerodynamics and vortex-sheet formation of a two-dimensional airfoil. J. Fluid Mech. **830**, 439–478 (2017)

Index

added mass tensor, 218, 221, 225
rotational, 147
rotational-translational, 147
translational, 146
translational-rotational, 147
aerodynamic center, 311
angular impulse, 38, 190

basis field, 5
basis vectors
body-fixed, 10
inertial, 10
basis complex potential field
due to unit body rotation, 132
due to unit body translation, 131
due to unit uniform flow, 131
unit vorticity-induced, 131
vorticity-induced, 131
basis scalar potential field
due to unit body rotation, 142
due to unit body translation, 141
due to unit uniform flow, 139
basis vector potential field
due to unit body rotation, 142
due to unit body rotation at infinity, 143
due to unit body translation, 141
due to unit uniform flow, 140
basis velocity field, 125
due to unit body rotation, 132, 142
due to unit body rotation at infinity, 143
due to unit body translation, 131, 141
due to unit uniform flow, 131, 139
two-dimensional flat plate, 276
unit vorticity-induced, 131
vorticity-induced, 130, 137
basis vortex sheet, 125
due to unit body rotation, 133, 141
due to unit body translation, 133, 140
due to unit uniform flow, 133, 139
two-dimensional flat plate, 276
unit vorticity-induced, 132, 138
vorticity-induced, 132, 137, 278
Bernoulli equation, 44
Biot–Savart integral, 58
Birkhoff–Rott equation, 255
blob, *see* vortex blob
blob function, 261
Body Motion
oscillatory Pitching, 318
oscillatory Plunging, 317
body reference point, 10
body-fixed frame, 8
bound vortex sheet, *see* vortex sheet, bound
bound vorticity, 92, 203
branch, *see* branch cut
branch cut, 51
Brown–Michael equation, 246, 256

Cauchy Integral, 280
Cauchy integral, 420
Cauchy Residue Theorem, 423
Chebyshev collocation points, 338
Chebyshev polynomial, 288
extended, 450
circle plane, 108
circle theorem, 113
circulation, 34
and the starting vortex, 123
constraint on total, 122
total, 35

J. D. Eldredge, *Mathematical Modeling of Unsteady Inviscid Flows*, Interdisciplinary Applied Mathematics 50, https://doi.org/10.1007/978-3-030-18319-6

circulation-preserving flow, 39
complex conjugate velocity, *see* complex velocity
complex potential, 49
complex velocity, 16, 49
  circle plane, $\hat{w}$, 108
conformal map, *see* conformal transformations
conformal transformations
  Jacobian of, 108
  Joukowski, 274, 429
  Joukowski and Kármán–Trefftz airfoils, 431
  power series transformation, 428
  Schwarz–Christoffel, 116, 436
corner flow, 52
  signed intensity, 55
corners, *see* edges
critical leading-edge suction parameter, 178

d'Alembert's paradox, 203
deformation velocity, 334
differentiation of integrals
  line integrals, 415
  surface integrals, 413
  volume integrals, 413
dipole, 48
discrete cosine transform, 290, 338
divergence theorem, generalized, 393
double-layer potential, 73
doublet, *see* dipole
drag
  on a flat plate in steady flow, 311
Duhamel's integral, 326

edge suction, 215
edge suction parameter, 178, 215, 300
edges
  behavior of velocity near, 117
  mathematical form, 429
ellipsoidal coordinates, 369
Euler equations, 29

flow contributors, 5, 97, 125
Fourier–Chebyshev expansion, 287
free vortex sheet, *see* vortex sheet, free
free vorticity, 203

Green's theorem, 397

Helmholtz decomposition, 26
higher-order singularities, 48
holomorphic, 49

impulse
  unit vorticity-induced, 219
impulse matching equation, 257
indicial response, 326
inertial frame, 8, 126
inverse surface operator, *see* surface operator

Jacobian
  see conformal transformations, Jacobian of, 108
Joukowski hypothesis, *see* Kutta condition

Küssner function, 331
Kelvin's circulation theorem, 36
Kelvin's minimum energy theorem, 31
Kirchhoff velocity, 246
Kutta condition, 162, 173
  Giesing–Maskell extension, 170
Kutta–Joukowski lift theorem, 297, 311
Kutta-Joukowski condition, *see* Kutta condition

leading-edge suction parameter, 302
leading-edge vortex, 3
lift
  on a flat plate in steady flow, 311
Lighthill vorticity creation mechanism, 2
line vortex, 63
linear impulse, 38, 190

material derivative, 29, 412
material point, 12, 412
monopole source
  three-dimensional, 57
  two-dimensional, 47, 50
multipole coefficients
  complex form in two dimensions, 158
  conformal mapping, 159
  plates, 159
  two-dimensional flows, 152
  vorticity form, 156
multipole expansion, 150

nascent vortex, 172, 303
no-penetration condition, 28
  direct form, 86
  tangency form, 86

panel method, 107, 364
partial circulation, 54
physical plane, 108
Plücker coordinates, 17

plate, 98
Plemelj formulae, 421
point vortex, 59
  three-dimensional, 56
  two-dimensional, 46, 50
potential flow, 43
principal value, 420

quasi-steady circulation, 129, 177, 220, 312
quasi-steady force, 220
quasi-steady lift, 315
quasi-steady moment, 220, 316

reduced frequency, 317
regularization, 60
rotation matrix, 10
rotation operator, 16
Routh correction, 246, 251
Runge–Kutta method, 343

scalar potential, 26
Schwarz reflection principle, 110
Sears function, 329
single-layer potential, 72
stagnation point, 283
starting vortex, 310
Stokes' Theorem, Generalized, 405
streamfunction, 26
  in a rigid body, 27
surface operator, 103
  inverse, 104

Theodorsen function, 318
thin airfoil theory, 302
total vorticity, 37
transformation operator
  body-to-inertial, 19
two-dimensional bodies
  geometric properties, 426

uniform flow, 43

vector potential, 26
velocity kernel, 47
vortex blob, 260
vortex cloud, 59, 253
vortex filament, 60
  cutoff method, 265
vortex force, 203
vortex line, 39
vortex sheet, 66
  basis, *see* basis vortex sheet
  bound, 2, 81, 96, 98
  complex strength, 87
  free, 4, 81, 98
  local circulation, 68
  local strain, 83
  strength, 66
vortex tube, 39
  strength, 39
vorticity flux, 256
vorticity transport equation, 30

Wagner function, 323
wedge flow, 56
windtunnel frame, 8, 126

MIX
Papier aus verantwortungsvollen Quellen
Paper from responsible sources
FSC® C105338

If you have any concerns about our products,
you can contact us on
ProductSafety@springernature.com

In case Publisher is established outside the EU,
the EU authorized representative is:
Springer Nature Customer Service Center GmbH
Europaplatz 3, 69115 Heidelberg, Germany

Printed by Libri Plureos GmbH
in Hamburg, Germany